CANCER SYSTEMS BIOLOGY

T0225480

CHAPMAN & HALL/CRC
Mathematical and Computational Biology Series

Aims and scope:

This series aims to capture new developments and summarize what is known over the entire spectrum of mathematical and computational biology and medicine. It seeks to encourage the integration of mathematical, statistical, and computational methods into biology by publishing a broad range of textbooks, reference works, and handbooks. The titles included in the series are meant to appeal to students, researchers, and professionals in the mathematical, statistical and computational sciences, fundamental biology and bioengineering, as well as interdisciplinary researchers involved in the field. The inclusion of concrete examples and applications, and programming techniques and examples, is highly encouraged.

Series Editors

N. F. Britton
Department of Mathematical Sciences
University of Bath

Xihong Lin
Department of Biostatistics
Harvard University

Hershel M. Safer

Mona Singh
Department of Computer Science
Princeton University

Anna Tramontano
Department of Biochemical Sciences
University of Rome La Sapienza

Proposals for the series should be submitted to one of the series editors above or directly to:
CRC Press, Taylor & Francis Group
4th, Floor, Albert House
1-4 Singer Street
London EC2A 4BQ
UK

Published Titles

Chapman & Hall/CRC Mathematical and Computational Biology Series

CANCER SYSTEMS BIOLOGY

EDITED BY

EDWIN WANG

CRC Press
Taylor & Francis Group
Boca Raton   London   New York

CRC Press is an imprint of the
Taylor & Francis Group an **informa** business

A CHAPMAN & HALL BOOK

CRC Press
Taylor & Francis Group
6000 Broken Sound Parkway NW, Suite 300
Boca Raton, FL 33487-2742

First issued in paperback 2017

ISBN-13: 978-1-4398-1185-6 (hbk)
ISBN-13: 978-1-138-11329-9 (pbk)

Library of Congress Cataloging-in-Publication Data

Cancer systems biology / editor, Edwin Wang.
    p. ; cm. -- (Chapman & Hall/CRC mathematical and computational biology series ; 32)
    Includes bibliographical references and index.
    ISBN 978-1-4398-1185-6 (alk. paper)
    1. Cancer--Research--Methodology. 2. Systems biology. I. Wang, Edwin. II. Series: Chapman and Hall/ CRC mathematical & computational biology series ; 32.
    [DNLM: 1. Neoplasms--genetics. 2. Gene Expression Regulation, Neoplastic--physiology. 3. Gene Regulatory Networks--physiology. 4. Genes, Neoplasm--physiology. 5. Systems Biology--methods. QZ 200 C21748 2010]

RC267.C375 2010
362.196'9940072--dc22   2009039065

Visit the Taylor & Francis Web site at
http://www.taylorandfrancis.com

and the CRC Press Web site at
http://www.crcpress.com

Table of Contents

Foreword

Cancer is a complex disease. In fact, it should be considered as a constellation of diseases that consist of various malfunctions of cellular systems that failed to maintain their coordination capability. While striking progress has been made in past decades on identification of genes that are involved in cancer outbreak and progression, only a handful of such discoveries resulted in effective clinical practices. The systems biology approach is essential in the battle against cancer because it emerges and survives through active modification of cellular and physiological host systems. A simple approach to interrupt one of molecules is limited in its successful applications. This reality has now been recognized and interest is growing.

This book is comprised of chapters that collectively discuss recent progress in the understanding of cancer systems biology, at a time when more and more researchers and pharmaceutical companies are looking into a systems biology approach to find drugs that can effectively be used to treat cancer patients. These chapters will also help readers appreciate the breadth of subjects that are involved in the study of cancer systems biology. This is particularly important because it illustrates the fact that cancers are a complex phenomena and mobilize every possible biological process and mechanism to secure their survival and proliferation.

Cancer systems biology is a field of study that is still in its infancy. Nevertheless, I believe that it will eventually become an indispensible part of the effective future treatment of cancer, and this book on cancer biology and its therapeutic applications is an important step toward this goal.

I hope the readers of this book will find it useful, and will be encouraged by it to join us in the battle against cancer.

Hiroaki Kitano
President, The Systems Biology Institute, Tokyo, Japan
Director, Sony Computer Science Laboratories, Inc.

Preface

The history of biology consists of "waves" that have profound impacts on thinking and advances in biological research. Biological research started from discovering, naming, classifying, and collecting specimens throughout the seventeenth and eighteenth centuries. Carolus Linnaeus, who introduced scientific names for species, is a key figure in the activities. Obervations of and thinking about species nurtured Darwin's theory of evolution in 1859, which has become the central event in the history of modern biology.

In the early nineteenth century, Theodor Schwann and Matthias Schleiden proposed the cell theory, which led to the cell being broken down into cellular components such as chromosomes, mitochondria, the nucleus, and so on. This is the reductionist approach, which is to break down a complex system into simpler or more fundamental parts. Reductionism proposes that the sum of these parts will explain the biological system. Since then the reductionist approach has dominated biology.

In the early twentieth century, Hans Krebs and Carl Cori led the movement of biochemistry. They worked out many of the central metabolic pathways, each of which contains a series of enzymes and linear reactions. Meanwhile, work on the fruitfly by Thomas Morgan recaptured the genetics of Gregor Mendel, who is recognized as the "father of modern genetics." In 1943 Oswald Avery proposed that DNA but not protein was the genetic material of the chromosome. Ten years later, James Watson and Francis Crick suggested that the structure of DNA was a double helix, the basis of molecular biology. Since then molecular biology has come to dominate every aspect of biology. In 1988 the Human Genome Project moved biology into the genome era.

The practices of functional genomics have led to the realization that the reductionist approach can't often capture the properties of the biological system. From the year 2000 onward, a paradigm shift has begun, from the reductionist approach to the holistic approach, which proposes that not all the properties of a biological system can be determined or explained by its component parts alone. Instead, the biological system often has emergent properties, which are either predictable or unpredictable and occur from the intricate causal relations across different scales and feedbacks, or the interconnectivities between parts. Systems biology represents such a paradigm shift. Systems biology is proposed to study the complex interplays between the biological components and how these interplays give rise to the emergent properties of that system. Furthermore, it seeks to integrate different types of information to advance the understanding of the biological whole, and develop models to uncover how biological systems change over time, such as with the onset of disease or in response to a perturbation.

Compared to other "waves," the current systems biology wave will dramatically change the reductionism that has dominated biology for the past centuries. Furthermore, it is the first time in the history of biology that components of computational science are required in the field of biology. Computing is critical for analysis and modeling of the integrated data at the systems level, such as developing a model of a cancer cell in order to find the "weak spots" in its signaling networks.

So far, cancer is the problem that has been studied the most using the systems biology approach. This book tends to reflect these efforts. It contains three parts: (1) basic concepts and theories of systems biology and their applications in cancer research; (2) basic cancer biology and cutting-edge topics of cancer research for computational biologists; and (3) computational tools and data resources for experimental biologists.

Finally, I would like to thank Mathieu Cloutier, Cong Fu, Jie Li, Chabane Tibiche, and Naif Zaman for their help in editing this book.

Edwin Wang
Montreal, Canada

About the Editor

Edwin Wang received his undergraduate training in computer science and obtained his PhD in experimental molecular genetics from the University of British Columbia in 2002. After working at FlyBase for a year, he moved to the Biotechnology Research Institute, National Research Council Canada, as a scientist working on bioinformatics and systems biology. At present he is a senior scientist at the National Research Council Canada and an adjunct professor at the McGill University Center for Bioinformatics. His work is currently centered around bioinformatics, computational, and experimental systems biology.

Contributors

Réka Albert
Department of Physics
Pennsylvania State University
University Park, Pennsylvania

Diego di Bernardo
Telethon Institute of Genetics and Medicine
Naples, Italy

Nicholas R. Bertos
Department of Medicine
Rosalind & Morris Goodman Cancer
    Centre
McGill University
Montreal, Canada

Mathieu Cloutier
Hamilton Institute
National University of Ireland
Maynooth, Ireland

Miroslava Cuperlovic-Culf
Institute for Information Technology
National Research Council Canada
Ottawa, Canada

Thamara K.J. Dayarathna
Department of Biochemistry
Siebens-Drake Medical Research Institute
Schulich School of Medicine
    and Dentistry
University of Western Ontario
London, Canada

Cong Fu
College of Life Science
Beijing Normal University
Beijing, China
and
Biotechnology Research Institute
National Research Council Canada
Montreal, Canada

Jing-Dong Han
Center for Molecular Systems Biology
Institute of Genetics and Developmental
    Biology
Chinese Academy of Science
Beijing, China

Zhenjun Hu
Program in Bioinformatics and Systems
    Biology
Boston University
Boston, Massachusetts

Maria Luz Jaramillo
Biotechnology Research Institute
National Research Council Canada
Montreal, Canada

Do Han Kim
Department of Life Science
Systems Biology Research Center
The Gwangju Institute of Science
    and Technology
Kwangju, Korea

Seon-Young Kim
Functional Genomics Research Center,
    KRIBB
Yuseong-gu, South Korea

Mario Lauria
Telethon Institute of Genetics
    and Medicine
Naples, Italy

Anne E.G. Lenferink
Biotechnology Research Institute
National Research Council Canada
Montreal, Canada

Jie Li
Biotechnology Research Institute
National Research Council Canada
Montreal, Canada

Shao Li
MOE Key Laboratory of Bioinformatics
    and Bioinformatics Division
TNLIST-Department of Automation
Tsinghua University
Beijing, China

Shawn S.-C. Li
Department of Biochemistry
Siebens-Drake Medical Research Institute
Schulich School of Medicine
    and Dentistry
University of Western Ontario
London, Canada

Thomas P. Loughran, Jr.
Penn State Hershey Cancer Institute
The Pennsylvania State University College
    of Medicine
Hershey, Pennsylvania

Yun Ma
College of Life Science
Tianjin Normal University
Tianjin, China

Wayne Materi
Institute for Nanotechnology
National Research Council Canada
Edmonton, Canada

Maria J. Moreno
Institute of Biological Science
National Research Council Canada
Ottawa, Canada

Dougu Nam
Division of Industrial Mathematics
National Institute for Mathematical
    Sciences
Yuseong-gu, South Korea

Morag Park
Department of Medicine and Departments
    of Oncology and Biochemistry
Rosalind & Morris Goodman Cancer
    Centre
McGill University
Montreal, Canada

Ally Pen
Institute of Biological Science
National Research Council Canada
Ottawa, Canada

Pradeep Kumar Shreenivasaiah
Department of Life Science
Systems Biology Research Center
The Gwangju Institute of Science
    and Technology
Kwangju, South Korea

Danica B. Stanimirovic
Institute of Biological Science
National Research Council Canada
Ottawa, Canada

Jinsheng Sun
College of Life Science
Tianjin Normal University
Tianjin, China

Chabane Tibiche
Biotechnology Research Institute
National Research Council Canada
Montreal, Canada

Edwin Wang
Biotechnology Research Institute
National Research Council Canada
and
Center for Bioinformatics
McGill University
Montreal, Canada

David S. Wishart
Department of Biological and Computing
    Science
University of Alberta
and
National Research Council
National Institute for Nanotechnology
Edmonton, Canada

Xuebing Wu
MOE Key Laboratory of Bioinformatics
    and Bioinformatics Division
TNLIST-Department of Automation
Tsinghua University
Beijing, China

Ranran Zhang
Penn State Hershey Cancer Institute
The Pennsylvania State University College
    of Medicine
Hershey, Pennsylvania

I

Cancer Systems Biology: Concepts and Applications

CHAPTER 1

A Roadmap of Cancer Systems Biology

Edwin Wang

CONTENTS

1.1 CANCER SYSTEMS BIOLOGY AND PERSONALIZED MEDICINE

1.1.1 Systems Biology Is Transforming Attitudes about Cancer Biology

When an accident occurs on a busy road during rush hour in a big city, such as Montreal or New York, traffic is blocked for a short time. Soon, however, drivers begin to turn around and use alternative roads to reach their destinations. A road map of a city is a web, a collection of intertwined roads that allows for identification of alternative routes. Increasing evidence (see Chapters 4 to 7) shows that, similar to roads, molecules in cells are also net-worked. This structure suggests that biochemical pathways are interconnected, which may allow cancer to bypass the effects of a drug.

Traditional approaches to biological studies rely mainly on linear verbal logic and illustrative descriptions without mathematical explanations. These approaches are only satisfactory for addressing mechanisms that involve a small number of elements or short chains of causality. Therefore, the approaches of traditional biology are unable to capture and unravel elaborate webs of molecular interactions. Most diseases, including cancer, involve a large number and variety of elements that interact via complex networks and, consequently, display highly nonlinear dynamics. Therefore, simply knocking out one target molecule in a biochemical pathway is not sufficient for treating a disease like cancer, because the cells often find alternative molecular routes to escape the blockage. This is one reason why current drug design strategies often fail. It is increasingly believed that a systems perspective, rather than the current gene-centric view, could solve these problems and open up entirely new options for cancer treatment.

The systems approach to biological studies combines empirical, mathematical, and computational techniques to gain an understanding of complex biological and physiological phenomena. For example, hundreds of proteins might be involved in signaling processes that ensure proper functioning of a cell. If such a signaling network is disturbed or altered, a cancer phenotype could be generated. As is discussed in Chapters 4 and 5, systems biology helps to shed light on these complex phenomena by generating detailed route maps of the various kinds of cellular networks and by developing sophisticated mathematical, statistical, and computational methods and tools to analyze these networks. Understanding the complex systems involved in cancer will make it possible to develop smarter therapeutic strategies, for example, by disrupting two or three key interactions in a biochemical network at the same time. These approaches could lead to significant advances in the treatment of cancer and help in transforming traditional reductionism-based approaches into unbiased systems-level approaches for drug discovery.

The birth and growth of the field of systems biology have been driven by technological innovation in high-throughput techniques targeted to life science applications. Over the past few years, high-throughput techniques, such as next generation genome sequencing, RNA-seq, chip-on-chip, large-scale immunoprecipitation (ChIP-seq), microarrays, and others, have been developed and used to measure gene expression and gene regulatory elements to identify genes that influence some interesting phenotype on a genome-wide scale. These technologies have triggered a dramatic change in the style of biological studies from a "one gene model" (i.e., focusing on the identification of individual genes and proteins and pinpointing their roles in the cell) to a "multiple gene model" (i.e., the belief that molecules almost never act alone and biological entities are *systems*—collections of interacting parts). These technologies have generated many "large-scale biology projects" and as they become more affordable and accessible, the implementation of large-scale biological projects is becoming more popular and routine.

With the emergence of systems biology, huge amounts of biological data have been produced and this trend is expected to continue in the future. The nature of high-throughput data is more comprehensive and unbiased than one-on-one biological data. This high-throughput approach to research has greatly altered the field of cancer research. Scientists have quickly realized that the combination of data management, interpretation, and our ability to obtain insights into these data are now the bottleneck in systems science, because "real signals" or molecular mechanisms and biological principles are buried in this flood of data.

The only way to deal with large amounts of data and the relationships within those datasets is through mathematical representation and computation. Systems biology tends to meet these challenges by integrating many types of -omic data and developing effective computational tools to decipher the complex systems. Network and graph theory have been developed to describe, analyze, and model the complexity of these biological systems using a mathematical language. As shown in Chapters 2, 4, 6, and 8, by applying network theory to biological systems, we are able to transform the biological language into a mathematical language, which is computable and can deal with the huge number of relations in a biological dataset. In fact, the fundamental framework of systems biology is network biology, which involves the use of networks to represent complexity, compute and model biological relationships, and seek to uncover biological principles and insights. A detailed discussion of network biology can be found in Chapter 2. Examples of cancer network studies can be found in Chapters 4 to 6 and 8.

This chapter illustrates strategies, procedures, and computational techniques for the study of cancer systems biology by focusing on network reconstruction, network analysis, and modeling. Meanwhile, to match the contents of these strategies and procedures, I will guide readers to the relevant chapters of this book. Finally, certain challenges and hurdles in cancer systems biology will also be discussed.

1.1.2 Systems Biology Is the Tool for Personalized Medicine

Recent studies have determined that many drugs work well for less than half of the patients for whom they are prescribed. Furthermore, nearly 3 million incorrect or ineffective prescriptions are written annually and more than 100,000 people in the United States die each year from drug-related adverse events (Kirk et al. 2008). These data strongly suggest that one-size-fits-all medicine and preventive care are not effective. Moreover, effective treatment of disease requires that the provider consider the effects of the patient's personal genetic background. Personalized medicine is a proposed approach to develop treatment regimes that take into account each patient's unique genetic profile, allowing the treatment to fit the specific needs of subpopulations of patients with different genetic backgrounds. Furthermore, this approach would help doctors to better evaluate the risk-to-reward scenarios and prescribe appropriate pharmaceuticals for different subpopulations of patients.

Over the past decade, cancer therapy has slowly begun to change from a one-size-fits-all approach to a more personalized approach. In a personalized approach, patients are treated based on the specific genetic defects present in their tumor. However, cancer is an extremely complex, heterogeneous disease. It is believed that crucial breakthroughs in the treatment of cancer, in the framework of personalized medicine, rely on the achievements of the powerful scientific approach of systems biology. Therefore, more efforts in "-omics" and systems biology have been made in the cancer research community. As a result, a tremendous amount of money has been poured into the field of cancer research over the past few years. Relatively speaking, more high-throughput data have been generated in cancer biology than in any other field of biology. However, the complexity of cancer is a major obstacle preventing a comprehensive understanding of the underlying molecular mechanisms of tumorigenesis. To crack the cancer code, network approaches have been developed and applied to cellular networks of cancer.

The examination of the entire genome of tumors (i.e., for the identification of cancer driver-mutating genes) and the global profiling of -omic data for cell signaling (i.e., gene expression, epigenetic and metabolomic profiles, and signaling data such as phospho-proteomic profiles) will aid in the construction of patient-specific cancer signaling networks. Analysis of such tumor signaling networks could help in making individualized risk predictions and treatment decisions. The cost of sequencing an entire human genome is rapidly falling. The continual development of faster and cheaper DNA sequencing technologies (for example, the next generation of DNA sequencing, which aims to decode a human genome for $1,000) could provide the ability to identify cancer driver-mutating genes in individual patients. Furthermore, profiling of tumor gene expression is also accessible and affordable.

Because these data can be generated in a routine clinical manner, it is possible to adopt a systems biology strategy for medical research and finally move forward into the era of personalized medicine. For example, construction and analysis of patient-specific tumor signaling maps will allow for the identification of key protein communication modules that are critical for development of a specific tumor. Modeling and simulation of such a patient-specific tumor signaling map will help to infer the molecular mechanisms responsible for the cancer and will aid in pinpointing the key targets of the tumor. Furthermore, the use of computational modeling and simulation would lessen the risk of therapeutic failure at clinical stages. Therefore, it is predicted that network analysis and modeling will become a mainstream tool in both the pharmaceutical and the biotech industries (Figure 1.1).

Three major aspects of cancer biology are expected to benefit from the application of a systems biology approach: (1) identification of prognostic and drug-response biomarkers of tumors by using a systems approach to link genomic data and medical records, such as blood samples, lifestyle questionnaires, and patient survival (see Chapter 4); (2) an understanding of network-oriented molecular mechanisms by building networks and computational models of different stages of cancer progression; and (3) an understanding of the network-based molecular mechanisms of metastasis and improved treatment of the later stages of tumors by comparative analysis of the networks of primary and metastatic tumors (see Chapter 5). Finally, cancer systems biology could provide new insights into the network-based molecular mechanisms that cause certain drugs to fail, thereby helping in the selection of multiple anticancer drugs and optimization of treatment strategies.

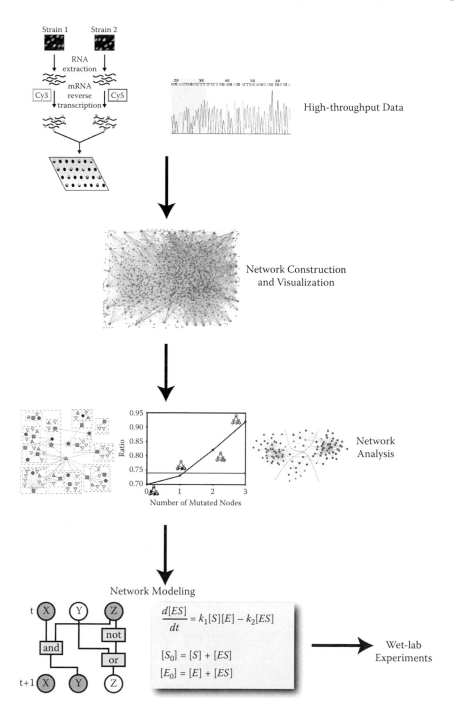

FIGURE 1.1 (See color insert following page 332.) The strategy and procedures for cancer systems biology research.

1.2 STRATEGIES FOR CANCER SYSTEMS BIOLOGY STUDY

Recent tumor genome sequencing efforts have shown that there may be thousands of cancer driver-mutating genes. Moreover, cancer driver-mutating genes are diverse and have little overlap between different tumors. This diversity is seen among different types of tumors and between tumors that originate from the same tissue. These observations suggest that cancer is a phenotype that can be caused by a collection of many genetic paths. However, several functional modules, the hallmarks of cancer (details of cancer hallmarks are described in Chapter 12), have been uncovered and documented. In general, cancer driver-mutating genes reflect cancer hallmarks or functional modules. Each hallmark, or functional module, is composed of a set of functionally linked pathways. Therefore, it is possible to map the functional modules and the mutating genes onto network modules, each of which is a subnetwork that contains the functionally linked pathways on the network. For example, an integrative analysis of the human signaling network and cancer driver-mutating genes has revealed network modules of this type (details of such modules are described in Chapter 5).

The systems approach to cancer studies must build realistic network models of tumors (network construction) and identify network modules, as well as the key genes and other network features in each module, from these networks (Pujana et al. 2007). Ultimately, the results derived from this systems biology approach must be experimentally validated in cancer cell lines and mouse models. Through this approach, cancer systems biology enables the integration of biological and clinical data at various levels and has the potential to provide insight into this complex disease (Figure 1.1).

Network construction focuses on reconstructing functional networks that reflect the relationships between genes and proteins under specific conditions, such as cancer gene signaling networks in metastasis. These networks can encode the links between the -omic data and the fundamental processes of cancer development and metastasis, that is, cancer hallmarks, cell cycle, apoptosis, and immunological response. Constructing a series of networks that incorporate time-course data may reveal the dynamics of biological processes such as tumor progression. Network approaches ease computational analysis, simplify and reduce complex interactions, and allow for the identification and quantification of relationships between inputs and outputs. Furthermore, network analysis aids in uncovering the general principles that underlie systems. To reach these goals, network construction relies heavily on integrative approaches to combining -omic data and accumulated knowledge.

Network visualization is the process of providing tools to build intuition that is unsurpassed by analysis tools. These intuitions may help in forming ideas regarding network exploration using analytical tools.

Network analysis focuses on computational analysis of the constructed networks using mathematical and statistical tools. Analysis may be performed on a single network to identify the important nodes, key network modules/subnetworks, and high-order relations between modules, such as collaboration, coexpression or coregulation of modules. Furthermore, functional principles of cancer can be inferred from this type of analysis. Therefore, hypotheses about mechanisms underlying cancer progression and metastasis can be generated through network analysis. Comparative analysis of time-course networks can highlight the dynamic nature of the functioning (or malfunctioning) of cells in the development and progression of diseases (i.e., cancer progression). Furthermore, comparative analysis can aid in the identification of key network components and their causal relationships during developmental stages. These analyses would capture the dynamic interactions between large numbers of components across different time scales, as well as the nonlinear nature of the systems. In addition, network analysis may lead to the identification of gene signatures that could be used for prognosis and drug response prediction through integration of gene expression or protein abundance profiles and clinical information about the cancer patients.

Network modeling involves the use of dynamic systems theory and mathematical tools to investigate complex biological systems in order to demonstrate nonlinear spatiotemporal behavior. However, the generation of experimental data that are suitable to parameterize, calibrate, and validate such models is often time consuming and expensive, or even impossible, with the technology available today. Regardless, the spatiotemporal dynamics of the system as a whole are of such complexity that understanding those dynamics challenges conventional approaches and makes mathematical modeling a necessity.

The behavior of complex cancer cell networks cannot be deduced by intuitive approaches. Instead, it requires sophisticated and elegant network models and computational analysis and simulation. Cancer cell network models will aid in the generation of experimentally testable hypotheses and discovery of the underlying mechanisms of tumorigenesis and metastasis. Network construction may provide insights into specifying the necessary components of a biological process, a subject that is highly related to the explicit hypotheses of cancer development, progression, and metastasis. Both network analysis and modeling may improve our understanding of the cancer system and reveal hidden patterns or counterintuitive mechanisms in cancer, uncovering critical points about which our understanding is still poor. Furthermore, both analysis and modeling may help generate hypotheses that, in turn, can be tested in a wet lab. Finally, network analysis may help identify biomarkers useful in the clinical practice of personalized medicine (Figure 1.1).

1.2.1 Requirements for Experimental Models to Perform Cancer Systems Biology

Before applying systems biology approaches, we should focus on the types or subtypes of cancer that have high clinical relevance and are well studied in terms of molecular pathology (i.e., pathological features can be mapped onto gene signatures). Breast cancer is one example of this type of cancer. Gene microarray profiles can stratify breast cancer samples into four subtypes (Sotiriou and Pusztai 2009). In fact, mammary epithelial cells of different origins can give rise to tumors with distinctly different phenotypes. Gene expression profiling of tumor types has shown that gene expression changes depend on the nature of the precursor cells (Ince et al. 2007). Therefore, it is important to know the clinical subtypes of tumors when performing analyses.

For a particular type or subtype of cancer, it is essential to have high quality experimental mouse models (i.e., progression models for different cancer stages, for instance, prostate cancer) and a set of targeted drugs available for treatment of the cancer type or subtype. In general, how closely the mouse models mimic human cancer types and subtypes will affect the usefulness of those models in understanding the molecular mechanisms and guiding therapeutic decisions. Therefore, it is desirable to have mouse models that mirror clinical outcomes in patients with specific types of cancer. Such models can be used to predict poor responses to chemotherapy in cancer patients and, therefore, might help in determining patient prognosis.

The selected cancer types or subtypes should have high quality cell line models that have been well characterized using genomic approaches. The breast cancer cell line MDA-MB-*231* is a good example of one of these models. The MDA-MB-*231* line has several derivative lines, which display the features of organ-specific (lung, bone, or brain) metastases (Nguyen, Bos, and Massague 2009).

For experimental models, large-scale -omic data, such as genome-wide gene expression, phosphoproteomics, epigenetics, cancer driver-mutating genes (i.e., via tumor genome sequencing), and metabolomics, can be generated. To produce the best data for network construction, analysis, and modeling, experimental biologists should interact with computational scientists to design experiments properly. Using this variety of data, construction, analysis, and modeling of tumor-specific networks can be conducted. The networks can also be applied to mathematical frameworks for modeling and simulation.

1.2.2 Data Integration and Cancer Gene Network Construction

1.2.2.1 Cancer Gene Network Construction

The major objective of cancer systems biology is to create dynamic models of biological processes closely related to cancer initiation, progression, and metastasis. Therefore, cancer networks should capture the important functional themes of cancer biology. Several fundamental biological processes play central roles in cancer. For example, the hallmarks of cancer are typical examples of these fundamental processes (detailed descriptions of cancer hallmarks can be read in Chapter 12). It should be noted that "new cancer hallmarks" might be added as understanding of cancer biology, especially cancer systems biology, increases. Indeed, some groups have recently proposed that inflammation may be a "new hallmark" for cancer (Mantovani 2009). Cancer metastasis depends on both intrinsic properties of the tumor cells and factors in the tumor microenvironment. These factors provide tumors with blood vessels and an inflammatory environment, consisting of immune cells and their secretory products, which promote tumor growth (the tumor microenvironment and blood vessels are described in detail in Chapters 14 and 15).

Cell cycle and division, differentiation, apoptosis, angiogenesis, insensitivity to inhibitors (robustness), tissue invasion and metastasis [i.e., epithelial to mesenchymal transition (EMT); more details about EMT are described in Chapter 13], and inflammation are among the important properties of tumors. All of these processes are associated with cell signaling. Integrative analysis of the human signaling network with data about cancer driver-mutating genes suggests that cell cycle and apoptotic signaling are essential in all types of cancers (Cui et al. 2007; see Chapter 5). Most of these biological themes should be captured in cancer gene networks. Furthermore, the subnetworks of these themes should be viewed as a priority for systems biology research. Separation of these subnetworks will allow different cellular processes to be studied in a relatively isolated manner by network analysis and modeling. It should be noted that at least some of these processes are potentially interlinked within the cell or the tumor microenvironment. Therefore, high-order relationships between these processes could be modeled after a careful and systematic study of each individual process.

There are three approaches for construction of cancer gene networks. The first approach is to infer or reverse engineer the cancer gene or signaling network using genome-wide datasets, such as gene expression profiles, RNAi knockout phenotype data, etc. For example, a gene regulatory network has been constructed using time course microarray profiles from a mouse epithelial breast cell line (BRI-JM01; Wang, Lenferink, and O'Connor-McCourt 2007), which undergoes an EMT when treated with transforming growth factor (TGF)-β (Lenferink et al. 2004). Notably, clusterin, one of the genes that is upregulated at the middle and late time points, shows many regulatory links to other genes in the network. During the EMT process, clusterin is secreted by the BRI-JM01 cells. Interestingly, application of anti-clusterin antibodies to the TGF-β-treated BRI-JM01 cells blocks the TGF-β-induced EMT (Wang, Lenferink, and O'Connor-McCourt 2007). Chapter 3 describes additional computational methods and examples of reverse engineering of networks.

The second network construction approach is to extend a protein interaction or signaling network using high-throughput experimental approaches, such as protein interaction measurements (Chapter 19 describes an approach for extending the human signaling network). This extension method also allows for the construction of gene regulatory networks using large-scale ChIP-seq or ChIP-on-chip data. This approach works especially well for the construction of gene regulatory networks. For example, the application of this type of approach has constructed a P53 regulatory network containing 98 novel direct target genes of P53 (Wei et al. 2006).

A cancer gene collaborative and mutually exclusive interaction network has been constructed by large-scale mutagenesis (i.e., retroviral insertional mutagenesis) to screen ARF- and P53-deficient as well as wild-type mice to identify genes that interact with one or the other of these tumor suppressors (Uren et al. 2008). For this kind of network, it should be noted that cancer gene collaboration has two levels: (1) a gene tends to collaborate with another gene; or (2) one gene may have distinct sets of collaborators, based on different mutations of that gene (i.e., Notch1).

The third approach to network construction is to integrate data from high-throughput studies or manually curated literature databases onto current networks. For example, current signaling networks are largely constructed using manually curated data from the literature (Awan et al. 2007; Cui et al. 2007; Ma'ayan et al. 2005). This data integration approach is widely used in network construction because more high-throughput data are easily accessible. Integration of high-throughput data through computational approaches provides a powerful method to address and dissect the complexity of cancer at various levels in a systems manner. Quality of data is very important in the data integration approach. The decision regarding particular data sources used for network construction must be based on the questions to be addressed by the network analysis. A discussion of useful data sources for systems biology can be found in Chapters 16 to 18 and 20.

Directly mapping genes of interest (i.e., modulated genes between normal and cancer tissues) onto protein interaction networks and signaling networks will lead to the construction of cancer-related subnetworks or network modules (i.e., modulated genes are connected together to form a subnetwork). Several such examples are described in Chapters 4 to 7.

Alternatively, we can map the genes of interest onto a network (i.e., a human signaling network) and extract all of the shortest paths between any two genes of interest. These shortest paths can then be merged to construct a network. The genes of interest used for this type of approach may be cancer-modulated genes, cancer driver-mutating genes, or other groups.

We can examine which shortest paths of the entire network are important for a particular cellular condition using functional genomics data and collect these shortest paths to build networks. For example, we have constructed cancer cell line-specific signaling networks by collecting the shortest paths that are significantly enriched in the cell line gene expression profile of the human signaling network. Tumor gene coexpression networks can be constructed using different types or subtypes of tumors. Weighted gene coexpression networks have been constructed using gene microarray profiles from glioblastoma samples (Horvath et al. 2006). Analysis of such networks provides a blueprint for leveraging genomic data to identify key control networks and molecular targets in cancer.

Cancer molecular networks can also be constructed by linking the information between genotypes and phenotypes. For example, Quigley and colleagues crossed mice of two species, *Mus spretus* and *M. musculus*, which were either resistant (*Mus spretus*) or susceptible (*M. musculus*) to skin tumor development. Following the cross, they combined gene expression profiling with linkage analysis to construct a "susceptibility network" of gene expression and regulation in normal skin (Quigley et al. 2009). This study highlights the power of a network approach for identifying genotype–phenotype relationships.

Cancer development and metastasis are dynamic processes with different timescales. Time series high-throughput data of cancer processes may be used to construct a series of networks that represent these scenarios, using the methods discussed above. To develop dynamic network models, we must clearly conceptualize the way time is encoded in networks. Analytically, time has two distinct forms: discrete and continuous. A discrete representation of time often consists of a series of snapshots of the network. Hence, longitudinal analysis focuses on the change from one network state to another. In such cases, a process is generally inferred from the total change in the network across time. A continuous representation of time consists of sequential events or interactions recorded with exact starting and ending times. Continuous representations of time enable the identification of overall network changes. However, most of the current experimental systems are only able to produce data for discrete representation of time in networks.

All of the approaches discussed can be used to build tumor sample or cell line-specific cancer networks. Construction of such networks simplifies and reduces complex interactions. In other words, it removes the "noisy information" from the global network and assembles various parts that are highly related in tumors. After constructing the cancer networks, it is necessary to check which cancer hallmarks have been captured. If none of the hallmarks can be found in the networks, it is worthwhile to check the data and network construction procedures.

1.2.2.2 Data Integration in Systems Biology Drives New Concepts for Bioinformatics Analysis

Bioinformatics provides essential tools for data integration for network construction, analysis, and modeling. Many bioinformatics methods and tools have been developed for large-scale data analysis. These methods and tools include statistical tests (i.e., for testing gene expression differences, genetic associations, and gene expression correlations), data extraction from literature and databases (i.e., text mining), and procedures for pattern recognition and machine learning (i.e., clustering analysis).

One trend in bioinformatics analysis is the movement away from consideration of genes and proteins in an isolated manner. In the early days of microarray analysis, genes studied in gene expression profiles were examined statistically in an individual manner. Currently, the statistical significance of gene expression changes is assessed in a gene set-dependent manner (i.e., taking into account pathway genes or a set of genes in a biological process such as a cell cycle). Indeed, these methods have led to new insights into cancer biology. For example, alteration of the expression correlations of protein network modules seems to be involved in cancer metastasis. Many methods following this trend have been developed and used, for example the Gene Set Enrichment Analysis (GSEA). More details on these methods are described in Chapter 18. However, there are still many challenges in developing systems-oriented bioinformatics methods. For example, the problem of how to dissect the multivariable factors in systems, given the fact that biological variables are highly intertwined and correlated, remains to be solved.

1.2.3 Network Visualization

Networks represent complex systems. Although we have developed (and are continuing to develop) mathematical concepts and computational tools for network analysis (more details in the next section), we are still unable to fully decode complex systems. In certain contexts, the human brain is still more powerful than an analytical method in forming intuitions that can help guide network analysis.

Network visualization allows us to exploit the human mind's capacity for building intuition that is unsurpassed by analysis tools. Before conducting network visualization, we must carefully formalize questions that can be used to extract meaning and implication from cancer molecular networks.

The effectiveness of network visualization differs depending on network size. Thus, visualization has different objectives based on size. Visualization of small networks focuses on detailed elements of the graph structure, whereas that of larger networks mainly captures a more global picture. Visual analysis of small networks allows for the formation of insights into molecularly interacting relationships, whereas the analysis of larger networks allows for judgments about high-order relations (more abstractive) between subnetworks (i.e., interplay between network modules or components such as cell cycle and cell death in cancer progression). It is possible to define network modules and subnetworks in large networks. The intrinsic structure and relationships in each network module can be examined using network visualization tools. In turn, the nodes in a network module can be collapsed into token nodes and used to explore high-order relationships between modules.

Color coding of nodes or arcs based on molecular function, information flows, or other related features closely related to the problems addressed often helps in discovering network patterns and forming hypotheses to guide in-depth analysis.

There are increasing efforts to produce network visualization beyond "static" representations of cellular states, toward a more dynamic model of cellular processes. These efforts strive to incorporate high-throughput and functional data, such as time-series gene expression data, gene ontology terms, and subcellular localization data. In this context, dynamic network visualization helps in capturing the dynamic features of a process, augmenting theoretical intuition and extracting meaningful patterns.

Two basic approaches to visualization have been developed. The first common visualization approach encodes all changes and transitivity between developmental stages (Amato et al. 2006) into a single network. For example, a signaling network encodes all the modulated genes from a time course dataset of cancer progression with different colors (i.e., representing different stages in cancer progression). Such a network consists of patterns of causal or collaborative gene relationships. However, it is necessary to identify the stages that substantively capture the nature of the relational events and the character of temporary cellular states that arise in the focal context. Changes in transitivity provide information about a single dimension of a network's structure. One might find that a network reaches a high or low transitivity level, suggesting the potential importance of some stages. The clusters and information flows obtained by mixing different colorful nodes and links might suggest new network modules. This approach is commonly used to evaluate cancer progression or identify causal relationships between network modules.

One of the most effective methods to implement dynamic network visualization is to present sparse networks in a way that shows how the network emerges over time (i.e., modulated genes in a time-series manner for cancer progression models) by adding and color coding nodes and relationships as they appear (i.e., different colors can be used for nodes and relationships in a time-dependent manner). It is important to organize the nodes and edges in the display plane based on the final stage of the network. The appearance of dynamic elements over time reveals key genes that play roles in different stages and suggests regulatory relationships between network modules at different stages.

The second common approach to network visualization is to explore separate networks at each time point. However, these networks are often difficult to interpret using pure visualization, because it is impossible to identify the sequential links between node positions from one network to the next. In this situation, comparative network analysis is a proper and powerful approach to these problems (see Section 1.2.4).

A number of network visualization tools, such as Cytoscape and VisANT, have been developed. VisANT has developed high-level abstraction of network relationships (more details about VisANT are described in Chapter 17). Chapter 20 lists many network visualization and analysis tools for systems biology studies.

1.2.4 Network Analysis

Genes and proteins are often used as nodes in networks, while relationships between them are usually represented by edges (undirected links) or arcs (directed links). A detailed description of network types is provided in Chapters 2 and 4. Reading Chapter 2 prior to this section is recommended to provide an understanding of basic concepts and terms in network biology.

1.2.4.1 Different Biological Properties Are Encoded in Different Network Types

Although a set of common questions can be asked of cancer molecular networks, different questions can be addressed using different types of molecular networks, in which different network characteristics and relationships are encoded. Moreover, different types of data are necessary to construct different types of networks.

In gene regulatory networks, the length of regulatory cascades is often short, normally three to five steps from the first layer to the last layer of the network (Wang and Purisima 2005), reflecting the quick regulation response of these systems. Hubs in gene regulatory networks play a major role in responding to stimuli and coordinating the regulated genes. In agreement with this mechanism, the transcripts of the hub transcription factors often display the property of rapid decay (Wang and Purisima 2005). Local transcription factors often encode genes that take part in one or a few biological processes, whereas intermediate hub transcription factors encode the collaborative relationships (i.e., coexpression) between a few biological processes. The rapidly decaying transcripts of global hub transcription factors might encode "switch" functions, which are used under different conditions and stimuli. Most of the target genes of transcription factors are "workers," which directly perform the tasks of biological processes and do not have regulatory roles. Collaborative relationships between transcription factors can be also found. Therefore, gene regulatory networks are useful for identifying key regulators, coexpression of genes, and sets of "workers" involved in cancer processes.

Nodes in the human signaling network are sparsely connected. The length of regulatory cascades in signaling networks (normally 7 to 14 steps from receptors to transcription factors; Cui, Purisima, and Wang 2009) is often longer than the length of cascades in gene regulatory networks. Along these signaling cascades in protein signaling networks, almost all of the nodes are "regulators." Therefore, logical regulatory relationships are extensively encoded in signaling networks. Cell signaling information flow propagates from a receptor to the nucleus. It is believed that a number of proteins scattered directly downstream of receptors are logical "organizers" that integrate signals. For example, hubs in signaling networks play a major role in integrating different signals and pathways. Therefore, signaling networks are useful in identifying cancer causal genes and regulatory logic involved in cancer processes.

In a signaling network, paths represent signaling information flow and regulatory logics. In a gene regulatory network, paths represent regulatory hierarchy. In contrast, paths in a protein interaction network have no clear biological implications. Network paths also have different evolutionary features (Cui, Purisima, and Wang 2009). For example, in the case of directed shortest paths, the more distance between two proteins, the less chance they share similar evolutionary rates. However, such a correlation was not observed with respect to the neutral shortest path. It has been shown that the evolutionary rate of proteins decreases along the signaling information flow from the extracellular space (input layer) to the intracellular space to the nucleus (output layer) (Cui, Purisima, and Wang 2009).

The expression levels of major regulators (i.e., kinases) in signaling networks do not necessarily change dramatically during cancer progression and metastasis. The major regulatory reactions are modulated via protein modification (i.e., phosphorylation and dephosphorylation), not via modulation of gene expression. Therefore, direct functional consequences of cancer driver-mutating genes are difficult to address in gene regulatory networks. Most of the cancer driver-mutating genes are signaling genes. Furthermore, these mutating genes do not simply increase expression levels of their targets, but increase or decrease the activity of their targets (Cui et al. 2007). Monitoring the dynamics of phosphorylation and dephosphorylation is essential to decode signaling networks.

Network motifs in protein interaction networks represent protein complexes, whereas they represent information processing units and regulatory loops in signaling and gene regulatory networks. In protein interaction networks, network modules represent protein interaction communities associated with particular biological processes, whereas in signaling networks, they represent blocks of regulatory logics and information processing.

Compared to the human signaling network, nodes in the human protein interaction network are densely connected. Regulatory logics are difficult to identify in protein interaction networks. However, network modules or network communities are encoded in protein interaction networks. Therefore, such networks are suitable for integration of gene expression profiles to determine subnetworks (teams of protein "workers") that perform certain functions at different stages of cancer development, progression, and metastasis. Furthermore, such subnetworks could be used as biomarkers in a clinical setting.

Generally, signaling networks are sparse and full of logical codes of regulation, whereas protein interaction networks are dense and do not code for logic of regulation. Gene regulatory networks encode both regulatory logic and gene "workers."

It should be noted that posttranscriptional and posttranslational regulation are both prevalent in cells. It is important to consider these aspects in terms of network construction, analysis, and modeling.

Ubiquitination is applicable to a wide range of human proteins (Yen et al. 2008; see Chapter 6). In the human signaling network, ubiquitin-mediated regulation is enriched in receptors and ligands, the signal initiating portion of the network (Fu, Li, and Wang 2009). Initiated signals can be immediately organized and processed in the upstream region of the network, which resides in intracellular space close to the cell membrane. This network region is enriched for many built-in negative feedback loops (Legewie et al. 2008). In contrast, MicroRNAs (miRNAs) regulation (negative regulation) focuses on the downstream regions of the network (Cui et al. 2006).

Posttranscriptional and translational modifications of genes and their products provide feedback mechanisms in gene regulatory networks. miRNAs tend to posttranscriptionally regulate transcription factors. Nearly half of the human transcription factors are regulated by miRNAs (Cui et al. 2006). Furthermore, hub transcription factors tend to regulate more miRNAs (Chapter 7).

1.2.4.2 Network Analysis Using Network Biology Methods

Evolution is the central law of cancer cells. Similar to the laws of physics and chemistry, the design principles that constrain cancer biology are all amenable to discovery and modeling. "Core design principles" in biology must be modeled to express the mechanistic rules easily and efficiently. Abstraction is the most critical process required to uncover the design principles of biological systems. A proper abstraction aids in data examination from different perspectives and helps to extract meaningful knowledge from the data. Graph theory allows for representation of the abstraction of biological relationships, analysis of the information, and extraction of insights. Evidence shows that biological insights have been encoded in network properties (Wang, Lenferink, and O'Connor-McCourt 2007). Therefore, network property analysis of integrated networks (i.e., signaling networks incorporated with cancer-related high-throughput -omic data) will provide new biological insights.

The core concepts of network analysis are nonlinear and network perspectives. Emergent biological properties may be discovered from nonlinear thinking. The results of linear thinking are often predictable and expected, whereas the results of network analyses are often nonlinear and unexpected. In theory, network analysis could lead to more unexpected and, therefore, exciting results.

Network properties range from local (i.e., single node or edge, network bottlenecks, network motifs, and modules) to global or network-wide (i.e., whether all nodes are connected, network diameter, shortest path, density, average links, clustering coefficient, network centrality, degree centrality, closeness centrality, radiality, betweenness and pageRank, minimum spanning trees, and network flows). A detailed survey of network measurements and properties has been described by Costa et al. (http://arxiv.org/abs/cond-mat/0505185). Intrinsic relationships exist between local and global properties, such that sometimes a perturbation of a small number of linked nodes can result in widespread consequences. For example, a collection of protein network modules with gene coexpression alterations leads to breast cancer metastasis (Taylor et al. 2009). Further details about network biology concepts such as graph theory, network measurements, and analysis are described in Chapters 2 and 4. The terms and concepts in Chapters 2 and 4 will be used in the following examples to illustrate network analysis in cancer biology.

Based on the specific questions being addressed, different methods of network analysis can be applied. The architectural structure of cellular networks provides a framework to illustrate the logics and mechanisms of cancer biology. Regarding global network features, for example, the following cancer biology questions could be addressed: extracting a subnetwork of cancer signaling that reflects functionality of cancer development or metastasis; uncovering the mechanisms by which genetic and epigenetic events affect cancer cell signaling and tumor progression; identifying central players (i.e., network hubs that are cancer genes) in cancer cellular networks; identifying subnetworks that are functionally targeted cancer hallmarks, such as cell cycle and apoptosis. These analyses allow for the narrowing down of the scale of these complex networks. Further, they capture the communications between the core molecular processes in cancer and uncover the molecular mechanisms responsible.

Many questions can also be addressed using local network features. For example, the enrichment of cancer driver-mutating genes (Futreal et al. 2004) in positive feedback signaling loops (network motifs) suggests that oncogenes gain function due to mutation, whereas the enrichment of cancer methylated genes in negative regulatory signaling loops suggests that loss of function by gene methylation promotes tumorgenesis (Cui et al. 2007). The cancer signaling network, extracted from the human signaling network by integrating cancer driver-mutated genes and cancer methylated genes, has been decomposed into 12 modules (network modules or communities). Furthermore, high-order collaborative relationships between these modules have been identified in different types of cancers (Cui et al. 2007).

Although cancer is considered a very heterogeneous disease, querying mutated genes in tumor samples using the network modules defined by a human cancer signaling map reveals that one common network module occurs in most tumor samples. Specifically, breast and lung cancers show more complex collaborative patterns of oncogenic signaling modules than the other cancer types examined, highlighting their heterogeneous nature (Cui et al. 2007). These examples demonstrate that network biology is a powerful tool that elegantly provides new insights into biology. Moreover, most of these insights cannot be drawn from traditional biological approaches, which are dominated by linear thinking.

Network analysis also provides a powerful tool for generating testable hypotheses. For example, Fu and co-workers found that ubiquitin-mediated proteins are enriched in positive loops in the human signaling network (Fu, Li, and Wang 2009). Gene ontology enrichment analysis of the ubiquitin-mediated proteins in these positive loops suggests that the biological process apoptosis is enriched in this group of proteins. Furthermore, more than 85% of the ubiquitin-mediated apoptotic proteins in these positive loops are cancer-associated genes. These observations led to the hypothesis that the ubiquitination machinery, such as the 26S proteasome, could be more highly expressed in tumor cells than in normal cells (high expression of ubiquitination machinery genes will block apoptosis, an essential block in cancer signaling). Using microarray data from both tumor and normal samples, Fu, Li, and Wang (2009) provide evidence that this hypothesis is true. More examples of cancer network analyses are discussed extensively in Chapters 4 to 7.

1.2.4.3 Network Dynamics Analysis and Gene Markers for Diagnosis and Prognosis

Genetic variation and somatic mutations in human populations, and even in tumor samples from the same individual, make tumors a very heterogeneous tissue type. The heterogeneous nature of tumors leads to different responses from different patients with the same type of cancer to treatment with the same drug. To address this problem, personalized medicine proposes to identify molecular markers for drug responses.

Another issue in cancer treatment is how to predict which patients with cancer should receive extra therapy after surgery. Currently, most cancer patients undergo surgery. If physicians knew which patients are in the earliest stages of cancer, they could better predict which patients might benefit from additional treatment. However, physicians cannot predict which patients with cancer should receive extra therapy after surgery. At present, it is very difficult to predict which patients will be cured by surgery alone, the single treatment most patients receive, and which patients might benefit from the addition of chemotherapy. Therefore, it is critical to identify those genes that can be used as tools to predict survival after diagnosis of cancer and those genes that can guide how oncologists should treat the cancer to obtain the best outcome.

Practically, we are facing the challenge of identifying robust and highly accurate molecular markers for drug response and survival prediction (prognosis). Enormous efforts have been made for more than 10 years to identify such biomarkers from gene microarray profiles. In fact, PubMed contains more than 3000 publications on this subject. However, no robust biomarkers have yet been identified for cancer. Specifically, the current so-called breast cancer biomarkers are ineffective when used in a different set of breast tumor samples.

It is expected that the use of a systems approach could extract more accurate and mechanism-based markers of patient response to drug treatment by capturing the system dynamics. This approach requires integration of cellular networks and the alterations of gene expression, genetic mutation, methylation, and protein modifications with clinical information such as drug response, patient outcomes, etc. These efforts will fundamentally change both the health care system and the management strategies for cancer patients.

Two recent studies demonstrated that the network biology approach offers promising results toward finding better markers for cancer prognosis. Mapping tumor-expressed genes onto a human protein interaction network allows for the identification of subnetworks as cancer biomarkers. The resulting subnetwork markers are more reproducible than individual marker genes, which are selected without protein interaction information. These subnetwork markers also achieve higher accuracy in classification of metastatic versus nonmetastatic tumors (Chuang et al. 2007). Alteration of the coexpression of genes that are organized as protein network modules is associated with cancer metastasis, suggesting that dynamic rewiring of protein interaction modules is implicated in metastasis. Based on this discovery, network module markers have been shown to reach higher survival prediction for breast cancer patients than other markers selected from gene clustering, an approach that does not consider genes as interacting modules (Taylor et al. 2009). We hope that network biology approaches will be applied to the discovery of drug response markers in the future.

1.2.5 Dynamic Network Modeling

Biology is currently experiencing a high level of interest in developing an understanding of system dynamics, specifically in studying systems made up of communicating parts and machines, information processing (cell signaling), and interconnected computational and functional units. In this perspective, organisms are viewed as information manipulators and processors.

Network modeling uses methods from dynamical systems theory to model and simulate networks to decode the information processing machines and test hypotheses about the mechanisms that underlie the function of cancer cells. In network modeling, the behavior of cancer cells is represented in terms of quantitative changes in the levels of gene transcripts or enzyme activities (i.e., kinase activity).

Network modeling provides conceptual and computational tools with which to perform and iterate dry-lab experiments. Network modeling enables simulation-based research within a quantitative reference framework that connects *in silico* replica and real systems by means of quantitative conceptual and computational tools. In the long term, network modeling approaches could replace many time-intensive or expensive wet-lab experiments. In this context, the growing field of systems biology is expected to lead to fundamental breakthroughs in cancer biology.

Two basic approaches (qualitative and quantitative) to dynamic network modeling are often used. Qualitative network modeling considers the states (i.e., gene expression values, protein concentrations, active, or nonactive) of the network nodes in a finite number of values (i.e., ON and OFF, higher or lower than threshold, etc.), whereas quantitative network modeling considers the states of nodes over a wide range of values. In addition, quantitative modeling also considers probabilistic, deterministic, or stochastic characteristics of the network.

Dynamic network models are composed of three basic elements: the cellular network (i.e., networks of gene regulation, protein interaction, or signaling in a given cellular context), the initial state of each node, and the transfer functions that describe the state dependencies of each node in terms of its regulators. Node states can be modeled in either a continuous or a discrete manner, whereas the transfer functions can be modeled in either a deterministic or a stochastic fashion. Therefore, there are four methods (continuous deterministic model, continuous stochastic model, discrete deterministic model, and discrete stochastic model) for dynamic network modeling. A discrete state approach may "precisely" describe the system's behavior, whereas continuous state equations describe the "average behavior" of the system

In theory, a continuous stochastic model describes the system more accurately and more closely reflects the real system. However, high-quality experimental data are required to apply this modeling method. The limited availability of high-quality quantitative data forms a major bottleneck for the application of continuous stochastic models. Compared with the continuous stochastic model, a continuous deterministic model describes the system without taking into account the stochastic (noise) nature of the system. A discrete stochastic model usually accounts for noise in the transfer functions (differential equations) and models the nodes with two or a few states.

A discrete deterministic model has a high level of abstraction of node states (i.e., node states are assigned into only a few categories, even two binary states such as ON and OFF). The transfer functions are encoded as logic functions, such as "and," "or," and "not." Boolean models are one representative group of the discrete deterministic modeling method. This method requires relatively little detailed input information. Therefore, this method is more attractive and feasible, because the data generated from current experimental systems are suitable for this modeling method. One of the disadvantages of this method is that the predictions are generally more macro-scale and less quantitative. A comprehensive survey of the methods and computational tools for dynamic network modeling can be found in Chapter 16. An application of Boolean models to cancer cell death signaling networks is documented in Chapter 8. Other modeling examples can be found in Chapters 9 to 11.

In summary, systems biology is leading to fundamental changes in cancer biology. In terms of dynamic network modeling, high-quality quantitative data (i.e., quantitative proteomic data for signaling networks) are still required. Current methods for producing high-quality data for modeling are still expensive and time consuming. Development of new methods for network analysis and modeling are also needed. Most important, more scientific investigators need to be trained to think in a network fashion, rather than in a traditional linear fashion.

REFERENCES

Amato, R., Ciaramella, A., Deniskina, N. et al. 2006. A multi-step approach to time series analysis and gene expression clustering. *Bioinformatics* 22: 589–596.

Awan, A., Bari, H., Yan, F. et al. 2007. Regulatory network motifs and hotspots of cancer genes in a mammalian cellular signaling network. *IET Syst Biol* 1: 292–297.

Chuang, H. Y., Lee, E., Liu, Y. T., Lee, D., and Ideker, T. 2007. Network-based classification of breast cancer metastasis. *Mol Syst Biol* 3: 140.

Cui, Q., Ma, Y., Jaramillo, M. et al. 2007. A map of human cancer signaling. *Mol Syst Biol* 3: 152.

Cui, Q., Purisima, E. O., and Wang, E. 2009. Protein evolution on a human signaling network. *BMC Syst Biol* 3: 21.

Cui, Q., Yu, Z., Purisima, E. O., and Wang, E. 2006. Principles of microRNA regulation of a human cellular signaling network. *Mol Syst Biol* 2: 46.

Fu, C., Li, J., and Wang, E. 2009. Signaling network analysis of ubiquitin-mediated proteins suggests correlations between the 26S proteasome and tumor progression. *Mol Biosyst* 5: 1809–1816.

Futreal, P. A., Coin, L., Marshall, M. et al. 2004. A census of human cancer genes. *Nature Rev Cancer* 4: 177–183.

Horvath, S., Zhang, B., Carlson, M. et al. 2006. Analysis of oncogenic signaling networks in glioblastoma identifies ASPM as a molecular target. *Proc Natl Acad Sci USA* 103: 17402–17407.

Ince, T. A., Richardson, A. L., Bell, G. W. et al. 2007. Transformation of different human breast epithelial cell types leads to distinct tumor phenotypes. *Cancer Cell* 12: 160–170.

Kirk, R. J., Hung, J. L., Horner, S. R., and Perez, J. T. 2008. Implications of pharmacogenomics for drug development. *Exp Biol Med (Maywood)* 233: 1484–1497.

Legewie, S., Herzel, H., Westerhoff, H. V., and Bluthgen, N. 2008. Recurrent design patterns in the feedback regulation of the mammalian signalling network. *Mol Syst Biol* 4: 190.

Lenferink, A. E., Magoon, J., Cantin, C., and O'Connor-McCourt, M. D. 2004. Investigation of three new mouse mammary tumor cell lines as models for transforming growth factor (TGF)-beta and Neu pathway signaling studies: identification of a novel model for TGF-beta-induced epithelial-to-mesenchymal transition. *Breast Cancer Res* 6: R514–R530.

Ma'ayan, A., Jenkins, S. L., Neves, S. et al. 2005. Formation of regulatory patterns during signal propagation in a mammalian cellular network. *Science* 309: 1078–1083.

Mantovani, A. 2009. Cancer: inflaming metastasis. *Nature* 457: 36–37.

Nguyen, D. X., Bos, P. D., and Massague, J. 2009. Metastasis: from dissemination to organ-specific colonization. *Nature Rev Cancer* 9: 274–284.

Pujana, M. A., Han, J. D., Starita, L. M. et al. 2007. Network modeling links breast cancer susceptibility and centrosome dysfunction. *Nature Genet* 39: 1338–1349.

Quigley, D. A., To, M. D., Perez-Losada, J. et al. 2009. Genetic architecture of mouse skin inflammation and tumour susceptibility. *Nature* 458: 505–508.

Sotiriou, C. and Pusztai, L. 2009. Gene-expression signatures in breast cancer. *N. Engl. J. Med.* 360: 790–800.

Taylor, I. W., Linding, R., Warde-Farley, D. et al. 2009. Dynamic modularity in protein interaction networks predicts breast cancer outcome. *Nature Biotechnol* 27: 199–204.

Uren, A. G., Kool, J., Matentzoglu, K. et al. 2008. Large-scale mutagenesis in p19(ARF)- and p53-deficient mice identifies cancer genes and their collaborative networks. *Cell* 133: 727–741.

Wang, E., Lenferink, A., and O'Connor-McCourt, M. 2007. Cancer systems biology: exploring cancer-associated genes on cellular networks. *Cell Mol Life Sci* 64: 1752–1762.

Wang, E. and Purisima, E. 2005. Network motifs are enriched with transcription factors whose transcripts have short half-lives. *Trends Genet* 21: 492–495.

Wei, C. L., Wu, Q., Vega, V. B. et al. 2006. A global map of p53 transcription-factor binding sites in the human genome. *Cell* 124: 207–219.

Yen, H. C., Xu, Q., Chou, D. M., Zhao, Z., and Elledge, S. J. 2008. Global protein stability profiling in mammalian cells. *Science* 322: 918–923.

CHAPTER 2

Network Biology, the Framework of Systems Biology

Jing-Dong Han

CONTENTS

2.1 INTRODUCTION

Systems biology studies the nature of biological systems that allow gene products to be linked together in nonlethal and even useful combinations and to understand the special properties that allow them to function together to generate different phenotypes (Kirschner 2005). In essence, network biology is the framework of systems biology. Full genome sequences have provided complete part lists for the molecular networks and have spurred the take-off of systems biology. The part lists allow construction of raw maps of the molecular networks at a genome-wide scale and in an unbiased way; these raw maps then form the backbone for further computational annotation and inference of the networks; both the experimentally and computationally derived networks then allow examination of the structural or topological properties of the networks, and their link to biological properties of the networks, such as the robustness and modularity of the networks. Ultimately the dynamic networks under a certain biological condition or within a certain biological context can be identified and

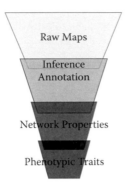

FIGURE 2.1   Information flow of network biology. See text for details.

studied to explain the biological responses and the steady states of the networks, which are reflected as phenotypic traits of the system (Figure 2.1). The phenotypic traits of a network can be its normal biological function, such as cell differentiation, or a pathological condition, such as cancer or diabetes mellitus (Han 2008). In this chapter, I will follow the hierarchical flow of information described above and discuss important concepts at each step of the flow.

2.2 CONSTRUCTION OF THE RAW MAPS OF THE MOLECULAR NETWORKS

The availability of the full gene list in an organism now enables high-throughput mapping of many types of networks.

Yeast two-hybrid (Y2H) screens and co-affinity purification followed by mass spectrometry identification (CoAP-MS) are two major approaches to mapping the protein–protein interaction (PPI) networks, where proteins are connected by direct PPI. The Y2H screens are done in an ex vivo system of the yeast nuclei for often a foreign Orfeome (the collection of open reading frames in an organism); therefore, it identifies all possible interactions without a biological context. However, it is much more sensitive than CoAP and can identify not only stable but also transient interactions. When used correctly, its accuracy can be similar to or better than CoAP, sometimes better than literature annotation (H. Yu et al. 2008a). CoAP-MS results have been more reproducible across different laboratories (von Mering et al. 2002); the PPIs identified can be specific to the particular cell context used for the experiments. However, CoAP-MS is less sensitive for detecting transient interactions than Y2H, and protein complexes identified by this approach do not necessarily represent direct PPIs between any two prey proteins (proteins identified by CoAP-MS), or even between the bait protein (a tagged protein used to pull down the protein complex) and its every prey protein. The edges (interactions) in the PPI networks identified by either Y2H or CoAP-MS are undirected, that is, they can only be detected as mutual binding or complex formation by these approaches. Directed enzyme–substrate interactions—for example, phosphorylation, dephosphorylation, ubiquitination, deubiquitination, methylation, and demethylation—can be represented as directed edges in PPI networks where the source nodes are enzymes and targets are substrates. However, networks of these interactions are still largely based on computational predictions (Linding et al. 2007; see below).

Chromatin immunoprecipitation followed by microarray identification (ChIP-chip) or sequencing (ChIP-seq) and yeast one-hybrid (Y1H) screens are commonly used approaches to mapping regulatory networks, which consist of directed transcription factor (TF)–target gene relationships. Similar to Y2H, Y1H is not context specific, whereas both ChIP-chip and ChIP-seq identify TF–target relationships for a specific state of a specific cell line or tissue. Compared to ChIP-chip, ChIP-seq can generate sharper binding signals at higher resolutions (Barski et al. 2007), given enough depth of sequencing, which is now carried out at many-fold coverage of a genome using the next-generation short sequencing technologies such as Solexa or Solid.

High-throughput synthetic lethal screens have been used to map genetic interaction networks, where the nodes are genes and the edges are undirected synthetic interactions. Directed networks of epistatic relationships have so far not been subjected to high-throughput mapping experiments, although genetic screens for key genes mediating a biological pathway or phenotype are routinely carried out in model organisms, and the genetic relationships between some of the players have been solved one at a time by epistatic analysis.

Similarly, small-scale biochemical and genetic experiments have delineated metabolic networks in great detail. These networks are well annotated in the Kyoto Encyclopedia of Genes and Genomes (KEGG) database, and are represented by network graphs where nodes are metabolites or small chemical molecules, and the edges are the reactions catalyzed by enzymes. High-throughput experiments are also becoming applicable to mapping small molecule-target networks and microRNA-target networks.

2.3 COMPUTATIONAL INFERENCE OF THE MOLECULAR NETWORKS

The large amount of high-throughput data has made it possible to computationally infer biological networks. Nondirectional networks can be inferred by correlations between nodes across many measurements, such as gene expression, evolution, or even literature citations. As correlation measurements, Pearson correlation coefficient (PCC), cosine correlation coefficient, mutual information, Hamming distance, or any other distance can be used. The choice among these distance metrics is largely dependent on the type and number of data. For example, PCC is more suitable for large datasets of continuous values, whereas Hamming distance can be used for relatively small datasets of binary data values. However, these similarity or correlation metrics are often based on data of a single type, and sometimes due to the difficulty of normalization among datasets, they have to be calculated based on a single dataset. This greatly reduces the coverage and accuracy of the inference. To solve these problems, the naïve Byes model or linear regression model can be enlisted to integrate heterogeneous datasets to make the strengths of inference from various datasets comparable based on the same set of gold standard positive (GSP) and gold standard negative (GSN) data (Xia, Dong, and Han 2006). Thus, multiple weak evidences can be turned into a strong prediction. Furthermore, networks inferred through these models have edges weighted by their likelihood of being true positive versus true negative relationships, which allows quantitative evaluation of functional associations of nodes inside such networks.

However, networks inferred simply based on correlations are not only nondirectional, but also not strictly structural in that they contain many transitive relationships although some pruning can be used to minimize them. Bayesian networks (BN) is a framework suitable for inferring network structure and sometimes directed causal relationships among the nodes. Nodes in a BN represent a set of variables, and a BN is a network representation of the joint probability distribution over the set of variables as a product of conditional probability distributions for each node. By evaluating conditional independencies between sets of variables, a Bayesian network structural learning algorithm tries to find the network structure that encodes the factorization of the joint probability distribution that best fits the data. The evaluation can be performed explicitly by testing various conditional independency relations or implicitly by using a scoring function to quantify the fitness of different networks with data. For real-world data, the latter approach typically yields better results by avoiding accumulating errors in multiple statistical testing. The Bayesian information criterion (Tsong et al. 2006) is a frequently used scoring function which contains a term to evaluate the likelihood that the data was generated by the model and another term to penalize the complexity of the model (Needham et al. 2006). To find the network with the highest score, current exact BN structural learning algorithms could scale up to 20–30 variables by using dynamic programming strategies, while heuristic search algorithms may scale up to hundreds of variables and yield close to optimal solutions.

The BN structure inferred is a directed acyclic graph (DAG), representing the conditional independencies between variables, which states the target node is conditionally independent of its nondescendants given its parents (Needham et al. 2006). Due to this constraint, BN must be acyclic; that is, no loops are allowed, even though in the real network they may exist. Such feedback relationships sometimes can be inferred by dynamic BN, which is essentially a large BN unfolded by a template unit over time or other dimensions. The additional temporal information can be exploited to resolve causal feedback relationships without violating the acyclic constraint. It is possible to identify potential causal relationships by finding the consistently directed edges (irreversible edges) within the whole set of equivalent BN structures (Chickering 1995; Kim et al. 2006; H. Yu et al. 2008b). For example, the BN formation can be used to infer the causal relationships among histone modifications and chromatin-binding factors and toward gene expression, whereas the correlations among the modifications and factors only hint at functional relatedness (H. Yu et al. 2008b). Interventions of the nodes can provide more direct evidence and stronger signals for learning causal relationships. For example, a signaling network of 11 molecules can be rather accurately inferred through thousands of single-cell flow cytometry measurements of the level of the molecules upon activating or inhibiting each of them in human primary T cells (Sachs et al. 2005).

Moreover, compared to networks derived from correlations, the conditional probability distribution for each node is able to identify both strong and weak, as well as linear and nonlinear dependencies.

However, due to the requirements of large amounts of incidences or data points for BN inference and noises in data, gene regulatory networks can only be robustly reverse engineered when a large number of gene expression measurements for each node (gene) exists (usually >1000), and is often impossible to learn when the number of nodes is also large. Recent chromatin-immunoprecipitation followed by microarray (ChIP-chip) or by deep sequencing (ChIP-seq) technologies has generated enormous amounts of genomic DNA-binding profiles for various regulatory factors, and made these data ideal for BN learning.

2.4 ANNOTATION OF THE RAW MAP AND COMPUTATIONALLY INFERRED NETWORKS BY DATA INTEGRATION

Large-scale data and computational predictions contain many false positives. These can be technical or biological false positives. Data integration when used to filter the results according to their confidence levels can minimize the technical false positives. To generate high confidence datasets, approaches from simple intersection analysis (Han et al. 2004; Said et al. 2004; von Mering et al. 2002) to more quantitative probability-based scoring systems, such as the naïve Bayes classifier, can be used where interactions are ranked by their likelihood of being true positives versus true negatives (see above). To remove biological false positives, other context information related to a certain biological process (e.g., early embryogenesis; Gunsalus et al. 2005) or compartment (e.g., mitochondria; Pagliarini et al. 2008) can be used to filter the raw maps and reinforce the biological relevance of the interactions.

2.5 TOPOLOGICAL PROPERTIES OF THE MOLECULAR NETWORKS

The topological features of a network can be described by statistical metrics, and by their unique components, such as subgraphs in the network. Many statistical metrics have been developed to characterize network graphs in general, some of which are especially relevant to biological networks.

2.5.1 Statistical Metrics to Characterize the Networks and Their Biological Relevance

The *shortest path length* (SPL) is the smallest number of edges in a network required to connect two nodes. It measures how close the nodes are to each other. Two proteins having shorter SPL in the PPI network are more likely to have similar phenotypes and gene expression profiles than those having longer SPL (Gunsalus et al. 2005). *Characteristic path length* (CPL) is the average SPL between any pair of nodes that are directly or indirectly connected in a network. It measures how tightly connected the whole network is. Nearly all biological networks have very short CPL, suggesting they are tightly connected. The importance of a node to the overall connectivity of a network can be reflected as the change of the network's CPL upon simulated removal of the node.

Node *degree* (k) is the number of links a node makes. The proportions of nodes of different degrees ($p(k)$) within a network are collectively called the *degree distribution* of the network. When $log(p(k))$ is linear to $log(k)$, the network can be called scale-free, which approximately fits all the known biological networks. Such a distribution has been suggested to maintain structural robustness of the networks (Albert et al. 2000). Hubs, or high degree nodes, in scale-free networks are more likely to be essential to the structural integrity of the network (Albert et al. 2000), and in yeast PPI networks, reflected as the essentiality to the survival of the yeast (Jeong et al. 2001).

Compared to this "degree centrality," "*betweenness* centrality" has been shown to be even better correlated with the essentiality of the node (H. Yu et al. 2007). Betweenness of a node or an edge is quantified as the number of shortest paths passing through the node or the edge.

2.5.2 Network Motifs

Other than the quantitative metrics to describe a network and its nodes and edges, the presence of unique subgraphs in a network also distinguishes biomolecular networks from other types of networks. Network motifs are small subgraphs that are statistically over-represented compared to random expectation. For example, feedback and feed-forward motifs are enriched in the regulatory networks (Milo et al. 2002), where negative feedback loops can stabilize the signal passing through it, or serve to terminate an input signal, and feed-forward loops suggest combinatorial effects or signal redundancy.

2.5.3 Modularity of the Networks

Modularity is another feature of biomolecular networks. A network module is a context-coherent subnetwork with defined inputs and outputs, as well as conditionally comparable temporal and spatial profiles (Papin et al. 2005). However, depending on the data collection process, not all the input, output, temporal, and spatial parameters are available for a system. In practice, modules are often evaluated by their functional homogeneity as annotated by benchmark standards (such as gene ontology, http://www.geneontology.org/) and can be dissected from the whole network based on the coherency of the nodes or edges under one or more contexts, such as structural or dynamic contexts.

Structural modules are defined, based on network configurations alone without consulting the temporal or spatial context, as subnetworks where nodes are more densely connected within the subnetwork than toward the outside of the subnetwork, or more than random expectation. The modularity metric Q has been developed to quantitatively evaluate this (Newman and Girvan 2004).

Network clusters or cliques are a typical type of structural module. The tightness of clustering or the cliquishness of a cluster can be measured in many ways. The most frequently used is the average clustering coefficient ($C(v)$) among nodes in a subgraph. The $C(v)$ of a node is the ratio of the number of observed edges over that of all possible edges among the interactors of a node (Watts and Strogatz 1998). In PPI networks, subgraphs that have high average $C(v)$ often correspond to protein complexes (Bader and Hogue 2003). Larger structural modules can also be identified based on similarity of shortest distances between node pairs (Rives and Galitski 2003), the enrichment or conservation of subgraphs compared to random expectation (Sharan et al. 2005), or the number and distance of interactions to reference genes (Kohler et al. 2008).

Other than structural coherence or connectivity density, edge type consistency has been used to find epistatic modules (Segre et al. 2005); the expression profile similarity and dissimilarity have been used to find dynamic modules active during a biological process (Xue et al. 2007).

Modularity of the molecular networks is also reflected by the subnetworks that are associated with disease development (Goh et al. 2007). This feature allows predicting potential disease-associated genes through the modular networks containing the known disease-associated genes. A breast cancer gene network has been dissected from an integrated functional interaction network through coexpression with multiple known breast cancer genes. The network module has served as a template for ranking novel breast cancer related genes (Pujana et al. 2007). Based on direct PPIs (Lage et al. 2007), shortest distance in the PPI network (Wu et al. 2008), or interaction strength (Kohler et al. 2008) to already mapped Mendelian disease-associated genes, algorithms have been developed to narrow down the disease-associated genes within known disease-associated genomic loci, which each contain many genes, sometimes over 100 genes.

Based on the coexpression levels of a hub with its interacting partners, hub proteins in the yeast PPI network can be categorized into "date" (intermodule) and "party" (intramodule) hubs, with low and high coexpression levels, respectively. Functional modules are found to be organized around party hubs, and connected at a higher level by the date hubs through more dynamic interactions. This hierarchical structure implies dynamic modularity in the yeast PPI network (Han et al. 2004). Interestingly, in the human PPI network the date, or intermodule, hubs that have low cross-tissue coexpression levels with their interactors are the best predictors for breast cancer prognosis, suggesting that the aggressiveness of breast cancer is associated with altered dynamic modular organization of the network in cancer cells (Taylor et al. 2009).

Although topological properties are important to the biological functions of the networks, they are not the only determinants of functions. The kinetic properties, for instance, are another important factor contributing to functions. In fact, some of the topological properties contribute to the biological functions through affecting the kinetic properties of the networks. For example, negative feedback loops have been shown to be required for ensuring the signal sensitivity and response fidelity of a pathway to its input signal concentrations (Bhalla, Ram, and Iyengar 2002; R.C. Yu et al. 2008).

2.6 EVOLUTION OF MOLECULAR NETWORKS

The nature of biological systems that allows gene products to be linked together in nonlethal and even useful combinations (Kirschner 2005) is achieved through evolution. Therefore, the ultimate explanation of the structure and dynamics of the network lies in the fitness of the organism to the environment and the selection pressures in an organism's evolutionary history.

When a protein complex or a pathway achieves an essential biological function, it is likely to be preserved through evolution and to experience negative selection pressures (Kirschner 2005; Sharan et al. 2005; Wuchty, Oltvai, and Barabasi 2003). However, the network motifs within different sets of paralogous genes in an organism are mostly not conserved (Conant and Wagner 2003; Teichmann and Babu 2004). Gene duplication has been a mechanism to generate new connectivity patterns in the new and old set of paralogous genes by relieving the negative selection pressure on singleton genes, thereby achieving new functions. Network structure analysis has suggested that the two paralogous gene groups derived from the whole genome duplication in yeast generated separate functional subnetworks (Conant and Wolfe 2006). Even without gene duplication, network diversification can occur while maintaining its functions. The evolutionary history of the yeast hormonal response elements illustrates a step-wise adaptation toward functionally identical structural diversification through an intermediate state where both new and old regulatory elements function simultaneously (Tsong et al. 2006).

Given that the structures of the network motifs or connectivity configuration per se are not direct targets of evolutionary selection on molecular networks, what are the direct target(s)? The structures of genes and proteins seem to contribute directly to the complexity of molecular networks, and are selected through evolution. Gene duplication, generation of gene and protein variants by alternative splicing, as well as thriving noncoding regulatory genes and elements all undoubtedly contribute to the increase in the number of nodes of molecular networks, and the temporally and spatially dynamic concentration of the nodes, and therefore linearly increase the complexity of the molecular networks in higher and more complex organisms.

However, increasing a node's connectivity in the network can exponentially increase the complexity of the network by adding edges without an increase in the number of nodes, or the size of the proteome or genome. The average PPI domain coverage (the percentage of total length of a protein occupied by PPI domains) of proteins in an organism is highly correlated with the complexity of the organism as estimated by the number of cell types of the organism. Interestingly in both human and yeast PPI networks, a protein's PPI domain coverage is also highly correlated with the protein's interaction degrees, suggesting an adaptation of the proteins toward higher compactness in domain structures occurs while organism and PPI network complexity increases through evolution (Figure 2.2A; Xia et al. 2008). The PPI domain organizations on a protein are also associated with the dynamics of the interactions. A protein containing a single PPI interface is likely a date hub and is involved in multiple dynamic interactions, whereas a protein containing multiple PPI interfaces tends to be a party hub and binds to its multiple interactors simultaneously (Kim et al. 2006; Figure 2.2B).

Therefore, the selection pressure on network evolution is reflected and preserved verbatim at the genome level through protein domain structures. In this sense, the genome sequences not only provide parts lists for network biology, when examined from the evolutionary perspectives, they also reveal the evolutionary history and selection pressure experienced by the molecular networks at the systems and organism level.

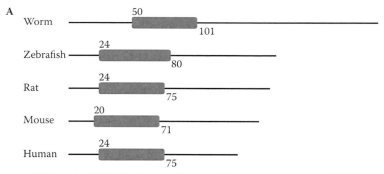

A

Worm

50

101

Zebrafish

24

80

Rat

24

75

Mouse

20

71

Human

24

75

■ PPI domain (IPR001092 Basic helix-loop-helix dimerisation region bHLH)

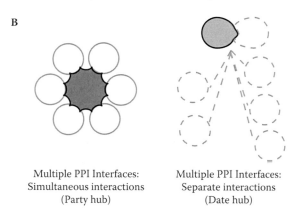

B

Multiple PPI Interfaces:
Simultaneous interactions
(Party hub)

Multiple PPI Interfaces:
Separate interactions
(Date hub)

FIGURE 2.2 Adaptations in PPI domain arrangement on proteins are associated with network connectivity and dynamic network connectivity. (A) Proteins that contain PPI domains tend to lose the domain-free regions to make the proteins more compact for PPI domains, and hence specializing in PPI functions. The ancestral protein of an exemplary human protein, myc associated factor x (MAX), gradually sheds the C-terminus domain-free tail from the nematode *C. elegans*, zebra fish, rat, mouse to human. Modified based on Xia et al. (Xia et al. 2008). (B) In PPI networks, hub proteins with multiple PPI interfaces tend to be party hubs, whereas those with a single PPI interface tend to be date hubs. Modified based on Kim et al. (Kim et al. 2006).

REFERENCES

Albert, R., Jeong, H., and Barabasi, A. L. 2000. Error and attack tolerance of complex networks. *Nature* 406: 378–382.

Bader, G. D. and Hogue, C. W. 2003. An automated method for finding molecular complexes in large protein interaction networks. *BMC Bioinformatics* 4: 2.

Barski, A., Cuddapah, S., Cui, K. et al. 2007. High-resolution profiling of histone methylations in the human genome. *Cell* 129: 823–837.

Bhalla, U. S., Ram, P. T., and Iyengar, R. 2002. MAP kinase phosphatase as a locus of flexibility in a mitogen-activated protein kinase signaling network. *Science* 297: 1018–1023.

Chickering, D. M. 1995. A transformational characterization of equivalent Bayesian network structures. *Proceedings of Eleventh Conference on Uncertainty in Artificial Intelligence*, San Francisco, CA: Morgan Kaufman Publishers, 87–98.

Conant, G. C. and Wagner, A. 2003. Convergent evolution of gene circuits. *Nature Genet* 34: 264–266.

Conant, G. C. and Wolfe, K. H. 2006. Functional partitioning of yeast co-expression networks after genome duplication. *PLoS Biol* 4: e109.

Goh, K. I., Cusick, M. E., Valle, D. et al. 2007. The human disease network. *Proc Natl Acad Sci USA* 104: 8685–8690.

Gunsalus, K. C., Ge, H., Schetter, A. J. et al. 2005. Predictive models of molecular machines involved in *Caenorhabditis elegans* early embryogenesis. *Nature* 436: 861–865.

Han, J. D. 2008. Understanding biological functions through molecular networks. *Cell Res* 18: 224–237.

Han, J. D., Bertin, N., Hao, T. et al. 2004. Evidence for dynamically organized modularity in the yeast protein-protein interaction network. *Nature* 430: 88–93.

Jeong, H., Mason, S. P., Barabasi, A. L., and Oltvai, Z. N. 2001. Lethality and centrality in protein networks. *Nature* 411: 41–42.

Kim, P. M., Lu, L. J., Xia, Y., and Gerstein, M. B. 2006. Relating three-dimensional structures to protein networks provides evolutionary insights. *Science* 314: 1938–1941.

Kirschner, M. W. 2005. The meaning of systems biology. *Cell* 121: 503–504.

Kohler, S., Bauer, S., Horn, D., and Robinson, P. N. 2008. Walking the interactome for prioritization of candidate disease genes. *Am. J. Hum. Genet.* 82: 949–958.

Lage, K., Karlberg, E. O., Storling, Z. M. et al. 2007. A human phenome-interactome network of protein complexes implicated in genetic disorders. *Nature Biotechnol* 25: 309–316.

Linding, R., Jensen, L. J., Ostheimer, G. J. et al. 2007. Systematic discovery of *in vivo* phosphorylation networks. *Cell* 129: 1415–1426.

Milo, R., Shen-Orr, S., Itzkovitz, S. et al. 2002. Network motifs: simple building blocks of complex networks. *Science* 298: 824–827.

Needham, C. J., Bradford, J. R., Bulpitt, A. J., and Westhead, D. R. 2006. Inference in Bayesian networks. *Nature Biotechnol* 24: 51–53.

Newman, M. E. and Girvan, M. 2004. Finding and evaluating community structure in networks. *Phys Rev E Stat Nonlin Soft Matter Phys* 69: 026113.

Pagliarini, D. J., Calvo, S. E., Chang, B. et al. 2008. A mitochondrial protein compendium elucidates complex I disease biology. *Cell* 134: 112–123.

Papin, J. A., Hunter, T., Palsson, B. O., and Subramaniam, S. 2005. Reconstruction of cellular signalling networks and analysis of their properties. *Nature Rev Mol Cell Biol* 6: 99–111.

Pujana, M. A., Han, J. D., Starita, L. M. et al. 2007. Network modeling links breast cancer susceptibility and centrosome dysfunction. *Nature Genet* 39: 1338–1349.

Rives, A. W. and Galitski, T. 2003. Modular organization of cellular networks. *Proc Natl Acad Sci USA* 100: 1128–1133.

Sachs, K., Perez, O., Pe'er, D., Lauffenburger, D. A., and Nolan, G. P. 2005. Causal protein-signaling networks derived from multiparameter single-cell data. *Science* 308: 523–529.

Said, M. R., Begley, T. J., Oppenheim, A. V., Lauffenburger, D. A., and Samson, L. D. 2004. Global network analysis of phenotypic effects: protein networks and toxicity modulation in *Saccharomyces cerevisiae*. *Proc Natl Acad Sci USA* 101: 18006–18011.

Segre, D., Deluna, A., Church, G. M., and Kishony, R. 2005. Modular epistasis in yeast metabolism. *Nature Genet* 37: 77–83.

Sharan, R., Suthram, S., Kelley, R. M. et al. 2005. Conserved patterns of protein interaction in multiple species. *Proc Natl Acad Sci USA* 102: 1974–1979.

Taylor, I. W., Linding, R., Warde-Farley, D. et al. 2009. Dynamic modularity in protein interaction networks predicts breast cancer outcome. *Nature Biotechnol* 27: 199–204.

Teichmann, S. A. and Babu, M. M. 2004. Gene regulatory network growth by duplication. *Nature Genet* 36: 492–496.

Tsong, A. E., Tuch, B. B., Li, H., and Johnson, A. D. 2006. Evolution of alternative transcriptional circuits with identical logic. *Nature* 443: 415–420.

von Mering, C., Krause, R., Snel, B. et al. 2002. Comparative assessment of large-scale data sets of protein-protein interactions. *Nature* 417: 399–403.

Watts, D. J. and Strogatz, S. H. 1998. Collective dynamics of "small-world" networks. *Nature* 393: 440–442.

Wu, X., Jiang, R., Zhang, M. Q., and Li, S. 2008. Network-based global inference of human disease genes. *Mol Syst Biol* 4: 189.

Wuchty, S., Oltvai, Z. N., and Barabasi, A. L. 2003. Evolutionary conservation of motif constituents in the yeast protein interaction network. *Nature Genet* 35: 176–179.

Xia, K., Dong, D., and Han, J. D. 2006. IntNetDB v1.0: an integrated protein-protein interaction network database generated by a probabilistic model. *BMC Bioinformatics* 7: 508.

Xia, K., Fu, Z., Hou, L., and Han, J. D. 2008. Impacts of protein-protein interaction domains on organism and network complexity. *Genome Res* 18: 1500–1508.

Xue, H., Xian, B., Dong, D. et al. 2007. A modular network model of aging. *Mol Syst Biol* 3: 147.

Yu, H., Braun, P., Yildirim, M. A. et al. 2008a. High-quality binary protein interaction map of the yeast interactome network. *Science* 322: 104–110.

Yu, H., Kim, P. M., Sprecher, E., Trifonov, V., and Gerstein, M. 2007. The importance of bottlenecks in protein networks: correlation with gene essentiality and expression dynamics. *PLoS Comput Biol* 3: e59.

Yu, H., Zhu, S., Zhou, B., Xue, H., and Han, J. D. 2008b. Inferring causal relationships among different histone modifications and gene expression. *Genome Res* 18: 1314–1324.

Yu, R. C., Pesce, C. G., Colman-Lerner, A. et al. 2008. Negative feedback that improves information transmission in yeast signalling. *Nature* 456: 755–761.

CHAPTER 3

Reconstructing Gene Networks Using Gene Expression Profiles

Mario Lauria and Diego di Bernardo

CONTENTS

3.1 INTRODUCTION

Gene expression microarrays, and more recently deep sequencing approaches, yield quantitative and semiquantitative data on cell status in a specific condition and time. Molecular biology is rapidly evolving into a quantitative science and, as such, it is relying increasingly on engineering and physics to make sense of high-throughput data. The aim is to infer, or reverse engineer, from gene expression data, the regulatory interactions among genes using computational algorithms. There are two broad classes of reverse-engineering algorithms (Gardner et al. 2005): those based on the physical interaction approach that aim at identifying interactions among transcription factors and their target genes (gene-to-sequence interaction), and those based on the influence interaction approach that try to relate the expression of a gene to the expression of the other genes in the cell (gene-to-gene interaction), rather than relating it to sequence motifs found in its promoter (gene-to-sequence). We will refer to the ensemble of these "influence interactions" as gene networks.

The interaction between two genes in a gene network does not necessarily imply a physical interaction, but can also refer to an indirect regulation via proteins, metabolites, and ncRNA that have not been measured directly. Influence interactions include physical interactions, if the two interacting partners are a transcription factor and its target, or two proteins in the same complex. Generally, however, the meaning of influence interactions is not well defined and depends on the mathematical formalism used to model the network. Nonetheless, influence networks do have practical utility for: (1) identifying functional modules, that is, identifying the subset of genes that regulate each other with multiple (indirect) interactions but have few regulations to other genes outside the subset; (2) predicting the behavior of the system following perturbations, that is, gene network models can be used to predict the response of a network to an external perturbation and to identify the genes directly "hit" by the perturbation (di Bernardo et al. 2005), a situation often encountered in the drug discovery process, where one needs to identify the genes that are directly interacting with a compound of interest; (3) identifying real physical interactions, by integrating the gene network with additional information from sequence data and other experimental data (i.e., chromatin immunoprecipitation, yeast two-hybrid assay, etc.).

In addition to reverse-engineering algorithms, network visualization tools are available online to display the network surrounding a gene of interest by extracting information from literature and experimental datasets, such as Cytoscape (Shannon et al. 2003; http://www.cytoscape.org/features.php) and Osprey (Breitkreutz, Stark, and Tyers 2003; http://biodata.mshri.on.ca/osprey/servlet/Index).

Here we will focus on gene network inference algorithms (the influence approach). A description of other methods based on the physical approach and more details on computational aspects can be found in Ambesi and di Bernardo 2006; Beer and Tavazoie 2004; Foat, Morozov, and Bussemaker 2006; Gardner and Faith 2005; Prakash and Tompa 2005; Tadesse, Vannucci, and Lio 2004. We will also briefly describe two "improper" reverse-engineering tools (MNI and TSNI), whose main focus is not inferring interactions among genes from gene expression data, but rather identification of the targets of the perturbation [point (2) above].

3.2 GENE NETWORK INFERENCE ALGORITHMS

In this section we will give an overview of different approaches to the inference of gene networks, and then we will describe in more detail the regression-based approach. The diagram in Figure 3.1 illustrates the relationship between the different algorithms and their specific domains of application.

In the following we will indicate gene expression measurement of gene i with the variable x_i, the set of expression measurements for all the genes with X, and the interaction between genes i and j with a_{ij}. X may consist of time-series gene expression data of N genes

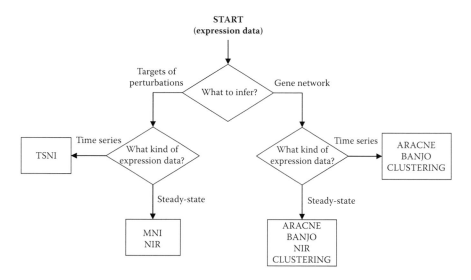

FIGURE 3.1 Diagram of the relationship between the different algorithms and their specific domains of application.

in M time points (i.e., gene expression changing dynamically with time), or measurements taken at steady state in M different conditions (i.e., gene expression levels in homeostasis). Some inference algorithms can work on both kinds of data, whereas others have been designed specifically to analyze one or the other.

Depending on the inference algorithm used, the resulting gene network can be either an undirected graph, that is, the direction of the interaction is not specified ($a_{ij} = a_{ji}$), or a directed graph specifying the direction of the interaction, that is, gene j regulates gene i (and not vice versa; $a_{ij} \neq a_{ji}$). A directed graph can also be labeled with a sign and strength for each interaction, signed directed graph, where a_{ij} has a positive, zero, or negative value indicating activation, no interaction, and repression, respectively.

3.2.1 Coexpression Networks and Clustering Algorithms

Correlation networks were among the first attempts to infer metabolic networks starting from time series measurements of species concentration (Arkin and Ross 1995). More advanced correlation-based methods were later developed and applied to the inference of genetic networks from high-throughput technologies (Rice, Tu, and Stolovitzky 2005; Ucar et al. 2007). Correlation among the expression profile of pairs of genes can be used alone to generate a network of interactions between genes, typically by setting a significance threshold to select meaningful interactions (Conant and Wolfe 2006), or it can be used as part of a clustering approach. Clustering, although not properly a network inference algorithm, is the current method of choice to visualize and analyze gene expression data. Clustering is based on the idea of grouping genes with similar expression profiles in clusters (Eisen et al. 1998). The similarity is measured by a distance metric, as for example the correlation coefficient among pairs of genes. The number of clusters can be set either automatically or by the user, depending on the clustering algorithm used (Amato et al. 2006;

Eisen et al. 1998). The rationale behind clustering is that coexpressed genes (i.e., genes in the same cluster) have a good probability of being functionally related (Eisen et al. 1998). This does not imply, however, that there is a direct interaction among coexpressed genes, since genes separated by one or more intermediaries (indirect relationships) may be highly coexpressed. It is therefore important to understand what can be gained by advanced gene network inference algorithms, whose aim is to infer direct interactions among genes, as compared to "simple" clustering for the purpose of gene network inference.

The most common clustering approach is hierarchical clustering (Eisen et al. 1998), where relationships among genes are represented by a tree whose branch lengths reflect the degree of similarity between genes, as assessed by a pairwise similarity function such as the Pearson correlation coefficient:

$$r_{ij} = \frac{\sum_{k=1}^{M}(x_i(k)x_j(k))}{\sqrt{\sum_{k=1}^{M} x_i^2(k)\sum_{k=1}^{M} x_j^2(k)}} \tag{3.1}$$

For a set of n profiles all the pairwise correlation coefficient r_{ij} are computed; the highest value (representing the most similar pair of genes) is selected and a node in the tree is created for this gene pair with a new expression profile given by the average of the two profiles. The process is repeated by replacing the two genes with a single node, and all pairwise correlations among the $n - 1$ profiles (i.e., $n - 2$ profiles from single genes plus 1 of the gene pair) are computed. The process stops when only one element remains. Clusters are obtained by cutting the tree at a specified branch level.

3.2.2 Bayesian Networks

A Bayesian network is a graphical model for probabilistic relationships among a set of random variables X_i, with $i = 1 \ldots n$. These relationships are encoded in the structure of a directed acyclic graph G whose vertices (or nodes) are the random variables X_i. The relationships between variables are described by a joint probability distribution $P(X_1,\ldots,X_n)$ that is consistent with the independence assertions embedded in the graph G and has the form:

$$P(X_1,\ldots, X_N) = \prod_{i=1}^{N} P(X_i = x_i \mid X_j = x_j,\ldots, X_{j+p} = x_{j+p}) \tag{3.2}$$

where the $p + 1$ genes on which the probability is conditioned are called the *parents* of gene i and represent its regulators, and the joint probability density is expressed as a product of conditional probabilities by applying the chain rule of probabilities and independence. This rule is based on the Bayes theorem: $P(A,B) = P(B|A)P(A) = P(A|B) P(B)$.

We observe that the JPD (joint probability distribution) can be decomposed as the product of conditional probabilities as in Equation 3.2 only if the *Markov assumption* holds, that is, each variable X_i is independent of its nondescendants, given its parent in the directed acyclic graph G.

In order to reverse engineer a Bayesian network model of a gene network we must find the directed acyclic graph G (i.e., the regulators of each transcript) that best describes the gene expression data X, where X is assumed to be a steady-state dataset. This is done by choosing a scoring function that evaluates each graph G (i.e., a possible network topology) with respect to the gene expression data X, and then searching for the graph G that maximizes the score.

The score can be defined using Bayes rule: $P(G|X) = P(X|G)P(G)/P(X)$, where $P(G)$ can either contain some a priori knowledge on network structure, if available, or can be a constant noninformative prior, and $P(X/G)$ is a function, to be chosen by the algorithm, that evaluates the probability that the data X has been generated by the graph G. The most popular scores are the Bayesian information criteria (Tsong et al. 2006) or Bayesian Dirichlet equivalence (BDe). Both scores incorporate a penalty for complexity to guard against overfitting of data.

Trying out all the possible combinations of interaction among genes, all the possible graphs of G, and choosing the G with the maximum Bayesian score is an NP-hard problem. Therefore, a heuristic search method is used, like the greedy-hill climbing approach, Markov chain Monte Carlo method, or simulated annealing.

In Bayesian networks, the learning problem is usually underdetermined and several high-scoring networks are found. To address this problem, one can use model averaging or bootstrapping to select the most probable regulatory interactions and to obtain confidence estimates for the interactions. For example, if a particular interaction between two transcripts repeatedly occurs in high-scoring models, one gains confidence that this edge is a true dependency. Alternatively, one can augment an incomplete dataset with prior information to help select the most likely model structure. Bayesian networks cannot contain cycles (i.e., no feedback loops). This restriction is the principal limitation of Bayesian network models. *Dynamic Bayesian networks* overcome this limitation. Dynamic Bayesian networks are an extension of Bayesian networks able to infer interactions from a dataset D consisting of time-series data rather than steady-state data. We refer the reader to Yu et al. (2004).

A word of caution: Bayesian networks model probabilistic dependencies among variables and not causality, that is, the parents of a node are not necessarily also the direct causes of its behavior. However, we can interpret the edge as a causal link if we assume that the *causal Markov condition* holds. This can be stated simply as: a variable X is independent of every other variable (except the targets of X) conditional on all its direct causes. It is not known whether this assumption is a good approximation of what happens in real biological networks.

For more information and a detailed study of Bayesian networks for gene network inference we refer the reader to Friedman et al. (2000).

Banjo is a gene network inference software that has been developed by the group of Hartemink (Yu et al. 2004). Banjo is based on the Bayesian networks formalism and implements both Bayesian and dynamic Bayesian networks; therefore, it can infer gene networks from steady-state gene expression data, or from time-series gene expression data. Banjo outputs a signed directed graph indicating regulation among genes. Banjo can analyze both steady-state and time-series data. In the case of steady-state data, Banjo, as well as the other Bayesian networks algorithms, is not able to infer networks involving cycles (e.g., feedback or feed forward loops). Other Bayesian network inference algorithms for which software is available have been proposed (Friedman and Elidan 2000; Murphy 2001).

3.2.3 Information-Theoretic Approaches

Information-theoretic approaches use a generalization of the pair-wise correlation coefficient in Equation 3.1, called mutual information (MI), to compare expression profiles from a set of microarrays. For each pair of genes, their MI_{ij} is computed and the edge $a_{ij} = a_{ji}$ is set to 0 or 1 depending on a significance threshold to which MI_{ij} is compared. MI can be used to measure the degree of independence between two genes. Mutual information MI_{ij} between gene i and gene j is computed as:

$$MI_{ij} = H_i + H_j - H_{ij} \qquad (3.3)$$

where, H_i and the joint entry H_{ij} are defined as:

$$H_k = -\sum_{k=1}^{n} p(x_k)\log(p(x_k)) \qquad (3.4)$$

$$H_{ij} = -\sum_{k=1}^{n}\sum_{l=1}^{m} P(x_j, x_i)\log(P(x_j, x_i))$$

The entropy H_k has many interesting properties; specifically it reaches a maximum for uniformly distributed variables, that is, the higher the entropy, the more randomly distributed are gene expression levels across the experiments. From the definition, it follows that MI becomes zero if the two variables x_i and x_j are statistically independent [$P(x_i x_j) = P(x_i)P(x_j)$], since their joint entropy $H_{ij} = H_i + H_j$. A higher MI indicates that the two genes are nonrandomly associated to each other. It can be easily shown that MI is symmetric, $M_{ij} = M_{ji}$; therefore, the network is described by an undirected graph G, thus differing from Bayesian networks (directed acyclic graph).

MI is more general than the Pearson correlation coefficient. This quantifies only linear dependencies between variables, and a vanishing Pearson correlation does not imply that two variables are statistically independent. In practical application, however, MI and Pearson correlation may yield almost identical results (Steuer et al. 2002).

The definition of MI in Equation 3.3 requires each data point, that is, each experiment, to be statistically independent from the others; thus information-theoretic approaches, as described here, can deal with steady-state gene expression datasets or with time-series data as long as the sampling time is long enough to assume that each point is independent of the previous points.

Edges in networks derived by information-theoretic approaches represent statistical dependencies among gene expression profiles. As in the case of Bayesian networks, the edge does not represent a direct causal interaction between two genes, but only a statistical dependency. A "leap of faith" must be made in order to interpret the edge as a direct causal interaction.

It is possible to derive the information-theoretic approach as a method to approximate the joint probability distribution (JPD) of gene expression profiles, as is done for Bayesian networks. We refer the interested reader to Margolin et al. (2006).

ARACNE (Basso et al. 2005; Margolin et al. 2006) belongs to the family of information-theoretic approaches to gene network inference first proposed by Butte and Kohane (2000) with their relevance network algorithm. ARACNE computes MI_{ij} for all pairs of genes i and j in the dataset. M_{ij} is estimated using the method of Gaussian kernel density (Steuer et al. 2002). Once M_{ij} for all gene pairs has been computed, ARACNE excludes all the pairs for which the null hypothesis of mutually independent genes cannot be ruled out ($H_0 : MI_{ij} = 0$). A p-value for the null hypothesis, computed using Monte Carlo simulations, is associated to each value of the mutual information. The final step of this algorithm is a pruning step that tries to reduce the number of false positives (i.e., inferred interactions among two genes that are not direct causal interactions in the real biological pathway). They use the data processing inequality (DPI) principle that asserts that if both (i, j) and (j, k) are directly interacting, and (i, k) are indirectly interacting through j, then $MI_{i,k} \leq min(MI_{ij}, MI_{jk})$. This condition is necessary but not sufficient, that is, the inequality can be satisfied even if (i, k) are directly interacting; therefore, the authors acknowledge that by applying this pruning step using DPI they may be discarding some direct interactions as well.

3.3 A MORE IN-DEPTH LOOK: ODE-BASED APPROACHES

3.3.1 Details of the ODE-Based Approach

In recent studies, algorithms based on Ordinary Differential Equations (ODE) have been developed that use a collection of steady-state RNA expression measurements (network identification by multiple regression, NIR; microarray network identifcation, MNI; sparse simultaneous equation model with Lasso, SSEMLasso), or time-series measurements (time-series network identification, TSNI) following transcriptional perturbations, to reconstruct gene-gene interactions and to identify the mediators of the activity of a drug (Bansal, Gatta, and di Bernardo 2006; Cosgrove et al. 2008; di Bernardo et al. 2005; Gardner et al. 2003). Other algorithms based on ODEs have been proposed in the literature (Bonneau et al. 2006; D'haeseleer et al. 1999; Tegner et al. 2003; van Someren et al. 2006).

In ODE-based approaches the gene network dynamics describing the time evolution of the mRNA concentration transcribed by each gene is modeled by a set of ordinary differential equations:

$$\frac{dx}{dt} = f(x,u) \tag{3.5}$$

where x represents the mRNA concentrations of the genes in the network and u is a set of transcriptional perturbations. Assuming that the cell under investigation is at equilibrium near a stable steady-state point, we can apply a small perturbation to each of its genes. A perturbation is small if it does not drive the network out of the basin of attraction of its stable steady-state point and if the stable manifold in the neighborhood of the steady-state point is approximately linear. With these assumptions the set of nonlinear rate equations can be linearized near their stable steady-state point.

Thus, for each gene i, in a network of N genes, we can write the above equations in the form:

$$\frac{dx_{il}}{dt} = \sum_{j=1}^{m} a_{ij} x_{jl} + u_{il} = a_i^T x_l + u_{il}, \quad i=1,\ldots,N, \quad l=1,\ldots M \qquad (3.6)$$

where x_{il} is the mRNA concentration of gene i following the perturbation in experiment l; a_{ij} represents the influence of gene j on gene i; u_{il} is an external perturbation to the expression of gene i in experiment l.

Identifying the gene interactions network means to derive the matrix A of the coefficient a_{ij} for each gene i in the model described above. This can be accomplished if we measure the mRNA concentration of all the N genes at steady state (i.e., $\dot{x}_l = 0$) in M experiments and then solve the system of equations:

$$AX = -U \qquad (3.7)$$

where X is an $N \times M$ matrix whose columns are the x_l vectors and U is an $N \times M$ matrix whose columns are the u_l vectors. This system can be solved only if $M \geq N$; however, the recovered weights A will be extremely sensitive to noise both in the data and in the perturbations and thus unreliable unless we overdetermine the system (increasing the number of experiments or assuming the number of regulators of each gene, k, is much smaller than M).

Having described the underlying model, we now briefly outline the method used by NIR to infer A, given U and X. In order to estimate the coefficients of the gene interaction network, NIR essentially solves a *linear regression* problem for each equation in (3.7) assuming an upper bound *of* $k = 10$ regressors for each predicted gene, that is, we assume that each gene can be regulated at most by 10 other genes (Figure 3.2). The value of 10 was found

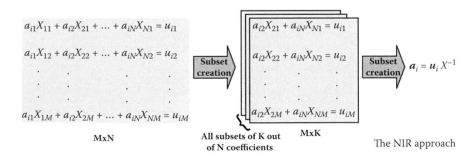

FIGURE 3.2 (See color insert following page 332.) Network identification by multiple regression algorithm.

empirically to be the best compromise between computational complexity and complete-ness of the results. The set of variables comprising the regressor set was chosen according to the *residual sum of square error* (Bonneau et al. 2006) minimization criterion. Due to the infeasibility of exhaustively searching the best set of regressors in the space of all the possible *n*-ples (for each value of *n*), we used the forward/step-wise method (di Bernardo et al. 2004) to solve the variable selection problem, for each cardinality k of regressors. We used the Akaike's final prediction error (FPE) to solve the model selection problem, that is, to select the final value of k within a range of values. In our experiments we used a range of (2, 10) unless specified otherwise.

The MNI algorithm (di Bernardo et al. 2005) is based on Equation 3.7 as well, and uses steady-state data like NIR, but importantly, each microarray experiment can result from any kind of perturbation, that is, we do not require knowledge of U. MNI is dif-ferent from other inference methods since the inferred network is used not per se but to filter the gene expression profile following a treatment with a compound to determine pathways and genes directly targeted by the compound. This is achieved in two steps: in the first step the network A is estimated using a training set of expression data. In the second step, the regulatory network is used as a filter to determine the genes affected by the test condition; Equation 3.7) and to compute the values u_{il} for each i, using the a_{ij} estimated in the previous step and X being the set of expression profiles obtained fol-lowing the treatment. The u_{il} different from 0 represent the genes that are directly hit by the compound. The output is a ranked list of genes; the genes at the top of the list are the most likely targets of the compound (i.e., the ones with the highest value of u_{il}).

The network inferred by MNI could be used per se, and not only as a filter. However, if we do not have any knowledge about which genes have been perturbed directly in each perturbation experiment in dataset X (right-hand side in Equation 3.7), then, differently from NIR, the solution to Equation 3.7 is not unique, and we can only infer one out of many possible networks that can explain the data. What remains unique are the predic-tions (Milo et al. 2002), that is, all the possible networks predict the same u_{il}.

The TSNI (time series network identification) algorithm (Bansal, Gatta, and di Bernardo 2006) identifies the gene network (a_{ij}) as well as the direct targets of the perturbations (Milo et al. 2002). TSNI is based on Equation 3.6 and is applied when the gene expression data are dynamic (Arkin and Ross 1995). To solve Equation 3.6, we need the values of $\dot{x}_i(t_k)$ for each gene i and each time point k. This can be estimated directly from the time-series of gene expression profiles. TSNI assumes that a single perturbation experiment is performed (e.g., treatment with a compound, gene overexpression, etc.) and M time points following the perturbation are measured (rather than M different conditions at steady state as for NIR and MNI). For small networks (tens of genes), it is able to correctly infer the network structure (i.e., a_{ij}). For large networks (hundreds of genes) its performance is best for predicting the direct targets of a perturbation (i.e., u_{il}) (for example, finding the direct targets of a transcription factor from gene expression time series following overexpression of the factor). TSNI is not described further here, but we refer the reader to Bansal, Gatta, and di Bernardo (2006).

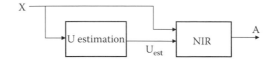

FIGURE 3.3   Block diagram of NIRest. Input X is the gene expression data, output A is the inferred network, and the internal variable U_{est} holds the estimate of the perturbation. The functional block labeled *U estimation* derives an estimate of the perturbations matrix U needed as second input to NIR.

3.3.2 The ODE-Based Approach When Perturbation Information Is Not Available

The information on which genes were perturbed in each microarray is not always available. In these cases, the regression-based procedure illustrated above needs to be modified. The approach used in NIRest to infer the network is described by a two-step procedure depicted in Figure 3.3 (Lauria, Iorio, and di Bernardo 2009):

1. Given X, infer the matrix U_{est} of the most likely set of gene perturbations (also called mode of action).

2. Using X and U_{est}, infer the matrix A using NIR.

We have implemented these two steps in a tool called NIRest. We have tested a number of different methods to obtain an estimate of U, including, for example, MNI. Based on the results of our tests (not shown) in the end we selected an indirect approach, which consisted of obtaining a first rough estimate A_{est} of A with a fast statistical method, then deriving the perturbation matrix as $U_{est} = -A_{est}X$.

Then for the second of the two steps we fed X and U_{est} to NIR to obtain a better estimate of A (Figure 3.3). In principle the newly found A could be used to compute a new estimate of U and the algorithm could be iterated one or more times. In practice we did not observe substantial improvements, due mainly to the accumulation of errors outweighing the benefits of additional passes.

3.4 EVALUATING THE PERFORMANCE OF REVERSE-ENGINEERING ALGORITHMS

3.4.1 Introduction

In this section we will describe how the different approaches to network inference are tested. Given the importance of the genetic network inference problem, many efforts have been devoted not only to devising new algorithms but also to evaluating their performance. The main issue in testing algorithms is the lack of suitable real networks to use as gold standards. Most biological networks are either completely determined but small, or large but incomplete. An example of the first type of network is a nine-transcript subnetwork of the SOS pathway in *Escherichia coli* often used as a test network and first adopted in Gardner et al. (2003), available at gardnerlab.bu.edu. Typically this is a subset of a well-known biological network, whose knowledge is the result of years of focused efforts by part of the scientific community on a specific model organism and model pathway. An example of the second type of network is the set of interactions between 844 transcriptional factors and their targets in *Saccharomyces cerevisiae* published by Lee et al. (2002). In this case the set of interactions obviously represents only a subset of the set of all real interactions among genes in *S. cerevisiae*; these networks are usually the result of large-scale studies focusing on a single aspect of gene interaction.

Small networks are not an ideal gold standard because in general the complexity of recovering a network from a set of expression data grows more than proportionally with size, both in terms of accuracy of the results and in terms of computational effort. The disadvantage of the large but incomplete networks is that it is difficult to measure the accuracy of a network recovered from expression data if the gold standard is faulty in the first place. In view of the above limitations, it is not surprising to find a widespread use of synthetic data ("in silico" data), that is, gene expression data generated by a computer model of gene regulation assuming an artificial network of gene interactions. While less realistic, synthetic data enables a performance comparison in a completely controlled testing environment that would not be possible with a breadth of scenarios and a level of detail that would be hardly feasible using real data. Unfortunately, there is no agreement on a standard set of test data, as is the case in other areas of computational biology (i.e., CASP for protein structure prediction); therefore direct comparison between new and existing approaches is often spotty. This situation is partially remedied by the recent introduction of a yearly competition called DREAM, described later in this section.

3.4.2 A Comparison of the Different Reverse-Engineering Approaches

A comparison of different mainstream algorithms for network inference is reported in Bansal et al. (2007). Here we report an excerpt from that work, integrated with results for NIRest from Lauria et al. (2009); for more details the reader is referred to the original papers. The performance is reported in terms of positive predictive value (PPV; also known as precision and sensitivity or recall). PPV represents the accuracy of the inferred network and is defined as TP/(TP + FP). Sensitivity is defined as TP/(TP + FN), where TP is true positive, FP is false positive, and FN is false negative. The in-house datasets were obtained by taking five of the synthetic networks (Net1 to Net5) described in the review paper by Bansal et al. (2007) and multiplying their inverse by a matrix P describing the desired set of perturbations, according to the formula $X = -A^{-1}U$. Each network contained 100 genes with an average number of connections k to other genes of 10; 100 different experiments were simulated for each dataset, generating gene expression datasets X of 100 genes × 100 experiments. Specifically, the presence of a perturbation to gene i in experiment j was represented by setting to 1 the corresponding element of U, that is, $U_{ii} = 1$. The single perturbation matrix U^1 coincided with the identity matrix I; the multiple perturbation matrices U^k were obtained by repeated shift-down and addition to itself of I. In other words, in the pth experiment (pth column of U^k) the genes $p, p+1, \ldots, (p+k) \bmod N$ were perturbed (elements in row $p, p+1, \ldots, (p+k) \bmod N$ of $U^k =$ were set to 1). Finally, noise was added to X as described in Bansal et al. (2007) to simulate experimental errors. Table 3.1 reports the results for synthetic datasets of size 10 × 10

TABLE 3.1   Performance Comparison of the Different Reverse-Engineering Approaches Using a Mix of Synthetic and Biological Datasets

Dataset	ARACNE		BANJO		NIR		NIRest		Clustering		Random
	Recall	Prec	Recall	Prec	Recall	Prec	Recall	Prec	Recall	Prec	Recall
10x10 (u)	0.53	0.61	0.41	0.50	0.63	0.96			0.39	0.38	0.36
10x10 (d)			0.25	0.18	0.57	0.93					0.20
100x100(u)	0.56	0.28	0.71	0.00	0.97	0.87	0.80	0.49	0.29	0.18	0.19
100x100(d)			0.42	0.00	0.96	0.86	0.47	0.43			0.10
E. coli (u)	0.69	0.34	0.78	0.44	0.80	0.88			0.8	0.63	0.71
E. coli (d)			0.67	0.24	0.74	0.67					0.63

Source: Data from Bansal et al. (2007) and Lauria et al. (2009).

and 100×100, and the *E. coli* dataset described before. Note that, as mentioned before, depending on the inference algorithm used, the resulting gene network can be either an undirected graph (the direction of the interaction is not specified), or a directed one (direction is specified). For algorithms like NIR that have the ability to recover both versions of the network the performance for both is reported [marked as (u) and (d), respectively].

3.4.3 The DREAM Competition

An annual competition called DREAM is organized with the purpose of assessing the current state of the art in genetic network inference. Every year a new set of synthetic and biological data is used to test the performance of the current version of the inference algorithms proposed by the contenders. In the words of the organizers (DREAM2007, 2007),

> DREAM is a Dialogue for Reverse Engineering Assessments and Methods. Its main objective is to catalyze the interaction between experiment and theory in the area of cellular network inference. The fundamental question for DREAM is simple: How can researchers assess how well they are describing the networks of interacting molecules that underlie biological systems? The answer is not so simple. Researchers have used a variety of algorithms to deduce the structure of very different biological and artificial networks, and evaluated their success using various metrics. What is still needed, and what DREAM aims to achieve, is a fair comparison of the strengths and weaknesses of the methods and a clear sense of the reliability of the network models they produce.

In the following we will describe the results of NIR using the 2007 DREAM2 challenge.

Since in the NIR model knowledge of the perturbation is required, we applied the NIR algorithm to the challenge number 4 of the DREAM competition (*The In-Silico-Network Challenge*). For this challenge, three networks were created by the organizers (*InSilico1, InSilico2, and InSilico3*) and they were endowed with dynamics that simulate biological interactions (DREAM2007 2007). Specifically, a nonlinear differential equations model with standard Hill kinetics was created reflecting the connections of each network; the data was then obtained by simulating the network with the COPASI software (Hoops et al. 2006; Mendes 2007). The data from *InSilico1* and *InSilico2* correspond to mRNA levels of gene networks with qualitatively different topologies. *InSilico3* corresponds to a full biochemical network, including metabolites, proteins, and mRNA concentrations. The challenge consisted of predicting the connectivity and some of the properties of one or more of these three networks. Each of these datasets was provided in three different versions:

- *Heterozygous*: steady-state levels for the wild type and 50 heterozygous knock-down strains for each gene (+/−). Values of gene expression were provided for a standard condition (steady state).

- *Null-mutants*: steady-state levels for the wild type and 50 null mutant strains for each gene (−/−). Values of gene expression were provided for a standard condition (steady state).

- Trajectories: time courses of the network recovering from several external perturbations. 23 different perturbations and 26 time points were available.

The predictions for the three datasets could be submitted in one or more of the following categories: *undirected-unsigned, undirected-signed, directed-unsigned, directed-signed (excitatory, inhibitory)*. The categories are self-explanatory; this flexibility in submitting the results was introduced to accommodate the different abilities of the various network prediction approaches to infer the direction and/or the sign of the relationship between each pair of genes. We submitted results for all the above categories.

The performance curves are shown in Figure 3.4. For the Undirected-Unsigned prediction we obtained a network with 148 edges, with a PPV equal to 1 until the 33rd most reliable connection, and a PPV of 0.5 if considering the first 100 most reliable connections. For the Directed-Unsigned prediction the performance was essentially the same in terms of PPV, with 328 total connections inferred. For the Directed Signed version, the challenge rules required specifying separately the set of excitatory connections and the set of Inhibitory connections. The 12 most reliable excitatory connections we predicted were all correct, and the same was true for the 25 most reliable inhibitory connections. We have not investigated the reason for this asymmetry in the prediction accuracy of the excitatory versus inhibitory connections; one possibility is that some difference in the nonlinear equations describing inhibitory vs. excitatory regulation used to simulate the network and generate the data might make it slightly harder to detect inhibition.

3.4.4 Synthetic Biology Meets Systems Biology

A recent development was the construction of a synthetic genetic circuit in yeast and its application to the comparison of reverse-engineering algorithms (Cantone et al. 2009). The novelty of this project was that for the first time a genetic circuit was designed from the ground up and then implemented for the explicit purpose of evaluating computational biology tools, namely, transcriptional modeling and genetic network inference algorithms. Despite the small scale, the network includes several representative interactions found in natural gene networks, such as transcriptional chains, positive and negative feedback

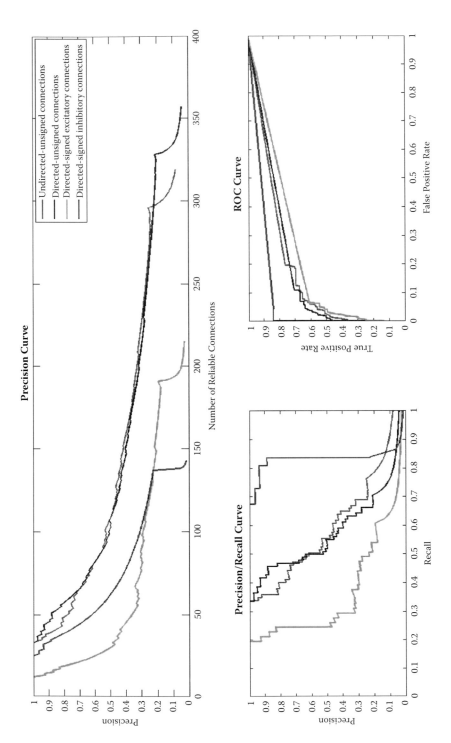

FIGURE 3.4 NIR performance assessment obtained using the DREAM2 Challenge Scoring Methodology, adapted from Lauria et al. (2009). *Precision Curve*: value of Precision (PPV) obtained when retaining only the *n* most reliable network connections (ranked by absolute value of the corresponding A element), for increasingly larger values of *n* (from left to right). *Precision/Recall Curve*: each point represents a (Precision, Recall) value pair obtained for a certain classification threshold on the values of the network coefficients (to classify network links as either null or not null); a curve represents the set of points obtained for all possible values of the threshold. *ROC Curve*: same as Precision/Recall curve, but with (True Positive Rate, False Positive Rate) value pairs plotted in correspondence of each value of the threshold.

loops, and indirect regulation mediated by protein-protein interactions. Furthermore, the network can be controlled through a small molecule, galactose, so that the circuit can be activated or inactivated through an external agent. The network was engineered to achieve complete isolation from the natural regulatory network of the yeast. The combination of these features made possible the analysis of the circuit using the perturbation-driven approach described in this chapter in a systematic and highly controlled manner. By combining the advantages of a realistic biological behavior of a living cell and detailed knowledge of the topology of the network of a synthetic model, this experiment pioneered a new approach to the study of genetic circuits and the computational tools employed to study them.

3.5 DISCUSSION

In this section we summarize the current state of the art, and the future directions for research. In silico analysis gives reliable guidelines on performance of algorithms in line with the results obtained on real datasets: ARACNE performs well for steady-state data and can be applied also when few experiments are available, as compared to the number of genes, but it is not suited for the analysis of short time-series data. This is to be expected due to the requirement of statistically independent experiments. BANJO is very accurate, but with a very low sensitivity, on steady-state data when more than 100 different perturbation experiments are available, independently of the number of genes, whereas it fails for time-series data. BANJO (and Bayesian networks in general) is a probabilistic algorithm requiring estimation of probability density distributions, a task that requires a large number of data points. NIR works very well for steady-state data, also when few experiments are available, but requires knowledge of the genes that have been perturbed directly in each perturbation experiment. NIR is a deterministic algorithm and if the noise on the data is small, it does not require large datasets, since it is based on linear regression. Clustering, although not a reverse-engineering algorithm, can give some information on the network structure when a large number of experiments is available, as confirmed by both in silico and experimental analysis, albeit with a much lower accuracy than the other reverse-engineering algorithms.

The different reverse-engineering methods considered here infer networks that overlap for about 10% of the edges for small networks, and even less for larger networks. Interestingly, if all algorithms agree on an interaction between two genes (an edge in the network), this interaction is not more likely to be true than the ones inferred by a single algorithm. Therefore, it is not a good idea to "trust" an interaction more just because more than one reverse-engineering algorithm finds it. Indeed, the different mathematical models used by the reverse-engineering algorithms have complementary abilities; for example, ARACNE may correctly infer an interaction that NIR does not find and vice versa. Hence in the intersection of the two algorithms, both edges will disappear, causing a drop in sensitivity without any gain in accuracy (PPV). Taking the union of the interactions found by all of the algorithms is not a good option, since this will cause a large drop in accuracy.

REFERENCES

Amato, R., Ciaramella, A., and Deniskina, N. et al. 2006. A multi-step approach to time series analysis and gene expression clustering. *Bioinformatics* 22: 589–596.

Ambesi, A. and di Bernardo, D. 2006. Computational biology and drug discovery: from single target to network drugs. *Current Bioinformatics*: 3–13.

Arkin, A. and Ross, J. 1995. Statistical construction of chemical reaction mechanism from measured time-series. *J. Phys. Chem.* 99: 970–979.

Bansal, M., Belcastro, V., Ambesi-Impiombato, A., and di Bernardo, D. 2007. How to infer gene networks from expression profiles. *Mol Syst Biol* 3: 78.

Bansal, M., Gatta, G. D., and di Bernardo, D. 2006. Inference of gene regulatory networks and compound mode of action from time course gene expression profiles. *Bioinformatics* 22: 815–822.

Basso, K., Margolin, A. A., Stolovitzky, G. et al. 2005. Reverse engineering of regulatory networks in human B cells. *Nature Genet* 37: 382–390.

Beer, M. A. and Tavazoie, S. 2004. Predicting gene expression from sequence. *Cell* 117: 185–198.

Bonneau, R., Reiss, D. J., Shannon, P. et al. 2006. The Inferelator: an algorithm for learning parsimonious regulatory networks from systems-biology data sets *de novo*. *Genome Biol* 7: R36.

Breitkreutz, B. J., Stark, C., and Tyers, M. 2003. Osprey: a network visualization system. *Genome Biol* 4: R22.

Butte, A. J. and Kohane, I. S. 2000. Mutual information relevance networks: functional genomic clustering using pairwise entropy measurements. *Pac Symp Biocomput*: 418–429.

Cantone, I., Marucci, L., Iorio, F. et al. 2009. A yeast synthetic network for *in vivo* assessment of reverse-engineering and modeling approaches. *Cell* 137: 172–181.

Conant, G. C. and Wolfe, K. H. 2006. Functional partitioning of yeast co-expression networks after genome duplication. *PLoS Biol* 4: e109.

Cosgrove, E. J., Zhou, Y., Gardner, T. S., and Kolaczyk, E. D. 2008. Predicting gene targets of perturbations via network-based filtering of mRNA expression compendia. *Bioinformatics* 24: 2482–2490.

D'haeseleer, P., Wen, X., Fuhrman, S., and Somogyi, R. 1999. Linear modeling of mRNA expression levels during CNS development and injury. *Pac Symp Biocomput*: 41–52.

di Bernardo, D., Gardner, T. S., and Collins, J. J. 2004. Robust identification of large genetic networks. *Pac Symp Biocomput*: 486–497.

di Bernardo, D., Thompson, M. J., Gardner, T. S. et al. 2005. Chemogenomic profiling on a genome-wide scale using reverse-engineered gene networks. *Nature Biotechnol* 23: 377–383.

DREAM2007. 2007. The In-Silico Challenge Description. http://wiki c2b2 columbia edu/dream/index php/The_In-Silico-Network_Challenges.

Eisen, M. B., Spellman, P. T., Brown, P. O., and Botstein, D. 1998. Cluster analysis and display of genome-wide expression patterns. *Proc Natl Acad Sci USA* 95: 14863–14868.

Foat, B. C., Morozov, A. V., and Bussemaker, H. J. 2006. Statistical mechanical modeling of genome-wide transcription factor occupancy data by MatrixREDUCE. *Bioinformatics* 22: e141–e149.

Friedman, N. and Elidan G. 2000. Bayesian network software libb 2.1.

Friedman, N., Linial, M., Nachman, I., and Pe'er, D. 2000. Using Bayesian networks to analyze expression data. *J. Comput. Biol.* 7: 601–620.

Gardner, T. S., di Bernardo, D., Lorenz, D., and Collins, J. J. 2003. Inferring genetic networks and identifying compound mode of action via expression profiling. *Science* 301: 102–105.

Gardner, T. S. and Faith J. J. 2005. Reverse-engineering transcription control networks. *Phys Life Rev* 2: 65–88.

Hoops, S., Sahle, S., Gauges, R. et al. 2006. COPASI: a COmplex PAthway SImulator. *Bioinformatics* 22: 3067–3074.

Lauria, M., Iorio, F., and di Bernardo, D. 2009. NIRest: a tool for gene network and mode of action inference. *Ann NY Acad Sci* 1158: 257–264.

Lee, T. I., Rinaldi, N. J., Robert, F. et al. 2002. Transcriptional regulatory networks in *Saccharomyces cerevisiae*. *Science* 298: 799–804.

Margolin, A. A., Nemenman, I., Basso, K. et al. 2006. ARACNE: an algorithm for the reconstruction of gene regulatory networks in a mammalian cellular context. *BMC Bioinformatics* 7 (Suppl 1): S7.

Mendes, P. 2007. The In-Silico network for the DREAM 2007 competition. http://www comp-sys-bio org/tiki-index php?page=DREAM2007.

Milo, R., Shen-Orr, S., Itzkovitz, S. et al. 2002. Network motifs: simple building blocks of complex networks. *Science* 298: 824–827.

Murphy, K. P. 2001. The Bayes net toolbox for Matlab. *Comput Sci Stat* 33: 331–350.

Prakash, A. and Tompa, M. 2005. Discovery of regulatory elements in vertebrates through comparative genomics. *Nature Biotechnol* 23: 1249–1256.

Rice, J. J., Tu, Y., and Stolovitzky, G. 2005. Reconstructing biological networks using conditional correlation analysis. *Bioinformatics* 21: 765–773.

Shannon, P., Markiel, A., Ozier, O. et al. 2003. Cytoscape: a software environment for integrated models of biomolecular interaction networks. *Genome Res* 13: 2498–2504.

Steuer, R., Kurths, J., Daub, C. O., Weise, J., and Selbig, J. 2002. The mutual information: detecting and evaluating dependencies between variables. *Bioinformatics* 18 (Suppl 2): S231–S240.

Tadesse, M. G., Vannucci, M., and Lio, P. 2004. Identification of DNA regulatory motifs using Bayesian variable selection. *Bioinformatics* 20: 2553–2561.

Tegner, J., Yeung, M. K., Hasty, J., and Collins, J. J. 2003. Reverse engineering gene networks: integrating genetic perturbations with dynamical modeling. *Proc Natl Acad Sci USA* 100: 5944–5949.

Tsong, A. E., Tuch, B. B., Li, H., and Johnson, A. D. 2006. Evolution of alternative transcriptional circuits with identical logic. *Nature* 443: 415–420.

Ucar, D., Neuhaus, I., Ross-MacDonald, P. et al. 2007. Construction of a reference gene association network from multiple profiling data: application to data analysis. *Bioinformatics* 23: 2716–2724.

van Someren, E. P., Vaes, B. L., Steegenga, W. T. et al. 2006. Least absolute regression network analysis of the murine osteoblast differentiation network. *Bioinformatics* 22: 477–484.

Yu, J., Smith, V. A., Wang, P. P., Hartemink, A. J., and Jarvis, E. D. 2004. Advances to Bayesian network inference for generating causal networks from observational biological data. *Bioinformatics* 20: 3594–3603.

CHAPTER 4

Understanding Cancer Progression in Protein Interaction Networks

Jinsheng Sun, Jie Li, and Edwin Wang

CONTENTS

4.1 NETWORKS OF CANCER GENES

4.1.1 Genes Do Not Work in Isolation

About 24,000 protein-encoding genes are contained in the human genome, and cellular processes are achieved by interactions among these genes. Diseases such as cancer are driven by dynamic interactions among sets of proteins involved in altering various normal cell activities. Several studies have illustrated the importance of gene and protein interactions in essential biological processes and cancer development.

Cell cycle activity is a hallmark of cancer. A recent study of the time dependence of protein complex assembly in the cell cycle showed that the whole picture of protein interactions is extremely dynamic and precisely controlled (de Lichtenberg et al. 2005). The authors constructed a protein interaction network and traced the pieces of each complex using public protein interaction data for yeast. After mapping diverse genome-wide cell cycle datasets (time course gene microarray data), they showed clear patterns in how the complexes are assembled: key components of the machines are assembled ahead of time, kept in stock, and prepared for use. When a new machine is needed, a few crucial pieces of the machine are produced and assembled to form the complete functional machine (Figure 4.1). This scenario suggests a few underlying principles: (1) cell cycle machines consist of two different subunits that are expressed both periodically and constitutively; (2) these subunits ensure that the machines can be built at the right times, which affords control of protein complex activity through "just-in-time assembly." The study indicates that biological systems have dynamic behavior, and it demonstrates that proteins not only interact but also must be produced at the right times and places in the cell. This example clearly tells us that genes not only work together but also are governed by certain underlying rules.

4.1.2 Cellular Networks

4.1.2.1 Pathways and Networks

During the past few decades, the concept of biological pathways has been developed and used to study the relationships among genes or proteins. The linear relationship between genes is the core of biological pathways, but the components and boundaries of signaling pathways are not clearly defined. For example, different numbers of proteins and relationships in the transforming growth factor (TGF-β) pathway are found in three public databases: BioCarta (http://www.biocarta.com/), KEGG (http://www.genome.jp/kegg/pathway.html), and Reactome (http://www.reactome.org/). Indeed, the linear relationships between proteins have been increasingly challenged by evidence for cross-talk between pathways, that is, evidence that many proteins are shared by multiple pathways (Wang, Lenferink, and O'Connor-McCourt 2007). It seems that cross-talk between pathways has become one of the underlying principles of evolution: fast-evolving proteins tend to form network components with other signaling pathways (Cui, Purisima, and Wang 2009), suggesting that apoptotic signaling proteins have proliferated in humans, and a significant portion of them have been integrated into the signaling processes for normal physiological conditions. For example, apoptotic proteins such as caspases are involved in many nonapoptotic signaling processes in humans and mice (i.e., cell proliferation and differentiation; Cui, Purisima,

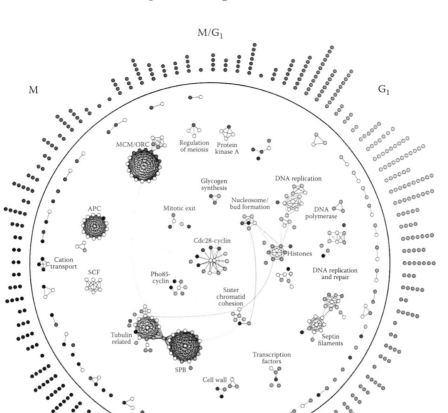

FIGURE 4.1 (See color insert following page 332.) Temporal protein interaction network of the yeast mitotic cell cycle. Circle nodes represent cell cycle proteins, which are members of protein complexes or other physically interacting partners. White nodes represent static proteins, while colored nodes represent the time of peak expression of the proteins. The proteins without interactions are outside the circle, and are positioned and colored according to their peak time. (From de Lichtenberg, U. et al. 2005. *Science* 307: 724–727. With permission.)

and Wang 2009; Kuida et al. 1998). In mice, caspase-9 is involved in both apoptosis and inner ear epithelium development (Cecconi et al. 2004), while caspase-8 is involved in critical signaling for cardiac and neural development during early embryogenesis (Sakamaki et al. 2002). Conversely, multiple normal signaling mechanisms have been recruited to cell death pathways either as backup or parallel mechanisms of apoptosis. Cytochrome c, a key electron carrier of mitochondrial complex III for respiration, is involved in apoptosis when mitochondria are damaged (Liu et al. 1996).

Pathway cross-talk and component sharing provide redundant and flexible mechanisms for cell signaling. Coevolution of the proteins from different pathways would enhance the rapid information transfer between these signaling pathways, and it significantly promotes new functionalities (Cui, Purisima, and Wang 2009). These studies suggest that extensive communication exists between pathways, to the point that many pathways lose their identities and the pathway concept gets fuzzy. Therefore, the concept of the network emerges as a necessity for biological systems.

In fact, both pathways and networks are concepts that reflect the efforts of humans to explore the activities with cells via abstraction. Cells are full of molecules and reactions; it's hard to say whether they really contain pathways or networks. In order to explore cellular activities, we must develop conceptual models and bring them closer to the real situation in cells. At the moment, a pathway may describe the real cellular situation in certain controlled conditions, but in most situations a network is probably more accurate.

4.1.2.2 Types of Cellular Networks

Cellular networks have been used to describe protein interactions, metabolism, gene regulation, signaling, genetic interactions, and gene co-expression (Wang et al. 2007). Each network can be placed into one of two categories—general or specific networks, based on whether it captures biological relationships on a genome-wide scale or only specific cellular activities such as metabolism or signaling.

A general cellular network encodes all the relationships among proteins or genes across all biological processes in a cell. The information in the network ranges from basic cellular machinery for DNA synthesis, metabolism, and transcription to protein complexes involved in cellular signaling. For example, a protein interaction network records all the physical interactions among proteins in a cell. To the category of general cellular networks, we can add gene regulation networks, which describe all the regulatory relationships between transcription factors and genes in a cell.

The second category of cellular networks is specific networks, which encompass metabolic networks, gene co-expression networks, and signaling networks. These networks describe relationships in specific cellular conditions or specific cellular activities such as signaling and metabolism. A signaling network encodes the flows of information and biochemical reactions for signal transduction, while a metabolic network collects all the metabolic reactions and metabolic flows. Co-expressed genes often represent a collection of genes that are involved in similar biological functions and activities.

4.2 BIOLOGICAL INSIGHTS INTO CANCER GENES IN THE PROTEIN INTERACTION NETWORK

4.2.1 Special Features of Cancer Genes in the Human Protein Interaction Network

During the past two decades, scientists have identified thousands of interactions among cellular proteins, and in the coming years they hope to comprehensively catalog all the interactions encoded in the human genome. Such information will be critical to understanding the basic mechanisms of cellular activities and how malfunctions in these activities contribute to cancer. As the high-throughput technologies for characterizing protein interactions have advanced, so has the feasibility of building large-scale protein interaction networks via sophisticated computational tools. Therefore, network-based approaches have been introduced for representing complex biological systems. Cellular networks are complex systems that are inherently nonlinear in nature. In a nonlinear system, simple changes in one part of the system produce complex effects throughout. However, biologists often treat biological systems as linear (e.g., linear cascades of biochemical reactions or gene regulations). Because many biological properties are encoded in cellular networks (Wang, Lenferink, and O'Connor-McCourt 2007), it is necessary to develop tools to treat such complex systems as nonlinear in order to discover common principles governing their organization and function. In the past few years, mathematical, statistical, and computational methods for characterizing and elucidating complex biological networks have been adopted from other fields, such as the social sciences. Network analysis methods have provided insights into many biological questions and have been applied to many types of biological networks, including transcriptional regulation, genetic interactions, protein interactions, expression correlation, and cell signaling (Wang, Lenferink, and O'Connor-McCourt 2007). Network theory has also been employed to discover the patterns of gene mutations that drive cancer on the human protein interaction network. Cancer driver-mutating genes promote cancer initiation, progression, or metastasis when mutated. There are two types of such genes: oncogenes and tumor suppressor genes. An oncogene, when mutated or expressed at high levels, helps turn a normal cell into a cancer cell, while a tumor suppressor gene protects a cell from progressing toward cancer. When a tumor suppressor gene is mutated to cause a loss or reduction of its function, the cell can progress to cancer, usually in combination with an oncogene's genetic changes. Cancer driver-mutating genes can be identified by either traditional genetic approaches or the tumor genome sequencing approach (Sjoblom et al. 2006).

4.2.1.1 Cancer Genes Occur Less Often in Duplicable Genes

Analysis of the human genome shows that cancer driver-mutating genes are significantly enriched in singletons and less so in duplicable genes (83.7% and 16.3%, respectively), regardless of their molecular functions (Rambaldi et al. 2008), suggesting that these genes are intrinsically fragile. Furthermore, there is evidence that targeted design of oncogene inhibitors could be more effective to block the onco-signal that is produced from that oncogene. Indeed, many cancer drugs (e.g., Epidermal Growth Factor Regulator [EGFR] inhibitors) can effectively prolong the survival of patients whose cancer is caused by a specific mutation. However, we must keep in mind that cancer is induced not by one mutating gene, but by several. Designing an inhibitor to target only one cancer driver-mutating gene cannot cure cancer, but it may prolong the lives of some patients.

4.2.1.2 Cancer Genes Have More Interacting Partners and Higher Network Interconnectivity

The human protein interaction network can be built using the data from several human protein interaction databases such as the Human Protein Reference Database (HPRD, http://www.hprd.org/), which is manually curated. More information about data sources for building protein interaction networks is presented in Chapter 20. Analysis of the human protein interaction network uncovered the fact that human duplicable proteins have significantly more interacting partners than singletons but have lower clustering coefficients (Liang and Li 2007; Liao and Zhang 2007), that is, they are associated with less network interconnectivity. Rambaldi et al. mapped a set of cancer driver-mutating genes by merging a manually curated census of human cancer genes (Futreal et al. 2004) and a set of genes derived from sequencing of tumor genomes. They then overlaid this map on the human protein interaction network. They found that cancer driver-mutating genes have significantly more interacting partners and higher clustering coefficients than human proteins (Rambaldi et al. 2008). Similar results have been shown in another study, which used another source of protein interaction data to characterize cancer driver-mutating genes (Jonsson and Bates 2006). In this study, Jonsson and Bates applied a computational method based on the principle of orthologous interactions (Jonsson et al. 2006) to build a human protein interaction network, and they mapped the human consensus cancer genes (Futreal et al. 2004) onto the network. The interesting point here is that the studies performed by both Rambaldi et al. and Jonsson and Bates used different datasets but obtained similar results, which suggests that cancer proteins form the backbone of the human proteome and become central interconnected hubs in the human protein interaction network. Their results led Rambaldi et al. to propose that cancer genes are intrinsically fragile and susceptible to perturbations: gene dosage modifications of highly interconnected hubs are likely to produce simultaneous effects on several processes (Rambaldi et al. 2008). Indeed, we previously showed that mutated oncogenes activate cancer signaling, while mutated and methylated (i.e., silenced) tumor suppressor genes release signaling brakes, in both cases increasing the dosage for cancer signaling (Cui et al. 2007). More details are described in Chapter 5.

4.2.1.3 Network Hot Spots of Cancer Gene Mutations Demonstrate Hallmarks of Cancer
Jonsson and Bates showed that cancer driver-mutating genes tend to form network clusters or communities; that is, when a few genes on the network do something, their "friends" (genes in the same module) probably do something similar. Detailed annotations indicate that these network clusters encode cancer hallmarks such as active cell cycles and immune systems. More information about cancer hallmarks and their associations with signaling pathways and genome science can be found in Chapter 12. Similarly, Rambaldi et al. showed that cancer genes are significantly enriched in network motifs (basic network building blocks) containing 3- and 4-connected proteins. Detailed information about network motif concepts is described in different contexts in Chapters 2 and 5. Furthermore, cancer proteins tend to co-occur within the same network motif.

4.2.1.4 Special Features of Cancer Genes in Cell Signaling Regulatory Loops
Results similar to those mentioned above have been observed when analyzing cancer genes on a manually curated human signaling network (Awan et al. 2007), that is, cancer genes are significantly enriched in network hub proteins. Furthermore, there exist substantial numbers of hot spots—11 and 9 of the 3- and 4-node network motifs, respectively, in which all nodes are cancer genes. These hot spots are potential biomarker clusters or anticancer

drug target clusters. However, in the signaling network, there are more interesting observations that have not been uncovered in protein interaction networks. Downstream regions of the signaling network are significantly enriched in cancer genes ($P < 2 \times 10^{-4}$), that is, 7.9%, 9.2%, and 18.1% of ligand-receptor, intracellular components, and nuclear genes, respectively, in contrast to 8.6%, the average rate at which cancer genes occur among signaling network proteins. These results suggest that the downstream regions of signaling pathways tend to be more *perturbed* in cancer signaling. Cancer genes are significantly enriched or depleted in some particular types of signaling network motifs; for example, they are enriched in positive feed-forward regulatory loops but depleted in bi-fan motifs.

A feed-forward regulatory loop consists of three proteins: A, B, and C. In the positive feed-forward loop, both A and B regulate C, while A also regulates B. This kind of loop could provide specific regulatory capacities and decode signal strength and process information (Wang, Lenferink, and O'Connor-McCourt 2007). Mutation of oncogenes in positive feed-forward regulatory loops could amplify and enhance the regulatory signals and therefore promote cancer signaling. A bi-fan regulatory loop consists of four proteins: A, B, C, and D. A regulates C and D, while B also independently regulates C and D. Therefore, gene mutation in the bi-fan regulatory loop does not amplify the underlying signal, and it is reasonable that cancer genes are less enriched in bi-fan loops. These results suggest that certain types of signaling regulatory network motifs are critical for cancer development and metastasis. Cancer genes are significantly enriched in the target nodes of most signaling motifs, especially the convergent target nodes that receive signal information consolidated from two or more source nodes. These findings indicate that the convergent nodes are critical and may be sufficient to activate other network nodes and induce cancer development. In signaling networks, multiple information flows could converge to produce a limited set of phenotypic responses (Prinz, Bucher, and Marder 2004) because convergence provides redundant cellular functions and robustness. Critical signaling nodes fall into two categories in the network: those that preserve homeostasis during perturbation and those that evoke phenotypic changes. Taken together, the above observations suggest that convergent nodes in the cancer-gene-enriched motifs could be crucial for preserving homeostasis, and perturbation of these nodes could lead to loss of cellular homeostasis and induction of cancer (Awan et al. 2007).

4.2.2 Protein Interaction Networks for Interpreting Cancer Microarray Data

4.2.2.1 Cancer Microarray Data Are Extremely Noisy

Microarray technology has been applied extensively to tumor gene expression profiling during the past decade. As of April 2009, more than 10,000 papers on cancer microarray studies have been published. However, it is well known that gene expression profiles differ greatly among tumor samples, even for the same type of cancer. Moreover, gene expression profiles of tumors are more complex than those of other disease samples. These facts could be rooted in the mechanisms of cancer initiation, progression, and metastasis. Through an integrative analysis of a human signaling network and the output of large-scale sequencing of tumor genomes, we have shown that alterations of the tumor suppressor gene p53 are essential to cancer development and progression (Cui et al. 2007). Mutation of tumor suppressor genes often increases genome instability, which, in turn, induces many genomic alterations such as rearrangements, chromosomal fragment amplifications, and deletions (Wang, Lenferink, and O'Connor-McCourt 2007). Thus, tumor cells often have many more "passenger signals" than other cells. The variability of the gene expression profiles of individual tumors is very high, and the "real" cancer gene expression signatures could be buried in these highly varied profiles. Microarray data contain a large number of genes, which indicate complex changes in genetic programs (i.e., driver mutations) and suggest the intertwining of unknown mechanisms. Advances in technology have made data generation much easier, but interpreting these data is a major hurdle in science today. Clustering methods are the most popular approach to microarray data analysis, but these methods do not make it easy to pinpoint molecular mechanisms or even particular pathways. The extremely noisy nature of gene expression data from tumors makes it very challenging to extract meaningful information and to obtain biological insights into the molecular mechanisms of cancer.

4.2.2.2 Networks Are Useful for Filtering Out the Noise in Microarray Data

A group of genes working together has intrinsic relationships; for example, cancer genes often form network communities. Therefore, we propose that cellular networks provide a platform to help filter out "noise" or "passenger signals," which are random signals unlikely to form any statistically significant patterns from tumor gene expression profiles. Furthermore, some lines of evidence suggest that tumorigenesis is rooted in coordinated reprogramming of molecular interactions in the context of highly connected and regulated cellular networks. Therefore, statistically significant patterns or systems-level reorganization of modulated genes during tumorigenesis observed in cellular networks could be interpreted as the "real signals" in tumor gene expression profiles. Several studies have applied cellular networks to interpret cancer microarray data, and novel biological insights have been drawn from these studies.

4.2.2.3 Properties and Organization of Cancer-Modulated Genes on Networks

Transformation of cells from normal to cancerous phenotypes requires dynamic changes in cancer signals within regulatory loops (i.e., sets of protein interactions), which could be modulated along with certain genes. To understand the systems-level properties and organization of cancer-modulated genes, Hernandez et al. (2007) performed an integrative analysis of the human protein network and the differentially expressed genes in tumors relative to healthy tissues. The genes differentially expressed in the presence of cancer were identified from microarray datasets of prostate, lung, and colorectal samples. Regardless of the tumor type, downregulated cancer genes have common properties: more interacting proteins and high betweenness (a measure of the number of paths along which signals can pass), suggesting that these genes are involved in multiple biological processes. These results further suggest that both cancer driver-mutating genes and downregulated cancer genes share common features in protein interaction networks, in that both take part in multiple biological processes or pathways. However, it was noted that upregulated cancer genes have no such patterns on the network, suggesting that downregulated genes might play major roles during tumorigenesis. Network analysis indicated that the dependence of a downregulated cancer gene node on its interacting neighbors is significantly lower, suggesting that cancer genes act independently and play dominant roles in information exchange and propagation within the network. Shortest-path analysis of the cancer-modulated genes suggests that downregulated genes or all modulated genes (including up- and downregulated genes) tend to form network clusters, which are involved in cancer-related processes. These results indicate that the coordinated downregulation of genes is involved in programmed cell death, cell adhesion, and cell communication processes, which could facilitate the metastatic behavior of cancer cells (Hernandez et al. 2007).

4.2.2.4 Dynamics of Network Rewiring and Functional Modules in Cancer Metastasis

4.2.2.4.1 Network Modules in Cancer Metastasis   Metastasis is a key process associated with cancer recurrence and the patient's death. Metastatic cancer cells have the ability to break away from the primary tumor and move to different organs and therefore must have properties such as increased motility and invasiveness. Detailed information about metastasis can be found in Chapter 13. Chapters 12, 14, and 15 have more information about factors that might affect metastasis from tumor-surrounding cells.

In networks, a module consists of a subset of nodes within which connections are dense, whereas connections are sparser between modules (Wang, Lenferink, and O'Connor-McCourt 2007). Modules are one of the higher-level topological properties of networks, and they contain correlations between the degree of a node and the degrees of the nearest neighbors.

Jonsson et al. (2006) applied the concept of network modules to cancer metastasis. They identified metastatic protein communities (network modules) by analyzing a protein interaction network in conjunction with up- and down-regulated genes related to cancer metastasis. By applying the clustering method, Jonsson et al. identified 37 network modules of highly interconnected proteins containing 313 proteins involved in 1094 interactions (Figure 4.2). Interestingly, most of the modules are associated with cancer metastasis, which indicates that key proteins are involved in metastasis.

The metastatic modules contain a high-order organization and include two modules representing TGF-β signaling and the cell cycle process. Both modules have been implicated in cancer metastasis, but this work suggests how these two modules are functionally linked. Most interestingly, among the identified modules, 17 are linked in a chain-like manner. The most upstream module represents intracellular signaling cascades, while the downstream modules include actinin, laminin, cell cycle regulation, NF-κB, hypoxia, nuclear hormone receptors, and metalloproteinases, which have been implicated in cancer metastasis. The importance of this type of analysis is that it not only captures the functional network modules but also suggests potential links between these modules and thus yields working hypotheses that can be tested experimentally.

4.2.2.4.2 Network Rewiring and Module Dynamics in Cancer Metastasis   Intuitively, the importance and roles of a network node are determined not only by the number but also the quality of its neighbors. Motivated by this idea, Guimerà, Sales-Pardo, and Amaral (2007) proposed a general framework to classify network hubs into "provincial hubs" and

FIGURE 4.2 (See color insert following page 332.) Protein communities of cancer metastasis. The communities were identified by *k*-clique analysis performed on the predicted genome-wide rat protein network. The communities are distinguished by different colors and labeled by the overall function or the dominating protein class. Note that proteins, particularly at community edges, can belong to more than two communities. (From Jonsson, P.F. et al. 2006. *BMC Bioinformatics* 7: 2.)

"connector hubs." Provincial hubs' neighbors tend to belong to the same module, while connector hubs efficiently link different modules. The provincial and connector hubs are similar to the concept of "party" and "date" hubs in protein interaction networks proposed by Han et al. (2004). Party hubs interact with most of their partners simultaneously, whereas date hubs prefer to meet their partners one at a time under different external or internal conditions. More detailed information about party and date hubs can be found in Chapter 2.

The dynamics of protein network modules have been investigated by examining the co-expression of hub proteins and their partners in the human protein interaction network (Taylor et al. 2009). Taylor and colleagues first classified the hub nodes as intermodular hubs (similar to party hubs or provincial hubs) and intramodular hubs (similar to date hubs or connector hubs), which display more highly correlated patterns of co-expression. In a functional network, the organization could be viewed as connecting network modules, which are comprised of intramodular hubs, via intermodular hubs.

To explore the relationships between network module dynamics and metastasis, Taylor et al. examined the co-expression correlation coefficients of hub proteins and their interacting partners in patients who were disease-free after extended follow-up (the samples had no metastasis) and in patients who died of disease (the samples had metastasis). They found that a substantial number of hubs (256), most of which were intramodular, had significantly altered co-expression correlation coefficients between metastatic and non-metastatic groups, but the hubs themselves were not significantly up- or downregulated between the two groups. For example, BRCA1, a protein that is mutated in a subset of familial breast cancers, was a hub in the network. The expression of BRCA1 was strongly correlated with the expression of its partners in tumors from non-metastatic patients, but it was not well correlated with their expression in tumors from metastatic patients.

These results suggest that gene expression reflects higher-level organization in the network; that is, network modules are altered during cancer metastasis. The alteration of gene co-expression produces changes in network modules, or changes in dynamic network modularity indicate that rewiring a network plays an important role in producing phenotypic changes.

Interestingly, further analysis of the two inter- and intra-modular hubs via integrating information about cancer driver-mutating genes showed that intermodular hubs were associated with cancer phenotypes more frequently than intramodular hubs. Taken together, these studies suggest that genetic changes in intermodular hubs drive the dynamic gene expression changes within intramodular hubs, which in turn take part in cancer metastasis.

4.3 A NETWORK-BASED APPROACH FOR IDENTIFYING PROGNOSTIC MARKERS

Early detection of various types of cancer is an important goal for clinicians and laboratory scientists. Prognostic biomarkers may help predict whether someone's cancer will come back after surgical removal. So far, several predictors, such as the intrinsic-subtype classifier (Hu et al. 2006; Perou et al. 2000; Sorlie et al. 2001), 70-gene signature, wound-response gene-expression signature (van 't Veer et al. 2002; van de Vijver et al. 2002), and ratio of the expression levels of two genes (Ma et al. 2004), largely based on an unsupervised analysis of breast-tumor gene-expression profiles (Paik et al. 2004), have been developed for breast cancer. However, these predictors cannot be used in other patient cohorts. Therefore, these reportedly predictive gene signatures lack reliability and robustness for cancer prognosis.

Factors such as tumor heterogeneity, limitations of microarray platforms, statistical methods used for marker discovery, and data overfitting (i.e., using a small number of samples to identify genes associated with patient survival from thousands of genes that display altered expression) have been identified as challenges to discovering robust biomarkers. However, statistical analysis of tumor microarray datasets suggests that the microarray platforms or the data analysis methods are not critical in marker discovery (Ein-Dor, Zuk, and Domany 2006). Our recent reanalysis of breast tumor microarray datasets by resampling tissues and genes suggests that data overfitting and tumor heterogeneity are the dominant barriers to marker discovery (Wang et al. 2009).

In fact, as mentioned above, too many "passenger gene expression signals" are buried in tumor gene expression data. Normally, identification of cancer biomarkers involves searching for differentially expressed genes and applying clustering methods in which a subset of genes can discriminate between different cancer diagnoses. However, these tools cannot remove the passenger gene expression signals, which definitely crush the reliability and robustness of markers. As proposed above, networks could help in filtering out gene expression noise. Therefore, it is reasonable to integrate protein interaction networks with tumor microarray data to identify prognostic biomarkers.

Recently, Ideker and colleagues improved the prognostic predictive performance of gene expression signatures by analyzing tumor microarray data that incorporated protein interaction data (Chuang et al. 2007). They mapped microarray data of breast tumor samples onto the human protein interaction network and identified subnetworks in which the expression patterns of the genes are coherent. They then searched each subnetwork for gene expression patterns that were able to distinguish whether a patient developed distant metastasis. The identified subnetwork components (proteins) are now used as prognostic markers for cancer metastasis and are more reproducible and accurate in classifying tumors as either metastatic or nonmetastatic.

Similarly, based on their analysis of tumor gene co-expression patterns in the human protein interaction network, Taylor et al. (2009) showed which genes had network neighbors that become dysregulated, and they therefore sought network signatures as prognostic markers. Toward this end, they computed the relative expression levels of hubs with each of their interacting partners, determined for which hubs the levels changed significantly between metastatic versus nonmetastatic patients, and then employed a clustering method. The accuracy of the resulting network signature has been improved.

These two examples support our notion that networks can help in filtering out gene expression noise, and biomarkers might be improved by overlaying tumor microarray data onto networks. Furthermore, the resulting network signatures also provide insights into the molecular mechanisms underlying metastasis.

Although both studies have improved the accuracy of network markers, the robustness of the markers has still not been tested extensively. For example, both studies only tested the markers in one independent dataset, although several public breast cancer datasets are available. It would be interesting to know how these network markers perform in other breast cancer datasets.

In addition, while both studies are more mechanism-based (network-based), breast cancer has distinct subtypes, ER+ and ER- (Sotiriou and Pusztai 2009), which have distinct molecular mechanisms. Recently we developed an algorithm using breast cancer microarray data and functional modules defined by Gene Ontology (Wang et al. 2009). Among the markers identified by applying the algorithm, there is a cell death-related gene signature (each containing 30 genes) for both ER+ and ER- patients, but no genes overlapping between the two signatures. These results suggest that ER+ and ER- cancers indeed have different molecular mechanisms. Along these lines, it is necessary to consider applying data from cancer subtypes to biological networks to identify subtype-specific markers, which might improve the performance of markers.

4.4 CONCLUSIONS

It is becoming increasingly clear that genes and their products do not function in isolation; they interact to form complex cellular networks. Deciphering how these networks operate and are rewired in order to achieve a deeper understanding of cancer molecular mechanisms requires integrative analysis of cellular networks and many types of -omic datasets from cancer studies. We have presented a series of examples of such studies and shown the power of network analysis and modeling to generate hypotheses for understanding the molecular mechanisms of cancer initiation, progression, and metastasis.

A global picture of a cancer protein network is emerging (Figure 4.3): (1) there are many small network modules containing intramodular hubs, and these modules are connected by intermodular hubs; (2) cancer driving-mutating genes are dominantly enriched in intermodular hubs; (3) their expression levels are not significantly changed between metastatic and nonmetastatic tumors; and (4) the co-expression relationships between intramodular hubs and their interacting partners are significantly changed between metastatic

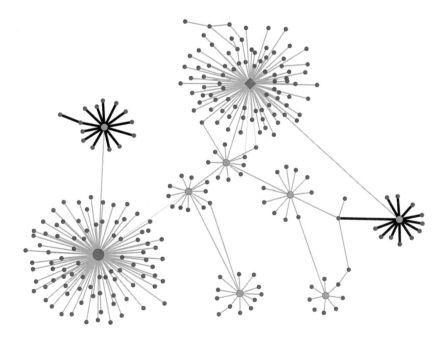

FIGURE 4.3 (See color insert following page 332.) A global picture of the cancer protein network encodes the driver-mutating information, network rewiring, and module dynamics. The cancer protein network architecture contains many small network modules containing intramodular hubs which are connected by intermodular hubs in which cancer driver-mutating genes are dominantly enriched. The expression levels of the intramodular and intermodular hub genes are not significantly changed between metastatic and nonmetastatic tumors; however, the coexpressions between intramodular hubs and their interacting partners are significantly changed between metastatic and nonmetastatic tumors. Nodes represent proteins while links represent physical interactions. Red, orange, and blue nodes represent intermodular and intramodular hubs, and nonhub nodes, respectively. Square nodes represent cancer driver-mutating genes while black links represent the co-expression changes between the two linked genes.

and nonmetastatic tumors, suggesting that network rewiring and module dynamics play important roles during metastasis.

Future studies in this direction rely on (1) a comprehensive catalog of reliable and high-quality protein interactions in the human genome collected by high-throughput approaches; (2) a more detailed understanding of cancer biology that acknowledges that cancer has subtypes with distinct molecular mechanisms (Sorlie 2009, Sotiriou and Pusztai 2009); (3) identification of key network modules, which orchestrate the fundamental processes of cancer, and understanding the dynamics of these modules using time-series data from mouse models of cancer initiation, progression, and metastasis; and (4) exploration of higher-order relationships in networks, such as modularity (Guimera, Sales-Pardo, and Amaral 2007), hierarchy (Clauset, Moore, and Newman 2008), and others. For example, a network can be seen as a hierarchical organization, where nodes cluster together to form modules, which themselves cluster into larger modules. This arrangement is similar to the organization of scientists (a professor, postdocs, and graduate students) into teams, teams into scientific departments, and departments into universities. Newman and colleagues proposed a direct but flexible model of hierarchical structure from which they could predict which interactions might have been missed (Clauset, Moore, and Newman 2008).

REFERENCES

Awan, A., Bari, H., Yan, F. et al. 2007. Regulatory network motifs and hotspots of cancer genes in a mammalian cellular signalling network. *IET Syst Biol* 1: 292–297.

Cecconi, F., Roth, K. A., Dolgov, O. et al. 2004. Apaf1-dependent programmed cell death is required for inner ear morphogenesis and growth. *Development* 131: 2125–2135.

Chuang, H. Y., Lee, E., Liu, Y. T., Lee, D., and Ideker, T. 2007. Network-based classification of breast cancer metastasis. *Mol Syst Biol* 3: 140.

Clauset, A., Moore, C., and Newman, M. E. 2008. Hierarchical structure and the prediction of missing links in networks. *Nature* 453: 98–101.

Cui, Q., Ma, Y., Jaramillo, M. et al. 2007. A map of human cancer signaling. *Mol Syst Biol* 3: 152.

Cui, Q., Purisima, E. O., and Wang, E. 2009. Protein evolution on a human signaling network. *BMC Syst Biol* 3: 21.

de Lichtenberg U., Jensen, L. J., Brunak, S., and Bork, P. 2005. Dynamic complex formation during the yeast cell cycle. *Science* 307: 724–727.

Ein-Dor, L., Zuk, O., and Domany, E. 2006. Thousands of samples are needed to generate a robust gene list for predicting outcome in cancer. *Proc Natl Acad Sci USA* 103: 5923–5928.

Futreal, P. A., Coin, L., Marshall, M. et al. 2004. A census of human cancer genes. *Nature Rev Cancer* 4: 177–183.

Guimera, R., Sales-Pardo, M., and Amaral, L. A. 2007. Classes of complex networks defined by role-to-role connectivity profiles. *Nature Phys* 3: 63–69.

Han, J. D., Bertin, N., Hao, T. et al. 2004. Evidence for dynamically organized modularity in the yeast protein-protein interaction network. *Nature* 430: 88–93.

Hernandez, P., Huerta-Cepas, J., Montaner, D. et al. 2007. Evidence for systems-level molecular mechanisms of tumorigenesis. *BMC Genomics* 8: 185.

Hu, Z., Fan, C., Oh, D. S. et al. 2006. The molecular portraits of breast tumors are conserved across microarray platforms. *BMC Genomics* 7: 96.

Jonsson, P. F. and Bates, P. A. 2006. Global topological features of cancer proteins in the human interactome. *Bioinformatics* 22: 2291–2297.

Jonsson, P. F., Cavanna, T., Zicha, D., and Bates, P. A. 2006. Cluster analysis of networks generated through homology: automatic identification of important protein communities involved in cancer metastasis. *BMC Bioinformatics* 7: 2.

Kuida, K., Haydar, T. F., Kuan, C. Y. et al. 1998. Reduced apoptosis and cytochrome c-mediated caspase activation in mice lacking caspase 9. *Cell* 94: 325–337.

Liang, H. and Li, W. H. 2007. Gene essentiality, gene duplicability and protein connectivity in human and mouse. *Trends Genet* 23: 375–378.

Liao, B. Y. and Zhang, J. 2007. Mouse duplicate genes are as essential as singletons. *Trends Genet* 23: 378–381.

Liu, X., Kim, C. N., Yang, J., Jemmerson, R., and Wang, X. 1996. Induction of apoptotic program in cell-free extracts: requirement for dATP and cytochrome c. *Cell* 86: 147–157.

Ma, X. J., Wang, Z., Ryan, P. D. et al. 2004. A two-gene expression ratio predicts clinical outcome in breast cancer patients treated with tamoxifen. *Cancer Cell* 5: 607–616.

Paik, S., Shak, S., Tang, G. et al. 2004. A multigene assay to predict recurrence of tamoxifen-treated, node-negative breast cancer. *N. Engl. J. Med.* 351: 2817–2826.

Perou, C. M., Sorlie, T., Eisen, M. B. et al. 2000. Molecular portraits of human breast tumours. *Nature* 406: 747–752.

Prinz, A. A., Bucher, D., and Marder, E. 2004. Similar network activity from disparate circuit parameters. *Nature Neurosci* 7: 1345–1352.

Rambaldi, D., Giorgi, F. M., Capuani, F., Ciliberto, A., and Ciccarelli, F. D. 2008. Low duplicability and network fragility of cancer genes. *Trends Genet* 24: 427–430.

Sakamaki, K., Inoue, T., Asano, M. et al. 2002. Ex vivo whole-embryo culture of caspase-8-deficient embryos normalize their aberrant phenotypes in the developing neural tube and heart. *Cell Death Differ* 9: 1196–1206.

Sjoblom, T., Jones, S., Wood, L. D. et al. 2006. The consensus coding sequences of human breast and colorectal cancers. *Science* 314: 268–274.

Sorlie, T. 2009. Introducing molecular subtyping of breast cancer into the clinic? *J Clin Oncol* 27: 1153–1154.

Sorlie, T., Perou, C. M., Tibshirani, R. et al. 2001. Gene expression patterns of breast carcinomas distinguish tumor subclasses with clinical implications. *Proc Natl Acad Sci USA* 98: 10869–10874.

Sotiriou, C., and Pusztai, L. 2009. Gene-expression signatures in breast cancer. *N. Engl. J. Med.* 360: 790–800.

Taylor, I. W., Linding, R., Warde-Farley, D. et al. 2009. Dynamic modularity in protein interaction networks predicts breast cancer outcome. *Nature Biotechnol* 27: 199–204.

van 't Veer, L. J., Dai, H., van de Vijver, M. J. et al. 2002. Gene expression profiling predicts clinical outcome of breast cancer. *Nature* 415: 530–536.

van de Vijver, M. J., He, Y. D., van't Veer, L. J. et al. 2002. A gene-expression signature as a predictor of survival in breast cancer. *N. Engl. J. Med.* 347: 1999–2009.

Wang, E., Lenferink, A., and O'Connor-McCourt, M. 2007. Cancer systems biology: exploring cancer-associated genes on cellular networks. *Cell Mol Life Sci* 64: 1752–1762.

Wang, E., Li, J., Lenferink, A. et al. 2009. Cancer gene markers and methods for their identification. *United States provisional patent* 12020-1.

CHAPTER 5

From Tumor Genome Sequencing to Cancer Signaling Maps

Cong Fu and Edwin Wang

CONTENTS

5.1 CANCER GENOME SEQUENCING

In recent years, the cost of genome sequencing technology has dropped rapidly, owing to the continual development of newer, faster, and cheaper DNA sequencing technologies. It is believed that within a few years, the cost of sequencing a human genome will fall to $1,000. Eventually, genome sequencing technology may allow doctors to decode the entire genetic code of any patient disease samples in a clinical setting. In this situation, it would be possible for governments to include the genome sequences of all individuals in healthcare systems.

At the root of all forms of cancer are genetic and epigenetic alterations, which are either inherited or acquired (i.e., mutated or methylated) during our lives. Somatic mutations are the major cause of cancer initiation, progression, and metastasis. Cancer genomes carry two classes of mutations: driver mutations, which are positively selected because they are essential for tumor growth and development, and passenger mutations, which are not subject to selection because they do not confer a growth advantage. Positive selection indicative of driver mutations is evidenced by a higher ratio, compared with that determined by chance, of amino acid-changing nonsynonymous mutations to synonymous mutations that do not involve amino acid changes.

Advances in the genetic understanding of many forms of cancer have been made during the past decades. However, we are no closer to uncovering the molecular underpinnings of the disease. If we could catalog cancer driver-mutating genes, we would be able to link these genes to biological pathways, biological processes, and cellular networks.

5.1.1 Sequencing All Coding Genes in a Limited Number of Tumor Samples

Genome sequencing technology makes it possible to search for cancer driver-mutating genes on a genome-wide scale. Furthermore, profiling cancer driver mutations could provide a molecular portrait of each individual tumor sample. Such a molecular portrait could help clinicians to improve their diagnosis and offer the most suitable therapy to each patient. In 2006, a team of researchers obtained a molecular portrait of individual tumors by completing an unbiased, large-scale sequencing study of protein-coding genes in tumors caused by breast and colorectal cancer (Sjoblom et al. 2006). The study found a surprising number of mutated genes, many of which have not been previously implicated in tumorigenesis. On average, each tumor had at least 14 to 15 protein-altering cancer driver-mutating genes.

The same team extended the genome sequencing to all of the genes in the Reference Sequence database in 11 breast and 11 colorectal tumor samples (Wood et al. 2007). They conducted a comprehensive assessment of the genomic landscapes of human breast and colorectal cancer. From this study, a global picture of the genomic landscape of cancer emerged. Only a few genes (i.e., P53) are repeatedly mutated in many tumors, and most of the cancer genes are mutated at relatively low frequencies, that is, in fewer than 5% of tumors.

These studies demonstrate that many of the cancer genes are important in a relatively small proportion of tumors. The studies suggest that low-frequency gene mutations are more relevant to directing tumorigenesis and survival than previously defined high-frequency gene mutations. The research confirms the notion that cancer is caused by an accumulation of mutations. Further analyses of these mutating genes provide additional evidence that pathways, especially signaling pathways, rather than individual genes, govern the course of tumorigenesis.

For each individual tumor, these studies demonstrated that different tumors have different profiles of gene mutations (i.e., different sets of mutated genes) or a unique signature of gene mutations. This observation raises interesting possibilities for developing biomarkers and novel, personalized treatment strategies.

Similar results related to genome-wide mutations in cancer have been obtained using the transposon-mediated forward genetic screen for colon cancer in mice (Starr et al. 2009). The screening method is adopted from the Sleeping Beauty transposon-based insertional mutagenesis system. Sleeping Beauty is able to insert itself into or near genes to either activate or deactivate a gene's normal function. Compared to other methods, this method is faster, more accurate, and more efficient for identifying groups of genes associated with specific cancers. It also provides information about the specific combinatory patterns of gene mutations in each individual tumor sample. By comparing the identified mutated genes with the genes from the genome sequencing approach mentioned above, the authors concluded that there is significant overlap between the mouse candidate genes and human genes that are altered in colon cancer. These results confirm that tumor genome sequencing is a powerful approach for identifying cancer driver-mutating genes.

5.1.2 Sequencing of Selected Genes in a Large Number of Tumor Samples

Large-scale sequencing of preselected genes, that is, known cancer genes, in a large population represents another approach to sequencing tumor genes. For example, 1000 samples derived from 17 different tumor types have been analyzed for mutations of 17 well-known oncogenes (Greenman et al. 2007; Thomas et al. 2007). They surveyed the number and pattern of somatic mutations in coding regions of 518 kinase genes, which are among the most commonly mutated genes in cancer, in 210 tumor samples of different origins. These studies showed that mutational signatures of tumors are affected by tissue origin, DNA-repair ability, and the chance of exposure to carcinogens. Lung cancer, for instance, has more mutations due to the direct exposure of lung cells to air.

5.1.3 Complete DNA Sequencing of a Human Cancer Genome

In the November 6 issue (2008) of the journal *Nature*, a report detailed the first sequencing of the entire genome of a patient with acute myeloid leukemia (AML), a woman in her 50s who died of the disease (Gridley 2003). The study identified cancer-related mutations specific to her cancer. The DNA for the reference genome was taken from a skin sample of the patient. The tumor and the reference samples were obtained before the patient received cancer treatment. By doing so, the mutations induced by anticancer agents could be avoided.

This study was the first to conduct a full genome comparison between normal cells and tumor cells from the same patient. Single base changes in the patient's tumor genome compared with her normal genome were scanned. Almost 98% of the nucleotide variants in the patient's tumor genome were identical to those from the patient's skin sample. Ten mutations (including the two previously known genetic mutations that are common to AML) were identified. Among the eight novel mutations, three were in genes that normally act to suppress tumor growth, for example, a mutation in the PTPRT tyrosine phosphatase gene, which is frequently altered in colon cancer. Four other mutated genes were involved in molecular pathways that promote cancer growth. In the near future, the study's authors may release more results on the mutations of noncoding DNA regions, which would be the major contribution of the project. One of the advantages of the full genome sequencing of tumor samples is that it provides the mutation information for noncoding DNAs, which has not yet been explored in tumor genomes.

Tumor samples from 187 additional AML patients were scanned for the eight novel mutations, but none of them were found. This result confirmed the conclusion of other tumor genome-sequencing studies: there is a tremendous amount of genetic diversity in cancer, even in one type or subtype of cancer. The unique nature of the mutation profiles for this patient strongly indicates the huge genetic complexity and diversity of cancer genomes.

It is likely that a full genome-sequencing approach will be applied to more samples and extended to other cancer types with the advance of the next generation of genome sequencing technology.

5.1.4 Genome Sequencing of Tumors in Different Developmental Stages

The tremendous amount of genetic diversity in cancer suggests that there are many ways to mutate a small number of genes to get the same result. Furthermore, it suggests that the mutations may occur sequentially. The first mutation gives the cell a slight tendency toward cancer, and subsequent mutations compound this tendency. The last mutation in a malignant tumor might represent a turning point at which the cancer cells become more dangerous and aggressive.

One example of such a metastatic mutation is a constitutive activating mutation in the 1 integrin subunit (T188I 1) associated with human squamous cell carcinoma. Transgenic cells with the T188I 1 mutation showed increased cell spreading; however, this did not affect epidermal proliferation, epidermal organization, or stem cell number. Further analysis suggests that integrin mutations may play a part in cancer malignancy (Ferreira et al. 2009).

Cooperation of oncogenic mutations leads to synergistic changes in downstream signaling pathways. Furthermore, a significant number of such synergistic changes are crucial for tumorigenesis (McMurray et al. 2008). These synergistic changes in gene expression profiles could be used as a metric to efficiently identify key players that function downstream of oncogenic mutations and might be viable therapeutic anticancer targets.

It is critical to pinpoint the driver mutations in different developmental stages of tumors so that malignant mutation and the cooperation between mutations can be identified. However, current efforts of tumor genome sequencing cannot distinguish the driver mutations for cancer initiation, progression, or metastasis. In the future, it will be possible to explore the sequencing efforts for different developmental stages of tumors and catalog the mutations in these stages. These efforts will help sort out the relationships between driving mutations and their contribution to tumorigenesis.

It is also important to follow the patients whose tumors are sequenced to gather clinical data such as survival rates, tumor recurrence, and drug responses. Such clinical information will help link certain mutations for the identification of gene markers and illustrate the molecular mechanisms associated with prognosis, diagnosis, and drug response.

5.2 FROM A CANCER GENOME SEQUENCING APPROACH TO A SYSTEMATIC MULTIDIMENSIONAL GENOMIC APPROACH

The participants in the Cancer Genome Atlas (TCGA) project, which aims to discover and catalog major cancer-causing genomic alterations by assessing multiple human tumor samples, have proposed an integrated and multidimensional genomic approach to cancer genomic study. TCGA has a long-term goal of systematically exploring the universe of genomic changes involved in all types of human cancer and demonstrating the values of such efforts in advancing cancer research and improving patient care.

Recently, TCGA reported a comprehensive study of 206 samples of primary glioblastoma, including analysis of DNA methylation status and copy number aberrations, as well as coding and noncoding RNA expression and the sequencing of 601 preselected cancer genes (TCGA 2008). This is the first summary of data from the $100 million TCGA pilot project.

TCGA researchers also searched for mutations of 623 known cancer genes in 188 lung adenocarcinoma patients by sequencing DNA from tumor samples and matching noncancerous tissue from the patients (Ding et al. 2008). They identified 26 driver mutating genes. Most of these genes had not previously been associated with lung adenocarcinoma. Most interestingly, the authors found that the number of genetic mutations detected in tumor samples from smokers was significantly higher than that in tumors from people who had never smoked. Tumors from smokers contained as many as 49 mutations, whereas none of the tumors from people who had never smoked had more than five mutations.

A similar analysis has been applied to 22 human glioblastoma samples and 24 advanced pancreatic adenocarcinoma samples. For these samples, 20,661 protein-coding genes have been sequenced. Furthermore, genomic changes such as DNA methylation as well as gene expression changes have been analyzed. The genetic mutations in different cellular pathways have been mapped. For example, 12 core signaling pathways and processes have been linked to pancreatic cancer. The studies suggest that drugs that target a pathway rather than a gene are most likely to be more effective for the treatment of cancer (Parsons et al. 2008).

All of these studies provide a comprehensive view of the complicated genomic landscape of cancer. Additionally, they clearly confirm that unbiased systematic and integrative approaches can lead to a more comprehensive understanding of the changes that occur during tumor development and treatment. These studies also illustrate how an unbiased and systematic cancer genome approach can lead to paradigm-shifting discoveries. For example, this research could reveal an important link between a methylation change in the glioblastoma cells and the drugs that should be used for treatment. Tumors containing the methylated MGMT gene are more susceptible to the cancer drug temozolomide.

5.3 DATA INTERPRETATION BECOMES INCREASINGLY CHALLENGING

All of the studies mentioned above point to an unexpected conclusion: tumor genomes are extremely complex in terms of the genetic alterations that drive tumorigenesis. There is a lot of diversity and little overlap in the different types of mutated genes. This diversity is seen among different types and subtypes of tumors and even between tumors that originate from the same tissue. The discoveries in these studies are only the tip of the iceberg. With the rapid development of high-throughput sequencing platforms, as well as other large data generation systems, we expect that more and more complex datasets for tracking all genetic/epigenetic and gene/noncoding-RNA expressional changes occurring within tumors, and even within a specific type of tumor, will be generated.

Future studies will eventually help to untangle the biological roots of cancer by applying advanced genomic tools to the complexities of cancer. If so, this information will accelerate efforts by the worldwide scientific community to improve outcomes for cancer patients. For example, it may help guide the design of new drugs and other cancer therapies. It may also lead to an individualized approach to cancer treatment that maximizes efficacy by tailoring the course of therapy for each patient. However, the current challenge is to integrate and interpret these datasets in a way that provides insight into the molecular basis of cancer.

Several of the cancer genome studies mentioned above tried to map genetic mutations onto signaling pathways. However, there are several limitations to such an approach. First, only a fraction of the mutated genes can be mapped onto the pathways. In this context, it is hard to say whether the "core pathways" identified using this approach are representative. Most importantly, the mutated genes derived from each study are far from comprehensive. Typically, only a couple of tumor samples were used to sequence all the coding genes, and such analyses reveal only a fraction of the mutating genes. In contrast, some studies used several hundred tumor samples, but only sequenced a limited number of genes (500 to 600).

These analyses cannot catalog a comprehensive list of mutated genes. Therefore, it leads to the question whether the so-called "core pathways" are representative.

To test whether we could obtain a set of "core pathways" of cancer signaling, we tried to map signaling pathways using a more comprehensive cancer mutated gene list from the COSMIC database, which collects data about the cancer driver-mutating genes from the literature and genome sequencing efforts. From this analysis, we found that most of the signaling pathways can be mapped, even for the mutation genes coming from one type of cancer. These results suggest that the so-called "core pathways" become less defined when we have data compiled for more cancer driver-mutating genes. The final limitation is the unclear definition of signaling pathways. For example, the boundaries of individual signaling pathways differ from one database to another.

The cancer genome studies reveal that "real signals" or biological insights, molecular mechanisms, and biological principles are concealed by the abundance of extremely complex data. These studies underscore the notion that in the "new biology" era, the bottleneck is no longer a lack of data but the lack of ingenuity and the computational means to extract biological insights and principles by integrating knowledge and high-throughput data.

5.4 THE SIGNALING NETWORK, AN EFFECTIVE FRAMEWORK FOR MODELING COMPLEX CANCER DATA

5.4.1 Biology Is a Science of Relationships

To develop effective computational tools, we must first understand what biology is. Biology deals with many kinds of relationships among genes, proteins, RNAs, cells, tissues, organs, and environmental factors. For example, biological relationships include those encompassing gene regulation, protein interaction, activation, genetic interaction, inhibitory action, and so on. Biology is a science of relationships. Traditionally, biologists describe the relationships between a limited number of genes or proteins.

As shown in the cancer genome literature, high-throughput techniques have become more affordable and accessible, a driving force in modern biology. As a result of the huge amount of data produced by high-throughput techniques, biologists have to account for thousands of biological relationships in a single experiment. In this situation, the traditional ways of describing biological relationships are not sufficient. The only way to analyze a large number of relationships is through mathematical representation and computation.

One of the most important models used to describe biological relationships is a pathway in which a linear relationship between genes or proteins is established. From an abstract point of view, a pathway is a model that represents the efforts of human beings to explore, describe, and organize the biological relationships in cells. Such an effort is important because it allows us to further understand biological systems and predict cell behaviors. However, such goals have not been attained because of the crosstalk between pathways, which has been documented extensively in recent years. In this context, the network model, in which the interactions and relationships between genes or proteins are described in a nonlinear manner, has been proposed. It is important to note that both pathways and networks are conceptual models. We are making projections of living cell molecules onto conceptual frameworks. Thus, we can study the models, which more closely mimic cellular reality.

5.4.2 Cancer and Cell Signaling

Cells use sophisticated communication between proteins to initiate and maintain basic functions such as growth, survival, proliferation, and development. Traditionally, cell signaling is described via linear diagrams and signaling pathways. As more crosstalk between signaling pathways has been identified (Natarajan et al. 2006), a network view of cell signaling has emerged: the signaling proteins rarely operate in isolation through linear pathways, but rather through a large and complex network. Because cell signaling plays a crucial role in cell responses like growth and survival, alterations of cellular signaling events, like those caused by mutations, can result in tumor development. Indeed, cancer is largely a genetic disease that is caused by the acquisition of genomic alterations in somatic cells. Alterations to the genes that encode key signaling proteins, such as RAS and PI3K, are commonly observed in many types of cancers. During tumor progression, it has been proposed that a malignant tumor arises from a single cell, which undergoes a series of evolutionary processes of genetic or epigenetic changes and selections. Thus, the cell can acquire additional selective advantages for cellular growth or survival within the population, resulting in progressive clonal expansion (Nowell 1976).

Genetic mutations of the signaling proteins may over-activate key cell-signaling properties such as cell proliferation or survival, giving rise to a cell with selective advantages for uncontrolled growth and the promotion of tumor progression. In addition, mutations may also inhibit the function of tumor suppressor proteins, resulting in a relief from the normal constraints on growth. Furthermore, epigenetic alterations by promoter methylation, resulting in transcriptional repression of genes that control tumor malignancy, is another important mechanism for the loss of gene function that can provide a selective advantage to tumor cells.

The cancer phenotype is the result of the collaboration between a group of genes. This notion provides a structured network knowledge-based approach to analyzing genome-wide data in the context of known functional interrelationships among genes, proteins, and phenotypes. Many lines of evidence suggest that biological relationships and complexity are encoded in cellular networks (Cui et al. 2007a). Therefore, a network or systems-level view of cellular events emerges as an important concept.

Signaling networks contain the most complicated relationships between proteins. For example, nodes can represent different functional proteins such as kinases, growth factors, ligands, receptors, adaptors, scaffolds, transcription factors, and so on, which all have different biochemical functions and are involved in many different types of biochemical reactions that characterize specific signal transduction machinery (Cui et al. 2007b). In signaling networks, hub proteins are the proteins most commonly used by multiple signaling pathways. They become the information exchange and processing centers of the network (Cui et al. 2007a). Signaling network motifs are the smallest functional modules in signal processing, amplification, and noise buffering in cell signaling (Cui et al. 2007a).

5.5 CANCER SIGNALING MAPS DERIVED FROM COMPLEX CANCER DATA

5.5.1 Addressing Questions about Cancer Signaling Networks

Enormous efforts have been made over the past few decades to identify mutated genes that are causally implicated in human cancer. A genome-wide or large-scale sequencing of tumor samples across many kinds of cancers represents a largely unbiased overview of the spectrum of mutations in human cancers (see sections above). Similarly, genome-wide identification of epigenetic changes in cancer cells has recently been conducted (Ohm et al. 2007; Schlesinger et al. 2007; Widschwendter et al. 2007). These studies showed that a substantial fraction of the cancer-associated mutated and methylated genes are involved in cell signaling. This information is consistent with the previous finding that the protein kinase domain is most commonly encoded by cancer genes.

Although there is a wealth of knowledge about molecular signaling in cancer, the complexity of human cancer genomes prevents us from gaining an overall picture of the mechanisms by which these genetic and epigenetic events affect cancer cell signaling and tumor progression. Where are the oncogenic stimuli embedded in the network architecture? What are the principles by which genetic and epigenetic alterations trigger oncogenic signaling events? Because so many genes possess genetic and epigenetic aberrations in cancer signaling, what is the architecture of cancer signaling? Do any tumor-driven signaling events represent "oncogenic dependence," the phenomenon by which certain cancer cells become dependent on certain signaling cascades for growth or survival? What are the central players in oncogenic signaling? Are there any signaling partnerships that are generally used to generate tumor phenotypes? To answer these questions, we conducted a comprehensive analysis of cancer mutated and methylated genes in a human signaling network, focusing on network structural aspects and quantitative analysis of gene mutations in the network.

5.5.2 Where Are the Oncogenic Stimuli Embedded in the Network Architecture?

The architecture and relationships among the proteins of a signaling network play a significant role in determining the sites at which oncogenic stimuli occur and through which oncogenic stimuli are transduced. Integration of the data about mutated and methylated cancer genes into the network could help identify critical sites involved in tumorigenesis and increase our understanding of the underlying mechanisms in cancer signaling.

Extensive signaling studies during recent decades have yielded an enormous amount of information regarding the regulation of signaling proteins for more than 200 signaling pathways, most of which have been assembled and collected in diagrams in public databases. We manually curated the data on signaling proteins and their relationships (activation, inhibitory, and physical interactions) from the BioCarta database and the Cancer Cell Map database. We merged the curated data with another literature-mined signaling network that contains ~500 proteins (Ma'ayan et al. 2005). As a result, we have created a human signaling network containing 1634 nodes and 5089 links. We also collected the cancer driver-mutating genes from both the literature and the large-scale sequencing of tumor samples. Additionally, we isolated the cancer-methylated genes from the genome-wide identification of the DNA methylated genes in cancer stem cells. Finally, 227 cancer mutated genes and 93 DNA methylated genes were mapped onto the network. Among the 227 cancer mutated genes, 218 (96%) and 55 (24%) genes were derived from the large-scale gene sequencing of tumors and the literature curation, respectively.

5.5.2.1 Cancer Mutated Genes Are Enriched in Signaling Hubs but Not in Neutral Hubs

Genes that result in tumorigenesis when mutated or silenced often lead to the aberrant activation of certain downstream signaling nodes, resulting in dysregulated growth, survival, and/or differentiation. The architecture of a signaling network plays an important role in determining the site at which a genetic defect is involved in cancer. To discover where the critical tumor signaling stimuli occur in the network, we explored the network characteristics of the mutated and methylated genes. The signaling network is presented as a graph in which nodes represent proteins. Directed links are operationally defined to represent effector actions such as activation or inhibition, whereas undirected links represent physical protein interactions that are not characterized as either activating or inhibitory. For example, scaffold proteins do not directly activate or inhibit other proteins, but provide regional organization for activation or inhibition through protein-protein interactions. In this case, undirected links are used to represent the interactions between scaffold and other proteins. On the other hand, adaptor proteins are able to activate or inhibit other proteins through direct interactions. In this situation, directed links are used to represent these relationships. There are two kinds of directed links, incoming and outgoing. An incoming link represents a signal from another node, and the sum of the incoming links of a node is called the indegree of that node. An outgoing link represents a signal to another node; the sum of the outgoing links of a node is called the outdegree of that node. We refer to incoming and outgoing links as signal links, whereas the physical links are neutral links. We initially examined the characteristics of the nodes that represent mutated genes on the network. We compared the average indegree of the mutated genes to that of the nodes in the network as a whole. We found that the average indegree and outdegree of the mutated nodes are significantly higher than the indegree and outdegree of the network nodes. In contrast, there is no difference in the average neutral degrees between the mutated nodes and other nodes in the network.

These results suggest that cancer mutations most likely occur in signaling proteins that act as signaling hubs (i.e., RAS), actively sending or receiving signals, rather than in nodes involved in passive physical interactions with other proteins. Because these hubs are focal nodes that are shared by and important to many signaling pathways, alterations of these nodes or signaling hubs may affect more signaling events, resulting in cancer or other diseases. In previous studies, we found that genes associated with cancer are enriched in hubs (Cui et al. 2007a). However, these results indicate that genes associated with cancer are enriched in signaling hubs but not neutral hubs.

Methylated gene nodes do not appear to differ significantly from the network nodes with regard to their indegree, outdegree, and neutral degree. These results suggest that cancer mutated genes and methylation-silenced genes have different regulatory mechanisms in oncogenic signaling.

5.5.2.2 The Output Layer of the Network Is Enriched with Mutating Genes

We hypothesized that the downstream genes of the network, especially the genes of the output layer of the network, would have a higher mutation frequency. To test this possibility, we compared the average gene mutation frequency of the nuclear proteins, which represent the members of the output layer of the network, with that of the other network genes. Indeed, the nuclear genes have a higher mutation frequency than others, which correlates with our previous finding that cancer-associated genes are enriched in nuclear proteins (Cui et al. 2007a). In contrast, the distributions of the methylated genes have no such preference, suggesting that DNA methylation does not tend to directly affect the output layer of the network. These results strongly suggest that the genes in the output layer of the network, which play direct and important roles in determining phenotypic outputs, are frequent targets for activating mutations. The importance of this output layer is reinforced by our previous observation that the expression of the output layer genes of the signaling network is heavily regulated by microRNAs (Cui et al. 2006) and is evolutionarily conserved (Cui et al. 2009).

5.5.3 What Are the Principles by Which Genetic and Epigenetic Alterations Trigger Oncogenic Signaling Events?

The complex architecture of signaling networks can be seen as consisting of interacting network motifs, which are statistically overrepresented subgraphs that recur in networks. A signaling network motif, also known as a regulatory loop, is a group of interacting proteins capable of signal processing. The proteins are characterized by specific regulatory properties and mechanisms (Babu et al. 2004; Wang and Purisima 2005). The structure and intrinsic properties of the frequently recurring network regulatory motifs provide a functional view of the organization of signaling networks. Thus, the study of the distributions of the mutated and methylated genes in the network motifs will provide insight into mechanisms that regulate cancer signaling.

5.5.3.1 Mutated and Methylated Genes Are Enriched in Positive and Negative Regulatory Loops, Respectively

We examined the mutated genes in all of the 3-node size network motifs. We classified the 3-node size network motifs into 4 subgroups (labeled 0 to 3) based on the number of nodes that represent mutated genes. We calculated the ratio (Ra) of positive (activating) links to the total directed (positive and negative) links in each subgroup and compared it with the average Ra in all of the 3-node size network motifs, which is shown as a horizontal line in Figure 5.1a. As the number of mutated nodes rises, the Ra for the corresponding group increases to a maximum of ~0.93 (Figure 5.1a). We obtained similar results when we extended the same analysis to all of the 4-node size network motifs. These motifs show a clear positive correlation between the positive link ratio and the number of mutated genes in the motifs. These results suggest that cancer gene mutations occur preferentially in positive regulatory motifs. In contrast, all of the 3-node and 4-node size motifs show an obviously negative correlation between the positive link ratio and the number of methylated genes in the motifs (Figure 5.1b). These results suggest that

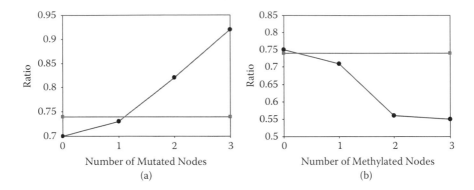

FIGURE 5.1 (See color insert following page 332.) Enrichment of mutated and methylation genes in network motifs. (a) Relations between the fractions of positive links in all 3-node size network motifs and the fractions of mutated genes in these motifs. (b) Relations between the fractions of positive links in all 3-node size network motifs and the fractions of methylated genes in these motifs. All network motifs were classified into subgroups based on the number of nodes that are either mutated genes or methylated genes, respectively. The ratio of positive links to total positive and negative links in each subgroup was plotted. The horizontal lines indicate the ratio of positive links to the total positive and negative links in all network motifs. (Adapted from Cui et al. 2007a.)

cancer gene methylation preferentially occurs in negative regulatory motifs. A similar trend was found among the 15 known tumor suppressors, which supports the notion that cancer-associated methylated genes act as tumor suppressors. Collectively, these facts suggest that mutated and methylated genes use different regulatory mechanisms in cancer signaling and support the notion that gene mutations and methylations are strongly selected in tumor samples.

5.5.3.2 Principles by Which Genetic and Epigenetic Alterations Trigger Oncogenic Signaling Events

Signaling information is propagated through a series of built-in regulatory motifs that contribute to cellular phenotypic functions (Ma'ayan et al. 2005). The transition from a normal cellular state into a long-term deregulated state such as cancer is often driven by prolonged activation of downstream proteins, which are regulated by upstream proteins or regulatory motifs or circuits. Positive regulatory loops (Ferrell 2002) could amplify signals, promote the persistence of signals, serve as sites for information storage, and evoke biological responses to generate phenotypes such as cancer. Cancer cells require constitutive activation of oncogenic signaling. The enrichment of gene mutations in positive regulatory loops suggests that the mutants in the motifs must have gain of function. Alternatively, compared with wild-type genes, they may increase their biochemical activities in order to constitutively activate downstream proteins. Indeed, a recent study showed that 14 of the 15 PI3K mutants in tumors have gain of function (Gymnopoulos, Elsliger, and Vogt 2007). Gain-of-function mutants in a positive regulatory loop amplify weak input stimuli and serve as information storage sites, which extend the duration of the activation of the affected downstream proteins. This might allow the downstream signaling cascades to persistently hold and transfer information, leading to tumor phenotypes.

Promoter gene methylation is a mechanism known to induce loss of function by inhibiting the expression of genes (Ohm et al. 2007; Widschwendter et al. 2007). Negative regulatory loops controlled by tumor suppressor proteins repress positive signals and play an important role in maintaining cellular homeostasis and restraining the cellular state transitions (Ma'ayan et al. 2005). A loss of function of gene methylation in a negative regulatory loop could break the negative feedback, thereby releasing the restrained activation signals and promoting oncogenic state transitions. Homeostasis relies on the balance between positive and negative signals in crucial components of the network. Both the gain-of-function mutated genes in positive regulatory loops and the loss-of-function methylated genes in negative regulatory loops could break this delicate balance, thus promoting state transitions and generating tumor phenotypes. Therefore, both mutated and methylated genes and their oncogenic regulatory loops are critical components of the network in which the oncogenic stimuli occur.

5.5.4 What Is the Architecture of Cancer Signaling?

5.5.4.1 The Overall Architecture of Cancer Signaling

Within a network, genes whose mutations or epigenetic silencing are crucial triggers for oncogenic signaling might link together as network components. Identification of these components will help us discover the relationship between and structural organization of the oncogenic proteins. To uncover the architecture of cancer signaling and gain insight into higher-order regulatory relationships among signaling proteins that govern oncogenic signal stimuli, we mapped all of the genetic mutations and epigenetically silenced genes onto the network. We found that most of these genes (67%) are linked, forming a giant network component. To build an oncogenic map, we included other mutated and methylated genes not present in the composition of the component in the giant network component based on node connectivity. The resulting oncogenic signaling map consists of 326 nodes and 892 links (Figure 5.2).

The emerging oncogenic signaling map represents a "hot area" where extensive oncogenic signaling events might occur. As a proof of concept, we found that the MAPK kinase and TGF-β pathways, well-known cancer signaling pathways, are embedded in the map. For example, 50 of 87 proteins in the MAPK kinase pathway and 22 of 52 proteins in the TGF-β pathway, respectively, are included in the map. More importantly, in addition to known oncogenic pathways, there are many other novel candidates for cancer signaling cascades present in the map. For a particular gene muation in a tumor, one could use this map to generate testable hypotheses to discover the underlying oncogenic signaling cascades in that tumor.

As mentioned above, events dependent on oncogenic signaling, which we define as the interactions between the cancer mutated or methylated genes, are frequently found in tumor samples and represent various oncogenic driving events that could play more critical roles in generating tumor phenotypes. To systematically identify such events and discover how they are organized in the map, we charted the gene mutation frequency onto the map and highlighted the signaling links between any two genes with high mutation frequencies. Most genes have mutation frequencies lower than 2%; however, a handful of genes have very high mutation frequencies, such as p53 (41%), PI3K (10%), and RAS (15%). Therefore, a gene mutation frequency equal to or greater than 2% was categorized as high. Interestingly, nearly 10% of the links in the map are dependent on oncogenic signaling. Certain signaling events, such as Pten-PI3K and RAS-PI3K in the map, are well-known oncogenic signaling-dependent events/cascades that frequently act as triggers for various cancers.

5.5.4.2 Do Any Tumor-Driven Signaling Events Represent "Oncogenic Dependence"?

Oncogenic dependence is the phenomenon by which certain cancer cells become dependent on certain signaling cascades for growth or survival. As shown in Figure 5.2, most

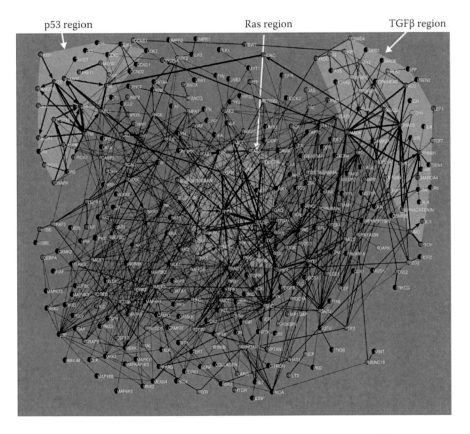

FIGURE 5.2 (See color insert following page 332.) Human oncogenic signaling map. The human cancer signaling map was extracted from the human signaling network, which was mapped with cancer mutated and methylated genes. The map shows three "oncogenic dependent regions" (background in light grey), in which genes of the two regions are also heavily methylated. Nodes represent genes, while the links with and without arrows represent signal and physical relations, respectively. Nodes in red, purple, brown, cyan, blue, and green represent the genes that are highly mutated but not methylated, both highly mutated and methylated, lowly mutated but not methylated, both lowly mutated and methylated, methylated but not mutated, and neither mutated nor methylated, respectively. (Adapted from Cui et al. 2007a.)

oncogenic signaling-dependent events are connected, and three major regions that contain densely connected oncogenic signaling-dependent events emerge in the map. The first region (p53 region) primarily contains tumor suppressors such as p53, Rb, BRCA1, BRCA2, and p14 (CDKN2A). The second region, RAS, primarily contains well-known oncogenes such as RAS, EGFR, and PI3K. The third region, TGF-β, contains SMAD3, SMAD4, and a few other TGF-β signaling proteins. Interestingly, genes in the p53 and TGF-β regions are also heavily methylated in cancer stem cells, suggesting that these regions are involved in the early stage of oncogenesis. Other methylated genes are intertwined with the mutated genes in the map, suggesting that they share some oncogenic signaling cascades and might be regulated to cooperate in cancer signaling via gene mutation and/or methylation. Notably, it seems that in cancer stem cells, the TGF-β signaling pathway is shut down, supporting its known role as a tumor suppressor in the early stages of tumorigenesis (Hanahan and Weinberg 2000; Siegel and Massague 2003). These results suggest that the crucial players in oncogenic signaling tend to be closely clustered and regionalized. This map uncovers the architectural structure of the basic oncogenic signaling process and highlights the signaling events that are involved in the generation of tumor phenotypes.

5.5.5 What Are the Central Players in Oncogenic Signaling?

The oncogenic signaling map can be broken down into several network communities, whereby each community contains a set of more closely linked nodes and ties to particular biological functions. To find such network communities in the map, we applied an algorithm that detects network communities. As a result, 12 network communities, referred to as "oncogenic signaling modules" and ranging in size from 11 to 65 nodes, were found in the map. Structurally, the nodes within each module have more links and signaling regulatory relationships to each other than to other nodes. The genes in each module share similar biological functions like cell proliferation, development, and apoptosis.

5.5.5.1 Genes Are Exclusively Mutated in the Same Cancer Signaling Modules

We investigated whether the genes in each module could operate in a compensatory or concerted manner to govern a set of similar functions. We surveyed the gene mutations in tumor samples in which at least two genes were screened for mutations. As a result, the co-occurrence in tumor samples of 25 mutated gene pairs was found to be statistically significant. Importantly, only three collaborative gene pairs came from the same module, whereas other collaborative gene pairs came from two different modules. One of the pairs came from Module 11 (defined as the p53 module), which contains p53, Rb, p14, BRCA1, BRCA2, and several other genes involved in the control of DNA damage repair and cell division. Collectively, these results suggest that the signaling genes from the same modules are exclusively mutated, but most likely work in a complementary way to generate tumor phenotypes.

5.5.5.2 p53-Apotopotic Signaling Module Plays a Central Role in Cancer Signaling

We surveyed the gene mutations in the tumor samples in which at least two gene mutations have been found. In total, 592 tumor samples fit this criterion. Notably, we found that at least one gene mutation in the p53 module had occurred in the tumor samples we examined, suggesting that the p53 module is involved in generating tumors for most cancers. This result suggests that the p53 module is a central oncogenic signaling player and plays an essential role in tumorigenesis. The p53 module is enriched with tumor suppressors and biological processes such as apoptosis and cell cycle. This finding is further supported by the following observations:

1. To become oncogenic, tumor suppressors require loss-of-function mutations, which occur more often than gain-of-function mutations (Gymnopoulos, Elsliger, and Vogt 2007). Indeed, the average gene mutation frequency in the p53 module is higher than that of other signaling modules, including the RAS module.

2. The methylation of genes in the cancer stem cells that result in long-term loss of expression represents the early stage of tumorigenesis. In fact, most of the members of the p53 module are methylated in cancer stem cells. These facts further support the possibility that the p53 module plays an important role in the earlier stages of oncogenesis.

3. Gene methylation or inactivating mutations of the DNA damage checkpoint genes like p53 induce genome instability. Consequently, these phenomena increase the chances of mutations occurring in other genes, including the genes of other oncogenic signaling modules that could functionally collaborate with the p53 module genes to generate tumor phenotypes.

5.5.5.3 Are There Any Signaling Partnerships That Often Generate Tumor Phenotypes?
We also investigated which oncogenic signaling modules work together to produce a tumor phenotype. To address this question, we used the 592 samples mentioned above to build a matrix (M) in which samples are rows and the signaling modules are columns. If a gene of a particular signaling module (b) is mutated in a tumor sample (s), we set $M_{s,b}$ to 1; otherwise we set $M_{s,b}$ to 0. A heat map was generated using the matrix (Figure 5.3a). As shown in Figure 5.3a, two of the signaling modules have a significantly greater number of gene mutations, suggesting that genes in these two signaling modules are predominantly used to generate tumor phenotypes. One oncogenic signaling module (Module 1, defined as the RAS module) contains genes like RAS, EGFR, and PI3K, which share similar biological functions such as cell proliferation, cell survival, and cell growth. The other oncogenic signaling module, the p53 module, shares similar biological functions such as cell cycle checkpoint control, apoptosis, and the ability to affect genomic instability. These two modules also represent the two oncogenic signaling-dependent regions (p53 and RAS regions) in Figure 5.2, respectively.

When a tumor sample has a mutation in a gene from the RAS signaling module, it is also more likely to contain a mutation in a gene from the p53 module ($P < 2 \times 10^{-4}$). To find out whether this phenomenon is primarily a result of the actions of a particular pair of genes, we calculated the likelihood of co-occurrence for each pair of the genes, in which one gene is mutated in one module and the other gene is mutated in the other module. We found that the P values for gene pairs always indicate greater significance compared

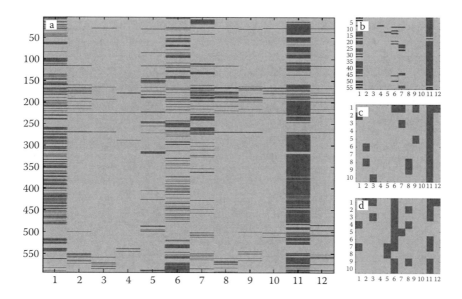

FIGURE 5.3 (See color insert following page 332.) Heatmaps of the gene mutation distributions in oncogenic signaling blocks. Twelve topological regions or oncogenic signaling blocks have been identified based on the gene connectivity of the human oncogenic signaling map. A heatmap was generated from a matrix, which was built by querying the oncogenic signaling blocks using tumor samples, in which each sample has at least two mutated genes. If a gene of a particular signaling block (b) gets mutated in a tumor sample (s), we set Ms,b to 1; otherwise we set Ms,b to 0. (a) A heatmap generated using the gene mutation data of the 592 tumor samples. (b) A heatmap generated using the gene mutation data of the NCI-60 cancer cell lines. (c) and (d) Heatmaps generated using the output from the genome-wide sequencing of breast and colon tumor samples, respectively. Rows represent samples, while columns represent oncogenic signaling blocks. Blocks with gene mutations are marked in red. (Adapted from Cui et al. 2007a.)

with the pair of modules RAS and p53. For example, the P value of co-occurrence of RAS (in module RAS) and p53 (in module p53) mutations is 0.01, which is greater than that of the two modules ($P < 2 \times 10^{-4}$). This indicates that these two oncogenic signaling modules collaborate to generate tumor phenotypes for most tumors. Experimental examples have shown similar gene collaboration in tumorigenesis: the activation of RAS (RAS module) and inactivation of p53 (p53 module) induce lung tumors (Meuwissen and Berns 2005), whereas the activation of RAS (RAS module) and inactivation of p16 (p53 module) induce pancreatic tumors (Obata et al. 1998). Generally, tumor cells exhibit either elevated cell proliferation or reduced differentiation or apoptosis when compared with normal cells. The oncogenic modules we have identified, especially the RAS and p53 modules, encode functions that are tumor related, such as cell cycle control, cell proliferation, and apoptosis. The activation of genes in the RAS module promotes cell proliferation, whereas the inactivation of genes in the p53 module prevents apoptosis. Thus, a functional collaboration between the genes in these two modules would promote synergistic cancer signaling and foster tumorigenesis.

Using the map as a framework, we benchmarked the mutated genes in the NCI-60 cell lines that represent a panel of well-characterized cancer cell lines and various cancer types. A systematic mutation analysis of 24 known cancer genes showed that most NCI-60 cell lines have at least two mutations among the cancer genes examined (Ikediobi et al. 2006). We built a matrix and constructed a heat map using these cell lines and their mutated genes, as described above (Figure 5.3b). Overall, the pattern obtained from the NCI-60 panel resembles that of the 592-tumor panel, with both the RAS and the p53 modules enriched with gene mutations and exhibiting statistically significant collaborations in these cell lines. These data are consistent with the earlier observations.

We also benchmarked the mutated genes derived from a genome-wide sequencing of 22 tumor samples (Sjoblom et al. 2006). Among these 22 samples, 10 breast and 10 colon tumor samples had at least two gene mutations in the map. As shown in Figure 5.3c-d, the p53 module is enriched with gene mutations. The 10 colon tumor samples reveal collaboration between Module 6 and Module p53. The 10 breast tumors establish collaborative patterns between multiple modules.

5.5.5.4 The Common Module Collaborates with Tumor Type-Specific Signaling Modules
To further examine the collaborative patterns of modules in individual tumor types at higher resolutions (Figure 5.3a), we extracted the sub-heat map from the heat map for several of the tumor types that occur more frequently within the group of 592 tumor samples (Figure 5.4). As shown in Figure 5.4, signaling module collaborative patterns are tissue dependent. The mutations of the genes in the common module (p53) are frequently accompanied by mutations in one or two other signaling modules. Different tumor types appear to achieve tumorigenesis via distinct mechanisms like collaboration between different signaling modules.

The signaling module collaborative patterns are classified into two groups. One group contains pancreatic, skin, central nervous system, and blood tumors that have simple module collaborative patterns. In these tumors, signaling collaborations mainly occur between module p53 and module RAS, with some minor contributions from modules 5, 6, or 7. This suggests that they predominantly use these oncogenic signaling routes to generate tumors, resulting in relatively homogeneous cancer cell types. The other group contains breast and lung tumors, which also contain large proportions of mutations from the p53 module but also reveal complex patterns of collaboration between assortments of multiple modules. This suggests that these tumors may have a larger variety of oncogenic signaling routes, which may explain, in part, the heterogeneous nature of the tumor subtypes in this category. These results might also explain why both lung and breast cancers are the most common types of human tumors.

In summary, the cancer signaling map allows complex mutations to be divided into a few common signaling modules, revealing the underlying logic of cancer signaling. Both common and tumor type-specific signaling modules are observed. The common module contains genes that are frequently mutated in most tumors, regardless of the tumor type. However, the common module is generally not sufficient for tumorigenesis, because mutations of the genes in the common module are frequently accompanied by mutations in one

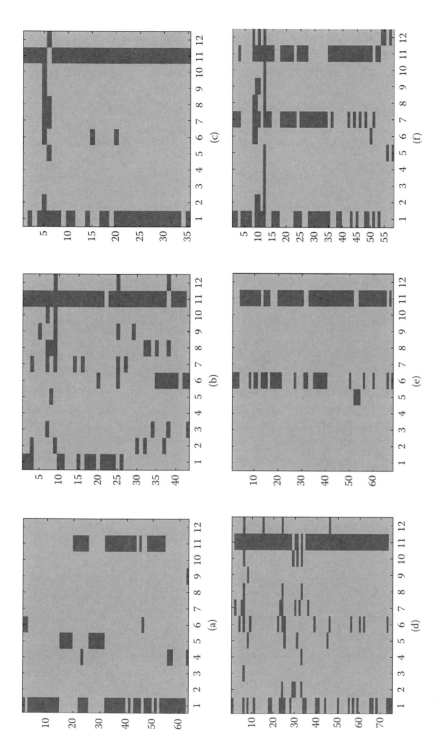

FIGURE 5.4 (See color insert following page 332.) Heatmaps of the gene mutation distributions in oncogenic signaling blocks for six representative cancer types. Twelve topological regions or oncogenic signaling blocks have been identified based on the gene connectivity of the human oncogenic signaling map. A heatmap was generated from a matrix, which was built by querying the oncogenic signaling blocks using tumor samples, in which each sample has at least two mutated genes. If a gene of a particular signaling block (b) gets mutated in a tumor sample (s), we set $M_{s,b}$ to 1; otherwise we set $M_{s,b}$ to 0. Heatmaps for (a) blood, (b) breast, (c) central nervous system, (d) lung, (e) pancreas, and (f) skin tumors were built using tumor samples of these cancer types, respectively. Rows represent samples, while columns represent oncogenic signaling blocks. Blocks with gene mutations are marked in red. (Figure is adapted from Cui et al. 2007a).

or two other signaling modules. Different tumor types appear to achieve tumorigenesis via distinct mechanisms like collaboration between different signaling modules. Taking a systems biology approach, this work presents a network view of the molecular mechanisms of cancer signaling that will shape our understanding of fundamental tumor cell biology.

5.5.6 Dissecting Dynamic Cancer Signaling Modules Using Gene Expression Profiles

Integrative analysis of signaling networks incorporating cancer driver-mutating genes leads to a static understanding of cancer, whereas integrative analysis of molecular networks using a combination of other-omic data such as gene expression profiles allows us to dissect dynamic signaling modules relevant to the individual tumor sample. One proposed approach has been to dissect oncogenic signaling networks into conditionally dependent signaling modules based on gene expression signatures. Furthermore, these modules have been shown to be useful for analyzing cancer patient outcomes and drug responses (Chang et al. 2009).

To dissect a Ras signaling module, Chang et al. mapped the proteins of a Ras pathway onto a human protein interaction network in order to collect the directly interacting proteins of the pathway proteins, defined as a core gene set. Statistical analyses of the Ras pathway genes using gene expression profiles of the NCI-60 cell line dataset showed similar variation in their expression to the core genes. This led to the identification of 20 gene signatures linked to the Ras pathway. Comparative analysis has been performed of these signatures with the signatures of mutants (i.e., tumors having specific mutation genes) that selectively activate downstream effectors of Ras or to signatures from cells that are sensitive to drugs that target specific pathway members or gene signatures for specific signaling effectors, such as Raf or phosphatidylinositol 3-kinase for Ras signaling. Such an approach allows us to define signaling modules used in a single tumor.

Chang et al. derived a set of 20 gene expression signatures for the EGFR signaling module based on the responses of cancer patients to an EGFR-specific drug, cetuximab. Further analysis of the gene signatures of the EGFR and Ras modules revealed that only the EGFR signatures could distinguish between the patients who are sensitive or resistant to cetuximab, indicating the specificity of each set of signatures for a particular oncogenic signaling module.

It is reasonable to expect that such an approach could be used for more accurate modeling of oncogenic information processing and transmission through the cancer signaling maps. Thus, it may be possible to uncover the collaborations between the modules and the correlations between modules and cancer phenotypes, such as cancer initiation, progression, and metastasis.

5.6 SIGNALING MAPS FOR INDIVIDUAL TUMORS AND PERSONALIZED MEDICINE

Technological advances in genomic and proteomic signaling capacity will enable us to determine individually relevant gene mutations and signaling events. It is expected that future opportunities for cancer management will involve individual target assessment (i.e., identifying key signaling events and genes involved in tumorigenesis and metastasis) and matching of individual targets to target-based therapy like small molecule and RNAi treatment.

It is increasingly clear that key signaling events rather than individual genes or pathways appear to be particularly important for tumorigenesis. Such an observation raises interesting possibilities for developing biomarkers and personalized treatment strategies.

An individual tumor-signaling map can be constructed using the data generated from one tumor sample by using multidimensional genomic approaches. Comparative analysis of the individual tumor-signaling maps with the general signaling map of the same tumor type will lead to the identification of key signaling events for each individual tumor-signaling map. Further network modeling of the individual tumor-signaling maps with known key signaling events will lead to the identification of key drug targets for individual patients (a roadmap for such an analysis and modeling has been described in Chapter 1). Data from the efforts of RNAi knockout cancer cells and the profiling of small molecules in cancer cells will help us find appropriate drugs for the key signaling events and drug targets identified using signaling map analysis. Finally, these combinations of large-scale analyses will speed up the process of identifying individual targets and target-based therapy.

REFERENCES

Babu M. M., Luscombe N. M., Aravind L., Gerstein M., and Teichmann S. A. 2004. Structure and evolution of transcriptional regulatory networks. *Curr Opin Struct Biol* 14:283–291.

Chang J. T., Carvalho C., Mori S. et al. 2009a. A genomic strategy to elucidate modules of oncogenic pathway signaling networks. *Mol Cell* 34:104–114.

Cui Q., Ma Y., Jaramillo M. et al. 2007a. A map of human cancer signaling. *Mol Syst Biol* 3:152.

Cui Q., Purisima E. O., and Wang E. 2009. Protein evolution on a human signaling network. *BMC Syst Biol* 3:21.

Cui Q., Yu Z., Purisima E. O., and Wang E. 2006. Principles of microRNA regulation of a human cellular signaling network. *Mol Syst Biol* 2:46.

Cui Q., Yu Z., Purisima E. O., and Wang E. 2007b. MicroRNA regulation and interspecific variation of gene expression. *Trends Genet* 23:372–375.

Ding L., Getz G., Wheeler D. A. et al. 2008. Somatic mutations affect key pathways in lung adenocarcinoma. *Nature* 455:1069–1075.

Ferreira M., Fujiwara H., Morita K., and Watt F. M. 2009. An activating beta1 integrin mutation increases the conversion of benign to malignant skin tumors. *Cancer Res* 69:1334–1342.

Ferrell J. E., Jr. 2002. Self-perpetuating states in signal transduction: positive feedback, double-negative feedback and bistability. *Curr Opin Cell Biol* 14:140–148.

Greenman C., Stephens P., Smith R. et al. 2007. Patterns of somatic mutation in human cancer genomes. *Nature* 446:153–158.

Gridley T. 2003. Notch signaling and inherited disease syndromes. *Hum Mol Genet* 12 Spec No 1:R9–R13.

Gymnopoulos M., Elsliger M. A., and Vogt P. K. 2007. Rare cancer-specific mutations in PIK3CA show gain of function. *Proc Natl Acad Sci USA* 104:5569–5574.

Hanahan D. and Weinberg R. A. 2000. The hallmarks of cancer. *Cell* 100:57–70.

Ikediobi O. N., Davies H., and Bignell G. et al. 2006. Mutation analysis of 24 known cancer genes in the NCI-60 cell line set. *Mol Cancer Ther* 5:2606–2612.

Ma'ayan A., Jenkins S. L., Neves S. et al. 2005. Formation of regulatory patterns during signal propagation in a mammalian cellular network. *Science* 309:1078–1083.

McMurray H. R., Sampson E. R., Compitello G. et al. 2008. Synergistic response to oncogenic mutations defines gene class critical to cancer phenotype. *Nature* 453:1112–1116.

Meuwissen R. and Berns A. 2005. Mouse models for human lung cancer. *Genes Dev* 19:643–664.

Natarajan M., Lin K. M., Hsueh R. C., Sternweis P. C., and Ranganathan R. 2006. A global analysis of cross-talk in a mammalian cellular signalling network. *Nature Cell Biol* 8:571–580.

Nowell P. C. 1976. The clonal evolution of tumor cell populations. *Science* 194:23–28.

Obata K., Morland S. J., Watson R. H. et al. 1998. Frequent PTEN/MMAC mutations in endometrioid but not serous or mucinous epithelial ovarian tumors. *Cancer Res* 58:2095–2097.

Ohm J. E., McGarvey K. M., Yu X. et al. 2007. A stem cell-like chromatin pattern may predispose tumor suppressor genes to DNA hypermethylation and heritable silencing. *Nature Genet* 39:237–242.

Parsons D. W., Jones S., Zhang X. et al. 2008. An integrated genomic analysis of human glioblastoma multiforme. *Science* 321:1807–1812.

Schlesinger Y., Straussman R., Keshet I. et al. 2007. Polycomb-mediated methylation on Lys27 of histone H3 pre-marks genes for *de novo* methylation in cancer. *Nature Genet* 39:232–236.

Siegel P. M., Massague J. 2003. Cytostatic and apoptotic actions of TGF-beta in homeostasis and cancer. *Nature Rev Cancer* 3:807–821.

Sjoblom T., Jones S., Wood L. D. et al. 2006. The consensus coding sequences of human breast and colorectal cancers. *Science* 314:268–274.

Starr T. K., Allaei R., Silverstein K. A. et al. 2009. A transposon-based genetic screen in mice identifies genes altered in colorectal cancer. *Science* 323:1747–1750.

TCGA 2008. Comprehensive genomic characterization defines human glioblastoma genes and core pathways. *Nature* 455:1061–1068.

Thomas R. K., Baker A. C., Debiasi R. M. et al. 2007. High-throughput oncogene mutation profiling in human cancer. *Nature Genet* 39:347–351.

Wang E. and Purisima E. 2005. Network motifs are enriched with transcription factors whose transcripts have short half-lives. *Trends Genet* 21:492–495.

Widschwendter M., Fiegl H., Egle D. et al. 2007. Epigenetic stem cell signature in cancer. *Nature Genet* 39:157-158.

Wood L. D., Parsons D. W., Jones S. et al. 2007. The genomic landscapes of human breast and colorectal cancers. *Science* 318:1108–1113.

CHAPTER 6

Ubiquitin-Mediated Regulation of Human Signaling Networks in Normal and Cancer Cells

Cong Fu, Jie Li, and Edwin Wang

CONTENTS

6.1 INTRODUCTION

Recently, posttranscriptional regulation has been recognized as an important aspect of gene expression in mammalian genomes. For example, microRNAs (miRNA) have been shown to participate in many cellular activities. Similarly, ubiquitination has been shown to contribute to the regulation of a broad range of cellular processes, including cell division, differentiation, and signal transduction. Aberrations in the ubiquitination system have been implicated in many kinds of diseases, including cancer.

Ubiquitination, which relies on the activity of the ubiquitin protein (Ub), is a reversible posttranslational modification of cellular proteins. Ubiquitin is a highly conserved protein, which can covalently attach to lysine residues of target proteins. Protein-attached

Ub acts as a substrate for the attachment of additional Ub residues, leading to the formation of a polyubiquitin chain. The polyubiquitin chain directs proteins to the proteasome, where the Ub is recycled and the protein is degraded (Ghaemmaghami et al. 2003). In general, protein ubiquitination is catalyzed by a cascade of enzymes, including a ubiquitin-activating enzyme E1, a ubiquitin-conjugating enzyme E2, and a ubiquitin ligase E3. E3 ubiquitin ligases are crucial for the selective recognition of target proteins and also function in subsequent protein degradation mediated by the 26S proteasome (Laney and Hochstrasser 1999). Protein ubiquitination, like protein phosphorylation, involves specific, diverse, and reversible modification of proteins. More than 600 Ubs are found in the human genome, suggesting that Ubs might potentially modify a large fraction of human proteins.

Many regulatory proteins, such as signaling proteins, are regulated or degraded in a temporally and spatially specific manner. These proteins are often tightly controlled by posttranslational modifications that are dependent on cell-signaling events. For example, ubiquitination is known to be involved in the internalization of signaling receptors and ligands. In fact, most regulated protein degradation in eukaryotes is controlled by the ubiquitin-proteasome system (Hershko and Ciechanover 1998; Hochstrasser 1996; Pickart and Eddins 2004).

Protein degradation occurs in either the lysosomal compartment or at the proteasome, which is the site of ubiquitin ligase-mediated protein degradation. Recently, Yen et al. (2008) conducted a survey of global protein stability (GPS) to identify the collective substrates of ubiquitin ligases. The GPS approach allows exploring the ubiquitin-proteasome system and identifying substrates of Ubs. Furthermore, the GPS method excludes the proteins that are functionally affected by modification of their enzymatic activity, cellular localization, or ability to physically interact with other cellular constituents. The GPS approach first determines comprehensive protein turnover rates then seeks to identify the substrates of a particular Ub. More than 8000 distinct human proteins have been scanned using the GPS approach. Based on turnover rates, each protein was assigned a protein stability index value that was further categorized into four groups, which have short (S), medium (M), long (L), and extra long half-lives (XL). In total, 6528 proteins have been assigned to these groups. In general, those proteins with short and medium half-lives are most likely to be degraded by Ubs.

Many extracellular stimuli evoke cellular responses by engaging intracellular signaling networks, which ultimately activate nuclear transcription factors and lead to a phenotypic response to the extracellular stimulation. Cellular decision making frequently requires dynamic regulation of signaling activity before the cell reaches a certain fate. Depending on the cellular context and stimulus, some signaling cascades deactivate in as little as a few minutes through posttranslational regulation (Legewie et al. 2008). In these situations, signaling proteins must be regulated in a temporally and spatially specific manner. It is known that most signaling protein degradation is carried out by the ubiquitin system. Therefore, it is desirable to understand the effects of ubiquitination on signaling networks in normal and cancer cells, and, in turn, to identify the potential implications of these insights to human diseases. In this chapter, we will summarize the results of ubiquitination on the human signaling network in normal cells (Fu, Li, and Wang 2009) and the insights of ubiquitination on a human cancer signaling network.

6.2 EXTRA LONG HALF-LIFE PROTEINS FORM A NETWORK BACKBONE

To systematically analyze the principles of ubiquitination on the human signaling network, we examined a previously literature-mined human signaling network that represents signal transduction processes from multiple cell surface receptors to various cellular machines and signaling outputs in humans (Cui et al. 2007a). The network contains more than 1600 nodes and 5000 links, including 2287 activating (positive) links, 651 inhibitory (negative) links, and 1914 neutral (protein physical interaction only) links. Nodes with positive links indicate that the node (protein) is required to sense and to transmit the signal (i.e., kinases), while a node with a negative link indicates that the node (protein) attenuates information transfer (i.e., by catalyzing kinase dephosphorylation). Neutral links represent physical interactions between proteins. We mapped the Ub-mediated half-lives of the proteins, which have been determined from the GPS survey, onto the human signaling network. In total, 570 proteins were mapped, of which 53 (9.3%), 122 (21.4%), 198 (34.7%), and 197 (34.6%) had short (S), medium (M), long (L), and extra long (XL) half-lives, respectively. We also mapped the Ub-mediated half-lives of the proteins onto the cancer signaling network obtained from Cui et al. (2007a). There are 126 cancer proteins, in which 11 (8.7%), 21 (16.7%), 46 (36.5%), and 48 (38.1%) proteins have S, M, L, and XL half-lives, respectively.

Hub regulators in gene transcriptional networks are characterized by unstable mRNAs; however, hubs in protein integration networks are characterized by stable mRNAs (Balaji, Babu, and Aravind 2007; Janga and Babu 2009; Wang and Purisima 2005). Therefore, we examined the enrichment of groups of proteins with different half-lives (S, M, L, and XL) in the groups of nodes categorized based on node degree. As shown in Table 6.1A,

TABLE 6.1A   Enrichment of Different Half-Life Protein Groups in the Groups of Nodes, Categorized Based on Degree (the Entire Human Signaling Network)

Subgroup (Node Number)	Ratio of Node Number of Each Half-Life Group to the Total Node Number of Each Subgroup Categorized by Node Degree (P-value)			
	S	M	L	XL
Degree <= 1 (136)	11.00% (0.967)	21.30% (0.981)	33.80% (0.981)	33.80% (0.981)
Degree <= 2 (241)	11.60% (0.342)	21.60% (0.981)	33.60% (0.981)	33.20% (0.981)
Degree <= 3 (322)	7.50% (0.151)	20.80% (0.981)	35.70% (0.967)	36.00% (0.967)
Degree <= 4 (251)	*6.40% (0.077)*	22.70% (0.967)	32.70% (0.999)	38.20% (0.342)
Degree <= 5 (200)	*4.50% (0.0056)*	22.00% (0.981)	33.00% (0.981)	40.50% (0.205)
Degree <= 6 (169)	*2.40% (0.0028)*	20.10% (0.981)	34.30% (0.981)	**43.20% (0.042)**
Degree <= 8 (119)	*0.80% (0.0028)*	19.30% (0.981)	32.80% (0.981)	**47.10% (0.028)**

Values in italics represent negative enrichment, while those in bold represent positive enrichment. S, M, L, and XL represent short, medium, long, and extra long half-life proteins, respectively.

Source: Adapted from Fu, Li, and Wang (2009).

proteins with more than three links are significantly less enriched for short half-life proteins, whereas proteins with more than five links are significantly enriched for extra long half-life proteins. Further, we observed a statistically significant positive correlation between node (protein) degree and protein stability index value (correlation coefficient = 0.14; Spearman correlation, $P = 0.015$). Proteins with fewer links that take part in fewer signaling events are more likely to be a Ub substrate. In contrast, proteins with more links or those involved in more signaling events are more likely to avoid being Ub substrates. The positive correlation between Ub-mediated protein half-life and node degree in the human signaling network agrees with the findings in an *Escherichia coli* protein inter-action network study (Janga and Babu 2009), but contradicts the observation that hubs in the *E. coli* gene regulatory network have short half-lives (Wang and Purisima 2005). This difference might reflect the different roles of the hubs in different types of cellular networks. Hubs in protein interaction networks and signaling networks play a major role in "integrating different signals and pathways," while hubs in gene regulatory networks are often transcription factors, which play a major role in "responding to the stimuli and coordinating the regulated genes."

To understand the distribution of different half-life protein groups in the network, we reconstructed a subnetwork using only the proteins whose half-lives were determined in the GPS survey (570 proteins). Interestingly, most of these proteins (342/570, 60%) are con-nected and form a subnetwork. We individually removed each group (S, M, L, and XL) of the proteins from the subnetwork. When Groups S, M, and L were removed, the subnet-work remained largely connected, suggesting that the XL proteins form the backbone of the network. This suggests that Ubs are not likely to degrade the backbone nodes of the sig-naling network. We further analyzed the XL proteins using a Gene Ontology tool, DAVID. The DAVID results indicated that the backbone nodes are mainly involved in intracellular signaling. When Groups XL, S, and M were removed, the subnetwork collapsed. Moreover, when Groups XL and L were removed, the subnetwork also collapsed. These results indi-cate that Ubs are targeted to degrade the periphery of the network. We also examined the GO terms and pathways enriched in the S and M protein groups. We found that the S proteins are enriched for receptors, especially G-protein-coupled receptors and plasma membrane proteins. The periphery of the network is known to sense signals for the cell and also undergoes rapid evolution (Cui, Purisima, and Wang 2009). Therefore, ubiquitination of the human signaling network might play a role in sensing cellular signals and enabling the system to adapt to environmental changes.

A similar result was obtained for the cancer signaling network (Table 6.1B). These analyses showed that the signaling network is composed of two major parts: the network backbone, a set of signaling proteins that are largely connected and generally stable, and a flexible portion, composed of a set of signaling proteins that are inducible and have high turnover rates. These results suggest that proteins in the network can either be flexible and rapidly responding (unstable proteins) or stable and relatively static (stable proteins). The stable portion of the network forms a subnetwork that is shared by many signaling path-ways (i.e., stable nodes have high degrees) to perform common signaling activities in the

TABLE 6.1B   Enrichment and Less Enrichment of Different Half-Life Protein Groups in the Groups of Nodes, Which Are Categorized Based on Node Degree (the Cancer Signaling Network)

Subgroup (Node Number)	Ratio of Node Number of Each Half-Life Group to the Total Node Number of Each Subgroup Categorized by Node Degree (P-value)			
	S	**M**	**L**	**XL**
Degree <= 1 (26)	19.20% (0.077)	7.70% (0.985)	42.30% (0.2691)	30.80% (0.736)
Degree <= 2 (39)	17.90% (0.057)	12.80% (0.948)	33.30% (0.635)	35.90% (0.486)
Degree >= 3 (87)	*4.60% (0.021)*	18.40% (0.806)	37.90% (0.279)	39.10% (0.203)
Degree>= 4 (69)	*2.90% (0.016)*	21.70% (0.527)	34.80% (0.544)	40.60% (0.164)
Degree >= 5 (55)	*3.60% (0.023)*	23.60% (0.391)	30.90% (0.784)	41.80% (0.149)
Degree >= 6 (48)	*4.20% (0.040)*	20.80% (0.605)	29.20% (0.841)	**45.80% (0.059)**
Degree >= 8 (36)	*2.80% (0.020)*	19.40% (0.686)	30.60% (0.760)	**47.20% (0.071)**

Ratio of node number in each half-life protein group to the total number of network nodes mapped with half-lives (126 proteins). Values in italics represent negative enrichment, while those in bold represent positive enrichment. S, M, L, and XL represent short, medium, long, and extra long half-life proteins, respectively.

organism. Nodes of the flexible part (short half-life proteins) are scattered throughout the network and have low degrees. Ub-mediated degradation of proteins in the flexible part of the network might provide a regulatory mechanism in a cell-type-specific or physiological manner.

6.3 SHORT AND MEDIUM HALF-LIFE PROTEINS DO NOT TEND TO CONNECT TO EACH OTHER IN BOTH NETWORKS OF NORMAL AND CANCER CELLS

To understand the interacting relationships among the proteins with different half-lives, we tested the interaction preferences of the different protein groups (S, M, L, and XL). We first extracted the neighbors of the XL proteins and counted the number in each group (S, M, L, and XL). To examine the statistical significance of the enrichment of certain groups of proteins in the XL proteins' neighbors, we randomly reassigned the half-lives of these proteins. We found that XL proteins are preferentially connected to each other (P = 0.0001). Similar analyses were extended to check the neighbors of the L, M, and S proteins, respectively. Interestingly, both L and M proteins are preferentially connected with the XL proteins (P = 0.035 and 0.042, respectively). However, both S and M proteins avoid connecting to the S proteins (P = 0.021 and 0.012, respectively). These results suggest that the XL proteins tend to interconnect to form basic signaling cascades or the network backbone. The L and M proteins tend to attach to the XL-formed network backbone to extend the signaling cascades and the network. Finally, Ub-mediated S and M proteins generally are not neighbors of each other. For example, it is difficult to find edges connecting S-S, S-M, and M-M proteins in a signaling cascade. Similar results have been obtained for the cancer signaling network: XL proteins are preferentially connected to each other (P = 0.0004) and S proteins avoid connecting to the S proteins (P = 0.0002). These results suggest that Ubs do not regulate consecutive proteins in a signaling cascade, but that they instead degrade key individual proteins in signaling cascades.

It should be noted that the above finding differs from the observation that the mRNA half-lives of interacting proteins tend to be similar in the *E. coli* protein interaction network (Janga and Babu 2009). In bacteria, genes belonging to one pathway can be arranged on one operon and tightly coregulated. These tightly coregulated genes often perform a single task (i.e., genes in a biosynthetic pathway used for a metabolic process). Therefore, it makes sense that the tightly coregulated genes should be degraded at the same time. However, in mammalian signaling networks, proteins in a chain are not tightly coregulated. This finding also appears to hold true for posttranscriptional regulation in human signaling networks. For example, miRNA regulated signaling proteins are not neighbors in a signaling cascade (Cui et al. 2007b). Such a regulatory mechanism might provide for better adaptation to the complexity of the signaling networks in mammalian genomes.

6.4 SHORT HALF-LIFE PROTEINS ARE ENRICHED IN THE UPSTREAM PORTION OF BOTH SIGNALING NETWORKS OF NORMAL AND CANCER CELLS

Cellular signaling information flow propagates from the extracellular space to the nucleus. Therefore, network components can be divided into ligands, cell surface receptors, intracellular signaling proteins, and nuclear proteins, based on their positions along the flow of signaling information. To discover which stages of the signal information flow are predominantly regulated by Ubs, we examined the enrichment of the different protein groups (S, M, L, and XL) along the signaling information flow and determined which groups (S, M, L, and XL) of the proteins are enriched in each signaling category (i.e., ligands, cell surface receptors, intracellular signaling proteins, and nuclear proteins). We found that ligands and receptors are significantly enriched for S and M proteins, but less enriched for XL proteins (Table 6.2A). Furthermore, intracellular signaling proteins are enriched for XL proteins (Table 6.2A). Similar results have been seen in the cancer signaling network (Table 6.2B). These results suggest that Ubs tend to regulate ligands and receptors. In cells, ligands and receptors are used for initial signal processing and specificity (i.e., ligands act in a specific manner). These results indicate that Ubs might act as rapid posttranslational mediation and could be most important for initial signal processing and specificity.

Ub-mediated regulation of the Notch signaling pathway is one example that illustrates this conclusion. The Notch pathway is an evolutionarily conserved signaling system that is absolutely required for normal embryonic development (Gridley 2003). Ligand-induced Notch signaling regulates a variety of cell types during specification, patterning, and morphogenesis through effects on differentiation, proliferation, survival, and apoptosis (Fiuza and Arias 2007). The Notch ligands Delta, Serrate, Jagged1, Jagged2, and Lag2 (DSL) are the major Notch signaling activators. It is known that two structurally distinct E3 ligases,

TABLE 6.2A   Enrichment of Different Half-Life Protein Groups along the Flow of Signaling Information (the Entire Human Signaling Network)

Signaling Stage	Ratio of Node Number in Each Half-Life Protein Group to the Total Node Number in Each Signaling Stage (P-value)			
	Short Half-Life Proteins	Medium Half-Life Proteins	Long Half-Life Proteins	Extra Long Half-Life Proteins
Ligands	*17.8% (0.05)*	*33.3% (0.04)*	40.0% (0.253)	**8.9% (0.001)**
Cell surface receptors	**23.6% (0.002)**	*30.3% (0.02)*	**25.8% (0.016)**	**20.2% (0.001)**
Intracellular signaling proteins	**6.3% (0.001)**	**17.3% (0.001)**	32.9% (0.083)	*43.6% (0.0002)*
Nuclear proteins	**3.1% (0.001)**	23.3% (0.317)	39.5% (0.112)	34.1% (0.596)
Ave*	9.3%	21.4%	34.7%	34.6%

* Ratio of node number in each half-life protein group to the total number of network nodes mapped with protein half-lives (570 proteins). Values in italics represent positive enrichment, while those in bold represent negative enrichment. S, M, L, and XL represent short, medium, long, and extra long half-life proteins, respectively.

Source: Adapted from Fu, Li, and Wang (2009).

Neuralized (Neur) and Mind bomb (Mib), influence Notch signaling by interacting with and ubiquitinating the DSL ligands. Neur1 ubiquitinated Jagged1 leads to degradation and attenuation of Jagged1-induced Notch signaling (Koutelou et al. 2008). However, Mib2 ubiquitinated Jagged2 is associated with activation of Notch signaling (Takeuchi, Adachi, and Ohtsuki 2005). These results suggest that different ubiquitination states of DSL ligands have different functional roles for Notch signaling, illustrating the signaling specificity mediated by Ubs. Furthermore, ubiquitination of different G-protein-coupled receptors has also been shown to be highly cell-type dependent (Dromey 2008).

TABLE 6.2B   Enrichment of Different Half-Life Protein Groups along the Signaling Information Flow (the Cancer Signaling Network)

Signaling Stage	Ratio of Node Number in Each Half-Life Protein Group to the Total Node Number in Each Signaling Stage (P-value)			
	Short Half-Life Protein	Medium Half-Life Protein	Long Half-Life Protein	Extra Long Half-Life Protein
Ligands	33.3% (0.115)	**50.0% (0.072)**	*0% (0.002)*	*16.7% (0.062)*
Cell surface receptors	14.3% (0.304)	23.8% (0.291)	38.1% (0.510)	*23.8% (0.053)*
Intracellular signaling proteins	7.0% (0.882)	14.0% (0.896)	*32.6% (0.058)*	**46.5% (0.005)**
Nuclear proteins	*3.4% (0.055)*	13.8% (0.759)	44.8% (0.201)	37.9% (0.616)
Ave*	8.7%	16.7%	36.5%	38.1%

* Ratio of node number in each half-life protein group to the total number of network nodes mapped with half-lives (126 proteins). Values in italics represent negative enrichment, while those in bold represent positive enrichment. S, M, L, and XL represent short, medium, long, and extra long half-life proteins, respectively.

Ubiquitination of the EGFR signaling pathway has also been documented in the literature. Upon ligand activation of many receptor tyrosine kinases, such as EGFR, there is a rapid decrease in the number of cell surface receptors and an eventual decrease in the cellular content of activated receptors, a process known as downregulation. This process can be divided into two distinct stages: internalization of the membrane receptor and degradation of the internalized receptor. Upon activation of the EGFR by a ligand, rapid ubiquitination of the EGFR occurs and leads to receptor internalization and degradation (Thien and Langdon 2001). For many growth hormone receptors, an active ubiquitination system is required for both uptake and degradation (Strous et al. 2004).

6.5 UBIQUITINATION OCCURS MORE FREQUENTLY IN POSITIVE NETWORK MOTIFS

A signaling network can be broken down into distinct regulatory patterns, or network motifs, typically comprised of three to four interacting components capable of signal processing. The function of a motif depends on whether the links are positive or negative. For example, positive feedback network motifs, also known as positive regulatory loops, lead to emergent network properties such as ultrasensitivity, disability, and switch-like behavior, whereas negative feedback network motifs, known as negative regulatory loops, permit adaptation, desensitization, and preservation of homeostasis (Babu et al. 2004; Barabasi and Oltvai 2004; Cui et al. 2007a, 2007b; Ferrell 2002; Ma'ayan et al. 2005). We asked if Ubs tend to regulate positively or negatively linked loops. Toward this end, we first extracted all 4-node network motifs from the subnetwork reconstructed using the proteins with half-life data from the GPS survey (570 proteins). Since XL proteins form the network backbone, we used the number of XL proteins (nodes) to classify the 4-node motifs into subgroups. Motifs containing 0, 1, 2, 3, and 4 XL proteins were defined as motif-subgroups 0, 1, 2, 3, and 4, respectively. Finally, we calculated the ratio (R) of positive links to the total positive and negative links in each motif-subgroup and compared it with the average R for all motifs, which is shown as a horizontal line in Figure 6.1. For the 4-node motifs, the values of R for motif-subgroups 0 and 1 are higher than the average R of all 4-node motifs ($P < 10^{-10}$, Wilcoxon rank-sum test; Figure 6.1a). The values of R for motif-subgroups 3 and 4 are lower than the average R of all the 4-node motifs ($P < 10^{-7}$, Wilcoxon rank-sum test; Figure 6.1a). In general, these motifs show a clear negative correlation between the positive link ratio and the number of XL proteins in the motif (Figure 6.1a). We extended this analysis to the 3-node motifs and similar results were obtained (Figure 6.1b). These results suggest that Ubs target negative regulatory motifs less frequently than positive regulatory motifs.

Ubiquitination in positive regulatory motifs allows faster inhibition of signaling pathways or cascades. Such quick attenuation might be advantageous for the cell to dynamically respond to external stimuli. In principle, such a process can be rapidly reversed (i.e., proteins can be de-ubiquitinated). Therefore, such a posttranslational mechanism provides robust and efficient regulation to speed up adaptation of signaling networks upon activation.

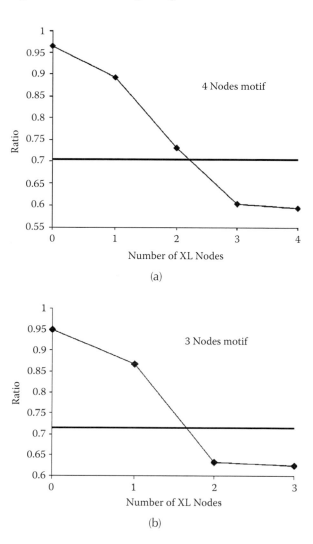

FIGURE 6.1   The relationship between the fraction of positive links and the fraction of extra long (XL) proteins in network motifs. All network motifs were classified into motif-subgroups based on the number of XL nodes. The ratio of positive links to total links in each motif-subgroup was plotted. The horizontal lines indicate the ratio of positive links to the total links in all network motifs: (a) 4-node motifs; (b) 3-node motifs. (Adapted from Fu, Li, and Wang 2009.)

6.6 HIGH EXPRESSIONS OF 26S PROTEASOME GENES PREDICT POOR OUTCOMES FOR BREAST CANCER PATIENTS

To understand which biological processes are primarily involved in Ub-mediated positive regulatory motifs, we extracted the S and M proteins (71 total proteins, 15 S and 56 M proteins) from motif-subgroups 0 and 1 and performed GO analysis. We found that the biological processes of inducing and positively regulating apoptosis were significantly enriched in the Ub-mediated positive regulatory loops. Apoptosis is one component of cancer signaling (Cui et al. 2007a; Wang, Lenferink, and O'Connor-McCourt 2007). Furthermore, analysis of cancer mutations in the human signaling network showed that, regardless of cancer types, almost all tumor samples bear at least one mutated gene in a network module dominated by apoptotic signaling (Cui et al. 2007a). These data suggest that disruption of apoptotic signaling is essential for cancer signaling, tumor progression, and metastasis. Twenty out of the 71 proteins extracted from motif-subgroups 0 and 1 are apoptotic proteins. To explore the relationship between cancer and these apoptotic proteins, we compared the 20 apoptotic proteins to our previously compiled list of cancer-associated proteins (Cui et al. 2007a), most of which have cancer promoting mutations. We found that 17 of the 20 apoptotic proteins are among the cancer-associated proteins. These results suggest that most of the Ub-mediated apoptotic proteins in the positive regulatory motifs are cancer-associated.

Functional links between cancer, apoptosis, and ubiquitination have been reported for both Smad3 and Siva-1, both of which are included in the 20 apoptotic proteins. Smad3 is an important component of the transforming growth factor-β (TGF-β) pathway, which regulates essential cellular functions ranging from cellular proliferation and differentiation to apoptosis. TGF-β-induced apoptosis is mediated through a Smad3-dependent mechanism (Ramesh et al. 2008). Moreover, 26S proteasome-dependent degradation of Smad3 has been implicated in cancer development (Inoue and Imamura 2008). Siva-1 induces apoptosis of T lymphocytes through a caspase-dependent mitochondrial pathway (Jacobs et al. 2007) or by negatively regulating NF-κB activity in T cell receptor-mediated activation-induced cell death (Gudi et al. 2006). Siva-1 can be ubiquitinated and degraded, resulting in downregulation of its proapoptotic activity (Lin et al. 2007).

If Ubs preferentially regulate positive regulatory motifs, which are predominately involved in positive regulation of apoptosis, which is itself implicated in cancer initiation and progression, we expect that increased expression of the genes of the common ubiquitination machinery (Ub machinery), such as genes encoding the 26S proteasome proteins, would block cell apoptosis and might be correlated with cancer progression and metastasis. To test this hypothesis, we compiled a set of known genes of the common Ub machinery, the 26S proteasome proteins (43 proteins). We tested whether the expression levels of the 26S proteasome genes are significantly higher in tumors than in normal tissues using Gene Set Enrichment Analysis (GSEA, http://www.broad.mit.edu/gsea/). GSEA is a computational tool that determines whether *a priori* defined set of genes shows statistically significant concordant differences between two biological phenotypes (i.e., tumor and normal samples). Twenty-five and twenty-four out of the 26S proteasome genes have been mapped onto the microarray platforms used for lung and bladder tumors, respectively. As shown in Figures 6.2a and 6.2b, 92% of the 26S proteasome genes are more highly expressed in tumor samples than in normal samples (P = 0.02 for lung tumors; P = 0.03 for bladder tumors; Figures 6.2a and 6.2b). These results suggest that elevated expression of the 26S proteasome is highly correlated with cancer development. Although some of the 26S proteasome genes have been implicated in cancer progression previously, this is the first report that the 26S proteasome genes, as a whole set, have higher expression levels in tumors than in normal tissues.

Presently, many cancer patients are over-treated. For example, 70% to 80% of lymph node-negative breast cancer patients may undergo adjuvant chemotherapy when it is, in

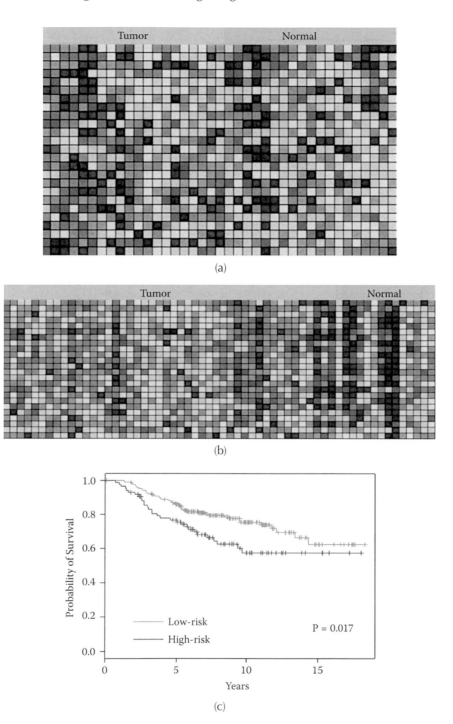

FIGURE 6.2   (See color insert following page 332.) The association of the 26S proteasome genes with tumor progression and metastasis. Heat map generated by GSEA based on the gene expression values of the 26S proteasome genes: (a) lung tumor, GSE2514, (b) bladder tumor, GSE3167. Kaplan–Meier survival analysis of breast cancer patient groups stratified by the 26S proteasomes genes' expression in tumors: (c) breast tumor, 295-set, (d) breast tumor, GSE349. (Adapted from Fu, Li, and Wang 2009.)

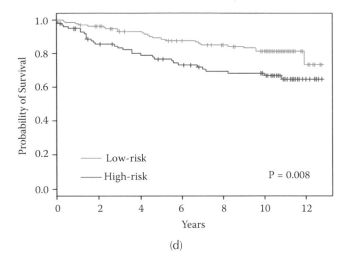

FIGURE 6.2 (Continued)

fact, unnecessary (van de Vijver et al. 2002). Unfortunately, there is currently no satis-factory approach to determine which cancer patients would benefit from extra therapy (such as chemotherapy) after surgery. Therefore, identification of genes and proteins that can be used for prognostic purposes would be greatly beneficial. Furthermore, prognostic biomarkers can be used as tools for predicting survival after a diagnosis of cancer and to guide oncologists in treatment of the cancer to obtain the best outcome. Cancer patients who have "bad" tumors have a higher chance of tumor recurrence and a short survival time, while the patients who have "good" tumors will not experience tumor recurrence after removal of the tumor.

We investigated whether the whole set of the 26S proteasome genes can be used as a prognostic signature. Using GSEA, we found that the expression levels of the 26S protea-some genes are significantly higher in "bad" breast tumors compared to "good" tumors (P = 0.02, P = 0.02, for both breast cancer datasets, respectively). Furthermore, we performed survival analysis, which was previously implemented (Cui et al. 2007a), using the gene expression profiles of two breast tumor cohorts in which patient clinical information (such as survival data) is available. Survival analysis showed that the 26S proteasome gene set can distinguish good and bad tumors (P = 0.017, P = 0.008, for the two breast cancer sets, respectively; Figures 6.2c and 6.2d). These results support our hypothesis that common ubiquitination machinery, specifically the 26S proteasome gene set, is significantly cor-related with cancer progression and metastasis. In agreement with our discovery, it has been reported that RNAi silencing of UbcH10, an E2 enzyme and a Ub machinery gene, enhances cell death in neoplastic human cells (Wagner et al. 2004). Our findings imply that the Ub machinery, and the 26S proteasome genes specifically, can be used as prog-nostic biomarkers to distinguish good and bad tumors. This is also the first evidence that the 26S proteasome genes, as a whole set, predict poor outcomes of breast cancer patients. Taken together, these results suggest that the expression of these 26S proteasome genes is highly correlated during cancer development and metastasis.

6.7 CONCLUDING REMARKS

We performed a comprehensive analysis of ubiquitin-mediated protein stability on a literature-mined human signaling network. We showed that the signaling network could be divided into two parts: the first consisting mainly of extra long half-life proteins connected to form a network backbone, and the second consisting of short and medium half-life proteins that preferentially attach to the network backbone and are scattered throughout the network. Nodes representing the extra long half-life proteins have significantly higher degrees of connection, suggesting that the network backbone might participate in many signaling pathways and perform basic signaling activities or integrate signals for different pathways. We also showed that short and medium half-life proteins avoid connecting to one another in the network, suggesting a rule of mutually exclusive ubiquitination of the proteins that are neighbors in a signaling cascade. This organization of the ubiquitination network may enable more efficient control of signaling.

We found that short half-life proteins are enriched in the upstream portion of the signaling network. The upstream portion of the network consists mainly of signaling receptors and ligands. Many studies have reported the functions of ubiquitination of signaling receptors and ligands in such pathways as Notch signaling and growth hormone factor signaling. Furthermore, ubiquitination of receptors and ligands helps initiate signal processing and specificity in the context of cell-specific or cellular conditions. We also showed that ubiquitination preferentially occurs in positive regulatory loops, which might enable faster breakdown of the amplification of signals in the positive regulatory loops and inhibit certain signaling pathways or cascades. Such a quick attenuation might be advantageous for the cell in dynamically altering responses to external stimuli and providing robust and efficient regulation for increased adaptation of signaling networks upon activation.

Interestingly, the ubiquitinated positive regulatory loops predominately take part in inducing or positively regulating cell apoptosis. Most of the apoptotic proteins in these loops have been implicated in cancer development. Thus, we expected that the common machinery of ubiquitination might be associated with tumor progression and metastasis. Indeed, we showed that high levels of expression of the ubiquitination machinery, specifically the 26S proteasome, are significantly associated with tumor progression and metastasis in several cancer types. We showed that high levels of expression of the 26S proteasome predicted poor outcomes of breast cancer patients. These findings have implications for the development of cancer treatments and prognostic biomarkers using the ubiquitination machinery.

REFERENCES

Babu M. M., Luscombe N. M., Aravind L., Gerstein M., and Teichmann S. A. 2004. Structure and evolution of transcriptional regulatory networks. *Curr Opin Struct Biol* 14:283–291.

Balaji S., Babu M. M., and Aravind L. 2007. Interplay between network structures, regulatory modes and sensing mechanisms of transcription factors in the transcriptional regulatory network of *E. coli. J. Mol. Biol.* 372:1108–1122.

Barabasi A. L. and Oltvai Z. N. 2004. Network biology: understanding the cell's functional organization. *Nature Rev Genet* 5:101–113.

Cui Q., Ma Y., Jaramillo M. et al. 2007a. A map of human cancer signaling. *Mol Syst Biol* 3:152.

Cui Q., Purisima E. O., and Wang E. 2009. Protein evolution on a human signaling network. *BMC Syst Biol* 3:21.

Cui Q., Yu Z., Purisima E. O., and Wang E. 2007b. MicroRNA regulation and interspecific variation of gene expression. *Trends Genet* 23:372–375.

Dromey, J. R., and Pfleger, K. D. 2008. G protein coupled receptors as drug targets: the role of beta-arrestins. *Endocr Metab Immune Disord Drug Targets* 166:1817–1826.

Ferrell J. E., Jr. 2002. Self-perpetuating states in signal transduction: positive feedback, double-negative feedback and bistability. *Curr Opin Cell Biol* 14:140–148.

Fiuza U. M. and Arias A. M. 2007. Cell and molecular biology of Notch. *J. Endocrinol.* 194:459–474.

Fu C., Li J., and Wang E. 2009. Signaling network analysis of ubiquitin-mediated proteins suggests correlations between the 26S proteasome and tumor progression. *Mol Biosyst* DOI: 10.1039/B905382D.

Ghaemmaghami S., Huh W. K., Bower K. et al. 2003. Global analysis of protein expression in yeast. *Nature* 425:737–741.

Gridley T. 2003. Notch signaling and inherited disease syndromes. *Hum Mol Genet* 12 Spec No 1:R9-13.

Gudi R., Barkinge J., Hawkins S. et al. 2006. Siva-1 negatively regulates NF-kappaB activity: effect on T-cell receptor-mediated activation-induced cell death (AICD). *Oncogene* 25:3458–3462.

Hershko A. and Ciechanover A. 1998. The ubiquitin system. *Annu Rev Biochem* 67:425–479.

Hochstrasser M. 1996. Ubiquitin-dependent protein degradation. *Annu Rev Genet* 30:405–439.

Inoue Y. and Imamura T. 2008. Regulation of TGF-beta family signaling by E3 ubiquitin ligases. *Cancer Sci* 99:2107–2112.

Jacobs S. B. R., Basak S., Murray J. I., Pathak N., and Attardi L. D. 2007. Siva is an apoptosis-selective p53 target gene important for neuronal cell death. *Cell Death Differ* 14:1374–1385.

Janga S. C., and Babu M. M. 2009. Transcript stability in the protein interaction network of Escherichia coli. *Mol Biosyst* 5:154–162.

Koutelou E., Sato S., Tomomori-Sato C. et al. 2008. Neuralized-like 1 (Neurl1) targeted to the plasma membrane by N-myristoylation regulates the Notch ligand Jagged1. *J. Biol. Chem.* 283:3846–3853.

Laney J. D. and Hochstrasser M. 1999. Substrate targeting in the ubiquitin system. *Cell* 97:427–430.

Legewie S., Herzel H., Westerhoff H. V., and Bluthgen N. 2008. Recurrent design patterns in the feedback regulation of the mammalian signalling network. *Mol Syst Biol* 4:190.

Lin F. T., Lai Y. J., Makarova N., Tigyi G., and Lin W. C. 2007. The lysophosphatidic acid 2 receptor mediates down-regulation of Siva-1 to promote cell survival. *J. Biol. Chem.* 282:37759–37769.

Ma'ayan A., Jenkins S. L., Neves S. et al. 2005. Formation of regulatory patterns during signal propagation in a mammalian cellular network. *Science* 309:1078–1083.

Pickart C. M. and Eddins M. J. 2004. Ubiquitin: structures, functions, mechanisms. *Biochim Biophys Acta* 1695:55–72.

Ramesh S., Qi X. J., Wildey G. M. et al. 2008. TGF beta-mediated BIM expression and apoptosis are regulated through SMAD3-dependent expression of the MAPK phosphatase MKP2. *EMBO Rep* 9:990–997.

Strous G. J., dos Santos C. A., Gent J. et al. 2004. Ubiquitin system-dependent regulation of growth hormone receptor signal transduction. *Curr Top Microbiol Immunol* 286:81–118.

Takeuchi T., Adachi Y. and Ohtsuki Y. 2005. Skeletrophin, a novel ubiquitin ligase to the intracellular region of Jagged-2, is aberrantly expressed in multiple myeloma. *Am. J. Pathol.* 166:1817–1826.

Thien C. B., and Langdon W. Y. 2001. Cbl: many adaptations to regulate protein tyrosine kinases. *Nature Rev Mol Cell Biol* 2:294–307.

van de Vijver M. J., He Y. D., and van 't Veer L. J. et al. 2002. A gene-expression signature as a predictor of survival in breast cancer. *N Engl J. Med.* 347:1999–2009.

Wagner K. W., Sapinoso L. M., El-Rifai W. et al. 2004. Overexpression, genomic amplification and therapeutic potential of inhibiting the UbcH10 ubiquitin conjugase in human carcinomas of diverse anatomic origin. *Oncogene* 23:6621–6629.

Wang E., Lenferink A., and O'Connor-McCourt M. 2007. Cancer systems biology: exploring cancer-associated genes on cellular networks. *Cell Mol Life Sci* 64:1752–1762.

Wang E. and Purisima E. 2005. Network motifs are enriched with transcription factors whose transcripts have short half-lives. *Trends Genet* 21:492–495.

Yen H. C., Xu Q., Chou D. M., Zhao Z., and Elledge S. J. 2008. Global protein stability profiling in mammalian cells. *Science* 322:918–923.

microRNA Regulation of Networks of Normal and Cancer Cells

Pradeep Kumar Shreenivasaiah,
Do Han Kim, and Edwin Wang

CONTENTS

7.1 INTRODUCTION

microRNAs (miRNAs) are endogenous ~22-nucleotide RNAs that suppress gene expression by selectively binding, through base-pairing, to the 3′-noncoding region of specific messenger RNAs. Growing evidence supports the idea that miRNA has a crucial role in key biological processes, such as cell growth, differentiation and proliferation, embryonic development, and apoptosis. Several hundred miRNAs have been identified through experimental and computational approaches. These miRNAs are estimated to regulate more than half of the human genes that are related to a wide range of the biological processes mentioned above. Therefore, it would not be surprising if miRNAs could regulate cellular networks and subsequently exert their functions. Cells use cellular networks such as signaling networks to make decisions whether to grow, differentiate, move, or die. By regulating such cellular networks, miRNAs could affect many kinds of cellular activity. miRNAs have also been found to contribute to a variety of disease states. Global expression profiling has demonstrated a wide spectrum of aberrations in miRNA expression profiles that are correlated with different stages of tumor progression. Several studies have deciphered molecular mechanisms of tumor development mediated by miRNAs. miRNAs could potentially modulate several components of cellular networks and thus attain broad influence on cellular activity, including cancer development and progression. A systems-level analysis of the interactions between miRNAs and cellular networks would enable us to understand the principles of miRNA regulation of cellular networks.

7.2 miRNA REGULATION OF NETWORKS OF NORMAL CELLS

It is intriguing and crucial to understand how miRNAs regulate different types of cellular networks, and what biological functions miRNAs could exert in biological systems. In order to understand miRNA functions in biological processes, we must learn how miRNAs function either in isolation or in concert with other factors to regulate biological systems. In this section, we will summarize the recent progress in computational studies of miRNA regulation of a few representative cellular networks, and the principles and implications of miRNA regulation of cellular networks.

7.2.1 miRNA Regulation of Cell Signaling Networks

Physiological decisions made by a living cell are a consequence of dynamic interactions of components in complex signaling networks. Protein components that interact in molecular pathways could be modulated by miRNAs. Such modulation may be an important part of the underlying regulatory mechanisms of physiological processes. Alternatively, it could be the manifestation of a disease that results from alterations in cell signaling (Irish, Kotecha, and Nolan 2006). Thus, it is imperative to understand the role of miRNAs in the regulation of cell signaling. In this section, we review studies that have addressed miRNA regulation of the human signaling network.

7.2.1.1 miRNAs Preferentially Regulate Positive Signaling Regulatory
Loops but Avoid Regulating Negative Regulatory Loops

Cui and co-workers pioneered the systems-level analysis of miRNA networks. They analyzed the interactions between miRNAs and human cellular signaling networks and uncovered several key miRNA regulatory principles (Cui et al. 2006). Analysis also revealed miRNA roles in the regulation of the strength and specificity of cellular signaling. Integrative analysis of the human signaling network and miRNAs showed that miRNAs preferentially regulate positive regulatory motifs and avoid affecting negative regulatory motifs. Positive and negative regulatory loops are the major building blocks of cellular networks; however, these loops affect network behavior in distinct ways. Positive feedback loops promote the amplification of transient signal/noise in a system, causing a shift of cellular states. Therefore, they are associated with converting a transient signal into a long-lasting cellular response, as with the case of developmental switches. In contrast, negative feedback loops act as a filter, buffering noise or fluctuations, and thus stabilizing the network.

The yeast galactose (GAL) network, which contains positive and negative loops, is a good example of the importance of negative feedback loops (Acar, Becskei, and van Oudenaarden 2005). When negative feedback loops were experimentally removed from the network, GAL genes in the network randomly switched ON and OFF (randomly switching between two phenotypes) over time, regardless of network induction (Acar, Becskei, and van Oudenaarden 2005). In this network, negative feedback is executed by transcription factors (TFs). Compared with transcriptional repressors, miRNAs can turn off mRNA translation faster than TFs. These data hint at the possibility that miRNA may provide fast feedback responses and filter noise effectively by regulating positive feedback motifs (Brandman et al. 2005; Ferrell 2002).

7.2.1.2 miRNAs Preferentially Regulate Downstream Components
but Avoid Regulating Common Components of the Network

Cui and co-workers also showed that miRNAs are frequently involved in regulation of downstream network components such as transcription factors; however, they less frequently regulate upstream network components, such as ligands and receptors. For example, only 9.1% of ligands are miRNA targets, whereas 50% of the nuclear proteins, mostly TFs, are miRNA targets. In other words, miRNA targets are enriched more than fivefold in the most downstream components (the output layer of the signaling network) compared with the most upstream proteins. These results suggest that miRNA tends to control the output layer of the signaling network.

Cui et al. (2007a) also studied the cancer-signaling network by mapping cancer driver-mutating genes, determined by large-scale genome sequencing, onto a human signaling network that contained more than 1600 nodes and 5000 signaling relations. The authors showed that cancer driver-mutated genes are enriched in the output layer and in the positive regulatory motifs of the network. In general, these two principles of cancer signaling are in agreement with miRNA regulatory principles. Thus, lower expression of miRNAs may generally promote cancer signaling, and therefore may promote cancer progression. Indeed, a recent report showed that low expression of two components of miRNA machinery, Drosha and Dicer, was significantly associated with advanced tumor stages. Cancer cases with high expression of both Dicer and Drosha were associated with increased survival (Merritt et al. 2008).

miRNAs avoid targeting common components of cellular machines in the networks. The common components are the signaling proteins, which are shared by shortest paths starting from any signaling input node to cellular signaling machines, such as transcription and translation machinery, the secretion apparatus, motility machinery, and electrical response. The common components targeted by miRNAs are significantly underrepresented among the total proteins of the network. For example, only 14.3% of all proteins derived from basic cellular processes, that is, transcription, translation, secretion, motility, and electrical response machineries, were miRNA targets, compared with 30% of all proteins in the network. These results indicate that miRNAs avoid disturbing the commonly used signaling components (Cui et al. 2006).

7.2.1.3 *miRNAs Preferentially Target the Downstream Components of the Adaptors, Which Have Potential to Recruit More Downstream Components*

Many types of intracellular signaling activity, such as the recruitment of downstream signaling components to the vicinity of receptors, are performed by adaptor proteins. Adaptors perform such jobs by activating, inhibiting, or relocalizing downstream components through direct protein–protein interactions. An adaptor is able to recruit distinct downstream components in different cellular conditions. Therefore, some adaptors (high-linking adaptors) could recruit more downstream components than others (low-linking adaptors). miRNA targets more proteins in the high-linking adaptor group (36.1%, 39/108) than in the low-linking adaptor group (24.2%, 22/91, $P < 0.015$). This result suggests that miRNAs preferentially target the downstream components of the adaptors, which have potential to recruit more downstream components (Cui et al. 2006).

If an adaptor can recruit more downstream components, these components should have higher dynamic gene expression. To accurately respond to extracellular stimuli, adaptors need to selectively recruit downstream components. As miRNA targets are more common among the downstream components of high-linked adaptors, miRNAs may play an important role in the precise selection of cellular responses to stimuli by controlling the concentration of adaptors' downstream components.

7.2.1.4 *Dynamic Regulation of Signaling Pathways by miRNAs*

General principles of miRNA regulation of signaling networks are understood, and several studies have been done to investigate the expression dynamics of miRNA regulation of signaling pathways. The loci and patterns of miRNA coexpression are conserved across various organisms (Altuvia et al. 2005; Megraw et al. 2007). This makes it easier to investigate the signaling pathways that are potentially regulated by miRNA clusters. Yen and co-workers (2008) surveyed conserved clustering of mammalian miRNAs on chromosomes, and then examined the potential signaling pathway genes (targets) of each miRNA cluster. Their analysis suggested that each member of the miRNA clusters could regulate one or more components of a signaling pathway. Thus co-ordination between members of the miRNA clusters controlling components of a signaling pathway can regulate signal flow. The authors also experimentally verified their predictions and showed that one of the miRNA clusters, mmu-mir-183-96-182, targets irs1, Rasa1, and Grb2. Interestingly, all of these components are known members of the insulin signaling pathway. The ability of miRNA clusters to modulate multiple components of a single pathway suggested that miRNA could act as an efficient regulator for certain signaling pathways.

In another study, Gusev (2008) presented computational analyses of cellular processes, functions, and pathways that are collectively targeted by differentially expressed miRNAs. Gusev linked miRNAs to different GO categories, disease and toxicological categories, physiological functions and pathways. Results of the study suggested that coexpressed miRNAs may provide systematic compensatory responses to the abnormal phenotypic changes in cancer cells by targeting different signaling pathways.

7.2.2 miRNA Regulation of Gene Regulatory Networks

Gene regulatory networks describe the regulatory relationships between a regulator, which could be a TF and/or RNA, and its target genes. In such a network a TF receives input from an upstream signal transduction cascade and binds to a *cis*-regulatory element in the promoter of a gene. The bound transcription factor can stimulate or repress gene transcription and facilitate information flow from TFs to downstream effectors, resulting in a phenotype (Cui et al. 2007b; Zhu, Gerstein, and Snyder 2007). Deciphering gene regulatory networks is a way to understand cellular processes.

7.2.2.1 miRNA Preferentially Regulates the Genes with High Transcriptional Regulation Complexity

Much is known about the primary components of transcriptional regulation, that is, TFs, transcriptional factor binding sites (TFBS) on a promoter region, and the relationships (combinatorial regulation, cooperation and antagonism, autoregulation, feed-forward, and so on) between them. However, study of the regulatory networks with miRNAs as regulators is just beginning. In particular, the ability of miRNAs to regulate multiple genes invites the question of whether miRNAs share TF functional paradigms, such as combinatorial regulation and regulation of whole genetic programs.

miRNAs have great diversity and an abundance of targets; however, it is **unclear** why some genes are regulated by miRNAs whereas others are not. In other words, what principles govern miRNA regulation in animal genomes? To answer this question, Cui et al. (2007b) examined the regulatory principles of miRNA within a gene regulatory network of the human embryonic stem cell. The network contained three transcription factors (OCT4, NANOG, and SOX2) and 2043 genes that are regulated by the three TFs. The analysis revealed that there are significantly more miRNA target genes that are regulated by more TFs (Cui et al. 2007b). This observation means that a gene that is regulated by more TFs is most likely to be regulated by miRNAs. To extend the observation to the entire genome, the authors extended the analysis onto entire human genes by correlating the TFBSs of miRNA target and nontarget genes. The analysis confirmed that genes targeted by more miRNAs have more TFBSs (Cui et al. 2007b). This observation led Cui et al. to propose a novel mechanism for spatiotemporal expression behavior of miRNA-regulated networks: genes that are part of some dynamic process likely require complex regulation under different temporal and spatial conditions. Analysis of developmental genes, which are known to be regulated dynamically by TFs, confirmed that these genes have more TFBSs (implying that they are regulated by more TFs) and are regulated by more miRNAs. Borneman et al. (2006) demonstrated a similar concept, showing that genes that extensively regulate crucial processes are often heavily regulated by miRNAs.

7.2.2.2 miRNA Tends to Team up with TFs to Form Regulatory Loops

The discovery mentioned above has led to the suggestion of co-evolution between TF-mediated transcriptional regulation and miRNA-mediated posttranscriptional regulation. Other studies have also suggested that miRNAs are most likely to regulate cellular processes in coordination with TFs (Shalgi et al. 2007; Tsang, Zhu, and van Oudenaarden 2007). A statistical analysis of combinatorial regulation between TFs and miRNAs showed support for evolutionary advantages of joining transcriptional and posttranscriptional regulatory mechanisms for regulatory control of genes (Zhou et al. 2007).

Analysis of signaling networks integrated with miRNAs showed that miRNAs preferentially regulate positive regulatory loops of signaling networks (Cui et al. 2006), and this conclusion has been detailed in Section 7.2.2.1. In this context, it is interesting to ask whether miRNAs also regulate the positive regulatory loops in gene regulatory networks. Tsang and co-workers used mammalian gene expression data to compute positive and negative transcriptional coregulation circuits of miRNA and its targets (Tsang, Zhu, and van Oudenaarden 2007). These miRNA-mediated feedback and feed-forward loops were shown to be prevalent in the human and mouse genome. Results of the analyses strongly suggested that coordinated transcriptional and miRNA-mediated regulation become a common mechanism in mammalian genomes to maintain the robustness of gene regulation (Tsang, Zhu, and van Oudenaarden 2007).

There are two prevalent classes of circuits (Type 1 and Type 2 circuits; more details about these two types of circuits can be found in Section 7.3.1) in mouse and human miRNA regulation networks. In Type 1 circuits (Figure 7.1), any deviation from the

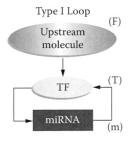

FIGURE 7.1　(See color insert following page 332.) Type-1 circuits of transcription factor and miRNA regulatory motifs (F), (T), and (m) represent upstream factor, target gene, and miRNA, respectively.

upstream factor (F)'s steady state would drive the target (T) and miRNA (m) away from their steady states in the same direction; thus, m could modulate the production rate of T in the direction opposite to F's fluctuation. Type 1 circuits could function as noise buffering that helps to maintain target protein homeostasis and ensures more uniform expression of T within a cell population. For example, in the human c-Myc/E2F1/miR-17-20 network (Acar, Becskei, and van Oudenaarden 2005; O'Donnell et al. 2005; Tsang, Zhu, and van Oudenaarden 2007), c-Myc and E2F1 can activate each other's transcription and form a positive regulatory loop, and both can activate the transcription of the miR-17 miRNA cluster. miRNA-17 mediates negative feedback to the E2F1 transcription factor. This negative loop, mediated by miRNAs, could prevent random activation of c-Myc/E2F1 by fluctuations in their expression. These motifs are overrepresented in animal genomes. Mathematical modeling has shown that miRNA in such motifs stabilizes the feedback loop to resist external perturbation, increasing network stability (Yu et al. 2008). Such robust action can be often observed in developmental processes that are regulated by miRNAs. The existence of extensive TF-like combinatorial interactions among miRNAs (Yuh, Bolouri, and Davidson 1998) and between miRNAs and TFs (Shalgi et al. 2007) has also been reported. Using evolutionarily conserved potential miRNA binding sites in human genes and conserved TF binding sites in gene promoter regions, the authors pointed out two global architectural features that are involved in recurring regulation networks consisting of miRNAs and TFs. The first feature is that the network consists of combinatorial interactions between pairs of miRNAs with many shared targets. Furthermore, networks encode several levels of hierarchy. The second feature is that the network included hundreds of miRNA-TF pairs that regulate a large set of common genes, which, in turn, tend to mutually regulate each other. The second global network character has been extensively observed in gene regulatory networks.

A recent genome-scale miRNA network motif analysis in *Caenorhabditis elegans* showed that miRNA-TF composite feedback loops, in which a TF controls a miRNA and, in turn, the TF, is also regulated by that same miRNA. Such network motifs were found in statistically significant numbers in the network (Martinez et al. 2008). Such miRNA-TF feedback loops are highly connected and heavily regulated (TFs and miRNAs regulate each other).

The data indicate that miRNAs and TFs regulate each other and form important regulatory motifs. These miRNA-TF motifs are prevalent in animal genomes; thus, they play important roles in many biological processes.

7.2.3 miRNA Regulation of Metabolic Networks

Cellular response to genetic and environmental perturbations is often reflected and/or mediated through changes in metabolism. It is accepted that mechanisms that control metabolic networks are complex and involve transcriptional, posttranscriptional, and translational regulation. Several studies have integrated expression with metabolic network data to analyze how the coordinated expression of enzymes shapes the metabolic network. Most of these studies have been successful in systematically characterizing transcriptional regulation of metabolic pathways. Since miRNAs could regulate many target genes, it is reasonable to believe that miRNA could extensively regulate metabolic networks.

Several studies have shown the effects of miRNA on a particular metabolic pathway. For example, miRNA may regulate entire metabolic pathways, such as cholesterol biosynthesis and triglyceride metabolism (Krutzfeldt and Stoffel 2006). miR-122 negatively regulates some transcriptional repressors that modulate a cluster of cholesterol-biosynthesis genes, including HMG-CoA reductase (Krutzfeldt and Stoffel 2006).

Tibiche and Wang recently analyzed the direct regulatory effect of miRNAs on enzymes in the human metabolic network. They explored miRNA regulatory principles in the network by exploring the relationships between the miRNA targets and the metabolic network nodes with distinct network structural features (Tibiche and Wong 2008). In the network, nodes represent either enzymes or reactions. In this study, the nodes were classified into five categories based on their structural features in the network (Figure 7.2). The nodes that uptake metabolites from extracellular space and have no incoming links were called upstream nodes (UPNs). The nodes with no outgoing links were called downstream nodes (DSNs), which were responsible for producing pathway output metabolites used in various cellular activities and biological processes. One crucial node in a network is a cut vertex node, or articulation or cut points (CPs). If the graph was connected before the removal of the vertex, it will be disconnected after its removal. Any connected graph with a cut vertex has a connectivity of 1. This means that a bottleneck node, when deleted, will disconnect at

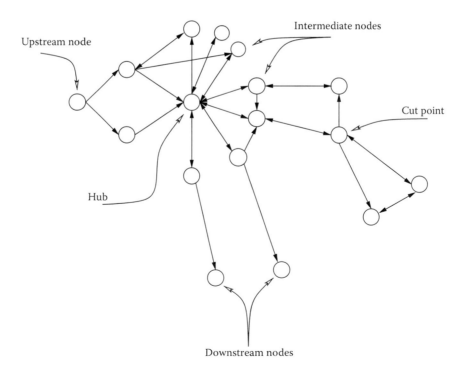

FIGURE 7.2 Node types with distinct network structural features. Nodes represent enzymes while the arcs represent chemical reactions.

least one component from the network. They are usually located in crucial network positions and, therefore, control metabolic flow from one part of the network to another. The nodes with a high degree of connectivity (top 5% of connection in network) were called hubs. All other nodes were called intermediate nodes.

Analysis of miRNA regulation of metabolic networks showed that miRNAs preferentially regulate network hubs. They avoid intermediate nodes, which is in agreement with the observation that miRNA avoids targeting common components of cellular signaling networks. miRNAs regulate CPs, allowing them greater control and targeted modulation of metabolic flow. miRNAs preferentially regulate two or more consecutive linear metabolic reactions but avoid regulating metabolic branches. Furthermore, certain metabolic pathways are predominately regulated by miRNA. For example, over 30 pathways, such as amino acid synthesis or degradation, glycan biosynthesis, pantothenate and CoA biosynthesis, and a group of lipid metabolism pathways, are extensively regulated by miRNAs. This is in agreement with previous studies reported in the mouse (Krutzfeldt and Stoffel 2006) and *Drosophila* (Stark et al. 2003). Taken together, the data show that miRNA regulates the central metabolic pathways and has a major role in cell metabolism.

7.2.4 miRNA Regulation of Protein Interaction Networks

Proteins interact with each other to perform specific cellular functions. Protein interaction networks or interactome maps provide a valuable framework for systematic study of protein function.

7.2.4.1 miRNAs Preferentially Regulate Network Hubs

Analysis of the human protein interaction network integrated with miRNAs showed that the average number of target-site types in the 3′ untranslated regions (3′UTR) of a set of genes encoding the protein is strongly correlated with the extent of protein connectivity in the network (Liang and Li 2007). Furthermore, the interacting proteins tend to share more miRNA target-site types compared with random pairs. In principle, proteins with more interacting partners suggest their participation in complex dynamic processes. Protein production in such processes is highly dynamic, requiring tighter regulatory control (Cui et al. 2007c). The analysis of microarray profiles of miRNA target genes in various tissues is consistent with this trend, showing that mRNAs with broad expression profiles have a tendency to be regulated by more miRNAs (Cui et al. 2007b). These data suggest that when a protein has many interacting partners, it will be regulated by many transcription factors and miRNAs. This is in agreement with the results from the gene network analysis, which showed that miRNAs preferentially regulate the genes regulated by many transcription factors (Borneman et al. 2006).

Based on the clustering coefficient (a ratio of the existing links among a node's neighbors and the maximum possible number of links between them) of network hubs, they can be classified into two categories, intermodular and intramodular. A hub protein with a high clustering coefficient is likely to be an intramodular hub, which interacts with most of its partners simultaneously to form a protein complex, and completes a coherent function. In contrast, a hub protein with a low clustering coefficient tends to be an intermodular hub, which tends to interact with other proteins at different times and in different places, and then coordinates different functional modules. The number of miRNA target-site types of a gene encoding a hub protein is significantly negatively correlated with the clustering coefficient (Liang and Li 2007). These results suggest that different types of hub proteins have different miRNA targeting propensities and the miRNAs regulate intermodular hubs more than intramodular hubs. Further analysis of hubs using betweenness centrality (a measure of the number of nonredundant shortest paths going through a node) showed that miRNAs preferentially regulate hubs with high betweenness, which are the "bottlenecks" of information flow in the network (Hsu and Huang 2008). These results indicate that miRNA tends to affect the "key nodes" in networks.

7.2.4.2 The Targets of Individual miRNAs Tend to Form Subnetworks
Decomposition of the network in a modular manner results in the finding that target genes of individual miRNA are significantly closer to each other, as shown by short characteristic path lengths between the target genes of the same miRNA, when compared to the whole network (Hsu and Huang 2008). The conclusion was drawn by comparing two subnetwork models: (1) a subnetwork formed by proteins directly regulated by one miRNA (level-0 proteins), and (2) an extended subnetwork involving interacting neighbors (level-1 proteins) of the proteins in the first network (level-0). Further statistical analysis led the investigators to conclude that formation of functional modules is rarer in the first type (level-0) of subnetwork than in the second (level-1). This result means that level-0 proteins do not interact closely with each other; instead, they interact with one another indirectly through some mediator proteins. These mediator proteins may correspond to the level-1 subnetwork proteins. Because level-1 proteins, along with level-0 proteins, form functional modules, miRNA could influence specific biological functions by regulating a small number of selected level-0 genes.

7.3 miRNA REGULATION OF CANCER GENE NETWORKS

Cancer is a disease that can be induced by multiple factors, such as gene mutation, methylation, and environmental factors. Certain genes induce cancer and promote oncogenesis (oncogenes), while others inhibit tumor development (tumor suppressors) (Cui et al. 2007a). Thus, pathogenesis of cancer is a delicate balance between oncogenes and tumor suppressors. Recently, miRNA expression was shown to be altered in cancer, compared with normal tissue, spanning a wide range of tumor types including lung, breast, brain, liver, and colon cancer, and leukemia. miRNAs regulate cancer-related processes like cell growth and tissue differentiation. Because abnormal proliferation and defective differentiation are characteristic features of tumor cells, and miRNAs are involved in regulation of these processes, miRNAs are considered to be key molecules that are likely to affect cancer gene networks. Global analysis of miRNA chromosomal locations suggested that miRNA genes were located at fragile sites in the genomic regions that are commonly amplified or deleted in human cancer (Calin et al. 2004). The functional roles of miRNA in cancer have been extensively reviewed elsewhere (Blenkiron and Miska 2007; Cui et al. 2007a; Sassen, Miska, and Caldas 2008; Wu et al. 2007). In summary, although miRNAs negatively regulate their targets, they perform more complex regulation by working with other regulatory genes, such as TFs. As explained above (see Section 7.2), it is expected that an integrative analysis of TFs, miRNAs, and their interactions with their target genes could shed light on transcriptional and posttranscriptional regulatory mechanisms of cancer gene networks.

7.3.1 miRNA Regulatory Circuits in Cancer

Gene regulatory loops are important components of cancer networks. Several lines of evidence have suggested that miRNA could interact with these cancer gene regulatory loops, and subsequently regulate cancer progression and metastasis. For example, mir-17-92 was implicated in regulating the Myc oncogenic pathway (Ma'ayan et al. 2005). Further studies of mir-17-92 in cancer suggested that a double feed-forward loop exists between E2F, Myc, and the miR-17/20 cluster, which plays a crucial role in regulation of cancer progression (O'Donnell et al. 2005; Sylvestre et al. 2007). Such a loop, containing TFs (E2Fs and Myc) and genes coding for miRNAs (miR-17-92 cluster), has been defined as a Type-I circuit, as described earlier (Figure 7.1; see Section 7.2.2). Briefly, the characteristics of this type of miRNA-mediated feedback and feed-forward circuits are: (1) the transcription of the miRNAs and their targets are positively coregulated; (2) they are found to be less abundant than type-II circuits (transcription of the miRNAs and their targets is oppositely regulated by common upstream factors); (3) they possess the ability to provide the host with some important regulatory and signal processing functions, that is, modulation and maintenance of the protein steady state.

Although we are able to infer general characteristics of the c-Myc/E2Fs and miR-17-92 regulatory loop based on the type of regulatory circuits, we still lack a full quantitative understanding of how the loop functions. In other words, when E2Fs/Myc are expressed at pathological levels, what quantitative levels induce these proteins to activate targets to control proliferation and apoptotic behavior, and what levels of miRNA are needed in tumor and normal states? To answer these questions, Aguda et al. (2008) proposed a simple mathematical model for this regulatory loop.

The authors explored the consequences of the network based on the steady states and dynamic states of miR-17-92 and the group of proteins that are its targets (Figure 7.3). The positive feedback loops between E2F and Myc can emerge into bistability (existence

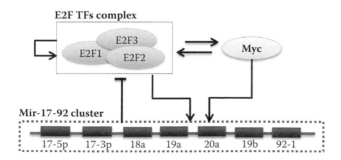

FIGURE 7.3 (See color insert following page 332.) The regulatory loop of c-Myc/E2Fs and miR-17-92. Arrows represent active regulation while T signs represent negative regulation.

of low and high levels of proteins, corresponding to bistable switches "OFF" and "ON," respectively, with sharp transitions between levels inducible by a triggering stimulus, such as a growth factor (Ferrell 2002)). The model described by Aguda et al. could exhibit bistability and predicted that miR-17-92 plays a critical role in regulating the position of the OFF-ON switch in E2F/Myc protein levels, and in determining the ON levels of these proteins. The model predictions agreed with current thinking about the regulatory mechanism by which c-Myc fine-tunes gene expression through the activation of E2F and miR-17-92 transcription. Upon activation, miR-17-92 simultaneously reduces the translation of E2F/Myc, preventing a rampant E2F/Myc feedback loop that could induce uncontrolled cell proliferation. Most importantly, the model predicted that miR-17-92 counteracts the cancer-associated decrease in growth factor requirement for cell proliferation. At non-steady-state conditions the model could also explain why miR-17-92 is anti-apoptotic. Initially the Myc/E2F protein levels were high, which corresponds to increased apoptosis. When the level of miR-17-92 was increased, the level of the target genes of miR-17-92 decreased, and, in turn, led to a decrease in the rate of apoptosis. The model demonstrated a delicate balance in the cellular system and accounted for a mechanism in which the same cellular components can cause or suppress the cancer depending on various influencing factors.

7.3.2 miRNA Regulatory Networks Associated with Cancer

As mentioned above, miR-17-92 has been studied extensively. Currently it is known that miR-17-92 has more than 20 genes, such as TGFBR2, PTEN, and THBS1, that are experimentally validated as its targets. Although many of these targets are known as cell cycle regulators, none of these interactions is sufficient to explain the oncogenic potential of this locus. The specific mechanisms of either the tumor suppressor or oncogenic activities of the miR-17-92 remain unknown, and may require investigation with the methods of systems biology. A recent study showed that miR-17-92 regulates a cell cycle specific network, a central component of cancer signaling (Cloonan et al. 2008).

Analysis of the miR-17-92 regulatory network predicted that miR-17-5p could act as an oncogene or a tumor suppressor by targeting both cell proliferation activators and proliferation inhibitors, respectively, in different cellular contexts. In situations where pro-proliferative genes dominate, miR-17-5p may stabilize the proliferation signal and maintain a net proliferative (oncogenic) outcome by removing proliferation inhibitors and increasing the mRNA levels of proliferation activators. On the other hand, in situations where proliferation inhibitors dominate, suppression of pro-proliferative signals is reinforced, leading to a net antiproliferative signal. Functional analysis of the network explained that miR-17-5p acts by suppressing the G1/S cell cycle check point, resulting in a sudden rise in cell proliferation rate by targeting a large genetic network of interacting proteins. This coordinated targeting allows miR-17-5p to efficiently decouple negative regulators of the MAPK signaling cascade, thus promoting growth in cancer cells. Withdrawal of miRNA-17-5p would result in increasing proliferation activators and decreasing mRNA levels of the proliferation inhibitors (pro-proliferative signal), possibly leading to tumor suppression (Cloonan et al. 2008).

Network approaches have been also used to construct a miR-21 regulatory network that may contribute to cancer progression. To uncover the molecular mechanisms of miR-21 in cancer development, Papagiannakopoulos, Shapiro, and Kosik (2008) first predicted all the potential targets of miR-21 using TargetScan, a computational program designed for miRNA target prediction (Lewis et al. 2003). They performed Gene Ontology (GO) analysis to identify a subset of the predicted targets that are involved in regulating cell growth or apoptosis. Their analysis showed that GO terms, such as cell growth and proliferation, cell death, cancer, and cell cycle, were statistically significant in the predicted targets. Finally, they mapped the predicted targets of these GO terms onto a protein interaction network and extracted a subnetwork mainly consisting of the predicted targets of the selected GO terms. The authors suggested that such a subnetwork is likely to be tightly regulated by miR-21 in what is also called the miR-21 regulatory network.

Predominant finds in the miR-21 regulatory network are the TGF-β pathway genes (TGFBR2, TGFBR3, and DAXX), p53 pathway genes (p53, TP73L, and TAp63), activating cofactors of the p53 pathway (JMY, TOPORS, HNRPK, and TP53BP2), and mitochondrial apoptotic pathway genes (APAF1, caspase-8, VDAC1, and PPIF). These genes, found in the subnetwork, are predicted to be regulated by miR-21. The TGF-β pathway is known to induce apoptosis in cancer cells, and also during development, in response to TGF-β ligand binding to its receptors, TGFBR2 and TGFBR3, which can, in turn, inhibit growth and activate apoptosis. Furthermore, genes involved in the p53 pathway are known to assist in the transcriptional activation of antiproliferative and proapoptotic genes in response to DNA damage. These results suggest that miR-21 may act as an oncogene to suppress p53 and the apoptotic pathways. Further experimental analysis of the regulatory effects of miR-21 on the components of the subnetwork suggested that changes in the level of miRNA-21 are likely to reduce the robustness of a highly interconnected tumor-suppressive network and result in global dysregulation of the network functions. This suggests that miRNA-21 upregulation may be a key step in oncogenesis.

7.3.3 miRNA Regulation of the p53 Network

The p53 network controls many pathways important for tumor suppression by regulating transcription. Recently, Sinha et al. (2008) have identified a cancer-signaling module (p53 module) that contains many tumor suppressors, such as p53, p16, and others that are commonly employed by tumors. The tumor suppressor p53 modulates the expression of target genes that promote growth arrest and apoptosis. Therefore, regulation of the p53 network may be critical for cancer progression and metastasis (Sinha et al. 2008).

A computationally systematic characterization of miRNAs that are components of the p53-miR transcriptional regulatory network was conducted by Sinha et al. (2008). They constructed a p53-centered regulatory network containing 23 TFs that regulate p53 and 48 TFs that are regulated by p53 (based on the p53 knowledgebase) (http://p53.bii.a-star.edu.sg). Furthermore, the authors integrated miRNAs that not only target the TFs in the p53-centered regulatory network, but may also be potentially regulated by p53, into the p53-centered regulatory network (Figure 7.4). In order to find the miRNAs that are potentially regulated by p53, they scanned the 10 kb flanking regions of each miRNA in the human genome using the p53MH. The latter

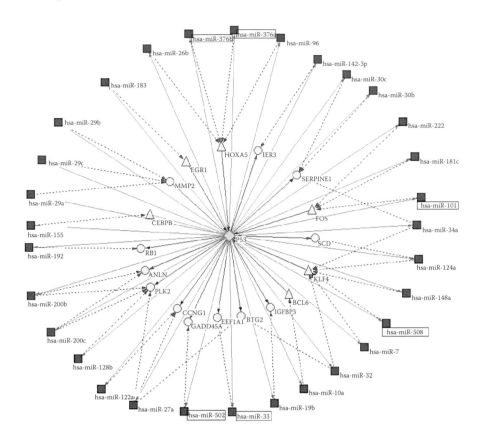

FIGURE 7.4   (See color insert following page 332.) The p53-centered miRNA regulatory cancer network. Red boxes represent miRs. Red dotted lines represent the regulation of miRs to TFs which either regulate p53 or are regulated by p53. Yellow triangles represent the TFs which regulate p53. Yellow circles represent the TFs which are regulated by p53. Red filled lines represent repression. Blue filled lines represent activation. Black filled lines represent repression/activation. Green filled lines represent uncharacterized regulation. Differentially expressed miRs in cancer are boxed. miRs whose expression 1.5-fold ($p < 0.05$) in cancer is triggered by the p53 activation are underlined. All of the TFs in this network were predicted as miRNA targets by two or more algorithms (picTar, miRtarget, microT, miRanda, TargetScanS).

is an algorithm that identifies p53-responsive genes in the human and mouse genome in order to identify putative p53-binding sites within miRNAs. Finally, a regulatory network centered on the p53 gene and miRNA was obtained. Notably, most of the miRNAs in the network had been reported to be modulated in either cancer cell lines or tumor samples. Further analysis of the potentially targeted network genes showed that miRNAs within the network are functionally enriched with biological processes, such as apoptosis and cell cycle. These data indicate that the p53-centered gene and miRNA regulatory network are involved in tumor progression. This case provides strong evidence for using integrative methods and data sources to tackle important problems regarding the roles of miRNA.

Using the p53 regulatory network, Sinha et al. identified four types of representative miRNA-TF regulatory circuits. They used these models to hypothesize the mechanisms of

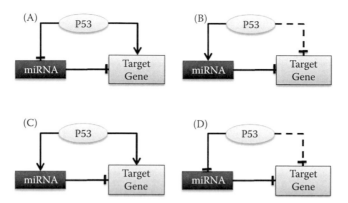

FIGURE 7.5 Types of representative miRNA-TF regulatory circuits in the p53-miRNA regulatory cancer network. (A) and (B): downstream incoherent models; (C) and (D): downstream coherent models.

action of p53 ("master transcriptional regulator") repression. The authors explained various scenarios of p53 repression as a direct or indirect effect of transactivating/transrepressing downstream targets (TFs/miRNA). For the downstream coherent models, the p53 action on downstream target genes and the miRNA is opposite (Figures 7.5A and 7.5B). In the downstream incoherent models, p53 acts on downstream target genes, as do the miRNAs (Figures 7.5C and 7.5D). By analyzing these circuits, the authors suggested that in the case of the p53 downstream targets, the foremost activated positive feedback in a coupled feedback circuit can rapidly induce the "ON" state transition of the signaling system (p53 activating downstream target genes), and then another delayed positive feedback (p53 suppressing miRNAs that target p53 downstream target genes) could robustly maintain this "ON" state. Finally, the most delayed negative feedback would reinstate the system to the original "OFF" state, preventing any further excessive response specific to the applied stimulus.

Using a computational and a systems biologic approach, it would be possible to perform a comprehensive survey of the miRNA-p53 network. Furthermore, the potential cross-talk between the miRNAs and the p53 transcriptional network could be investigated to illustrate the potential molecular mechanisms of p53-miRNA in cancer progression. Along this line, using an experimental approach, several studies showed that a part of the p53 network is mediated through transcriptional activation of a cluster of miRNAs. In turn, these miRNAs repress a number of target genes to induce growth arrest. All these studies compared the expression of miRNAs in cells with or without p53 expression. Sinha et al. compared the miRNA expression profiles of wild-type and p53-null mouse embryonic fibroblasts, which also expressed a variety of oncogenes. Three miRNAs, miR-34a, miR-34b, and miR-34c, were found to positively correlate with p53 status. The interaction of p53 and miR-34s has also been documented in the p53-centered gene and miRNA regulatory network.

Apoptosis and growth arrest are common consequences of p53 activation. Experiments have shown that the cyclin-dependent kinase inhibitor p21Waf1/Cip1 is an immediate target of p53. This inhibitory protein can affect most, but not all, of the antiproliferative capabilities of p53. This suggests the possibility that additional signaling molecules may participate in p53-mediated growth arrest. Furthermore, chromatin immunoprecipitation and luciferase assays (Blenkiron and Miska 2007; Raver-Shapira et al. 2007) confirmed that the activation of p53 resulted in binding to miR-34a promoters and activation of transcription in HCT116 cells. Additionally, Oren et al. and Mendell et al. were able to show that p53 regulates miR-34a in the same cancer cells (reported in Chang et al. 2007; Raver-Shapira et al. 2007). Hermeking (2007) and Raver-Shapira et al. (2007) have both reported that miR-34b and miR-34c are direct transcriptional targets of p53. Their studies have also shown that the ectopic expression of miR-34a, miR-34b, and miR-34c induced cell cycle arrest in primary human fibroblasts and in four other tumor cell lines. Expression of miR-34 inhibited proliferation by inducing cellular senescence and cell cycle arrest at G1. p53 upregulates miR-34a in response to DNA damage and subsequently leads to G1 cell cycle arrest and apoptosis in various cancer cell lines. All these studies substantiate the miRNA-p53 relationship in various mouse tissues and cancer cell lines. Interestingly, miR-34-mediated growth arrest can occur in the absence of p21waf1/Cip1. Taken together, the data highlight an important role of miRNAs in promoting p53 signaling through the p21-independent pathway for p53-dependent tumor suppression.

REFERENCES

Acar M., Becskei A., and van Oudenaarden A. 2005. Enhancement of cellular memory by reducing stochastic transitions. *Nature* 435:228–232.

Aguda B. D., Kim Y., Piper-Hunter M. G., Friedman A., and Marsh C. B. 2008. MicroRNA regulation of a cancer network: consequences of the feedback loops involving miR-17-92, E2F, and Myc. *Proc Natl Acad Sci USA* 105:19678–19683.

Altuvia Y., Landgraf P., Lithwick G. et al. 2005. Clustering and conservation patterns of human microRNAs. *Nucleic Acids Res* 33:2697–2706.

Blenkiron C. and Miska E. A. 2007. miRNAs in cancer: approaches, aetiology, diagnostics and therapy. *Hum Mol Genet* 16 Spec No 1:R106-13.

Borneman A. R., Leigh-Bell J. A., Yu H. et al. 2006. Target hub proteins serve as master regulators of development in yeast. *Genes Dev* 20:435–448.

Brandman O., Ferrell J. E., Jr., Li R., and Meyer T. 2005. Interlinked fast and slow positive feedback loops drive reliable cell decisions. *Science* 310:496–498.

Calin G. A., Sevignani C., Dumitru C. D. et al. 2004. Human microRNA genes are frequently located at fragile sites and genomic regions involved in cancers. *Proc Natl Acad Sci USA* 101:2999–3004.

Chang T. C., Wentzel E. A., Kent O. A. et al. 2007. Transactivation of miR-34a by p53 broadly influences gene expression and promotes apoptosis. *Mol Cell* 26:745–752.

Cloonan N., Brown M., Steptoe A. et al. 2008. The miR-17-5p microRNA is a key regulator of the G1/S phase cell cycle transition. *Genome Biol* 9:R127.

Cui Q., Ma Y., Jaramillo M. et al. 2007a. A map of human cancer signaling. *Mol Syst Biol* 3:152.

Cui Q., Yu Z., Pan Y., Purisima E. O., and Wang E. 2007b. MicroRNAs preferentially target the genes with high transcriptional regulation complexity. *Biochem Biophys Res Commun* 352:733–738.

Cui Q., Yu Z., Purisima E. O., and Wang E. 2006. Principles of microRNA regulation of a human cellular signaling network. *Mol Syst Biol* 2:46.

Cui Q., Yu Z., Purisima E. O., and Wang E. 2007c. MicroRNA regulation and interspecific variation of gene expression. *Trends Genet* 23:372–375.

Ferrell J. E., Jr. 2002. Self-perpetuating states in signal transduction: positive feedback, double-negative feedback and bistability. *Curr Opin Cell Biol* 14:140–148.

Gusev Y. 2008. Computational methods for analysis of cellular functions and pathways collectively targeted by differentially expressed microRNA. *Methods* 44:61–72.

Hermeking H. 2007. p53 enters the microRNA world. *Cancer Cell* 12:414–418.

Hsu C. W., Juan H. F., and Huang H. C. 2008. Characterization of microRNA-regulated protein-protein interaction network. *Proteomics* 8:1975–1979.

Irish J. M., Kotecha N. and Nolan G. P. 2006. Mapping normal and cancer cell signalling networks: towards single-cell proteomics. *Nature Rev Cancer* 6:146–155.

Krutzfeldt J. and Stoffel M. 2006. MicroRNAs: a new class of regulatory genes affecting metabolism. *Cell Metab* 4:9–12.

Lewis B. P., Shih I. H., Jones-Rhoades M. W., Bartel D. P., and Burge C. B. 2003. Prediction of mammalian microRNA targets. *Cell* 115:787–798.

Liang H. and Li W. H. 2007. MicroRNA regulation of human protein protein interaction network. *RNA* 13:1402–1408.

Ma'ayan A., Jenkins S. L., Neves S. et al. 2005. Formation of regulatory patterns during signal propagation in a mammalian cellular network. *Science* 309:1078–1083.

Martinez N. J., Ow M. C., Barrasa M. I. et al. 2008. A *C. elegans* genome-scale microRNA network contains composite feedback motifs with high flux capacity. *Genes Dev* 22:2535–2549.

Megraw M., Sethupathy P., Corda B., and Hatzigeorgiou A. G. 2007. miRGen: a database for the study of animal microRNA genomic organization and function. *Nucleic Acids Res* 35:D149–D155.

Merritt W. M., Lin Y. G., Han L. Y. et al. 2008. Dicer, Drosha, and outcomes in patients with ovarian cancer. *N. Engl. J. Med.* 359:2641–2650.

O'Donnell K. A., Wentzel E. A., Zeller K. I., Dang C. V., and Mendell J. T. 2005. c-Myc-regulated microRNAs modulate E2F1 expression. *Nature* 435:839–843.

Papagiannakopoulos T., Shapiro A., and Kosik K. S. 2008. MicroRNA-21 targets a network of key tumor-suppressive pathways in glioblastoma cells. *Cancer Res* 68:8164–8172.

Raver-Shapira N., Marciano E., Meiri E. et al. 2007. Transcriptional activation of miR-34a contributes to p53-mediated apoptosis. *Mol Cell* 26:731–743.

Sassen S., Miska E. A., and Caldas C. 2008. MicroRNA: implications for cancer. *Virchows Arch* 452:1–10.

Shalgi R., Lieber D., Oren M., and Pilpel Y. 2007. Global and local architecture of the mammalian microRNA-transcription factor regulatory network. *PLoS Comput Biol* 3:e131.

Sinha A. U., Kaimal V., Chen J., and Jegga A. G. 2008. Dissecting microregulation of a master regulatory network. *BMC Genomics* 9:88.

Stark A., Brennecke J., Russell R. B., and Cohen S. M. 2003. Identification of *Drosophila* microRNA targets. *PLoS Biol* 1:E60.

Sylvestre Y., De Guire V., Querido E. et al. 2007. An E2F/miR-20a autoregulatory feedback loop. *J. Biol. Chem.* 282:2135–2143.

Tibiche C. and Wang, E. 2008. MicroRNA regulatory patterns on the human metabolic network. *The Open Syst Biol. J.* 1:1–8.

Tsang J., Zhu J., and van Oudenaarden A. 2007. MicroRNA-mediated feedback and feedforward loops are recurrent network motifs in mammals. *Mol Cell* 26:753–767.

Wu W., Sun M., Zou G. M., and Chen J. 2007. MicroRNA and cancer: current status and prospective. *Int. J. Cancer.* 120:953–960.

Yen H. C., Xu Q., Chou D. M., Zhao Z., and Elledge S. J. 2008. Global protein stability profiling in mammalian cells. *Science* 322:918–923.

Yu X., Lin J., Zack D. J., and Mendell J. T., Qian J. 2008. Analysis of regulatory network topology reveals functionally distinct classes of microRNAs. *Nucleic Acids Res* 36:6494–6503.

Yuh C. H., Bolouri H., and Davidson E. H. 1998. Genomic cis-regulatory logic: experimental and computational analysis of a sea urchin gene. *Science* 279:1896–1902.

Zhou Y., Ferguson J., Chang J. T., and Kluger Y. 2007. Inter- and intra-combinatorial regulation by transcription factors and microRNAs. *BMC Genomics* 8:396.

Zhu X., Gerstein M., and Snyder M. 2007. Getting connected: analysis and principles of biological networks. *Genes Dev* 21:1010–1024.

CHAPTER 8

Network Model of Survival Signaling in T-Cell Large Granular Lymphocyte Leukemia

Ranran Zhang, Thomas P. Loughran, Jr., and Réka Albert

CONTENTS

8.1 INTRODUCTION

8.1.1 Signaling Studies in Humans: The Dilemma

The original idea of signaling pathways, as presented in the early studies of G-protein signaling (Gilman 1987; Hardman, Robison, and Sutherland 1971; Rodbell 1995) was heavily influenced by concepts of signal transduction in computer science. Each pathway was considered as an independent responding unit that coupled extracellular stimulations to specific outputs through defined signaling cascades. However, as more signaling pathways were identified, it became obvious that crosstalk was a common phenomenon among pathways, and the same pathway could participate in cellular responses against different environmental stimulations (Eungdamrong and Iyengar 2004; Ma'ayan 2008). It is now well established that molecular events in the cell occur in the context of a complex signaling network that is temporally, spatially, and concentration-wise dynamic.

Understanding this enormously dynamic signaling network is a long-term challenge for our experimental studies, particularly in humans because of technical and ethical concerns. Due to technical limitations, conventional signaling studies can only address signaling events among a few molecules. This approach tends to provide artificially separated pathways simply because the "connectors" among pathways are overlooked. Results from conventional signaling studies in turn constrain later studies to focus on interactions of the components in one pathway while neglecting the broader effects in the same or other pathways. As discussed in other chapters, the development of experimental techniques greatly facilitates qualitative and quantitative signaling studies in humans. However, studying temporal and spatial dynamics of multiple signaling components usually relies on *in vitro* tissue culture systems or animal models. Neither system can reproduce the exact molecular events under the same *in vivo* environment as seen in humans. On the other hand, limited availability and notorious variation curb the use of primary human samples. Taken together, some degree of approximation, simplification, generalization, and deduction is inevitable for most human signaling studies, no matter how capable the experimental systems are to extract quantitative information. In other words, what we obtain biologically from signaling pathway studies in humans are usually models rather than realities.

There is a long-lasting debate about the usefulness and possibility of integrating and analyzing these human signaling "models" with tools of systems biology. In conventional signaling studies, researchers focus more on qualitative changes in terms of a few signaling inputs and outputs, such as the relationship between overexpression of one protein and cell survival. It was widely believed that simple logic is sufficient to make qualitative predictions in well-defined signaling pathways without formal modeling efforts. However, with the exponential increase of signaling pathway information, it is becoming less intuitive to infer signaling events in complex signaling networks such as T-cell activation (Saez-Rodriguez et al. 2007). Moreover, although "low-throughput" per experiment, conventional signaling studies typically yield higher quality data compared to most high-throughput methods. They can also provide detailed signaling information of a particular pathway region from primary human samples, which cannot be replaced by high-throughput data obtained from *in vitro* systems or animal models. Thus, integrating the existing pathway knowledge is as important as obtaining high-throughput data to better our understanding of the human signaling network at a systematic level.

On the other hand, there is a need for tools that can aid this integration. Conventional signaling studies in humans are not naturally compatible with systems biology tools. Most of those studies were not equipped to ask quantitative questions such as reactant/product turnover rate, subcellular localization, and reaction kinetics. This limits the applicability of quantitative mathematical methods, such as the use of differential equations, in modeling existing data. Meanwhile, the topological analysis of signaling networks, although insightful in identifying motifs and key signaling components (Ma'ayan et al. 2005), has limited application in predicting the dynamic behaviors of specific signaling components.

8.1.2 Boolean Modeling: A Natural and Powerful Translator between Signaling Studies and Systems Biology

It is often observed that changes in signaling behaviors can be attributed to the qualitative or digital changes of certain components in a pathway, such as the activation or inhibition of certain proteins (Bornholdt 2008). To some extent, these observations serve as "the zeroth law" of conventional signaling research, based on which pathways can be studied. Recent quantitative research did indeed confirm the digital behavior of certain signaling pathway components (Altan-Bonnet and Germain 2005). These approximate or exact digital behaviors in biological processes triggered the initial exploration of using discrete dynamic models such as Boolean models to capture the dynamics of biological systems (Glass and Kauffman 1973).

Retrospectively, Boolean logic is a way of thinking implicitly embedded in conventional signaling pathway studies. Boolean models have no parameters, and only assume two possible states: ON, meaning above threshold, and OFF, meaning below threshold. This correlates well with the qualitative concepts of presence/activation and absence/inhibition in signaling studies. In Boolean models, relationships among components are indicated by three Boolean operators: AND (conjunction), OR (disjunction), and NOT (negation). Similar language is widely used in depicting regulatory relationships among signaling pathway components. The conditional dependency of upstream regulators to achieve a downstream effect, such as regulations exerted by proteins that form a complex, resembles the AND relationship. The combined effect of independent upstream regulators on a downstream node, such as regulations exerted from independent signaling pathways on the same target, resembles the OR relationship. NOT describes inhibitory effects. By using Boolean operators in combination, it is intuitive to translate biological regulation from upstream regulators to downstream targets into Boolean rules.

The establishment of Boolean rules from signaling pathways enables computational simulation of signaling events (Chaves, Albert, and Sontag 2005; Davidich and Bornholdt 2008; Kaufman, Andris, and Leo 1999; Li, Assmann, and Albert 2006; Saez-Rodriguez et al. 2007). In this chapter, we present our study in which we applied Boolean modeling to study the survival signaling network of T-cell large granular lymphocyte (T-LGL) leukemia (Zhang et al. 2008). This study proves that, as in other experimental systems, Boolean modeling retains its power of dynamic analysis and prediction when it comes to signaling studies in human cancers. By analyzing existing signaling information, Boolean modeling as a tool is capable of bringing insights into the pathogenesis and potential therapeutic targets in a rare leukemia, which would otherwise be inaccessible to intuitive logic deduction.

8.2 CASE STUDY: NETWORK MODEL OF THE SURVIVAL SIGNALING IN T-CELL LARGE GRANULAR LYMPHOCYTE LEUKEMIA

8.2.1 Studying Signaling Abnormalities in T-LGL Leukemia: Can We Make a Wiser Guess?

The name large granular lymphocyte (LGL) refers to a morphologically distinct subpopulation that normally comprises 10% to 15% of peripheral blood mononuclear cells (PBMC). They are characterized by a high cytoplasmic to nuclear ratio and abundant azurophilic granules (Loughran 1993; Timonen, Ortaldo, and Herberman 1981). Normal circulating LGL are mainly comprised of CD3− natural killer (NK) cells. Only about 15% are derived from CD3+ cytotoxic T lymphocytes (CTL) (Sokol and Loughran 2006). LGLs serve as the main executors of cell-mediated cytotoxicity. They are essential for eliminating infected somatic cells and tumor cells (Russell and Ley 2002). LGL leukemia is a rare disorder of cytotoxic lymphocytes. It was first described as a clonal proliferation of LGL involving blood, marrow, and spleen. In LGL leukemia patients, leukemic LGL usually form the major cell type in circulating PBMC, and bone marrow and spleen infiltration is frequently observed (Loughran et al. 1985). Based on the lineage of leukemic LGL, LGL leukemia is further divided into T-LGL leukemia and NK-LGL leukemia. T-LGL leukemia features an abnormal clonal expansion of antigen-primed, competent CTL. It occurs predominantly in the elderly. The majority of T-LGL leukemia cases follow an indolent course with median survival of more than 10 years, although acute cases have been reported. Currently, there is no curative therapy for this leukemia (Loughran 1993; Sokol and Loughran 2006). T-LGL leukemia is frequently associated with autoimmune diseases and autoimmune-mediated bone marrow disorders (Rose and Berliner 2004). This places T-LGL leukemia at the intersection of cancer and autoimmunity (Shah et al. 2008).

CTL activation normally involves an initial expansion of antigen-specific CTL clones and their acquisition of cytotoxic activity. Subsequently, the activated CTL population undergoes contraction mediated by activation-induced cell death (AICD), resulting in final stabilization of a small antigen-experienced CTL population (Klebanoff, Gattinoni, and Restifo 2006). This process requires a delicate balance between cell proliferation, survival, and apoptosis. Leukemic T-LGL exhibit constitutive activation of multiple survival signaling pathways that are only transiently activated during normal CTL activation, including the Janus kinases (JAK), signal transducers and activators of the transcription (STAT) pathway (Epling-Burnette et al. 2001), the mitogen-activated protein kinase (MAPK) pathway (Epling-Burnette et al. 2004; Schade et al. 2006), the phosphoinositide-3-kinase (PI3K)-v-akt murine thymoma viral oncogene homolog (AKT) pathway (Schade et al. 2006), and the mitochondria-related apoptosis pathways (Epling-Burnette et al. 2001). Most of these chronic activations can be attributed to the potential existence of long-term immunostimulation (Loughran 1993) and abnormal AICD of leukemic T-LGL *in vivo*. It has been shown that leukemic T-LGL are not sensitive to Fas-induced apoptosis (Lamy et al. 1998), a process essential for AICD (Krueger et al. 2003). Recent molecular profiling data further suggests that normal CTL activation and AICD are coupled by signaling components that are deregulated in leukemic T-LGL (Shah et al. 2008). In summary, T-LGL leukemia provides a unique opportunity to decipher the key mediators that bridge CTL activation and AICD in humans. This knowledge is essential for identifying potential therapeutic targets for T-LGL leukemia as well as generating long-term competent CTL necessary for tumor and cancer vaccine development.

Despite its importance, studying the signaling abnormalities in T-LGL leukemia is difficult. LGL leukemia is a rare disease (Sokol and Loughran 2006), and there is no animal model or cell line available for study of T-LGL leukemia signaling. The necessary reliance on primary lymphocytes from patients or healthy donors restricts the quantitative and even qualitative information that can be obtained in signaling studies of T-LGL leukemia. On the other hand, signaling perturbations in T-LGL leukemia, as discussed above, involve multiple signaling pathways. It is possible and likely that these perturbations result from deregulation of only a subset of pathway components. However, identifying such a subset is by itself difficult. Thus, in order to guide future signaling investigations in T-LGL leukemia, it is becoming increasingly important to synthesize the results of existing signaling studies as well as to discern the subset of key regulators hiding among them.

8.2.2 Constructing the Survival Signaling Network of T-LGL Leukemia

As the first step toward systematically understanding the long-term survival of leukemic T-LGL, we started to assemble the survival signaling network of T-LGL leukemia with the signaling network of CTL activation–AICD as a framework. The hypothesis behind this was that leukemic T-LGL were likely to arise from abnormal immune responses, and most of the deregulated signaling components were likely to interact with other signaling components as they do in normal CTL activation and AICD signaling. Due to the lack of an existing signaling network to depict signaling events during human CTL activation and AICD, we first constructed the human general CTL activation–AICD signaling network through an extensive literature search. As in many other signaling networks (Christensen, Thakar, and Albert 2007), proteins, mRNAs, and small molecules (such as lipids) were represented as nodes. In addition, "Cytoskeleton signaling," "Proliferation," and "Apoptosis" were included as nodes to summarize the biological effects of a group of components in the signaling pathways and serve as the indicators of cell fate. Interactions among nodes were denoted as edges, and the direction of edges followed the direction of the information flow, from the upstream (source) node to the downstream (product or target) node. Edges were characterized by signs, where a positive sign indicated activation and a negative sign indicated inhibition.

As expected, after constructing the general CTL activation–AICD signaling network, we did indeed find the majority of known deregulations in T-LGL leukemia involved in this network. However, we also noticed that a few deregulations, such as deregulations in platelet derived growth factor (PDGF) and sphingolipid signaling, were not involved. This observation by itself highlighted the topologically unique signal wiring in leukemic T-LGL as compared to normal CTL. To evaluate the relationships between these deregulations and other components normally involved in CTL activation as well as AICD, we augmented the general CTL activation–AICD network with nodes and edges corresponding to these deregulations and their signaling activities. We also included the node "Stimuli" to represent the unknown initial trigger of the CTL activation in T-LGL leukemia (Sokol and Loughran 2006).

To incorporate the most unique interactions through which all known deregulations in leukemic T-LGL were connected as well as to simplify the later simulation of the network dynamics, the augmented network was subjected to network simplification using NET-SYNTHESIS, a signal transduction network inference and simplification tool (Albert et al. 2007; Kachalo et al. 2008). NET-SYNTHESIS constructs the sparsest network that maintains all the causal (upstream-downstream) effects incorporated in a redundant starting network based on combinatorial optimization of graph algorithms for binary transitive reduction and pseudo-vertex collapse. Nodes and edges can be marked as "critical" so they will not be removed during the simplification process. In our study, we marked nodes known to be deregulated in T-LGL leukemia in the augmented network to be critical. We also marked the direct interactions or biochemical reactions between critical nodes as critical. Edges connecting two noncritical nodes are not critical regardless of whether or not they represent direct interactions. It is worth noting that the sparsest network may not be the most realistic network to recapitulate a biological process. Hence, we performed additional adjustment (manual curation) of the NET-SYNTHESIS output to obtain the T-LGL leukemia survival signaling network (Figure 8.1). This network contains 58 nodes and 123 edges.

Here is a brief description of the T-LGL leukemia survival signaling network. "Stimuli" initiates the CTL activation via engaging the T-cell receptor (TCR) complex (Samelson 2002). Consequently, Src kinase family members such as lymphocyte-specific protein tyrosine kinase (LCK) and FYN oncogene related to SRC, FGR, YES (FYN) are triggered (Veillette, Latour, and Davidson 2002) and stimulate the MAPK pathway (Samelson 2002), the PI3K pathway (Kane and Weiss 2003), and transcription factors such as NFκB and nuclear factor of activated T-cells (NFAT) (Feske 2007). These transcription factors facilitate the expression of IL-2 (Hayden, West, and Ghosh 2006; Kane, Lin, and Weiss 2000), amplifying the T-cell activation signal mainly through the JAK-STAT pathway (Leonard and O'Shea 1998). The negative regulation of T-cell activation from ZAP70 and FYN reflects the negative feedback loop that regulates the normal CTL activation at the TCR level. T-cell activation results in clonal expansion and the production of inflammatory factors such as interferon-γ, tumor necrosis factor family members, granzyme B, and perforin (Bouwmeester et al. 2004; Brueckmann et al. 2004; Glimcher et al. 2004; Grove and Plumb 1993; Hoffmann and Baltimore 2006).

Meanwhile, T-cell activation elevates the expression of Fas and Fas ligand, which prepare the competent CTL for AICD (Hoffman et al. 2002; Hsu et al. 1999; Macian 2005; Rengarajan et al. 2000). Fas-induced apoptosis is positively regulated by ceramide

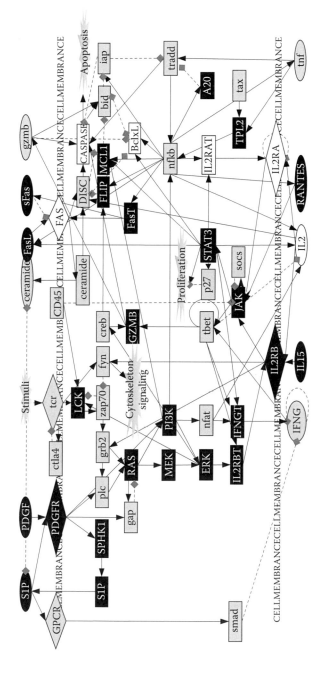

FIGURE 8.1 The T-LGL survival signaling network. Node and edge color and shape represent the current knowledge of the signaling abnormalities in T-LGL leukemia: upregulated or constitutively active nodes are in black; downregulated or inhibited nodes are in white; nodes that have been suggested to be deregulated (either upregulation or downregulation) are in grey with upper-case labels; and the states of grey nodes with lower-case labels are unknown or unchanged compared to normal. Black solid edge indicates activation and grey dashed indicates inhibition. The shape of the nodes indicates the cellular location: rectangular indicates intracellular components; ellipse indicates extracellular components; diamond indicates receptors. Star-shaped conceptual nodes (Stimuli, Cytoskeleton signaling, Proliferation, and Apoptosis) are also included. (Adapted from Zhang et al. 2008.)

(Grassme et al. 2001), and negatively regulated by FLICE inhibitory protein (FLIP) as well as inhibitor of apoptosis proteins (IAP) (Hoffmann and Baltimore 2006). Leukemic T-LGL overexpress soluble Fas (sFas), which inhibits Fas signaling by interfering Fas-FasL interaction (Lamy et al. 1998). In addition to the Fas-induced apoptosis, AICD is also regulated by mitochondrial pathways which are mainly governed by the balance between pro- and anti-apoptotic Bcl-2 family members (Siegel 2006). T-cell activation can be positively regulated by IL-15 (Weng et al. 2002), and negatively regulated by sphingosine-1-phosphate (S1P) through small mothers against decapentaplegic (SMAD) (Xin et al. 2004) as well as by PDGF (Daynes, Dowell, and Araneo 1991).

8.2.3 Studying the Signal Propagation in Motion: A Boolean Dynamic Model Based on the T-LGL Leukemia Survival Signaling Network

Constructing the T-LGL leukemia survival signaling network integrated the existing signaling knowledge about leukemic T-LGL in the context of CTL activation and AICD in the format of a network. However, due to the complexity of this network, it remains difficult to intuitively analyze the causal relationships among the signaling deregulations or the key regulators that determine the long-term survival of leukemic T-LGL. As the second step of systematically understanding the dynamics of survival signaling in T-LGL leukemia, we translated the T-LGL leukemia survival signaling network into a Boolean dynamic model. Each of the nodes was associated with a Boolean function using the three Boolean operators that describe the relationship between the state of the node and the states of the nodes regulating it. As in the biological system, there is a time lag between the state change of the regulators and the state change of the targets in the Boolean model. For example, the Boolean rule "PI3K*= (PDGFR OR RAS) AND NOT Apoptosis" indicates that the next state of PI3K (denoted by PI3K*) will be ON if one of its upstream regulators, PDGF receptor (PDGFR) or RAS, is currently ON, and the state of Apoptosis is currently OFF.

The kinetics of signal propagation are rarely known from experiments in T-LGL leukemia studies. Thus, to equally sample the space of all possible timescales, we used a random-order asynchronous update algorithm (Chaves, Albert, and Sontag 2005; Li, Assmann, and Albert 2006), which samples differences in the speed of signal propagation. In this algorithm, the unit of time (also called timestep) is a round of updating during which all nodes are updated in a randomly selected order. Thus, the timestep corresponds to the longest duration required for a node to respond to a state change of its regulators. The general updating scheme of the random-order asynchronous algorithm is written as

$$S_i^t = B_i\left(S_j^{mj}, S_k^{mk}, S_l^{ml}, \ldots\right),$$

where S_i^t is the state of component i at timestep t, B_i is the Boolean function associated with the node i and its regulators $j, k, l, \ldots$ and $mj, mk, ml, \ldots \in (t-1, t)$ are the timesteps when the last status change occurred for the regulators, which can be either the current or previous timestep.

To reproduce how a population of cells responds to the same signal, and to simulate cell to cell variability, we performed multiple simulations with the same initial conditions but different updating orders (i.e., different timing). The model was allowed to update for multiple rounds until the node "Apoptosis" became ON in all simulations (recapitulating the death of all CTL) or stabilized in the OFF state in a fraction of simulations (recapitulating the stabilization of the long-term surviving CTL population). The frequency of activation for each node, that is, the fraction of simulations where the node is ON, represents the trends of signaling events in the cell population studied. The onset of apoptosis removes cells from the CTL population. This effect was reproduced by terminating simulations in which the node "Apoptosis" was activated and obtaining the frequency of node activation from ongoing simulations only. The frequency of "Apoptosis" activation, however, is obtained from the whole population as in the beginning of simulations.

Before simulation, the states of nodes were set according to their states in resting CTL. The state of Stimuli was set to ON at the beginning of every simulation, recapitulating the activation of CTL by an antigen. At the end of the simulation, if the state of a node stabilized at ON although it was in the OFF state at the beginning of the simulation, we consider it as constitutively active. If the state of a node stabilized at OFF although it was in the ON state at the beginning of the simulation, or it was experimentally shown to be active after normal CTL activation, we consider it as downregulated/inhibited. During simulations, the state of a node can be fixed to reproduce signaling perturbations.

8.2.4 Revealing the Causal Relationships among the Known Signaling Abnormalities in T-LGL Leukemia

To investigate signaling abnormalities underlying the long-term survival of the leukemic T-LGL, we first tested whether our model could reproduce the uncoupling of CTL activation and AICD using all known deregulations. To reproduce these deregulations, we simultaneously set all known deregulations according to their state in T-LGL leukemia (either ON or OFF) throughout the simulations and tracked the state of "Apoptosis." "Apoptosis" turned OFF in all simulations after the first timestep. This suggests that the known deregulations are sufficient to reproduce the long-term survival of leukemic T-LGL in the model.

To determine the causal relationships within the known deregulations, we simulated the T-LGL survival signaling network by constantly setting the state of the known deregulated nodes individually and examined the other nodes that reached steady (time-independent) states during simulation. IL-15, PDGF, and Stimuli are three nodes that have been suggested to be abnormal in T-LGL leukemia without known upstream regulators in the T-LGL survival signaling network. To recapitulate the effect of their deregulations without masking the effect of the perturbation tested, the states of IL-15, PDGF, and Stimuli were randomly set at ON or OFF at every round of updating, that is, in a random state, except when probing for the effect of their own deregulations.

After setting the state S_i of node i, nodes achieving steady states (termed a fixed point in dynamical systems terminology) consistent with what is experimentally observed in T-LGL leukemia correspond to nodes whose state is determined by the deregulation of node i. All

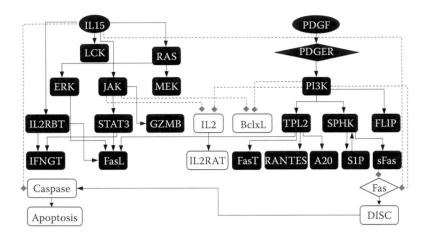

FIGURE 8.2 Hierarchy among known signaling deregulations in T-LGL leukemia. Color code for nodes and edges is the same as in Figure 8.1. (Adapted from Zhang et al. 2008.)

the nodes that reach their T-LGL-like fixed points due to the deregulation of node i form the induced node set of i, written as I_j. One can represent this causal relationship by $j \Rightarrow I_j$. For nodes a and b, if $I_a \subset I_b$ (the induced node set of a is a subset of the induced node set of b) and $a \in I_b$ (a belongs to the induced node set of b), the causal relationship between nodes b, a and I_a can be compressed into $b \Rightarrow a \Rightarrow I_a$. We present these relationships as a hierarchical network in Figure 8.2, with upstream deregulations as the potential cause of downstream deregulations.

These causal relationships among known deregulations, which are difficult to address through individual experiments, are particularly important to identify potential mechanisms of pathogenesis. Surprisingly, we found that keeping the state of IL-15 at ON was sufficient to reproduce all known deregulations in leukemic T-LGL when setting a random state for PDGF and Stimuli. To understand the effect of PDGF and Stimuli individually upon the constant presence of IL-15, we probed all the possible states of PDGF and Stimuli. We determined that the presence of PDGF is needed for the long-term survival of leukemic T-LGL. In contrast, the presence of Stimuli is not required after its initial activation (Figure 8.3). We concluded that, based on the available signaling information regarding T-cell activation and AICD, the minimal condition required for our model to reproduce all known signaling abnormalities in T-LGL leukemia (i.e., a leukemic-T-LGL-like state) is IL-15 constantly ON, PDGF intermittently ON, and Stimuli ON in the initial condition.

IL-15 has been shown to be important for CTL activation and generation of long-lived CD8+ memory cells (Liu et al. 2002; Weng et al. 2002). On the other hand, the need of PDGF for CTL long-term survival was not reported in humans. To validate this surprising prediction, we experimentally tested the effect of inhibiting PDGF signaling on the survival of leukemic T-LGL. Indeed, a specific chemical inhibitor of PDGFR effectively induced apoptosis in T-LGL leukemia patient PBMC but not in PBMC from healthy donors. This finding confirmed the possibility that the constitutive presence of IL-15 and PDGF is an upstream cause that can explain all known signaling abnormalities in T-LGL leukemia.

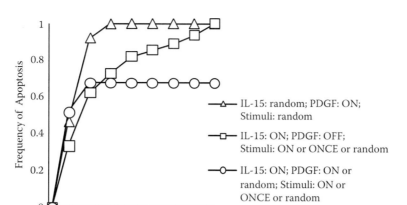

FIGURE 8.3   The Boolean model of the T-LGL survival signaling network predicts that the con-stitutive presence of IL-15 and PDGF is sufficient to induce all the known deregulations in T-LGL leukemia. The effects of IL-15, PDGF, and stimuli on the frequency of apoptosis during simulation were shown. Keeping PDGF ON does not prevent the onset of apoptosis (△). While keeping IL-15 ON, keeping PDGF OFF from the first round of updating delays but cannot prevent the onset of apoptosis (□). Setting Stimuli ON at the beginning of the simulation and then keeping it OFF ("ONCE") does not alter the inhibition of apoptosis upon keeping IL-15 ON in the presence of PDGF (○). (Adapted from Zhang et al. 2008.)

In addition, it suggested that provision of IL-15 and PDGF may be a strategy to generate long-lived CTL necessary for the development of virus and cancer vaccines.

8.2.5 Identifying Potential Key Regulators That Contribute to Long-term Survival of Leukemic T-LGL and Predicting Signaling Events

By constantly setting only two nodes, IL-15 and PDGF, in the ON state, we dynamically reproduced a leukemic-T-LGL-like state. This provided us a platform to examine the impor-tance of network components in determining the long-term survival of leukemic T-LGL. In experiments, the importance of a protein, small molecule, or complex in the survival of leukemic T-LGL can be evaluated by reducing its amount or interrupting its function and tracking the changes of apoptosis. In our model, the importance of a corresponding node can be evaluated by resetting and maintaining its state. For example, if the node would normally stabilize at ON, we reset its state to OFF and vice versa. We can then track the change in the activation frequency of the node "Apoptosis." Since during the *in vitro* experiments the assumption is that only the manipulated components were changed dur-ing treatment while the other components remain in their *in vivo* status, we maintained the setting of "IL15," "PDGF," and "Stimuli" as ON.

Intuitively, a key regulator of leukemic T-LGL survival should fulfill two criteria during model simulation. First, its state stabilizes once a leukemic-T-LGL-like state is achieved. Second, altering its state increases the activation frequency of "Apoptosis." Based on these criteria, we systematically simulated the effect of individually altering the states of all nodes that stabilize when a leukemic-T-LGL-like state is achieved. Out of 58 nodes in the network, 18 nodes were suggested to be key nodes for the survival of leukemic T-LGL. Among them, 11 were experimentally identified in previous studies (including PDGF and PDGFR), and 7 were novel predictions (SPHK1, NFκB, S1P, SOCS, GAP, BID, and IL2RB). As model verification, we tested 2 of the 7 predicted key regulators: sphingosine kinase 1 (SPHK1) and nuclear factor kappa-B (NFκB). Specific chemical inhibitors for SPHK1 and NFκB induced apoptosis in T-LGL leukemia patient PBMC but not in PBMC from healthy donors, validating the model's predictions.

In addition to predicting key regulators, our model is also capable of predicting additional signaling pathways that can connect and explain the known deregulations in T-LGL leukemia. As shown in Figure 8.4, the model predicted that NFκB functions downstream

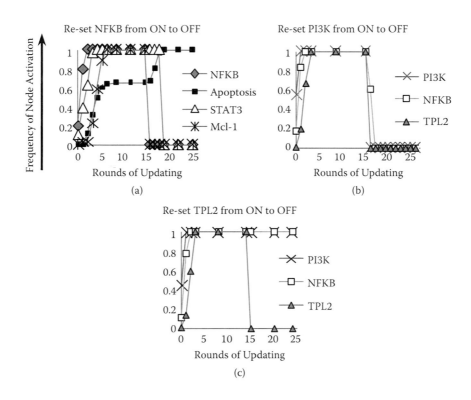

FIGURE 8.4 Examples of model predictions. (a) Model predictions of the effects of NFκB inhibition. Apoptosis (■) was rapidly induced after inhibiting NFκB (♦). The induction of apoptosis was tightly coupled with the downregulation of Mcl-1 (✳). In contrast, the state of STAT3 (Δ) remained unchanged until the simulation was terminated. (b, c) Analysis of the potential cause(s) of the constitutive activation of NFκB. The Boolean logical rule governing the state of NFκB is "NFKB*= ((TPL2 or PI3K) or (FLIP and TRADD and IAP)) and not Apoptosis". When a T-LGL-like state is achieved, the state of TRADD stabilizes at OFF. Thus the node that activates NFκB can only be TPL2 or PI3K, which are known to be constitutively active in T-LGL leukemia. Rapid inhibition of NFκB (□) was observed after inhibiting PI3K (X) but not after inhibiting TPL2 (▲). Figure is adapted from Zhang et al. (2008).

of PI3K and prevents apoptosis through maintaining the expression of myeloid cell leukemia sequence 1 (Mcl-1). In addition, the initial Mcl-1 downregulation upon NFκB inhibition was predicted to be independent of STAT3 activity, another transcriptional factor that was shown to regulate Mcl-1 in leukemic T-LGL (Epling-Burnette et al. 2001). We validated these predictions experimentally.

8.2.6 Boolean Modeling: You Can Do It, We Can Help

Our work on the signaling pathways in T-LGL leukemia illustrates the benefits of network-based dynamic modeling. Summarizing signaling information from cancer and its normal counterparts in the form of a graph/network, such as what we did for survival signaling pathways in T-LGL leukemia, facilitates knowledge integration and visualization. Translating the network into a Boolean model enables the initial systematic analysis of the signaling dynamics. The simulation process might appear less accessible to biologists. To overcome this obstacle, software tools are rapidly being developed to support modeling attempts at different levels. Network structures can be visualized by using software tools such as Graphviz (http://www.graphviz.org/), Cytoscape (http://www.cytoscape.org/), and yEd (http://www.yworks.com/en/products_yed_about.html). Some of these tools will be further reviewed in Part 3 of this book. Suites such as ProMoT (Saez-Rodriguez et al. 2006) and CellNetAnalyzer (Klamt, Saez-Rodriguez, and Gilles 2007) minimize the modeling efforts to just inputing the signaling network. They can automatically extract regulatory relationships depicted in the signaling network and offer outputs such as the final state based on given inputs, node state dependency (whether the state of one node is determined by the state of another node), and minimum node set to achieve a desired final state in the model. As a common drawback of comprehensive software tools, these programs usually have limited capacity for customization.

An intermediate modeling path is offered to biologists with a certain programming experience. Toolboxes that fall into this category provide basic "building blocks" for Boolean modeling, such as synchronous updating (the states of all network nodes are updated at the same time) and asynchronous updating (as we used in modeling the T-LGL leukemia survival signaling network). With intuitive text input of Boolean rules for network nodes and initial node states, network dynamics can be analyzed as desired by calling different simulation "building blocks" individually or in combination. An example of these toolboxes is BooleanNet (Albert et al. 2008), which serves as a mini-language under the programming language Python (http://www.python.org). Given the Boolean rules for individual network nodes, it is capable of analyzing network dynamics upon a given node-state input (including the constitutive activation or inhibition of certain nodes) using different simulation algorithms. It can also integrate available time-scale information by the use of piecewise linear differential equations. Although having a slightly steeper learning curve compared to the comprehensive modeling software tools, this type of toolbox greatly increases the flexibility and exploratory capacity of the model and provides a nice alternative to model simulation entirely by custom-made code.

8.3 DISCUSSION

8.3.1 Boolean Modeling in Cancer Systems Biology: A Way of Thinking in Addition to a Way of Modeling

Qualitative data are usually obtained from experiments in human cancer research, and their suitability for systematic level studies has been seriously questioned. Our study serves as an example that as long as we ask the right questions and use the right tools, qualitative data can be used to dynamically model complex signaling processes. In our study of a rare human cancer, T-LGL leukemia, Boolean modeling was proven to be powerful in predicting clinically relevant signaling events. It greatly facilitated our understanding of the pathogenesis and highlighted potential therapeutic targets in this leukemia. In addition, it provided feasible strategies for generating long-term competent CTL, which are necessary for tumor and cancer vaccine development. Moreover, in light of the lack of cell lines and animal models for T-LGL leukemia, Boolean modeling is likely to be the only way through which the existing qualitative data can be summarized and dynamically analyzed.

However, the power of Boolean modeling to benefit cancer systems biology studies may well exceed what was shown from the perspective of modeling. Conventional signaling studies rely heavily on intuitive pathway analysis. Researchers derive their hypotheses and deduce the causal relationships among pathway components from looking at the signaling pathway diagrams. The misconception is that signaling pathway dynamics are easy to predict as long as the pathways are well defined. A simple reality check shows that this cannot be further from the truth: in a signaling network containing n nodes, the number of potential initial states for signaling propagation is 2^n. It will not take more than half a dozen initial conditions to cause one to lose count, not to mention keeping track of the network dynamics after each signaling event. Instead of "staring and wondering," Boolean formalism is one of the most natural ways to translate a signaling diagram into a dynamic model, through which the effects of different initial conditions, interrupted signaling components, and node state dependency can be evaluated. Taken together, Boolean modeling may benefit wet-bench signaling studies if used as a routine research tool to aid hypothesis formulation. This is particularly true in light of the intuitiveness of the Boolean rules and availability of various modeling tools.

8.3.2 Boolean Modeling: Limitations

As all modeling tools, Boolean modeling has its own limitations. By nature, the efficacy and accuracy of Boolean models are heavily influenced by what we know now. Boolean models established from signaling data can be informative only if the amount of data is relatively abundant. Otherwise, it is difficult to bring insights into signaling dynamics beyond intuitive analysis. Boolean models are biased toward our current knowledge. A downside of Boolean modeling, as compared to large-scale high-throughput studies, is that it is difficult to predict the involvement or dynamics of a completely new pathway by Boolean models if no previous information is available. The sensitivity of Boolean modeling depends on whether qualitative input changes can induce qualitative output changes. Despite attempts at incorporating quantitative changes and time-scale differences into

Boolean models (Albert et al. 2008; Klamt, Saez-Rodriguez, and Gilles 2007), at present only limited signaling details (rates, kinetic parameters) can be reliably incorporated compared to differential equation models.

Moreover, as a supervised approach, Boolean modeling has been accused of subjectivity. Signaling components are usually affected by more than one upstream regulator, and the relationships among them may not be well studied. With the lack of experimental data, researchers are required to make an "educated guess" about the usage of Boolean operators when modeling. Suspicions may arise regarding the potential practice of altering the use of Boolean operators in order to obtain "desired" model predictions. In addition to emphasizing academic ethics, this concern can be addressed by testing the robustness of the Boolean models proposed. As models for signaling networks that are biologically robust, Boolean models are expected to exhibit robustness against small perturbations in model composition, such as altering a small portion of Boolean rules in the models. The accuracy of Boolean models may be questioned only if they are fragile even under the slightest rule changes. In fact, the involvement of decision making from researchers is a unique feature of Boolean modeling, which allows it to facilitate our intuitive understanding of biological systems but not to replace it.

In summary, Boolean modeling is a unique tool in systems biology. It is particularly useful in facilitating cancer systems biology studies because of its compatibility with qualitative data. Despite its relative coarseness compared to more sophisticated modeling frameworks such as differential equations, Boolean modeling is an ideal starting point for summarizing existing conventional signaling studies, for systematically analyzing signaling dynamics under a specified condition such as a particular cancer, for guiding future experiments, and for serving as the basis of more sophisticated models. Because of its robustness and straightforwardness, Boolean modeling has the potential to become a platform that directly facilitates the dialogue between wet- and dry-bench researchers. It provides a path through which conventional signaling studies and systems biology can be bridged.

REFERENCES

Albert I., Thakar J., Li S., Zhang R., and Albert R. 2008. Boolean network simulations for life scientists. *Source Code for Biology and Medicine* 3:16.

Albert R., DasGupta B., Dondi R. et al. 2007. A novel method for signal transduction network inference from indirect experimental evidence. *J. Comput. Biol.* 14:927–949.

Altan-Bonnet G. and Germain R. N. 2005. Modeling t cell antigen discrimination based on feedback control of digital ERK responses. *PLoS Biology* 3:e356.

Bornholdt S. 2008. Boolean network models of cellular regulation: prospects and limitations. *J. R. Soc. Interface* 5:S85–S94.

Bouwmeester T., Bauch A., Ruffner H. et al. 2004. A physical and functional map of the human TNF-[alpha]/NF-[kappa]B signal transduction pathway. *Nature Cell Biol* 6:97–105.

Brueckmann M., Hoffmann U., Dvortsak E. et al. 2004. Drotrecogin alfa (activated) inhibits NF-kappa B activation and MIP-1-alpha release from isolated mononuclear cells of patients with severe sepsis. *Inflammation Research* 53:528–533.

Chaves M., Albert R., and Sontag E. D. 2005. Robustness and fragility of Boolean models for genetic regulatory networks. *J. Theor. Biol.* 235:431–449.

Christensen C., Thakar J., and Albert R. 2007. Systems-level insights into cellular regulation: inferring, analysing, and modelling intracellular networks. *IET Syst Biol* 1:61–77.

Davidich M. I. and Bornholdt S. 2008. Boolean network model predicts cell cycle sequence of fission yeast. *PLoS ONE* 3:e1672.

Daynes R. A., Dowell T., and Araneo B. A. 1991. Platelet-derived growth factor is a potent biologic response modifier of T cells. *J. Exp. Med.* 174:1323–1333.

Epling-Burnette P. K., Bai F., Wei S. et al. 2004. ERK couples chronic survival of NK cells to constitutively activated Ras in lymphoproliferative disease of granular lymphocytes (LDGL). *Oncogene* 23:9220–9229.

Epling-Burnette P. K., Liu J. H., Catlett-Falcone R. et al. 2001. Inhibition of STAT3 signaling leads to apoptosis of leukemic large granular lymphocytes and decreased Mcl-1 expression. *J. Clin. Invest.* 107:351–362.

Eungdamrong N. J. and Iyengar R. 2004. Modeling cell signaling networks. *Biol Cell* 96:355–362.

Feske S. 2007. Calcium signalling in lymphocyte activation and disease. *Nature Rev Immunol* 7:690–702.

Gilman A. G. 1987. G proteins: transducers of receptor-generated signals. *Annu Rev Biochem* 56:615–649.

Glass L. and Kauffman S. A. 1973. The logical analysis of continuous, non-linear biochemical control networks. *J. Theor. Biol.* 39:103–129.

Glimcher L. H., Townsend M. J., Sullivan B. M., and Lord G. M. 2004. Recent developments in the transcriptional regulation of cytolytic effector cells. *Nature Rev Immunol* 4:900–911.

Grassme H., Jekle A., Riehle A. et al. 2001. CD95 signaling via ceramide-rich membrane rafts. *J. Biol. Chem.* 276:20589–20596.

Grove M. and Plumb M. 1993. C/EBP, NF-kappa B, and c-Ets family members and transcriptional regulation of the cell-specific and inducible macrophage inflammatory protein 1 alpha immediate-early gene. *Mol Cell Biol* 13:5276–5289.

Hardman J. G., Robison G. A., and Sutherland E. W. 1971. Cyclic nucleotides. *Annu Rev Physiol* 33:311–336.

Hayden M. S., West A. P., and Ghosh S. 2006. NF-[kappa]B and the immune response. *Oncogene* 25:6758–6780.

Hoffman B., Amanullah A., Shafarenko M., and Liebermann D. A. 2002. The proto-oncogene c-myc in hematopoietic development and leukemogenesis. *Oncogene* 21:3414–3421.

Hoffmann A. and Baltimore D. 2006. Circuitry of nuclear factor{kappa}B signaling. *Immunol Rev* 210:171–186.

Hsu S.-C., Gavrilin M. A., Lee H.-H. et al. 1999. NF[kappa]B-dependent Fas ligand expression. *Eur. J. Immunol.* 29:2948–2956.

Kachalo S., Zhang R., Sontag E., Albert R., and DasGupta B. 2008. NET-SYNTHESIS: a software for synthesis, inference and simplification of signal transduction networks. *Bioinformatics* 24:293–295.

Kane L. P., Lin J., and Weiss A. 2000. Signal transduction by the TCR for antigen. *Curr Opin Immunol* 12:242–249.

Kane L. P. and Weiss A. 2003. The PI-3 kinase/Akt pathway and T cell activation: pleiotropic pathways downstream of PIP3. *Immunol Rev* 192:7–20.

Kaufman M., Andris F., and Leo O. 1999. A logical analysis of T cell activation and anergy. *Proc Natl Acad Sci USA* 96:3894–3899.

Klamt S., Saez-Rodriguez J., and Gilles E. 2007. Structural and functional analysis of cellular networks with CellNetAnalyzer. *BMC Syst Biol* 1:2.

Klebanoff C. A., Gattinoni L., and Restifo N. P. 2006. CD8+ T-cell memory in tumor immunology and immunotherapy. *Immunol Rev* 211:214–224.

Krueger A., Fas S. C., Baumann S., and Krammer P. H. 2003. The role of CD95 in the regulation of peripheral T-cell apoptosis. *Immunol Rev* 193:58–69.

Lamy T., Liu J. H., Landowski T. H., Dalton W. S., and Loughran T. P., Jr. 1998. Dysregulation of CD95/CD95 ligand-apoptotic pathway in CD3+ large granular lymphocyte leukemia. *Blood* 92:4771–4777.

Leonard W. J. and O'Shea J. J. 1998. JAKs and STATs: biological implications. *Ann Rev Immunol* 16:293–322.

Li S., Assmann S. M., and Albert R. 2006. Predicting essential components of signal transduction networks: a dynamic model of guard cell abscisic acid signaling. *PLoS Biology* 4:e312.

Liu K., Catalfamo M., Li Y., Henkart P. A., and Weng N.-P. 2002. IL-15 mimics T cell receptor cross-linking in the induction of cellular proliferation, gene expression, and cytotoxicity in CD8+ memory T cells. *Proc Natl Acad Sci USA* 99:6192–6197.

Loughran T. P., Jr., Kadin M., Starkebaum G. et al. 1985. Leukemia of large granular lymphocytes: association with clonal chromosomal abnormalities and autoimmune neutropenia, thrombo-cytopenia, and hemolytic anemia. *Ann Internal Med* 102:169–175.

Loughran T. P., Jr. 1993. Clonal diseases of large granular lymphocytes. *Blood* 82:1–14.

Ma'ayan A. 2008. Network integration and graph analysis in mammalian molecular systems biology. *Syst Biol, IET* 2:206–221.

Ma'ayan A., Jenkins S. L., Neves S. et al. 2005. Formation of regulatory patterns during signal propagation in a mammalian cellular network. *Science* 309:1078–1083.

Macian F. 2005. NFAT proteins: key regulators of T-cell development and function. *Nature Rev Immunol* 5:472–484.

Rengarajan J., Mittelstadt P. R., Mages H. W. et al. 2000. Sequential involvement of NFAT and EGR transcription factors in FasL regulation. *Immunity* 12:293–300.

Rodbell M. 1995. Signal transduction: evolution of an idea. *Environ Health Perspect* 103:338–345.

Rose M. G. and Berliner N. 2004. T-cell large granular lymphocyte leukemia and related disorders. *Oncologist* 9:247–258.

Russell J. H. and Ley T. J. 2002. Lymphocyte-mediated cytotoxicity. *Annu Rev Immunol* 20:323–370.

Saez-Rodriguez J., Mirschel S., Hemenway R. et al. 2006. Visual setup of logical models of signaling and regulatory networks with ProMoT. *BMC Bioinformatics* 7:506.

Saez-Rodriguez J., Simeoni L., Lindquist J. A. et al. 2007. A logical model provides insights into T cell receptor signaling. *PLoS Comput Biol* 3:e163.

Samelson L. E. 2002. Signal transduction mediated by the T cell antigen receptor: the role of adapter proteins. *Annu Rev Immunol* 20:371–394.

Schade A. E., Powers J. J., Wlodarski M. W., and Maciejewski J. P. 2006. Phosphatidylinositol-3-phosphate kinase pathway activation protects leukemic large granular lymphocytes from undergoing homeostatic apoptosis. *Blood* 107:4834–4840.

Shah M. V., Zhang R., Irby R. et al. 2008. Molecular profiling of LGL leukemia reveals role of sphin-golipid signaling in survival of cytotoxic lymphocytes. *Blood* 112:770–781.

Siegel R. M. 2006. Caspases at the crossroads of immune-cell life and death. *Nature Rev Immunol* 6:308–317.

Sokol L. and Loughran T. P., Jr. 2006. Large granular lymphocyte leukemia. *Oncologist* 11:263–273.

Timonen T., Ortaldo J. R., and Herberman R. B. 1981. Characteristics of human large granular lym-phocytes and relationship to natural killer and K cells. *J. Exp. Med.* 153:569–582.

Veillette A., Latour S., and Davidson D. 2002. Negative regulation of immunoreceptor signaling. *Annu Rev Immunol* 20:669–707.

Weng N.-P., Liu K., Catalfamo M., Li Y. U., and Henkart P. A. 2002. IL-15 is a growth factor and an activator of CD8 memory T cells. *Ann NY Acad Sci* 975:46–56.

Xin C., Ren S., Kleuser B. et al. 2004. Sphingosine 1-phosphate cross-activates the SMAD signal-ing cascade and mimics transforming growth factor-β-induced cell responses. *J. Biol. Chem.* 279:35255–35262.

Zhang R., Shah M. V., Yang J. et al. 2008. Network model of survival signaling in large granular lym-phocyte leukemia. *Proc Natl Acad Sci USA* 105:16308–16313.

Cancer Metabolic Networks

Metabolic Pathways Modeling and
Metabolomics in Cancer Research

Miroslava Cuperlovic-Culf

CONTENTS

9.1 INTRODUCTION

Several hypotheses are being explored in an effort to understand the causes of malignant transformation. Currently, a principal view is that the formation of a cancer cell requires a series of mutations in the nuclear DNA sequence of its ancestral cell, either giving rise to oncogenes or impairing the action of tumor suppressor genes. Common features of cancer cells thus developed are dynamic changes in the genome, that is, genomic instability leading to a large variety of clones (Griffiths and Stubbs 2005; Hanahan and Weinberg 2000). This cellular diversity gives a basis for the selection of the most aggressive and resilient subpopulation. In other words, inherent genetic flexibility of these mutated cells allows cancer cells to progressively evolve functions that promote cell growth, disable cell death mechanisms, and evade immune surveillance and therapy. With more than 100 distinct types of cancer and many more subtypes, the starting point as well as the path that a cell can take on its way to become malignant are highly variable. However, in all cancers the endpoints that are ultimately reached appear to be the same and are termed hallmarks of cancer (Hanahan and Weinberg 2000). They include: (1) self-sufficiency in growth signals, (2) insensitivity to antigrowth signals, (3) ability to evade apoptosis, (4) limitless replicative potential and sustained angiogenesis, (5) tissue invasion and metastasis, (6) genome instability, and, recently added, (7) an inflammatory microenvironment (Mantovani 2009) as well as (8) cancer metabolic phenotype (Young and Anderson 2008). Regulation of cellular processes such as growth, replication, and apoptosis, which are all closely linked with metabolic processes, is the ultimate concern of the majority of oncogenes and oncogenic mutations. Most of these oncogenic mutations are clustered in a few pathways across different cancers. Therefore, the ultimate effects on the metabolism are similar across different cancers and this is described as the cancer metabolic phenotype. Altered metabolism of cancer cells gives them an advantage in survival and proliferation and is a necessary part of cancer development. This unique metabolic phenotype is in general characterized by (1) high glucose uptake, (2) increased glycolytic activity, (3) decreased mitochondrial activity for energy production, (4) low bioenergetic expenditure, (5) increased phospholipid turnover, altered lipid profile, and increase of *de novo* lipid synthesis, (6) increased amino acid transport and protein as well as DNA synthesis, (7) increased hypoxia (a pathological condition in which the body as a whole or a region of the body is deprived of adequate oxygen supply), (8) increased tolerance to reactive oxygen species (ROS), that is, highly reactive ions or small molecules with an unpaired valence shell electron.

The increased metabolic needs as well as altered metabolic pathways of cancer cells are regularly utilized in clinical practice for diagnosis using positron emission tomography (PET) and single-photon emission computed tomography (SPECT) as well as noninvasive magnetic resonance imaging (MRI) and the related method of magnetic resonance spectroscopy (MRS) (DeGraaf 2007; Vallabhajosula 2007). Some molecules used for PET and MRS in relation to the general metabolic changes observed in cancers are shown in Table 9.1. These diagnostic applications clearly show consistent alterations in cancer cells in energy production, lipid synthesis and turnover, amino acid and protein synthesis, as well

TABLE 9.1 Examples of PET Tracers and NMR or MRS Observed Metabolites Utilized for the Diagnosis of Different Cancers

Biochemical Process	PET Tracer	NMR Observed Metabolites
Energy metabolism: Glycolysis; incomplete TCA	[19F]Fluorodeoxyglucose	Citrate; glucose; acetate; glutamine; creatine; lactate; pyruvate; succinate
Membrane and lipid synthesis	[11C]Choline; [18F]Flurocholine; [18F]Fluoroacetate	Choline containing compounds; glycerol; various lipids; triglycerides; creatine
DNA synthesis	[11C]Thymidine; [18F]Fluorothymidine	
Amino acid transport and protein synthesis	[18F]FDOPA[1] [11C]-L-methionine; [18F]FMT[2]	Alanine, phenyalanine; threonine; tryptophan; valine; glycine; aspargine; aspartate; leucine; glutamate; glutamine; tyrosine; histidine
Hypoxia	[18F]FMISO[3]	

[1] [18F]FDOPA, 3,4-Dihydroxy-6-[18F]fluoro-l-phenylalanine.
[2] [18F]FMT, [18F]-l-*m*-tyrosine (FMT).
[3] [18F]FMISO, fluoromisonidazole.
Source: Data from DeGraaf (2007) and Vallabhajosula (2007).

as DNA synthesis. The causes and the consequences of these metabolic changes in cancers are still a point of contention primarily due to the still prevailing attempts to describe the changes individually rather than describing the cancer metabolic phenotype in a more holistic manner. The development of high-throughput approaches in biology as well as the increased popularity of quantitative systems biology analysis provide new methodologies for a more complete and clear understanding of both the causes and the consequences of the cancer metabolic alterations. Systems biology in this context involves an iterative interplay between more or less high-throughput and high-content "wet" experiments, data analysis, and theoretical and computational modeling (Alon 2007; Kell and Knowles; Kell 2006; Palsson 2006). The potential for improvement in treatment and diagnosis through increased understanding of the significance of metabolic changes in cancers makes systems biology analysis of cancer metabolism necessary and timely. Several different levels of analysis are needed for the description of these systems. On the first level, it is necessary to quantitatively measure changes in different metabolites. For many years measurements of metabolite concentrations and fluxes as well as enzyme kinetics were performed primarily in focused-hypothesis driven and often ex vivo experiments. Data provided by these approaches is a crucial starting point. However, ex vivo enzyme kinetics measured under optimal rather than physiological conditions cannot result in the most accurate information. Novel high-throughput, nonselective measurements provided by omics methodologies can supply *in vivo* data about a larger number of molecules in parallel. Although omics technologies are complementary, analysis of the metabolome (termed metabolomics) is an especially useful approach for identifying actual metabolic network transformations in pathologies. Metabolomics measurements can be performed *in vivo*, ex vivo, and *in vitro*, in body fluids as well as tissues and cell lines with different foci. The results of a metabolomics experiment can be analyzed independently or in conjunction with other omics methods, leading to a range of system-level information.

The next layer of systems biology analysis of metabolic changes in tumors involves the development of mathematical and computer models of changes in metabolic pathways and networks that can be exploited in different cellular situations, environments, and treatments. A mathematical model is an important tool used to integrate different data leading to an understanding of the system. Quantitative models of individual pathways and the complete metabolic network are based on various data, including qualitative metabolic pathways, that is, wiring diagrams, as well as enzyme kinetic information and concentration measurements for the molecules involved. Metabolism is subject to thermodynamic and stoichiometric constraints and there is a large body of data describing enzyme biochemistry, kinetics, and thermodynamics (as in Rojas 2007). This information provides a good starting point for the development of models that can subsequently be perfected based on new information.

Several texts provide detailed descriptions of general procedures used both in metabolomics and quantitative modeling and this will not be the focus of this chapter. Rather, we will provide an overview of the current knowledge of the metabolic network changes observed in cancers, followed by a description of the application of metabolomics analysis as well as metabolic pathway and network modeling and analysis that is related to cancer research. Although systems biology can be viewed in terms of the global analysis of both cellular systems and complete organisms, our primary focus in this chapter will be on the description of a cancer cell.

9.2 METABOLIC CHANGES IN CANCER

The metabolism of a cancer cell differs from normal cell metabolism in terms of the rate and the avenue for energy production, biosynthesis of lipids and other macromolecules. Virtually all cancers show metabolic changes that result in upregulated glycolysis and glucose consumption as well as upregulated and altered lipid, protein, and nucleotide synthesis. In the 1920s, Otto Warburg published the seminal observation that tumor cells consume glucose at a surprisingly high rate compared to normal cells. Further, in these early experiments it was also observed that cancer cells secrete most of the glucose-derived carbon as lactate rather than oxidizing it completely in mitochondrial respiration (Warburg 1925, 1956). The phenomenon is since termed the Warburg effect and it describes the anaerobic glycolysis of cancers. This anaerobic glycolysis is a highly inefficient, wasteful form of energy generation in comparison to mitochondrial respiration and, although in normal circumstances it only happens under hypoxia, in cancers it appears to be present even in an oxygenated environment (DeBerardinis et al. 2008a). Currently the prevalent hypothesis is that glycolysis is the major source of energy in cancer cells. Recent experimental and theoretical analysis of tumors *in vivo* (Griffiths and Stubbs 2005) show that even though glucose intake is indeed highly increased in cancer tissues, the production of ATP is higher than can be provided only through glycolysis. Sonveaux and co-workers (2008) have recently proposed a very interesting explanation for this apparent disagreement between observed high glucose intake in tumors both *in vivo* and *in vitro* and the observed discrepancy in ATP production *in vivo*. In this work the authors performed systematic analysis of the complete tumor tissue. They observed that hypoxic tumor cells do indeed follow the Warburg scenario and produce large amounts of lactate (2 mol for each 1 mol of glucose). The surrounding, well oxygenated, tumor cells that are adjacent to blood vessels appear to have a very different metabolic behavior. It was observed that these cells express proteins which allow them to take up lactate (e.g., monocarboxylate transporters, MCT) and use it in the presence of O_2 as their principal substrate for mitochondrial oxidative phosphorylation, generating in the process 36 mol of ATP per 2 mol of lactate. This "metabolic symbiosis" appears crucial for cancer and thus provides a new avenue for treatment. In fact, experiments on mice have shown that the inhibition of MCT significantly increased radiation-induced tumor retardation.

Apart from quenching energy needs, a replicating cell must duplicate its genome, proteins, and lipids and assemble the components into daughter cells, while at the same time avoiding apoptosis. For this, the cancer cell must take up extracellular nutrients like glucose and glutamine as well as essential amino acids and use them in metabolic pathways that convert them into biosynthetic precursors (DeBerardinis et al. 2008b). Tumor cells achieve this in a self-sufficient way while being insensitive to outside signals and nutrient concentrations (Hanahan and Weinberg 2000) through changes in the expression, activation, and sequence of enzymes that determine metabolic flux rates (Dang et al. 1997; DeBerardinis et al. 2008a, 2008b; Rashid et al. 1997).

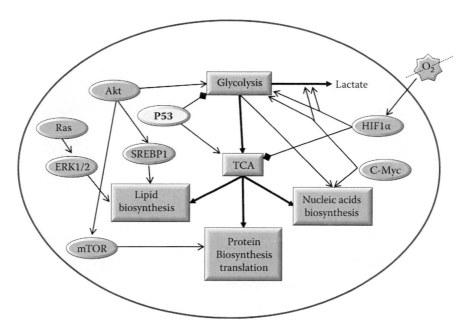

FIGURE 9.1 Schematic representation of some known effects of oncogenes and tumor suppressors on metabolic processes in cancer. Oncogenes are in white letters with dark grey background; tumor suppressors are in black letters with light background. Metabolic processes are represented by boxes.

With many different molecular changes that are happening during oncogenesis the ongoing challenge in describing tumor cell metabolism is to understand how individual pathways fit together into the global metabolic phenotype of the cell and how the metabolic changes are related to the other oncogenic processes. Although the answers to these questions are still not available, a schematic of the connection between the known oncogenes and the metabolic changes is starting to emerge. A brief summary of current knowledge related to the effects of major oncogenes and tumor suppressors on the cancer metabolome is given in Figure 9.1. One of the initial triggers leading to carcinogenic metabolism appears to be the low oxygen environment of the developing cancer cell, that is, hypoxia, caused by insufficient blood supply to the fast growing cancer tissue. The cell's primary answer to hypoxia is the activation of hypoxia-inducible transcription factor 1 (HIF-1), which is involved in transcription of many regulatory genes. In addition, particularly in cancers, HIF-1 accumulation can be caused by factors other than hypoxia. Known cancerogenic mutations that activate the mTOR pathway (PTEN, TSC2, PL3K$_\alpha$ or AKT1) promote transcription and translation of HIF-1 (Brugarolas and Kaelin 2004; Kaelin and Ratcliffe 2008). Activation of the oncogene Ras leads to suppression of HIF-1 hydroxylation. Further, HIF-1 also appears to be stabilized and its accumulation is stimulated by lactate as well as succinate, both observed in cancer cells as a result of increased glycolysis and incomplete TCA cycle (Vizan, Mazurek, and Cascante 2008). The resulting increased accumulation of HIF-1 has as its major function the initiation of a transcriptional program that provides possible multiple solutions for hypoxic stress (Kaelin and Ratcliffe 2008) as well as transcriptional activation of several oncogenes and enzymes. HIF-1 is additionally an effector of oncogenes such as Ras, Src, and Her-2. Also, HIF-1 upregulates expression and activation of genes and growth factors like VEGF and several glycolytic enzymes (Vizan, Mazurek, and Cascante 2008).

In addition to the indirect effect of oncogenes on metabolism through the activation of HIF-1, various oncogenes have a direct effect on the expression and activation of several enzymes. The alterations in the expression of oncogenes such as Akt, Ras, v-SRC, c-Myc, and tumor suppressors p53 and pVHL cause changes in the expression and activation of several enzymes involved in glycolytic or pentose phosphate as well as the biosynthesis pathway (Bensaad and Vousden 2007; Bertout, Patel, and Simon 2008; Costello and Franklin 2005; Dang et al. 1997, 2008; Dang and Semenza 1999; DeBerardinis et al. 2008a, 2008b; Gillies and Gatenby 2007a, 2007b; Gordan, Thompson, and Simon 2007; Hsu and Sabatini 2008; Kuhajda 2006; Ma et al. 2007b; Matoba et al. 2006; Menendez and Lupu 2007; Yeung, Pan, and Lee 2008; Young and Anderson 2008). The tumor suppressor p53 has a major role in the cellular response to a wide and diverse range of stress signals, such as DNA damage, hypoxia, or oncogenic activation (Bensaad and Vousden 2007; Vogelstein, Lane, and Levine 2000). The best understood is p53 control of cell cycle arrest and cell death. However, p53 functions appear to also include regulation of other processes, such as the response to and regulation of glucose metabolism. Tumor suppressor p53 was shown to be sufficient for induction of SCO2 expression, which ensures the maintenance of the cytochrome c oxidase complex. This complex is essential for mitochondrial respiration and the utilization of oxygen to produce energy (Matoba et al. 2006). Lack of functional p53 leads to lower oxygen consumption through mitochondrial respiration and the shift to glycolysis for energy production even in highly oxygenated environments, that is, the Warburg effect. In addition, p53 appears to downregulate the expression of phosphoglycerate mutase (PGM), an enzyme that is part of the glycolytic pathway (Kondoh et al. 2005). Although the mechanism for this downregulation is not clear, the loss of p53 is associated with increased PGM expression and thus enhanced glycolysis. Additionally, p53 influences glycolysis through the expression of TIGAR (T53-induced glycolysis and apoptosis regulator) which lowers the intracellular levels of fructose-2,6-bisphosphate, a substrate that promotes glycolysis and leads to an alternative pentose phosphate pathway (PPP).

Other metabolic changes in tumors involve altered biosynthesis of macromolecules such as lipids, proteins, and nucleic acids. Once again the oncogenes are the effectors of these metabolic changes with mTOR pathway alterations leading to modifications in protein biosynthesis and translation and Akt and Ras oncogenes leading to alterations in lipid biosynthesis. In the case of fatty acid biosynthesis the major effect is the highly increased expression of fatty acid synthase (FASN). The complete mechanism responsible for the tumor-associated FASN over-expression is not fully understood; however, there is experimental evidence linking FASN over-expression with the increased translation of known oncogenes Akt and Ras as well as hormonal stimulations (in breast and prostate tumors). Additionally, there is evidence for posttranscriptional activation of FASN once again by Akt as well as a PI3K and mTOR dependent mechanism. A number of different mechanisms leading to increased FASN expression in tumor cells highlight the importance of FASN in tumor progression. Other experimental evidence, primarily through inhibition of FASN activity, indicates that a change in FASN expression is not only a consequence of carcinogenesis but is actively contributing to the development, maintenance, and promotion of the cancer phenotype. The inhibition of FASN *in vivo* leads to the blocking of tumor growth as well as cell death through apoptosis in tumor cells. These experiments suggest that changes in the lipid biosynthesis pathway are crucial for the blocking of apoptosis in tumors (Little and Kridel 2008). In fact, FASN inhibition has been linked to p53 status but the actual correlation between FASN and p53 is still unclear. FASN is the key lipogenic enzyme that catalyzes terminal steps in the *de novo* biogenesis of fatty acids. In normal cells FASN expression depends on the levels of extracellular lipids; however, in tumors FASN is highly over-expressed regardless of the extracellular lipid concentration. FASN uses acetyl-CoA as a primer, malonyl-CoA as a carbon donor, and NADPH as a reducing equivalent and leads to the synthesis of the saturated fatty acid palmitate (Menendez and Lupu 2007; Figure 9.2). The activation of FASN leading to *de novo* synthesis of lipids as well as other changes in the expression of enzymes involved in the generation of lipids lead to the modified lipid profiles of tumor cells. In clinical applications, FASN expression correlates with poor patient prognosis and reduced survival. Anti-FASN drugs have successfully inhibited tumor growth in several animal models and their development is a point of active research. The lipid molecular fingerprint as well as concentration analysis through spectroscopy and the analysis of choline uptake through PET scanning are already used in diagnosis (Table 9.1).

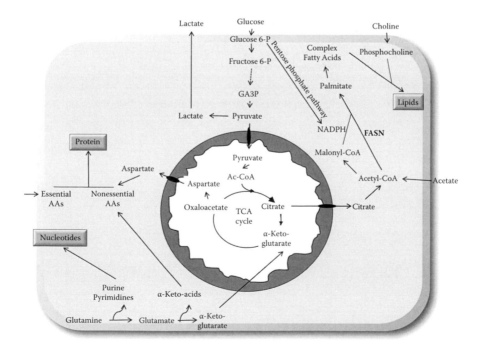

FIGURE 9.2 Brief outline of pathways suspected to be utilized in cancer cells for energy production, and biosynthesis of lipids, proteins, and nucleotides.

Another significant aspect of tumor metabolism is the specific expression of isoenzymes. These proteins have a different amino acid sequence than their normal counterparts but still catalyze the same reaction, although with altered kinetic and regulatory properties (Vizan, Mazurek, and Cascante 2008). Although the changes in the kinetic and regulatory properties of isoenzymes contribute to a tumor's metabolic changes, there is still no clear model explaining the cause and the role of isoenzymes. Therefore, further research is essential before the effects of isoenzymes can be included in cancer metabolic phenotype models as well as clinical applications.

The hypothesis developed thus far about the metabolic changes in carcinogenesis provides an illustrative description of some aspects of cancer metabolic phenotype. Further development of novel and focused treatment and diagnosis targeting metabolic changes requires more holistic and quantitative descriptions and understanding of the processes involved.

9.3 METABOLIC PATHWAY MODELING AND NETWORK DEVELOPMENT

Most biological characteristics arise from complex interactions between the cell's numerous constituents, such as proteins, DNA, RNA, and small molecules. The key challenge for biology is to understand and describe the structure and the dynamics of this interaction network (Barabasi and Oltvai 2004). Metabolism is a major part of this cellular network and various changes observed in metabolism in different cell phenotypes or stages of development can result from changes elsewhere in the network. A schematic model that describes the development of a cancer metabolic phenotype outlined above clearly shows that several seemingly unrelated genes strongly influence the changes in the metabolic processes. Thus, only a very large network of diverse biological molecules can lead to even a qualitative understanding of the process. Furthermore, even when only the metabolic processes are of direct interest it is crucial to keep in mind that metabolic reactions are highly interconnected. In human cells biological molecules, including metabolites, are involved in many different processes as products, substrates, or regulators and thus accurate kinetic information can only be achieved with the inclusion of all these different reactions. Once the network of stimulations leading to the cancer metabolic phenotype as well as the network of metabolic reactions is described, it will be possible to find focused and specific drug and diagnostic targets. The metabolic pathways can present a highly accurate diagram for the individual well studied processes, and metabolic pathway maps such as the Boehringer map (available on http://www.expasy.ch/cgi-bin/show_thumbnails.pl) can provide a very extensive diagram of many processes. However, pathways are largely artificial constructs of molecular interactions primarily resulting from *in vitro* analysis of individual enzymes. Networks that can be developed from omics data as well as automated literature searches can provide more information about the multifunctionality of biological molecules and can also give an indication of the function of unknown molecules (Barabasi and Oltvai 2004; Goodacre 2004). Furthermore, the analysis of the network and neighborhoods can lead to information about the structural properties of the network such as the determination of the most connected members as well as the distance between different members. Current efforts are focused on outlining the general metabolic network for a human cell and this general road map can subsequently be used for construction of specific routes that a cancer cell takes under different conditions. The availability of such a network will also allow direct visualization of various high-throughput experiments leading to more information about the activated neighbourhoods.

The reconstruction of a metabolic network is approached in several different ways. In one approach detailed networks and simulations are initially focused on the well understood, simple, model system such as *Escherichia coli* and yeast. These networks can then be upgraded and extended to describe more "complicated" cellular systems such as the human cell. The other approach involves the development of partial models and networks directly for a human cell, ultimately aiming at connecting them into a complete network. Several first round metabolic networks for a number of organisms including *Homo sapiens* are already available (Duarte et al. 2007; Ma et al. 2007a; Paley and Karp 2006) and can be utilized for the analysis of measurements in cancers. The development of such networks is a result of literature searches as well as genomics and protein-protein interaction information. Very detailed networks are available through the BIGG database (Recon 1 model available at http://bigg.ucsd.edu/), Reactome (http://www.reactome.org), and MetaCyc (http://metacyc.org). These tools in conjunction with other analysis tools (such as Cytoscape, http://cytoscape.org) allow the mapping of omics data and investigation of relationships between metabolites. Further optimization of such metabolic networks can be achieved from quantitative metabolomics measurements using methods such as comparative analysis of the correlation of concentration changes between different metabolites (Steuer et al. 2003) The network development is ultimately aimed at the building of the complete stoichiometric matrix, that is, a stochiometrically accurate network that would be able to describe all biochemical transformations of members in a self-consistent and chemically accurate mathematical model (Jamshidi and Palsson 2008; Figure 9.3). The development of a stoichiometric matrix is a prerequisite for the development of quantitative, mathematical models of the cellular processes. The main applications for quantitative models include possibilities: (1) to test whether a hypothesized model can describe known experimental facts and which changes in the network/model are necessary (hypothesize verification and generation); (2) to determine major components of the system for particular applications (e.g., biomarkers, drug targets, or external markers discovery); and (3) to provide a system for rapid testing of various system manipulations without costly and complex experiments (in silico experimentation). The development of network models as well as still prevailing partial pathway models intended for interpretation of biological data and in silico cell development can be approached in many different ways. These approaches include the development of detailed kinetic models, cybernetic models, stochastic models, metabolic control analysis, biochemical systems theory, and constraint-based methods. Constraint-based modeling is currently one of the most popular approaches as these models provide tools that can be used for genome-scale model construction. These models can include a large number of genes and reactions and were even proven to be predictive in some cases (Reed and Palsson 2003). In the constraint-based models parameter optimization is based on all allowed solutions for a set of equations. In this method detailed kinetic information is not included and rather than looking at individual reactions, the model calculates from the data major constraints for the whole network. In other words, the resulting models are made to fit a complete system of reactions rather than any individual reaction. The constraints used for model optimization are stoichiometric (mass balance), thermodynamics and enzymatic capacity (using the appropriate enzyme load values). The advantage of these models is that they do not require accurate kinetic data. In this type of model it is assumed that the network is in a steady state and therefore total concentration of each substance does not change. Following this assumption the system of reactions can be described with a set of linear equations that can be solved using linear programming. Constraint-based analyses of metabolic networks have gained considerable popularity and have been used to analyze genome-scale reconstructions of several organisms as well as the effect of various perturbations, such as gene deletions or drug inhibitions in silico. Genome-scale constraint-based models have an immense potential for building and testing hypotheses, as well as guiding experiments (reviewed in Raman and Chandra 2009 and Karlebach and Shamir 2008). The problem with such models is that they only provide an overall picture of the system and do not give any insight into cellular substrate concentrations. Furthermore, the steady-state assumption is problematic for metabolites that are exported from the cell. Also, their overall representation is only as good as the network used.

The growing availability of metabolomic and fluxomic datasets as well as methods for the determination of the thermodynamic properties of biochemical reactions *in vivo* has opened the possibility to formulate large-scale kinetic models (Jamshidi and Palsson 2008). In this type of model the goal is to determine accurate kinetic rules for each reaction and then to combine these individual reaction kinetics into a model of the complete network (Figure 9.3). The kinetic model of a metabolic process considers the cellular network as

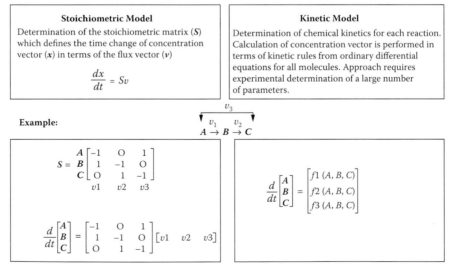

FIGURE 9.3 Outline of two extreme approaches to quantitative modeling of networks. Examples present stoichiometric and kinetic model approaches. The stoichiometric model is presented in detail in Palsson (2006). The functions used in kinetic models can be Michaelis–Menten, Hill, ping-pong, etc., and are described in detail in Demin and Goryanin (2009).

the combination of processes catalyzed by enzymes. The model consists of a system of nonlinear differential equations that provide time change information for each reaction species. These equations can sometimes be solved explicitly but in the majority of cases have to be solved approximately using many different, often computationally intensive, solvers. So far only small-scale models have been made and solved that describe only some pathways and only under some conditions. In cancer research most kinetic models have been aimed at describing signaling pathways with several kinetic models developed for the MAPK pathway (reviewed in Rangamani and Iyengar 2008 and Orton et al. 2005). Large-scale kinetic models as well as models of metabolic pathways for cancer cells have not been successfully constructed to date (Jamshidi and Palsson 2008). Kinetic models of metabolic networks are still unavailable at least in part due to the inherent complexity of cellular and regulatory networks as well as gaps in knowledge concerning system organization and kinetic rules. A large number of kinetic parameters required to define this system are still undetermined. Furthermore, even known kinetic parameters were largely determined from *in vitro* experiments on isolated enzymes under optimal rather than physiological conditions and thus they might not accurately represent the *in vivo* situation. Few attempts have been made at the development of detailed kinetic models of metabolic pathways in small healthy cells, such as simulation of the metabolism of the red blood cell (Nakayama, Kinoshita, and Tomita 2005) with good results. Although kinetic models can provide the most accurate information about the system, the problems with large numbers of kinetic parameters and equations that need to be determined as well as computationally intensive calculations makes them still only applicable to partial, relatively small and well-studied systems.

Several authors have made initial attempts to determine kinetic parameters on a network scale for both constrained as well as kinetic models. The availability of metabolomics data (Wishart 2008) and approaches for the extrapolation of thermodynamic quantities using computational approaches (Mavrovouniotis 1991; Zamboni and Kümmel 2008) are being explored for this application. The framework for metabolic modeling is also being proposed (Jamshidi and Palsson 2008), based on the combination of genomic and literature data for network design and thermodynamic and metabolomic data (for parameter determination). A great advancement in further development of models is the development of Systems Biology Markup Language (Hucka et al. 2003). Thanks to SBML the models are interoperable among many different software tools and users can upload them and further work on improvements as well as combining existing models, or alternatively use existing models for comparison and analysis of experimental data without going into mathematical details (http://www.sbml.org; Hucka et al. 2003). Many commercial and academic software tools are now available for the development of models and parameter determination from experimental data. The next major hurdle in the development of metabolic network models is the collection of high quality and quantity data that can show metabolite concentrations in different conditions and at different times. Once these data become available to modelers and data analysts it will be possible to make major contributions to the development of metabolomics network models for various systems, including cancer.

9.4 METABOLIC PROFILES OF CANCER

The human metabolome comprises thousands of endogenous molecules, many with still unknown functions and structures. The large diversity of molecular types in our metabolome and their significance both as participants and regulators of many cellular processes make their analysis both challenging and prudent. The analysis of metabolites should clearly be a major part of functional genomics as well as systems biology. An attraction of the metabolome is that, although more diverse in classes of molecules present, it is likely numerically smaller than the transcriptome or proteome (Oliver et al. 1998). Even more importantly, as metabolites are downstream of both transcription and translation they are potentially a better indicator of enzyme activity and actual phenotype defining changes (ter Kuile and Westerhoff 2001). In other words, metabolic changes are showing activated processes rather than the potential for process activation. Also, in catalyzed reactions in general and in enzyme-catalyzed metabolic processes in particular larger concentration changes are observed in metabolites (reaction substrates or products) than in catalysts, that is, either enzymes gene or protein form. However, thus far most of the focus in cancer research is still on observing changes in genes and proteins and the metabolomics applications in cancer research have been primarily aimed at fingerprinting or at diagnosis using only a handful of metabolites. The emerging field of metabolomics explores the structure and function of low molecular weight compounds beyond DNA and protein. Metabolic profiling has been performed for several decades. However, only recent technical innovations have allowed metabolite profiling to be carried out on a large scale in terms of both numbers of different classes of metabolites simultaneously observed as well as number of samples measured (Fernie et al. 2004). Also, the data-driven rather than the hypothesis-driven research initiated by omics methodologies makes metabolomics a conceptually different approach to previous metabolite analyses. Similar to other omics methodologies, metabolomics is an interesting platform for diagnostic biomarker discovery, functional genomics, and systems biology. In this context, metabolomics attempts to measure the complete set of metabolites that vary according to the physiology, development, or pathologic state of the cell, tissue, organ, or organism (Kim and Maruvada 2008; Oliver 2002). Several different, primarily spectroscopic methods are used for the high-throughput measure of metabolites. A brief outline of the most popular methods is given in Table 9.2. These technologies provide simultaneous measurements of mixtures and the resolution of different molecules arises from the dispersion of observed properties in one or more dimensions (Table 9.2). The major problem with current metabolomics analysis is that even the most involved technologies can individually allow observation of under about 300 different metabolites. Future efforts in experimental metabolomics should clearly be aimed at devising thoroughly high-throughput methodologies.

Metabolomics and metabolic profiling are extensively explored for direct clinical application in cancer patient stratification. Early efforts were aimed at finding cancer metabolic markers in the blood from the analysis of NMR spectral lines (Fossel, Carr, and McDonagh 1986). However, this initially highly praised result has since proven inaccurate (Chmurny, Hilton, and Halverson 1988; Herring et al. 1990) and this led to general

TABLE 9.2 Major Methods Used in Metabolomics Analysis

	NMR	MRS	MS	FTIR	Raman Spectroscopy
Metabolic fingerprinting	Yes	Yes	Yes	Yes	Yes
Quantitative analysis of metabolites	Yes	Yes	Yes	No	No
Observed molecular property	Nuclear spin's chemical shift	Nuclear spin's chemical shift	Mass to charge ratio	Chemical bond vibrations	Chemical bond vibration (complementary to FTIR)
Equipment cost	High	High	High	Low	High
Maintenance cost	High	High	High	Low	Low
Per sample cost	Low	Low	High	Low	Low
Reproducibility	High	High	Low	High	High
Identification of new metabolites	Yes	Difficult	Yes	Difficult	Difficult
Main advantage	Nondestructive	Noninvasive	High sensitivity	Cheap and fast	Very low water peak

The labels are NMR, nuclear magnetic resonance spectroscopy; MRS, *in vivo* NMR spectroscopy; MS, mass spectroscopy; FTIR, Fourier transform infrared spectroscopy.
Source: Data from Cuperlovic-Culf et al. (2008), Griffin and Shockcor (2004), and Lenz and Wilson (2007).

scepticism toward the application of metabolic profiling in cancer diagnosis. At the same time, both MS and NMR have been highly successfully utilized for the analysis of cancer cell lines and tissue extracts. Furthermore, magnetic resonance spectroscopy (MRS) allows noninvasive, *in vivo* fingerprinting of metabolic profiles of tumors. Despite limitations in sensitivity and resolution, MRS metabolic profiling is successfully used for diagnosis of tumors particularly of the central nervous system (CNS) and prostate as well as breast cancers (Table 9.3). In these measurements the most visible, that is, the most diagnostic are the alterations and increase in the energy and fatty acid production with many examples of cancer phenotype determination from metabolites such as lactate, lipids, phosphocholine, choline, citrate, various amino acids, and triglycerides. The direct analysis of cancer cells has also many clinical applications in diagnostics as well as development of novel treatments. Metabolomics of cells can be aimed at measuring the total complement of individual metabolites—metabolic fingerprint analysis, or as a measure of a particular class of metabolites. Both approaches lead to the determination of a metabolic fingerprint of the phenotype. Metabolic profiling of cancer cells can be further combined with available data on pathways and networks as well as other omics data such as transcriptomics (Cuperlovic-Culf et al. 2008) and proteomics with the ultimate goal in developing highly detailed, quantitative models.

In metabolic fingerprint analysis complete metabolomics spectra are used in the examination. In this chemometric approach the compounds are not initially identified. That is, only spectral patterns and intensities of the mixtures are recorded. Complete spectra

TABLE 9.3   Major Metabolites Observed Thus Far by NMR- and
MS-Based Metabolomics Studies of Various Cancer Tissues
and Cells

Cancer Type	Major Differentiating Metabolite
Breast	Choline metabolites
Liver	High lacate, high amino acids, low carbohydrates
Pancreatic	Decreased levels of phosphocholine, glycerophosphochiline, and phosphatidylinositol
Cervical	Low glucose, high cholines, high amino acids
Prostate	Altered amino acids levels; increased lactate; phospholipids, choline, decrease in citrate concentration, polyamines
Renal	High triglycerides and cholesteryl esters
Brain	Altered lipids; inositol, N-acetyl aspartate/choline ratio creatine/choline ratio

Source: From Gowda, G. A. et al. 2008. *Expert Rev. Mol. Diagn.* 8: 617–633. With permission.

are compared statistically and used to identify the relevant spectral features or areas that distinguish similar sample classes. Once these major features are identified different approaches can be used to assign them to the corresponding metabolites (Holmes, Wilson, and Nicholson 2008; Trygg, Holmes, and Lundstedt 2007). Alternatively, in another approach, termed quantitative metabolomics, compounds are identified and quantified initially. Once these compounds are identified and quantified, the data can be used for various applications including the development of more accurate systems biology models (Weljie et al. 2006; Wishart 2008). If the quantitative measurement of metabolite concentrations is performed over time the method is called fluxomics.

The chemometrics approach has a range of clinical applications for sample classification and in this context has some advantages over the quantitative approach (Serkova and Niemann 2006). At the same time this method has several inherent weaknesses caused by variations in the data due to the experimental conditions, overlapping peaks as well as the unresolved issue of result normalization. Utilization of complete spectra leads to the inclusion in the analysis of spectral regions with only background noise. This leads to the application of unnecessarily large datasets and can possibly also result in inaccuracies in classification. A major concern in quantitative metabolomics is therefore the problem of spectral assignment. Unlike transcriptomics where gene assignment is trivial thanks to highly specific hybridization of genes to specially designed, unique probes, in the high-throughput analysis of metabolic mixtures the peak assignment as well as measurement of intensity for each metabolite is highly challenging. Some of the major problems in quantitative metabolomics are the same as in the chemometric approach and include problems with overlapping spectra as well as changes in spectra of various compounds under different conditions (i.e., pH). Additionally, unidentified metabolites present a great challenge to quantitative metabolomics analysis. Possible experimental solutions are the utilization of more complex experiments (multidimensional NMR) or more involved preprocessing procedures (particularly for MS; Villas-Boas et al. 2005) and also combined utilization of different methods, for example MS and NMR methods or a combination of high-through-put liquid- and gas-based chromatography (MS) (Sreekumar et al. 2009). Another highly beneficial experimental method is stable isotope labeling followed by either MS or NMR measurement. This approach allows pathway tracing and easier metabolite assignment as well as metabolic flux measurements (Lane, Fan, and Higashi 2008). Isotope labeling was used in many different analyses of pathways and networks in various systems (Chikayama et al. 2008) including cancers (Yang et al. 2007). Mass spectrometry-based approaches that are coupled with chromatography provide high sensitivity for targeted compound analysis (Nordstrom et al. 2006; Sreekumar et al. 2009). NMR spectroscopy, on the other hand, does not require extensive sample preprocessing and separation and provides a number of different experimental protocols optimized for mixture analysis, as well as molecular formula or structure determination. Thus far the quantification of metabolites with or without isotope labeling requires comparison with standard measurements on individual metabolites. Several such databases are under development and some major noncommercial examples are outlined in Table 9.4. The spectral assignment is performed using methods for line comparison of the pure compound measurements and mixture spectra (such as Lewis et al. 2007; Sreekumar et al. 2009; Weljie et al. 2006) with either manual or semiautomatic spectral assignment (for example, Sreekumar et al. 2009; Xia et al. 2008;

TABLE 9.4   Some Major Noncommercial Databases of Metabolomics Standard Data for Quantification and Assignment

Name and Availability	Metabolomics Experiments	Additional Functionality
Human Metabolome Project (Wishart et al. 2007); http://www.hmdb.ca	NMR; MS	Biological data; chemical and clinical data specific to humans
BMRD http://www.bmrb.wisc.edu	NMR	Database search for NMR peaks assignment
Prime (Akiyama et al. 2008) http://prime.psc.riken.jp	MS, NMR	
Golm metabolome database http://csbdb.mpimp-golm.mpg.de	MS	Specific to plants
METLIN metabolite database http://metlin.scripps.edu	MS	Drug and drug metabolites; specific to humans
NIST Chemistry WebBook http://Webbook.nist.gov/chemistry	NMR, MS, IR	
Madison metabolomics database (Cui et al. 2008) http://mmcd.nmrfam.wisc.edu	MS, NMR	
NMR Lab of Biomolecules http://spinportal.magnet.fsu.edu	NMR	Database search for NMR peaks assignment

Zhang et al. 2008). Several tools for metabolic data processing including quantification are presented in reviews of the field (Katajamaa and Oresic 2007; Wishart 2008).

Although quantitative metabolomics is still under active development, there were several interesting applications in the analysis of metabolic pathways and networks in cancers. GC-MS and HPLC analysis of metabolites was used for the investigation of the effects of oncogenesis on metabolite profiles. In this three-dimensional screening experiment the authors analyzed four cell lines serially transduced with four different oncogenes and five small-molecule inhibitors of metabolic and nutrient-sensing pathways (Ramanathan, Wang, and Schreiber 2005). The resulting quantitative metabolomic data have shown the connection and consonance between the effects of oncogenes and metabolic changes. The conclusion from this study was that metabolic changes are likely the result of gene changes in cancers. The quantitative metabolic data clearly showed increased glucose consumption and lactate production (indicative of anaerobic glycolysis), increased consumption of oxygen, high levels of nucleotide biosynthesis, changes to the citric acid cycle metabolite concentrations, and changes in mitochondrial biogenesis. These data were in good agreement with the cancer metabolite model emerging from other methods described previously. However, these experiments led to some unexpected and as yet unexplained observations, including the observation that cells with greater tumorigenic potential consume more oxygen and yet exhibit diminished oxygen-dependent ATP synthesis.

Other research using metabolic profiling of cancer tissues has shown that in ovarian cancers there are significant quantitative metabolic differences in different tumor types (Denkert et al. 2006). In this work the authors investigated the fold change of metabolites in relation to known pathways. The high-throughput metabolite information additionally provided data for further investigation of pathways and networks. An example of such work came two years later from the same group and was focused on the analysis of metabolite profiling of human colon carcinoma (Denkert et al. 2008). In this case 206 metabolites were measured using time-of-flight mass spectrometry resulting in the determination of 82 significantly different metabolites between colon cancer and normal tissues. The list of identifiable metabolites was mapped on the cluster of closest neighbors determined from the KEGG pathway database. The neighbors were determined using the metabolite-metabolite distance calculations, based on the number of pathways needed to connect two metabolites. This analysis showed increased concentration in cancers of metabolites involved in amino acid as well as nucleic acid synthesis and downregulation of metabolites from the TCA cycle as well as fatty acid metabolism. These data are in general agreement with other published results, and additionally it shows a very interesting approach for the visualization of quantitative metabolomics results on the KEGG pathway network (Denkert et al. 2008). Several other groups have investigated the application of isotope labeling for quantitative metabolomics analysis of cancers. Many such studies were reviewed by Lutz (2005) and Boros and co-workers (2002, 2004). Recent efforts utilized more comprehensive high-throughput approaches (combined NMR and MS; 2D NMR) as well as isotopomer modeling approaches (Richardson et al. 2008; Yang et al. 2007). From these measurements it was possible to determine changes in the flux of different parts of energy production pathways (glycolysis and TCA cycle) as well as quantitative metabolic changes in breast cancer relative to breast normal cell lines. Richardson et al (2008) gained similar information, that is, carbon flux through central metabolism, in a cellular model of breast cancer progression.

Metabolomics and quantitative metabolomics produce voluminous data that can lead to highly useful information about cancers with the ultimate goal in clinical applications. However, this as well as other omics methods should be seen as data aimed at providing understanding or knowledge about biological processes, in this case understanding of metabolism in cancers (Kell 2004). Such understanding can only come from the development of, ultimately, quantitative models of metabolic networks that will one day become part of the complete cellular network.

9.5 CONCLUSION

Cancer is currently viewed as a disease resulting from cancer-causing genes that deregulate cellular proliferation, differentiation, and death. Although the relationship between genetic changes and the deregulation of energy production as well as biosynthesis is only partially understood, the significance of metabolic changes for cancer development and progression is now generally accepted. This recognition has led to many new ideas for various clinical applications of specific cancer metabolic phenotypes. The metabolic and mitochondrial changes are now being proposed as possible primary targets for cancer therapeutics. One example of a promising cancer drug is dichloroacetate, recently proposed and tested by researchers at the University of Alberta, Canada (Michelakis, Webster, and Mackey 2008). DCA is a known activator of pyruvate dehydrogenase and the effect of DCA on cancer cells appears to be increased delivery of pyruvate into the mitochondria, followed by increased mitochondria-based glucose oxidation and apoptosis, all resulting in the shrinking of tumors. DCA is currently undergoing clinical trials and promises to be widely applicable to a number of solid tumors. The applications in diagnostic are already clear from examples shown by PET and MRS methods. However, all of these applications are still primarily the result of observations rather than focused analysis due to the lack of quantitative *in vivo* metabolic data as well as quantitative models and networks of a cancer metabolic phenotype. System-level analysis and modeling can lead to more focused discovery and optimization of targets and markers. Metabolome measurements provide more information about the efficacy of cancer treatment and disease progression as well as the analysis of toxicity of the treatment. Quantitative models will allow in silico testing of various possible drugs and inhibitors and also lead to more optimal biomarkers. In silico modes of action studies on network and pathway models can lead to the discovery of the most significant enzymatic targets for treatment. Large progress is being made in all the individual fields of analysis of cancer metabolic phenotypes and only a combined effort from these different approaches can lead to a truly quantitative model of this important part of systems biology analysis.

REFERENCES

Akiyama, K., Chikayama, E., Yuasa, H. et al. 2008. PRIMe: a Web site that assembles tools for metabolomics and transcriptomics. *In Silico Biol.* 8: 339–345.

Alon, U. 2007. *An Introduction to Systems Biology.* Chapman & Hill/CRC, Boca Raton, FL.

Barabasi, A. L. and Oltvai, Z. N. 2004. Network biology: understanding the cell's functional organization. *Nature. Rev. Genet.* 5: 101–113.

Bensaad, K. and Vousden, K. H. 2007. p53: new roles in metabolism. *Trends Cell Biol.* 17: 286–291.

Bertout, J. A., Patel, S. A., and Simon, M. C. 2008. The impact of O_2 availability on human cancer. *Nature Rev. Cancer* 8: 967–975.

Boros, L. G., Cascante, M., and Lee, W. N. 2002. Metabolic profiling of cell growth and death in cancer: applications in drug discovery. *Drug Discov. Today* 7: 364–372.

Boros, László G., Serkova, Natalie J., Cascante, Marta S., and Lee, Wai Nang Paul. 2004. Use of metabolic pathway flux information in targeted cancer drug design. *Drug Discovery Today: Therapeutic Strategies* 1: 435–443.

Brugarolas, J. and Kaelin, W. G., Jr. 2004. Dysregulation of HIF and VEGF is a unifying feature of the familial hamartoma syndromes. *Cancer Cell* 6: 7–10.

Chikayama, E., Suto, M., Nishihara, T. et al. 2008. Systematic NMR analysis of stable isotope labeled metabolite mixtures in plant and animal systems: coarse grained views of metabolic pathways. *PLoS. One.* 3: e3805.

Chmurny, G. N., Hilton, B. D., and Halverson, D. et al. 1988. An NMR blood test for cancer: a critical assessment. *NMR Biomed* 1: 136–150.

Costello, L. C. and Franklin, R. B. 2005. "Why do tumour cells glycolyse?": from glycolysis through citrate to lipogenesis. *Mol. Cell Biochem.* 280: 1–8.

Cui, Q., Lewis, I. A., Hegeman, A. D. et al. 2008. Metabolite identification via the Madison Metabolomics Consortium Database. *Nature Biotechnol.* 26: 162–164.

Cuperlovic-Culf, Miroslava, Belacel, Nabil, and Culf, A. S. 2008. Integrated analysis of transcriptomics and metabolomics profiles. *Expert Opin. Med. Diagnostics* 497–509.

Dang, C. V., Kim, J., Gao, P., and Yustein, J. 2008. The interplay between MYC and HIF in cancer. *Nature Rev. Cancer* 8: 51–56.

Dang, C. V., Lewis, B. C., Dolde, C., Dang, G., and Shim, H. 1997. Oncogenes in tumor metabolism, tumorigenesis, and apoptosis. *J. Bioenerg. Biomembr.* 29: 345–354.

Dang, C. V. and Semenza, G. L. 1999. Oncogenic alterations of metabolism. *Trends Biochem. Sci.* 24: 68–72.

DeBerardinis, R. J., Lum, J. J., Hatzivassiliou, G., and Thompson, C. B. 2008a. The biology of cancer: metabolic reprogramming fuels cell growth and proliferation. *Cell Metab.* 7: 11–20.

DeBerardinis, R. J., Sayed, N., Ditsworth, D., and Thompson, C. B. 2008b. Brick by brick: metabolism and tumor cell growth. *Curr. Opin. Genet. Dev.* 18: 54–61.

DeGraaf, R. A. 2007. *In Vivo NMR Spectroscopy.* Wiley, Chichester, England.

Demin, O. and Goryanin, I. 2009. *Kinetic Modelling in Systems Biology.* Chapman & Hill/CRC, Boca Raton, FL.

Denkert, C., Budczies, J., Kind, T. et al. 2006. Mass spectrometry-based metabolic profiling reveals different metabolite patterns in invasive ovarian carcinomas and ovarian borderline tumors. *Cancer Res.* 66: 10795–10804.

Denkert, C., Budczies, J., Weichert, W. et al. 2008. Metabolite profiling of human colon carcinoma: deregulation of TCA cycle and amino acid turnover. *Mol. Cancer* 7: 72.

Duarte, N. C., Becker, S. A., Jamshidi, N. et al. 2007. Global reconstruction of the human metabolic network based on genomic and bibliomic data. *Proc. Natl. Acad. Sci. USA* 104: 1777–1782.

Fernie, A. R., Trethewey, R. N., Krotzky, A. J., and Willmitzer, L. 2004. Metabolite profiling: from diagnostics to systems biology. *Nature Rev. Mol. Cell Biol.* 5: 763–769.

Fossel, E. T., Carr, J. M., and McDonagh, J. 1986. Detection of malignant tumors: water-suppressed proton nuclear magnetic resonance spectroscopy of plasma. *N. Engl. J. Med.* 315: 1369–1376.

Gillies, R. J. and Gatenby, R. A. 2007a. Adaptive landscapes and emergent phenotypes: why do cancers have high glycolysis? *J. Bioenerg. Biomembr.* 39: 251–257.

Gillies, R. J. and Gatenby, R. A. 2007b. Hypoxia and adaptive landscapes in the evolution of carcinogenesis. *Cancer Metastasis Rev.* 26: 311–317.

Goodacre, R. 2004. Metabolic profiling: pathways in discovery. *Drug Discov. Today* 9: 260–261.

Gordan, J. D., Thompson, C. B., and Simon, M. C. 2007. HIF and c-Myc: sibling rivals for control of cancer cell metabolism and proliferation. *Cancer Cell* 12: 108–113.

Gowda, G. A., Zhang, S., Gu, H. et al. 2008. Metabolomics-based methods for early disease diagnostics. *Expert Rev. Mol. Diagn.* 8: 617–633.

Griffin, J. L. and Shockcor, J. P. 2004. Metabolic profiles of cancer cells. *Nature Rev. Cancer* 4: 551–561.

Griffiths, J. R. and Stubbs, M. 2005. Cancer metabolic phenotype in the *Oncegenesis Handbook*, LaRochelle W. J., Shunketes, R. A. (eds), Humana Press, Totowa, New Jersey.

Hanahan, D. and Weinberg, R. A. 2000. The hallmarks of cancer. *Cell* 100: 57–70.

Herring, F. G., Phillips, P. S., Pritchard, H., Silver, H., and Whittal, K. P. 1990. The proton NMR of blood plasma and the test for cancer. *Magn Reson. Med.* 16: 35–48.

Holmes, E., Wilson, I. D., and Nicholson, J. K. 2008. Metabolic phenotyping in health and disease. *Cell* 134: 714–717.

Hsu, P. P. and Sabatini, D. M. 2008. Cancer cell metabolism: Warburg and beyond. *Cell* 134: 703–707.

Hucka, M., Finney, A., Sauro, H. M. et al. 2003. The systems biology markup language (SBML): a medium for representation and exchange of biochemical network models. *Bioinformatics* 19: 524–531.

Jamshidi, N. and Palsson, B. O. 2008. Formulating genome-scale kinetic models in the post-genome era. *Mol. Syst. Biol.* 4: 171.

Kaelin, W. G., Jr. and Ratcliffe, P. J. 2008. Oxygen sensing by metazoans: the central role of the HIF hydroxylase pathway. *Mol. Cell* 30: 393–402.

Karlebach, G. and Shamir, R. 2008. Modelling and analysis of gene regulatory networks. *Nature Rev. Mol. Cell Biol.* 9: 770–780.

Katajamaa, M. and Oresic, M. 2007. Data processing for mass spectrometry-based metabolomics. *J. Chromatogr. A.* 1158: 318–328.

Kell, D. B. 2004. Metabolomics and systems biology: making sense of the soup. *Curr. Opin. Microbiol.* 7: 296–307.

Kell, D. B. 2006. Systems biology, metabolic modelling and metabolomics in drug discovery and development. *Drug Discov. Today* 11: 1085–1092.

Kell, D. B. and Knowles, J. D. 2006. The role of modeling in systems biology. In *Systems Modelling in Cellular Biology from Concepts to Nuts and Bolts*. Szallasi, Z. et al. (eds.), MIT Press, pp. 3–18.

Kim, Y. and Maruvada, P. 2008. Frontiers in metabolomics for cancer research: Proceedings of a National Cancer Institute workshop. *Metabolomics* 4: 105–113.

Kondoh, H., Lleonart, M. E., Gil, J. et al. 2005. Glycolytic enzymes can modulate cellular life span. *Cancer Res.* 65: 177–185.

Kuhajda, F. P. 2006. Fatty acid synthase and cancer: new application of an old pathway. *Cancer Res.* 66: 5977–5980.

Lane, A. N., Fan, T. W., and Higashi, R. M. 2008. Stable isotope-assisted metabolomics in cancer research. *IUBMB Life* 60: 124–129.

Lenz, E. M. and Wilson, I. D. 2007. Analytical strategies in metabonomics. *J. Proteome. Res.* 6: 443–458.

Lewis, I. A., Schommer, S. C., Hodis, B. et al. 2007. Method for determining molar concentrations of metabolites in complex solutions from two-dimensional 1H-13C NMR spectra. *Anal. Chem.* 79: 9385–9390.

Lindon, J. C., Nicholson, J. K., and Holmes, E. (Editors) 2007. *The Handbook of Metabonomics and Metabolomics.* Elsevier Science, New York.

Little, J. L. and Kridel, S. J. 2008. Fatty acid synthase activity in tumor cells. *Subcell. Biochem.* 49: 169–194.

Lutz, N. W. 2005. From metabolic to metabolomic NMR spectroscopy of apoptotic cells. *Metabolomics* 1: 251–268.

Ma, H., Sorokin, A., Mazein, A. et al. 2007a. The Edinburgh human metabolic network reconstruction and its functional analysis. *Mol. Syst. Biol.* 3: 135.

Ma, W., Sung, H. J., Park, J. Y., Matoba, S., and Hwang, P. M. 2007b. A pivotal role for p53: balancing aerobic respiration and glycolysis. *J. Bioenerg. Biomembr.* 39: 243–246.

Mantovani, A. 2009. Cancer: inflaming metastasis. *Nature* 457: 36–37.

Matoba, S., Kang, J. G., Patino, W. D. et al. 2006. p53 regulates mitochondrial respiration. *Science* 312: 1650–1653.

Mavrovouniotis, M. L. 1991. Estimation of standard Gibbs energy changes of biotransformations. *J. Biol. Chem.* 266: 14440–14445.

Menendez, J. A. and Lupu, R. 2007. Fatty acid synthase and the lipogenic phenotype in cancer pathogenesis. *Nature Rev. Cancer* 7: 763–777.

Michelakis, E. D., Webster, L., and Mackey, J. R. 2008. Dichloroacetate (DCA) as a potential metabolic-targeting therapy for cancer. *Br. J. Cancer* 99: 989–994.

Nakayama, Y., Kinoshita, A., and Tomita, M. 2005. Dynamic simulation of red blood cell metabolism and its application to the analysis of a pathological condition. *Theor. Biol. Med. Model.* 2: 18.

Nordstrom, A., O'Maille, G., Qin, C., and Siuzdak, G. 2006. Nonlinear data alignment for UPLC-MS and HPLC-MS based metabolomics: quantitative analysis of endogenous and exogenous metabolites in human serum. *Anal. Chem.* 78: 3289–3295.

Oliver, S. G. 2002. Functional genomics: lessons from yeast. *Philos. Trans. R. Soc. Lond. B Biol. Sci.* 357: 17–23.

Oliver, S. G., Winson, M. K., Kell, D. B., and Baganz, F. 1998. Systematic functional analysis of the yeast genome. *Trends Biotechnol.* 16: 373–378.

Orton, R. J., Sturm, O. E., Vyshemirsky, V. et al. 2005. Computational modelling of the receptor-tyrosine-kinase-activated MAPK pathway. *Biochem. J.* 392: 249–261.

Paley, S. M. and Karp, P. D. 2006. The Pathway Tools cellular overview diagram and Omics Viewer. *Nucleic Acids Res.* 34: 3771–3778.

Palsson, B. O. 2006. *Systems Biology: Properties of Reconstructed Networks.* Cambridge University Press, New York.

Raman, K. and Chandra, N. 2009. Flux balance analysis of biological systems: applications and challenges. *Brief. Bioinform.* 10: 435–449.

Ramanathan, A., Wang, C., and Schreiber, S. L. 2005. Perturbational profiling of a cell-line model of tumorigenesis by using metabolic measurements. *Proc. Natl. Acad. Sci. USA* 102: 5992–5997.

Rangamani, P. and Iyengar, R. 2008. Modelling cellular signalling systems. *Essays Biochem.* 45: 83–94.

Rashid, A., Pizer, E. S., Moga, M. et al. 1997. Elevated expression of fatty acid synthase and fatty acid synthetic activity in colorectal neoplasia. *Am. J. Pathol.* 150: 201–208.

Reed, J. L. and Palsson, B. O. 2003. Thirteen years of building constraint-based in silico models of *Escherichia coli. J. Bacteriol.* 185: 2692–2699.

Richardson, A. D., Yang, C., Osterman, A., and Smith, J. W. 2008. Central carbon metabolism in the progression of mammary carcinoma. *Breast Cancer Res. Treat.* 110: 297–307.

Rojas, I., Golebiewski, M., Kania, R., Kreb, O., Mir, S., Weidemann, A., Wittig, U. 2007. SABIO: a database for biochemical reactions and their kinetics. *BMC Sys. Biol.* (supplement 1): 56.

Serkova, N. J. and Niemann, C. U. 2006. Pattern recognition and biomarker validation using quantitative 1H-NMR-based metabolomics. *Expert. Rev. Mol. Diagn.* 6: 717–731.

Sonveaux, P., Vegran, F., Schroeder, T. et al. 2008. Targeting lactate-fueled respiration selectively kills hypoxic tumor cells in mice. *J. Clin. Invest.* 118: 3930–3942.

Sreekumar, A., Poisson, L. M., Rajendiran, T. M. et al. 2009. Metabolomic profiles delineate potential role for sarcosine in prostate cancer progression. *Nature* 457: 910–914.

Steuer, R., Kurths, J., Fiehn, O., and Weckwerth, W. 2003. Observing and interpreting correlations in metabolomic networks. *Bioinformatics* 19: 1019–1026.

ter Kuile, B. H. and Westerhoff, H. V. 2001. Transcriptome meets metabolome: hierarchical and metabolic regulation of the glycolytic pathway. *FEBS Lett.* 500: 169–171.

Trygg, J., Holmes, E., and Lundstedt, T. 2007. Chemometrics in metabonomics. *J. Proteome. Res.* 6: 469–479.

Vallabhajosula, S. 2007. (18)F-labeled positron emission tomographic radiopharmaceuticals in oncology: an overview of radiochemistry and mechanisms of tumor localization. *Semin. Nucl. Med.* 37: 400–419.

Villas-Boas, S. G., Mas, S., Akesson, M., Smedsgaard, J., and Nielsen, J. 2005. Mass spectrometry in metabolome analysis. *Mass Spectrom. Rev.* 24: 613–646.

Vizan, P., Mazurek, S., and Cascante. 2008. Robust metabolic adaptation underlying tumour progression. *Metabolomics* 4: 1–12.

Vogelstein, B., Lane, D., and Levine, A. J. 2000. Surfing the p53 network. *Nature* 408: 307–310.

Warburg, O. 1925. Uber den Stoffwechsel der Carcinomzelle. *Klin. Wochenschr. Berl.* 4: 534–536.

Warburg, O. 1956. On the origin of cancer cells. *Science* 123: 309–314.

Weljie, A. M., Newton, J., Mercier, P., Carlson, E., and Slupsky, C. M. 2006. Targeted profiling: quantitative analysis of 1H NMR metabolomics data. *Anal. Chem.* 78: 4430–4442.

Wishart, D. S. 2008. Quantitative metabolomics using NMR. *Trends Anal. Chem.* 27: 228–237.

Wishart, D. S., Tzur, D., Knox, C. et al. 2007. HMDB: the Human Metabolome Database. *Nucleic Acids Res.* 35: D521–D526.

Xia, J., Bjorndahl, T. C., Tang, P., and Wishart, D. S. 2008. MetaboMiner: semi-automated identification of metabolites from 2D NMR spectra of complex biofluids. *BMC Bioinformatics* 9: 507.

Yang, C., Richardson, A. D., Smith, J. W., and Osterman, A. 2007. Comparative metabolomics of breast cancer. *Pac. Symp. Biocomput.* 181–192.

Yeung, S. J., Pan, J., and Lee, M. H. 2008. Roles of p53, MYC and HIF-1 in regulating glycolysis: the seventh hallmark of cancer. *Cell Mol. Life Sci.* 65: 3981–3999.

Young, C. D. and Anderson, S. M. 2008. Sugar and fat—that's where it's at: metabolic changes in tumors. *Breast Cancer Res.* 10: 202.

Zamboni, N. and Kümmel. 2008. anNET: a tool for network-embedded thermodynamic analysis of quantitative metabolome data 9: 199.

Zhang, F., Bruschweiler-Li, L., Robinette, S. L., and Bruschweiler, R. 2008. Self-consistent metabolic mixture analysis by heteronuclear NMR: application to a human cancer cell line. *Anal. Chem.* 80: 7549–7553.

CHAPTER 10

Warburg Revisited

Modeling Energy Metabolism for Cancer Systems Biology

Mathieu Cloutier

CONTENTS

10.1 INTRODUCTION

This chapter on mathematical modeling and energy metabolism in cancer systems biology will pursue two major objectives. First, the importance of energy, energy signaling, and energy metabolism in cancer will be emphasized, because a major phenotype characteristic of cancer tumors is the loss of a very important systems property in energy metabolism. The second objective of this chapter is to show how the mathematical modeling of biological *processes* is relevant for cancer systems biology. Here, the emphasis on *processes* is extremely important. In engineering terms, a process is the sum of operations by which an input (energy, material, or information) is transformed into an output with specific properties. The concept of process is also well established in biology. For example, we could define glycolysis as the process by which glucose (GLC) is converted to pyruvate (PYR) with the regeneration of two molecules of ATP and two NADH. As will be seen, the deregulation of this process is very important in cancer and its mathematical modeling will yield significant insights on the disease development and potential therapeutic targets.

Thus, we will examine not only static interactions between cell components (a very important aspect of systems biology), but also the dynamic properties of the cell as a multi-input, multi-output system (the other important aspect of systems biology).

Otto Warburg, in his breakthrough study (Warburg 1930), observed a key phenotype difference between normal tissues and cancer tumors, that is, glycolytic overflow to lactate (LAC) in the presence of sufficient oxygen (O_2) and reduced (but not null) mitochondrial oxidation rate. This uncoupling of glycolysis and oxidative phosphorylation (OP) leads to a much less efficient energy metabolism, as glycolysis yields only 2 ATP per molecule of GLC consumed, while the complete oxidation of GLC yields approximately 30 molecules of ATP. Thus, the tumor cells require a much higher inflow of GLC and end up producing significant amounts of LAC, something that is not normally observed in mammalian metabolism in the presence of sufficient O_2 inflow. This came to be known as the Warburg effect. Relying on lactic fermentation is not optimal for mammalian cells, but it has been shown to provide "local" advantages for the tumor: increased availability of growth precursors, acidification of the surroundings, and prolonged resistance to hypoxia or anoxia. Moreover, tumors have access to a plentiful supply of GLC, especially with our modern, carbohydrates-rich alimentation. Thus, it might be that the "Warburg phenotype"—even though energetically suboptimal—provides mostly advantages to tumors, hence allowing their proliferation.

As observed by Pedersen in a recent review, the Warburg effect (and metabolism in general) received very little attention in the cancer literature (Pedersen 2007). Being more of a phenotype difference, the Warburg effect obviously cannot be the root cause of cancer and thus, the most valiant efforts were put into finding genetic and chromosomal factors involved in carcinogenesis. However, something very important might have been forgotten along the way. Even though Warburg had access to limited experimental techniques and incomplete knowledge of genetics (the DNA structure was not even known at the time), he managed to show something critical: cancer cells lack a very important *systems* property, that is, the process of OP is not coordinated with glycolysis and energy metabolism is thus suboptimal in tumors. If we are to develop efficient therapies for cancer, it will be important to assess not the specific properties of one cancer type, but the *systems* properties that induce cancer. In that regard, the aberrant energy metabolism of tumors might provide some cues.

Needless to say, the modern tools of molecular and systems biology can be used to reassess the Warburg effect and analyze the potential of targeting energy metabolism to cure cancer. Regarding the molecular biology behind the Warburg effect, the interested reader is directed to the review by Pedersen (2007), which offers an insightful, chronological report of some scientific advances since the days of Warburg. Elsewhere in this book,

Chapter 9 covers metabolic pathway modeling and presents a review of the current state of knowledge on the implications of metabolism in cancer.

The huge steps forward in our understanding of tumor development were possible because of the painstaking work of biologists, biochemists, and molecular biologists. As of now, the automation of biological measurements and high-throughput methods allows scanning an organism at practically all levels: genome, transcriptome, proteome, and metabolome. And yet, we won't completely elucidate cellular functioning (and malfunctioning) in the foreseeable future. Thus, new approaches to existing problems are sought. In that regard, systems biology offers unique opportunities to provide meaning to data and cellular processes that were, in the past, analyzed separately. According to Franklin M. Harold (2001 p. 65): "The time has come to put the cell together again." This chapter will thus present a possible way to do so, using energy as the underlying, common factor to all cellular processes.

10.2 ENERGY AND ENERGY METABOLISM IN CANCER

As was mentioned previously, the major phenotype particularity of a cancer tumor, the Warburg effect, seems to be a crisis in energy management. Energy is involved in cellular division, maintenance, and death, three modes between which a cancer cell will not switch normally. Most signaling pathways will include protein phosphorylation cascades, which require ATP. It would probably be impossible to identify a cellular process that does not involve, directly or indirectly, energy. The most fundamental function of a living organism, that is, the transcription of its genome and further protein synthesis, can account for a significant portion (up to 40% to 50%) of the energy budget (Buttgereit, Burmester, and Brand 2000). These considerations are presented in Figure 10.1, where energy is seen as being central in the multi-input, multi-output system that is the cell.

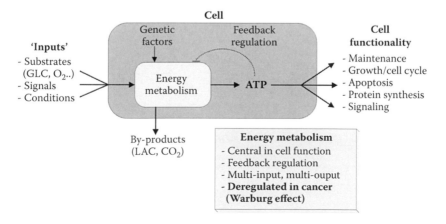

FIGURE 10.1   Energy in cellular processes.

Thus, it is not preposterous to assume that energy metabolism and energy-related interactions are, at some point, critical in cancer development. Recent findings (Pastorino, Shulga, and Hoek 2002; Pollak 2009; Heiden et al. 2001) have pointed at energy metabolism as being extremely important in cancer development. Mitochondrial dysfunctions because of impaired mtDNA are important in many, if not all, cancer types (Carew and Huang 2002). This implies that cancer cells are gradually deprived of their major energy generators (the mitochondria) and have to reorganize their metabolic activity. Analyzing the processes by which cells produce and manage energy will thus be crucial.

Another reason to consider energy metabolism is the diversity of experimental systems and cellular species that are studied in cancer research. Cell extracts and organelles (i.e., mitochondria) are used for molecular biology works, yeasts are used to analyze the cell cycle, tissues are used to study tumor development, animal models are used for genetic and *in vivo* studies, and, finally, humans are involved in clinical trials. In other words, cancer research and development of therapies has to deal with biological mechanisms that span many orders of magnitude both in time and space (Butcher, Berg, and Kunkel 2004). This concept is shown in Figure 10.2.

This layout presents, roughly, the available experimental systems for cancer studies (horizontal axis) and the relevant cellular mechanisms (vertical axis), from simple (*in vitro*

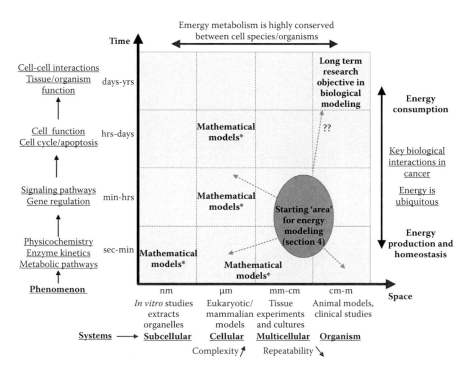

FIGURE 10.2 Energy and cancer systems biology. Horizontal axis shows the various experimental models on a physical scale. Vertical axis shows the various cellular processes involved in cancer on a time scale.

enzyme kinetics) to complex (cell function in the human body) that can be involved in cancer research. In that scheme, energy is definitely the conserved factor (horizontally and vertically). The most preserved biological mechanisms and pathways usually related to energy metabolism as glycolysis, TCA cycle, energy signaling, etc., are found in virtually all cells. It is thus hypothesized here that energy metabolism and energy-related signaling will be crucial in integrating the wide array of cellular processes relevant to cancer systems biology.

10.3 MATHEMATICAL MODELING OF CELLULAR PROCESSES IN CANCER SYSTEMS BIOLOGY

10.3.1 Integration with Traditional Approaches

Since the vertical (time) and horizontal (space) dispersion of biological mechanisms in Figure 10.2 is complicating our experimental investigations, a mathematical modeling approach can help to cope with this research challenge. Mathematical modeling is a powerful tool to organize knowledge and hierarchy in biological systems (Haefner 1996). Known biological processes (i.e., enzyme reactions, signaling, and gene regulation) can be translated into equations (i.e., differential equations) and solved analytically or numerically. The simulation results, in the form of time profiles of reaction rates, concentrations, and various protein/cellular functions, can then be compared to actual experimental data to improve the modeling. This comparison of experimental and theoretical works is actually extremely important in systems biology and in biology in general, as it improves the classical approach of formulating a hypothesis and trying to confirm or disprove it through experimental work alone. This is exemplified in Figure 10.3, where mathematical modeling of biological processes is included in the loop.

As mentioned by Khalil and Hill (2005), Stransky et al. (2007), and Ribba, Colin, and Schnell (2006), the integration of available biological knowledge through dynamic mathematical modeling will be crucial for cancer research. The physiome project is another example of such integration of biological knowledge through modeling (Crampin et al. 2004).

10.3.2 Modeling and Differential Equations in Systems Biology

As reviewed elsewhere in this book (Chapter 16, Modeling Tools for Cancer Systems Biology), several approaches and tools are available when modeling biological systems. Depending on the desired level of abstraction that is to be achieved, the model can take many forms. When the emphasis is on the dynamics of biological processes, ordinary differential equations (ODE) are frequently used. Several textbooks on the modeling of biological systems also use differential equations as the common ground for the mathematical description of the dynamics of biological processes. Haefner (1996) and Heinrich and Schuster (1996) are two good examples for the interested reader.

The use of ODE also allows many links with other fields of research, such as dynamical systems analysis and control engineering. The ODE model developed by Lotka and

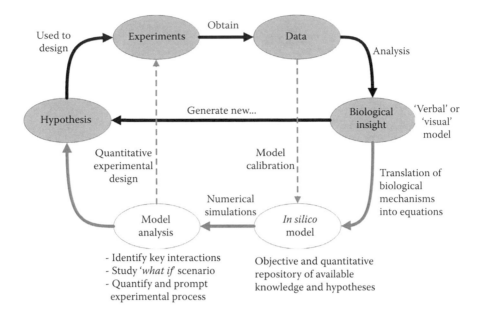

FIGURE 10.3 Mathematical modeling (grey lines) and interactions with the traditional approach (black lines).

Volterra to describe predator-prey interactions was one of the first attempts at a systems analysis of a biological process (Israel 1988). To emphasize the power of ODE models for systems analysis of biological problems, a simple, generic example can be used. Let's look at a pathway where a signal (S) is converted to a response (R) through a set of linear reactions with intermediates (X_1, X_2) and with feedback of the system's response (R) on the first step of the pathway. This pathway is shown in Figure 10.4.

This motif is found in various pathways for signaling, gene regulation, or metabolism. In glycolysis (a linear pathway) the third reaction, catalyzed by phosphofructokinase (PFK), is regulated by one of the end products of glycolysis (ATP) and also by citrate, one of the products of the TCA cycle. The example in Figure 10.4 could also serve as an analogy with gene regulation networks, where one signal upregulates a first transcription factor, which in turn upregulates a second transcription factor and so on until the desired response is achieved and the first step is inhibited (see Goldbeter et al. 2001 for further examples on gene regulation). "Visual" models (or cartoons) such as the one presented in Figure 10.4 are very common in biology and biologists will rapidly understand the implications and hypotheses behind such a graphical model.

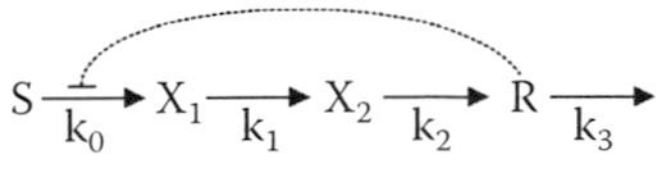

$$S \xrightarrow[k_0]{} X_1 \xrightarrow[k_1]{} X_2 \xrightarrow[k_2]{} R \xrightarrow[k_3]{}$$

FIGURE 10.4 A linear pathway with feedback inhibition.

However, the dynamic response of a system with feedback regulation is not necessarily intuitive and the transcription of the model in Figure 10.4 into an ODE form is necessary for a rigorous analysis of the interactions between the signal (S) and response (R). As the objective of this chapter is not to explain the details of ODE model construction, the interested reader is encouraged to read textbooks such as the one by Haefner (1996) for further details on how to build such models. ODE models can be built and solved using various software packages. MATLAB® (The Mathworks Inc.) and Mathematica® (Wolfram Research) are the most common commercial packages. Various free software packages such as Gepasi (http://www.gepasi.org) are also available to develop ODE models of biological systems. Chapters 9 and 16 in this book also offer a good overview of possible approaches for the modeling of metabolic pathways and links with experimental metabolomics studies. Moreover, many databases and repositories are now available to store and exchange ODE models, with the possibility of integrating cellular processes at a relatively large scale. Biomodels (http://www.biomodels.org) and CellML (http://www.cellml.org) are two very good examples of such databases.

Concerning our example in Figure 10.4, we can first derive the mass balances for X_1, X_2, and R:

$$\frac{\partial X_1}{\partial t} = k_0 \cdot S \cdot \left[\frac{1}{1 + (R/K_I)^{nH}} \right] - k_1 \cdot X_1$$

$$\frac{\partial X_2}{\partial t} = k_1 \cdot X_1 - k_2 \cdot X_2$$

$$\frac{\partial R}{\partial t} = k_2 \cdot X_2 - k_3 \cdot R$$

To the untrained eye, this mathematical description looks rather "dry" and will not reveal much about the underlying biology. These equations are nevertheless a transcription of the model presented in Figure 10.4. Each equation shows the rate of change of species involved in the system ($\partial X_1/dt$ ... in units of X_1 per unit of time). In each case, this rate of change is the difference between synthesis (first term on the right-hand side) and consumption or degradation of the molecule (second term on the right-hand side). Here, we assume simple mass action kinetics for all reactions ($k_i \cdot X_i$...) and the inhibition of the first step by the end product is described by the Hill equation: $1/(1+(R/K_I)^{nH})$.

When numerical values for parameters (k_0, k_1, k_2, K_I, nH) and initial conditions (amounts of X_1, X_2 and R at $t = 0$) are defined, the system can be solved analytically or numerically (the latter being more common with large systems). Figure 10.5 shows simulation results where an input ($S = 1$) was imposed after 5 units of time and the response (R) is shown with three values for the Hill coefficient ($nH = 1$, 10, and 20).

As can be seen, variations in parameter nH will induce qualitative changes in the system's behavior. At low values (black line in Figure 10.5) the system behaves as is normally

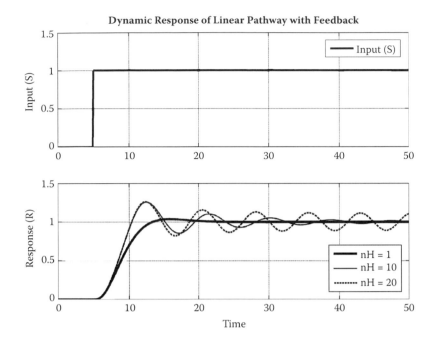

FIGURE 10.5 Dynamic response of the linear pathway with feedback inhibition. Parameters for the model were: $k_0 = 1$; k_1, k_2, $k_3 = 0.5$; $K_1 = 1$. A signal ($S = 1$) is imposed after 5 units of time.

expected in dose-response experiments, as a given input signal gives a stable response (R) after a short time. However, if the strength of the feedback inhibition is increased an unstable behavior develops, with damped oscillations for $nH = 10$ (grey line in Figure 10.5) or sustained oscillations for $nH = 20$ (dashed grey line). Of course, the system presented here is generic, but it must be mentioned that oscillations are observed in simple biological networks with negative feedback. Glycolytic oscillations have been studied for decades (Chance and Gosh, 1964) and genetic oscillators are also observed, even in simple systems (Guantes and Poyatos 2006). What we observe here is a mix of negative feedback and delay, that is, because of the three steps, the signal S is not instantly converted to response R. This combination is known to induce oscillations in dynamical systems. Again, the interested reader can find a more elaborate treatment of necessary conditions for biological oscillations in textbooks such as Heinrich and Schuster (1996).

First, these simulation results show that it might not be possible to guess a system's behavior by simply looking at the topology of the network. Here the topology is simple and does not change, but a variation in one parameter can change the qualitative behavior (from stable to unstable response). Second, it shows the importance of correctly analyzing the *dynamics* of the system, that is, the time-profile response after a perturbation. At the functional level, the dynamic response of a biological system is extremely important (Csete and Doyle 2002), because the coordination between cell components and the processing between signal and response are often critical (Tyson, Chen, and Novak 2003). Moreover, if oscillations were observed the usual approach of analyzing dose-response curves would be prone to intrinsic errors (i.e., the response is not stable) and a dynamic time-profile analysis would be more appropriate.

Finally, the example presented in Figures 10.4 and 10.5 shows that the *regulation structure* and regulatory interactions are important in determining the behavior of a biological system. Here, the key component in Figure 10.4 is not a specific reaction or topology, but the interaction between the response (R) and the first step in the pathway. As will be shown in the next section, the correct description of this regulation structure in tissue energy metabolism can be used to reassess the Warburg effect and identify potential therapeutic targets (i.e., sensitive parameters in the system).

10.4 CASE STUDY: MATHEMATICAL MODELING OF TISSUE ENERGY METABOLISM

10.4.1 Modeling Approach for Tissue Energy Metabolism

The dynamic modeling of metabolism is an established method in various fields such as biomolecule production, environmental engineering, food processing, and, more recently, in health sciences. This approach has been applied (specifically for energy metabolism) to many cellular systems such as *Escherichia coli* (Chassagnole et al. 2002), yeasts (Duarte, Herrgard, and Palsson 2004), muscle cells (Korzniewski and Zoladz 2001; Lambeth and Kushmerick 2002), blood cells (Holzhütter 2004), and brain tissue (Aubert and Costalat 2005; Cloutier et al. 2009), just to name a few. In that regard, a considerable amount of tools, kinetic equations, parameters, etc., are available from the literature and allow the rigorous development and application of these models. A common approach is to first develop the model based on available knowledge (reaction stoichiometries, kinetic properties of enzymes, and overall systems properties) and then to refine the modeling by comparing it to first-hand experimental data in an iterative loop (see Figure 10.3). In this section, the method will be presented for a generic modeling of tissue energy metabolism.

Figure 10.6 presents an overview of the major pathways and mechanisms that we will consider here for energy metabolism and energy-related signaling in mammalian tissues. This model describes:

- Glycolysis (reduced to four reactions, with regulation mechanisms on HK, PGI, PFK, and PK)

- Mitochondrial oxidation of pyruvate (lumped in one reaction, but accounting for the supply of O_2, PYR, and ADP)

- Phosphocreatine (PCr) buffering

- Exchanges of GLC, LAC, and O_2 with the blood flow

- Energy consuming processes (lumped in one ATPase reaction)

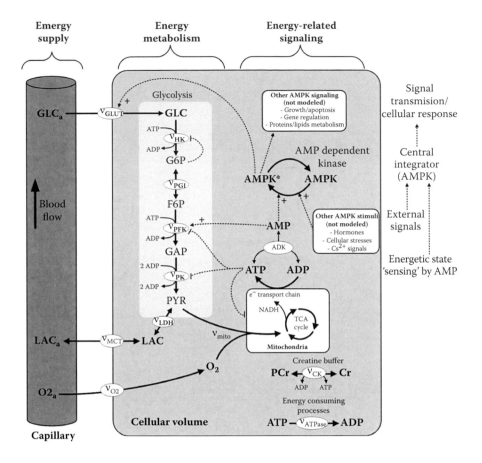

FIGURE 10.6 Tissue energy metabolism and signaling. An ODE implementation is presented in the Appendix.

- Adenylate kinase (ADK) equilibrium

- AMP activated protein kinase (AMPK) signaling for glucose transport (GLUT) and phosphofructokinase (PFK)

Further discussion of modeling hypotheses and complete presentation of the model's equation is found in the Appendix to this chapter.

As was the case for our simple example in the previous section, the mass balances of metabolic reactions are represented as a set of ODE. As an example, the differential equation for the rate of change of GLC will be the difference between uptake (ν_{GLUT}) and consumption by the hexokinases (ν_{HK}):

$$\frac{\partial GLC}{\partial t} = \nu_{GLUT} - \nu_{HK}$$

In a dynamic modeling approach, we will consider the kinetic equations and regulation of the transporter and hexokinases rates by the following equations:

$$v_{\text{GLUT}} = V_{\text{max,GLUT}} \cdot \left[\frac{GLC}{GLC + K_{T,\text{GLC}}} - \frac{GLCa}{GLCa + K_{\text{T,GLC}}} \right] \cdot \left[\frac{AMPK^{\star}}{K_{\text{AMPK}}} \right]^{0.5}$$

$$v_{\text{HK}} = V_{\text{max,HK}} \cdot \left[\frac{GLC}{GLC + K_{\text{m,GLC}}} \right] \cdot \left[\frac{ATP}{ATP + K_{\text{m,ATP}}} \right] \cdot \left[1 + \left(\frac{G6P}{K_{\text{I,G6P}}} \right)^{\text{nH}} \right]^{-1}$$

where the GLC transport is represented by the reversible Michaelis–Menten kinetic with an activation (i.e., similar to an increase in enzyme concentration) represented by the AMPK* term. This term represents the action of phosphorylated AMPK in releasing GLUT4 transporters during high demand periods. The hexokinase rate is a multiplicative Michaelis–Menten kinetic that accounts for the two substrates of the reaction (GLC and ATP). The last term represents the inhibition of hexokinases by their product ($K_{\text{I,G6P}} \approx$ 0.6 mM). Inhibition of HK by its product is an important mechanism to consider here, as it is affected in cancer cells, where HK2 binds to mitochondria, which removes the product inhibition.

This process of building mass balances and kinetic equations can thus be expanded to all the species and fluxes in Figure 10.6, which results in a model with 10 differential equations and 12 kinetic fluxes equations. The complete set of ODE, kinetic equations, and parameters is found in the Appendix. With well-defined kinetic equations, parameters, and initial conditions (i.e., resting metabolite concentrations) representative of the system, it is possible to start a numerical integration routine and obtain time profiles of GLC, v_{GLUT}, v_{HK} … and so on. The abundant literature and databases on enzyme kinetics, cell physiology, and other experimental and theoretical insights can be integrated in the modeling. Any information on steady-state metabolite concentrations and fluxes can also be implemented in the model. If quantitative experimental time profiles of metabolites or fluxes are available, the kinetic parameters of the model can be finely tuned to produce a realistic representation of specific conditions. Simulation results presented in this study were obtained using the Systems Biology Toolbox for Matlab (Schmidt and Jirstrand 2006).

There is recent interest in energy-related signaling, as links between energy and disease are reported and reviewed more and more frequently (Buttgereit, Burmester, and Brand 2000; Pedersen 2007). Interestingly, no mathematical modeling work has been done on the links between energy metabolism and cellular signaling. However, it is plausible that this kind of modeling will appear at some point, especially as mathematical models are now available for apoptosis signaling (Eissing et al. 2004), or cell cycle signaling (Novak and Tyson 2004). Links between energy metabolism and cancer-related processes will also be possible, as ODE models are available for tumor development (Spencer et al. 2004) or the apoptotic pathway (Bentele et al. 2004).

In that regard, the AMP dependent protein kinase (Figure 10.6: AMPK and AMPK* in its active form) will be key in connecting the cellular subsystems for energy management and the underlying biochemical pathways. AMPK signaling could potentially be a central player in disease development, as it is involved in many stress responses such as apoptosis, low glucose, and hypoxia, and in turn it up- or downregulates many signals related to cell growth, protein metabolism, and carbohydrates metabolism (Hardie 2007, 2008). This central cellular signal is thus included here for the regulation of glucose transport and will allow description of important regulatory effects in energy homeostasis.

10.4.2 Biochemical Regulation, Signaling, and Energy Homeostasis

Using the model presented in Figure 10.6, simulations show how the modeling can describe normal dynamic behavior when the energy load is increased (i.e., increase in ν_{ATPase} rate). Figure 10.7 presents simulated time profiles of metabolites and fluxes in response to a 1-, 5-, or 10-fold increase in energy demand for 30 minutes (i.e., similar to the effect of exercise in muscle cells). The ν_{ATPase} rate is thus used here as an "input" signal and we will analyze how the biochemical system responds to that input in order to maintain energy homeostasis.

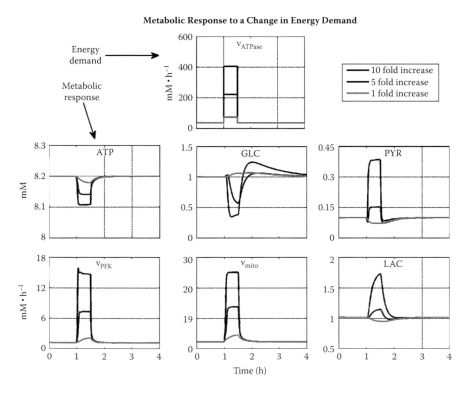

FIGURE 10.7 Metabolic response to a change in energy demand. +100% (light grey), +500% (dark grey), and +1000% (black). Fluxes are mM·h⁻¹ and concentrations are mM.

As can be seen, the model can represent typical observations from cell physiology. First, even in the presence of large changes in energy demand, the ATP concentration remains fairly stable, with only a ≈ 1% to 2% drop in the case of a 10-fold increase in ν_{ATPase} rate. This is reported in many studies (Korzeniewski and Zoladz 2001; Lambeth and Kushmerick 2002). This almost perfect energy homeostasis is achieved here by the sharp regulation of the PFK, a known "pace-making" enzyme in glycolysis and also by regulation of the mitochondrial oxidation rate (ν_{PFK} and ν_{mito} in Figure 10.7). In this model, the regulation of v_{PFK} was achieved by considering AMP signaling (simulated profiles not shown here) and the results (fluxes and concentrations in Figure 10.7) are physiologically realistic.

The observation of an overshoot in LAC during intense metabolic activity is consistent with observations of mammalian tissues submitted to large changes in energy demand such as muscle cells (see black line for LAC profile). Interestingly, this model can also predict LAC *consumption* by the cells if the increase in energy demand is low (see light grey line for LAC in Figure 10.7). From the biochemical regulation, mitochondrial activity is more sharply regulated at low energy loads, thus being able to consume PYR, favoring equilibrium toward LAC consumption. This type of regimen is observed in the brain (Pellerin and Magistretti 1994) and in the heart (Stanley 1991) and is suspected to be important for energy homeostasis.

The timescale for the simulation, with a 30 minute increase in ATPase rate (from 1 to 1.5 hours) is relatively short and represents what would happen at the biochemical regulation level (i.e., no genetic regulation). These results thus show the importance of energetic regulation for the fast response of the organism to an energy stress. The AMPK signaling (included in this model) does increase the amount of GLUT4 transporters. However, this increase is not necessarily at the genetic level, as the GLUT4 are stored in vesicles to allow a fast adaptation. Obviously, genetic regulation could also be included in longer simulations, as it is reported that AMPK signaling can increase the expression of mitochondrial enzymes and other genes related to energy stress.

10.4.3 The Warburg Effect Revisited

From the previous remarks on the Warburg effect, a simulation can be implemented to try and reproduce this typical phenotype. As the mtDNA and overall mitochondrial activity are known to be affected in cancer cells, a possible way to implement a cancer phenotype is to reduce the maximum reaction rate of mitochondrial oxidation (see $\nu_{max,mito}$ in the Appendix, Table A2). This will induce a lower *maximal* capacity for mitochondrial oxidation, while keeping the possibility of observing mitochondrial regulation by external effectors (PYR, ADP/ATP, O_2). Thus, it must be stated here that, for example, a 50% reduction of $\nu_{max,mito}$ will not necessarily result in a 50% decrease of the *actual* mitochondrial rate, as the reduction of the rate will induce PYR and/or ADP accumulation which will partly compensate for the loss of maximal capacity.

This approach thus represents what would happen *in vivo* when the mtDNA is damaged and mitochondrial oxidation cannot be performed correctly, with a cascade of events that leads to a reorganization of metabolites and fluxes. Results for a gradual lowering of

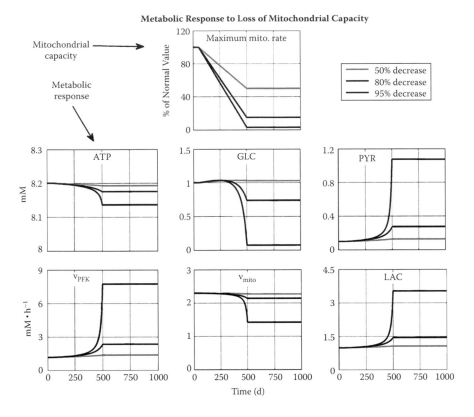

FIGURE 10.8 The Warburg effect at the metabolic level. Mitochondrial maximum rate was reduced by 50% (light grey lines), 80% (dark grey lines), and 95% (black lines) over a period of 500 days.

mitochondrial capacity are presented in Figure 10.8. In that regard, three simulations are performed, with a gradual reduction of 50%, 80%, and 95% of mitochondrial maximal capacity over a period of 500 days.

As can be seen from these simulations (and consistent with previous observations) the system is relatively robust to perturbations. A 50% loss of maximal mitochondrial capacity results in relatively small changes in ATP and other metabolite levels. As was the case for changes in energy demand (Figure 10.7) the adjustment of mitochondrial activity is sharp enough to react to small changes in ATP/ADP (see v_{mito} in the Appendix, Table A2). This could explain why some cancer tumors will not necessarily express a detectable Warburg phenotype, as impaired mitochondrial activity, if it is low enough, will have undetectable effects. However, the simulation shows (quantitatively) that this cellular adaptation has its limits. For 80% loss of mitochondrial capacity (dark grey lines in Figure 10.8), significant changes in ATP, PYR, LAC, and fluxes are observed. These are consistent with the Warburg effect, that is, glycolytic flux (v_{PFK}) is increased, LAC is produced in higher concentrations, and O_2 consumption is decreased, regardless of sufficient O_2 availability, as v_{mito} is lower. The black lines, showing the effect of a 95% loss of maximal mitochondrial capacity, reproduce what would be a "worst case" scenario, where the glycolytic flux produces 40% to 50% of the energy output of the cell. A similar proportion of glycolytic derived ATP was reported by Warburg for fast growing tumors. In that situation, the GLC inflow is even becoming limiting, as simulated GLC level (black line for GLC in Figure 10.8) is lower than the K_m value of hexokinase for GLC (0.1 mM).

As was mentioned previously, the expression and binding of a hexokinase isoform (HK2) to mitochondria was identified as a key mechanism in the development of the Warburg effect (Pedersen 2007). This mechanism increases the glycolytic flux by reducing or removing the inhibition of HK2 activity by its product, G6P. This inhibition is important in glycolysis, as it limits the inflow of GLC if G6P is accumulated. It thus acts as an "upper limit" to avoid pushing too much GLC in the cell if it is not needed.

Since the model used in this study does describe this inhibition mechanism, removing the inhibition—the $(1 + (G6P/K_I)^{nH})^{-1}$ in v_{HK} kinetics—would emulate the switch to HK2 isoform. Simulations of this transition are shown in Figure 10.9 with potential contributions of both impaired mitochondria and HK2 isoform in the Warburg effect.

Figure 10.9 shows that HK binding to mitochondria and removal of G6P inhibition alone can induce the Warburg metabolic phenotype, as it increases glycolytic rate (light grey line for v_{PFK}) while the mitochondrial rate is slightly reduced because of a higher energetic state (light grey lines for v_{mito} and ATP). As a comparison, the 80% loss of mitochondrial activity is also shown in Figure 10.9 (dark grey lines).

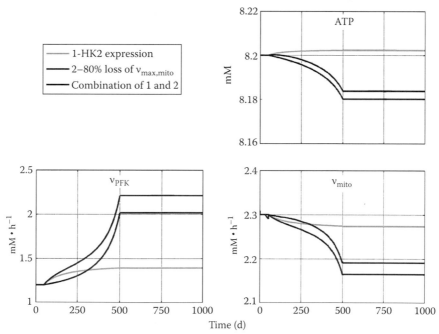

FIGURE 10.9 Combination of HK2 expression and mitochondrial defects to induce the Warburg effect. Light grey lines show expression of HK2, dark grey lines show 80% loss of mitochondrial maximum rate, and black lines show a combination of both factors.

Interestingly, the combination of both factors (black lines in Figure 10.9) will not necessarily add up in a linear fashion to deregulate tissue metabolism. The Warburg phenotype of higher glycolytic flux is indeed worse with a combination of the two factors (black line for v_{PFK}). However, the ATP level is in fact slightly higher with HK binding (black line for ATP) when compared to the case of mitochondrial loss only (dark grey line). Thus, the expression and utilization of the HK2 isoform might induce an energetic advantage to cancer tumors which would promote their survival and proliferation. This observation from model simulations would, however, require further validation at the experimental level.

Interestingly, in all the simulations presented in Figures 10.8 and 10.9, the tissue was able to metabolically adapt to the imposed perturbations, as a physiological steady state is reached in each case. So, for example, even with 95% damage to mitochondria, the tissue is still able to produce sufficient energy output (ATP concentration is reduced, but only slightly). The same can be said for the switch to HK2 where the metabolic state is perturbed but still within physiological range. A case where the tissue would exhibit significant metabolic problems would show ATP levels going to 0 and/or fluxes being reduced significantly (≈ 0), which is clearly not the case here.

Simulation results (not shown here) show that "metabolic death" can be induced if mitochondrial activity is impaired further (>95%). In that case the cells reach a limit in GLC transport capacity and a general degradation of metabolic state ensues (fluxes and metabolites going to 0). However, it is most probable that tumor cells have regulation mechanisms to avoid such a fate. Acidification of the tissue and increased cell proliferation, which are not modeled here, would probably be key mechanisms to promote tumor survival. However, the model presented here is still consistent with the observation of acidification, as LAC concentration is increased, and higher concentrations of glycolytic intermediates are observed (PYR in Figure 10.8), which would allow faster cell growth from increased availability of precursors. Thus, even though the energy modeling does not include growth and proliferation mechanisms, the simulation results observed here are consistent with current observations in tumor proliferation. Long-term simulations could also be improved by considering glycolytic gene expression levels, which are known to change during tumor development. However, the current study shows that the purely biochemical and signaling "management" of energy homeostasis is far from negligible in cancer.

10.4.4 Tumor Metabolic Control Analysis

Metabolic control analysis (MCA; Kascer and Burns 1973; Fell 1997) is a framework to analyze how a small perturbation on a metabolic system (i.e., change in enzyme level or metabolite concentration) can affect the metabolic system locally and globally. In that framework, the percent changes in fluxes (F_i) after a percent change in enzyme concentration (E_j) can be calculated as follows:

$$FCC_j^i = \frac{\Delta F_i / F_{i,0}}{\Delta E_j / E_{j,0}} \quad \text{with } i,j \; \varepsilon [1 \ldots m]; \; m = \text{number of reactions}$$

where $\Delta F_i/F_{i,0}$ is the relative change in flux i and $\Delta E_j/E_{j,0}$ is the relative change in enzyme j concentration. Thus, FCC_j^i is the control coefficient that enzyme j has on reaction i. An $FCC \approx 1$ would mean that a 10% increase in enzyme concentration would induce a 10% increase in flux. Note that this framework can be implemented experimentally, by using gene knock-out/over-expression or inhibitors. At the theoretical level, the implementation is straightforward, as model parameters can be easily changed and simulated fluxes can be compared to the original (unperturbed) behavior.

While this approach might seem highly theoretical, it is actually used extensively in bioprocess research (see Fell 1997 for a few examples) and, more recently, in health sciences. Effectively, the process of finding molecular targets for drug development is quite similar to the MCA framework, as a good drug target is a mechanism that induces a locally sensitive response in the cell (high FCC on a target reaction) with minimal effect on the rest of the system (low FCC on other cellular processes). Potential applications of MCA to disease and drug discovery have been discussed by Cascante et al. (2002) and success in applying MCA to limit tumor proliferation has been reported (Comin-Anduix et al. 2001).

As we have established models of "healthy" and "cancerous" phenotypes, applying the MCA framework to both would allow identifying possible differences in FCC. In that regard, an enzyme that has a high FCC in the disease phenotype, but a low FCC in the healthy phenotype would constitute an excellent drug target, as it would be more likely to affect cancer cells without inducing undesirable effects in normal cells.

Since tumor cells rely on glycolytic energy much more than normal cells, we will calculate the control coefficients of model parameters (enzymes and physiological parameters) on the glycolytic flux (i.e., v_{PFK}). Thus, a high FCC_j^{PFK} would mean that a change in the parameter j will affect the glycolytic flux, while a low value indicates that the flux is not affected. Results for this MCA assessment of the difference between normal and cancerous cells are presented in Figure 10.10.

First, we see that some parameters have positive FCC (increase in parameter leads to a rise in glycolytic flux), while others have negative FCC (inverse effect). As tumors rely on glycolysis, a decrease in the glycolytic flux would be desirable (i.e., negative FCC); however, it must be kept in mind that a parameter that has a positive FCC could also be interesting, in the sense that it shows that this parameter might have an aggravating effect in cancer development (as it helps increase the glycolytic flux). Also, it is clear that parameter values could also be *reduced* rather than *increased* (−1% instead of +1% here). In that case, a parameter that has a positive FCC (in Figure 10.10) could still be manipulated to induce a reduction in glycolytic flux.

As shown in Figure 10.10, the metabolic sensitivity of cancer cells is quite different than that of normal cells. First, cancer cells will be much more sensitive to arterial GLC, which is expected as they rely on high GLC inflow. Interestingly, low GLC diets are used to treat brain tumors in some situations (Seyfried and Mukherjee 2005), and many reports indicate a positive correlation between obesity (i.e., more likely to have high arterial GLC) and cancer. Thus, without being a root cause in carcinogenesis, high arterial GLC might be an aggravating factor and the MCA framework shows it.

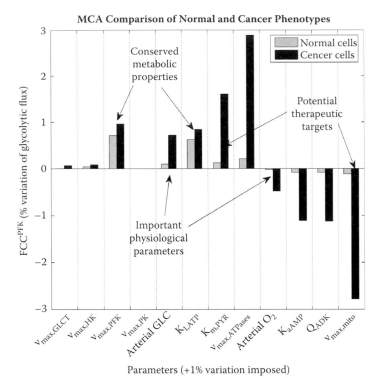

FIGURE 10.10 MCA of cancer cells (black) and normal cells (grey). A 1% variation in parameters was imposed and the vertical axis shows the resulting percent variation in glycolytic rate at steady state.

Not surprisingly, the mitochondrial reaction of the model is also a sensitive parameter. The maximum reaction rate of mitochondrial oxidation ($v_{max,mito}$) has a highly negative effect on the glycolytic flux. Interestingly, the healthy phenotype is much less affected by a change in $v_{max,mito}$. This shows that the mitochondria are indeed a good target to attack tumors, an observation that is coherent with the recent findings on dichloroacetate, a versatile cancer tumor "killer" (which has side effects, however). The sensitivity of mitochondrial affinity for PYR ($K_{m,PYR}$) also exhibits that difference between phenotypes. Parameters related to energy signaling, such as the activation constant for AMP ($K_{a,AMP}$) or the adenylate kinase equilibrium constant (Q_{ADK}), also affect tumor metabolism more aggressively, showing that energy signaling might also be a good target for drug design.

Finally, the glycolytic flux can also be either insensitive to certain parameters ($v_{max,HK}$, $v_{max,PK}$) or it can show similar sensitivities in both healthy and disease cases ($v_{max,PFK}$, $K_{I,ATP}$). These parameters would thus probably be much less promising targets for treatment.

10.4.5 Observations from the Modeling of Energy Metabolism

All the observations reported here from the simulation of energy metabolism are consistent with the current knowledge on tumor metabolism. The modeling presented here is among the first detailed mathematical description of the Warburg effect. Previous attempts concentrated on spatiotemporal aspects, with little detail at the metabolic level (Astanin and Preziosi 2009; Venkatasubramanian, Henson, and Forbes 2006). Metabolically speaking, it seems that cancer cells are more adapted for growth in an environment where GLC is plentiful and they do so without requiring a high inflow of O_2 (but tumors still require a minimal amount of O_2). Moreover, the modeling and MCA results showed that tumors might be more prone to environmental perturbations. Being more adapted for a specific condition (i.e., growth in a GLC-rich environment) usually comes at the price of losing flexibility. This shows that targeting energy metabolism and energy signaling will offer new possibilities to develop efficient and versatile therapies. Again, caution will be required as it is necessary to find targets that affect cancer cells and not normal cells, which is not straightforward given the complexity of energy regulation. As was shown here, the MCA framework could be used to that end.

10.5 CONCLUSIONS AND FUTURE OPPORTUNITIES

When studying a complex dynamical system it is extremely relevant to consider the energy balance. It allows analysis of the limits in terms of stability and operability. In engineering sciences, many examples can be found with energy as the underlying factor for a rigorous systems analysis (Wellstead 1979). So far, this approach is nowhere near being mainstream in any field of biological sciences and bioenergetics is still considered as a separate and relatively minor field. With large-scale, high-throughput experimental techniques and systems biology, there is a need for tools to integrate diverse biological processes. In that regard, it is clear that energy must be considered as the universal, underlying factor that links all the cell's components.

As was shown in this chapter, an energy-based modeling approach can be applied to gain a better understanding of cancer development. The simulation results show the importance of the energetic context in carcinogenesis. Mathematical modeling of energy metabolism allows identifying fundamental differences between healthy and disease phenotypes. This modeling could thus be further used to assess the importance of energy metabolism in specific cellular processes involved in cancer such as apoptosis, cell cycle, signaling pathways, or genetic regulation.

The approach presented here relies on our current understanding of energy metabolism in cancer cells and a dynamic modeling framework, using ODE to describe the major cellular processes in energy management. As the ODE model becomes a repository of available biological knowledge (see Figure 10.3) its power as an analytical tool will increase and it will be possible to make predictions and improve experimental design. Thus, the modeling approach presented here is not an end in itself, but rather a tool that can be refined depending on the specific issue to be addressed. Future work should thus be organized around the mathematical modeling of the energy balance in cellular processes involved in cancer. This approach will eventually offer unique opportunities to develop new and efficient therapies.

REFERENCES

Astanin S. and Preziosi L. (2009) Mathematical modelling of the Warburg effect in tumour cords. *J. Theor. Biol.* **258**: 578–590.

Aubert A. and Costalat R. (2005) Interactions between astrocytes and neurons studied using a mathematical model of compartmentalized energy metabolism. *J.Cerebral. Blood. Flow. Metab.* **25**: 1476–1490.

Bentele M., Lavrik I., Ulrich M., Stosser S., Heermann D. W., Kalthoff H., Krammer P H., and Eils R. (2004) Mathematical modeling reveals threshold mechanism in CD95-induced apoptosis. *J. Cell. Biol.* **166:** 839–851.

Butcher E. C., Berg E. L., and Kunkel E. K. (2004) Systems biology in drug discovery. *Nature Biotechnol* **22:** 1253–1259.

Buttgereit F., Burmester G. R., and Brand M. D. (2000) Bioenergetics of immune functions: fundamental and therapeutic aspects. *Immunol Today* **21:** 192–199.

Carew J. S. and Huang P. (2002) Mitochondrial defects in cancer. *Mol Cancer* **1:** 9–14.

Carling D. (2004) The AMP-activated protein kinase: a unifying system for energy control. *Trends Biochem Sci* **29:** 18–24.

Cascante, M., Beros, L. G., Comin-Andiux, B., Atauri, P., Centelles, J. J., Lee, P. 2002. Metabolic control analysis in drug discovery and disease. *Nat. Biotech.* **20:** 243–249.

Chance B. and Gosh A. (1964) Oscillations of glycolytic intermediates in yeast cells. *Biochem Biophys Res Commun* **16:** 174–181.

Chassagnole C., Noisommit-Rizzi N., Schmid J. W., Mauch K., and Reuss M. (2002) Dynamic modeling of the central carbon metabolism of *Escherichia coli*. *Biotechnol Bioeng* **79:** 53–73.

Cloutier M., Bolger F. B., Lowry J. P., and Wellstead P. (2009) An integrative dynamic model of brain energy metabolism using *in vivo* neurochemical measurments. *J. Comput. Neurosci.* doi: 10.1007/s10827-009-0152-8.

Comín-Anduix B., Boren J., Martinez S., Moro C., Centelles J. J., Trebukhina R., Petushok N., Lee, W. N. P., Boros L. G, and Cascante M. (2001) The effect of thiamine supplementation on tumour proliferation. *Eur. J. Biochem.* **268:** 4177–4182.

Crampin E. J., Halstead M., Hunter P., Nielson P., Noble D., Smith N., and Tawhai M. (2004) Computational physiology and the physiome project. *Exp Physiol* **89:** 1–26.

Csete M. E. and Doyle J. (2002) Reverse engineering of biological complexity. *Science* **295:** 1664–1669.

Duarte N. C., Herrgard M. J., and Palsson B. O. (2004) Reconstruction and validation of *Saccharomyces cerevisiae* iND750, a fully compartmentalized genome-scale metabolic model. *Genome Res* **14:** 1298–1309.

Edwards J. S., Palsson B. O. (2000) The *Escherichia coli* MG1655 in silico metabolic genotype: its definition, characteristics, and capabilities. *Proc Natl Acad Sci USA* **97:** 5528–5533.

Eissing T., Conzelmann H., Gilles E. D., Allgöwer F., Bullinger E., and Scheurich P. (2004) Bistability analyses of a caspase activation model for receptor-induced apoptosis. *J. Biol. Chem.* **279:** 36892–36897.

Fell D. (1997) *Understanding the Control of Metabolism*. London: Portland Press.

Goldbeter A., Gonze D., Houart G., Leloup J-C., Halloy J., and Dupont G. (2001) From simple to complex oscillatory behavior in metabolic and genetic control networks. *Chaos* **11:** 247–260.

Guantes R. and Poyatos J. R. (2006) Dynamical principles of two-component genetic oscillators. *PLoS Comput Biol* **2:** e30.

Haefner J. W. (1996) *Modelling Biological Systems: Principles and Applications*. New York: Chapman & Hall.

Haiden M. G. V., Plas D. R., Rathmell J. C., Fox C. J., Harris M. H., and Thompson C. B. (2001) Growth factors can influence cell growth and survival through effects on energy metabolism. *Mol Cell Biol* **21:** 5899–5912.

Hardie G. D. (2007) AMP-activated/SNF1 protein kinases: conserved guardians of cellular energy. *Nature Rev Mol Cell Biol* **8:** 774–785.

Hardie D. G. (2008) Role of AMP-activated protein kinase in the metabolic syndrome and in heart disease. *FEBS Lett* **582:** 81–90.

Harold F. M. (2001) *The Way of the Cell: Molecules, Organisms, and the Order of Life*. Oxford: Oxford University Press.

Heinrich R. and Schuster S. (1996) *The Regulation of Cellular Systems*. New York: ITP Chapman & Hall.

Heldt H. W., Klingenberg M., and Milovancev M. (1972) Differences between the ATP/ADP ratios in the mitochondrial matrix and in the extramitochondrial space. *Eur. J. Biochem.* **30:** 434–440.

Holzhütter H. G. (2004) The principle of flux minimization and its application to estimate stationary fluxes in metabolic networks. *Eur. J. Biochem.* **271**: 2905–2922.

Israel G. (1988) On the contribution of Volterra and Lotka to the development of modern biomathematics. *History Philos Life Sci* **10**: 37–49.

Jorgensen S. B., Rose A. J. (2008) How is AMPK activity regulated in skeletal muscles during exercise? *Frontiers Biosci* **1**: 5589–5604.

Kacser H. and Burns J. A. (1973) The control of flux. *Symp Soc Exp Biol* **27**: 65–104.

Khalil I. G. and Hill C. (2005) Systems biology for cancer. *Curr Opin Oncol* **17**: 44–48.

Korzeniewski B. and Zoladz J. A. (2001) A model of oxidative phosphorylation in mammalian skeletal muscle. *Biophys Chem* **92**: 17–34.

Lambeth M. J. and Kushmerick M. J. (2002) A computational model for glycogenolysis in skeletal muscle. *Ann Biomed Eng* **30**: 808–827.

Liguzinski P. and Korzeniewski B. (2006) Metabolic control over the oxygen consumption flux in intact skeletal muscle: in silico studies. *Am. J. Physiol. Cell. Physiol.* **291**: C1213–C1224.

Novak B. and Tyson J. J. (2004) A model for restriction point control of the mammalian cell cycle. *J. Theor. Biol.* **230**: 563–579.

Oliveras-Ferraros C., Vazquez-Martin A., Fernández-Real J. M., and Menendez J. A. (2008) AMPK-sensed cellular energy state regulates the release of extracellular fatty acid synthase. *Biochem Biophys Res Commun* **378**: 488–493.

Pastorino J. G., Shulga N., and Hoek J. B. (2002) Mitochondrial binding of hexokinase 2 inhibits bax-induced cytochrome c release and apoptosis. *J Biol Chem* **277**: 7610–7618.

Pedersen P. L. (2007) Warburg, me and hexokinase 2: multiple discoveries of key molecular events underlying one of cancers' most common phenotypes, the "Warburg Effect", i.e., elevated glycolysis in the presence of oxygen. *J Bioenerg Biomembr* **39**: 211–222.

Pellerin L. and Magistretti P. (1994) Glutamate uptake into astrocytes stimulates aerobic glycolysis: a mechanism coupling neuronal activity to glucose utilization. *Proc Natl Acad Sci USA* **91**: 10625–10629.

Pollak M. (2009) Energy metabolism, cancer risk, and cancer prevention. In *Cancer Prevention II*, pp 51–54. Senn, H. J., Kapp. U. Otto F., (Eds).

Ribba B., Colin T., and Schnell S. (2006) A multiscale mathematical model of cancer, and its use in analyzing irradiation therapies. *Theor Biol Med Model* **3**: 7.

Schmidt H. and Jirstrand M. (2006) Systems Biology Toolbox for MATLAB: a computational platform for research in systems biology. *Bioinformatics* **22**: 514–515.

Seyfried T. N. and Mukherjee P. (2005) Targeting energy metabolism in brain cancer: review and hypothesis. *Nutrition Metab* **2**: 30–39.

Spencer S. L., Berryman M. J., Garcia J. A., and Abbott D. (2004) An ordinary differential equation model for the multistep transformation to cancer. *J. Theor. Biol.* **231**: 515–524.

Stanley W. C. (1991) Myocardial lactate metabolism during exercise. *Med Sci Sports Exercise* **23**: 920–924.

Stransky B., Barrera J., Ohno-Machado L., and Souza S. J. D. (2007) Modelling cancer: integration of "omics" information in dynamic systems. *J Bioinformatics Comput Biol* **5**: 977–986.

Tyson J. J., Chen K. C., and Novak B. (2003) Sniffers, buzzers, toggles and blinkers: dynamics of regulatory and signalling pathways in the cell. *Curr Opin Cell Biol* **15**: 221–231.

Venkatasubramanian R., Henson M. A., Forbes N S (2006) Incorporating energy metabolism into a growth model of multicellular tumor spheroids. *J. Theor. Biol.* **242**: 440–453.

Warburg O. (1930) *The Metabolism of Tumours*. London: Constable and Company.

Wellstead P (1979) *Introduction to Physical System Modelling*. London: Academic Press.

APPENDIX

MODEL DESCRIPTION

Mass balances for the 10 metabolic species of the model are presented in Table A1, with reported steady-state concentrations taken for muscle tissue. Note here that conservation relations are used to reduce the number of model states. So for example, the mass balance on phosphocreatine (PCr) is described and creatine is calculated from the conservation of a total PCr + Cr = 35 mM. Table A2 presents the 11 kinetic equations of the model and Table A3 contains the model parameters that were either taken from the literature and/or adjusted to represent available knowledge on dynamic metabolic behavior (i.e., response to muscle contraction). All fluxes and names refer to Figure 10.6.

TABLE A1   Variables, Steady-State Concentrations, and Mass Balances of the Model

	Variable	Steady-State Concentration (mM)	Differential Equation
GLC	Glucose	1	$\dfrac{\partial GLC}{\partial t} = v_{GLUT} - v_{HK}$
G6P	Glucose-6-P	0.8	$\dfrac{\partial G6P}{\partial t} = v_{HK} - v_{PGI}$
F6P	Fructose-6-P	0.2	$\dfrac{\partial F6P}{\partial t} = v_{PGI} - v_{PFK}$
GAP	Glyceraldehyde-3-P	0.03	$\dfrac{\partial GAP}{\partial t} = 2 \cdot v_{PFK} - v_{PK}$
PYR	Pyruvate	0.1	$\dfrac{\partial PYR}{\partial t} = v_{PK} - v_{mito} - v_{LDH}$
LAC	Lactate	1	$\dfrac{\partial LAC}{\partial t} = v_{LDH} - v_{MCT}$
O_2	Intracellular O_2	0.13	$\dfrac{\partial O_2}{\partial t} = v_{O2} - 3 \cdot v_{mito}$
ATP	Adenosine triphosphate	8.2	$\dfrac{\partial ATP}{\partial t} = \begin{bmatrix} 15 \cdot v_{mito} - v_{ATPase} - v_{HK} + v_{PFK} \\ +2 \cdot v_{PK} + v_{CK} - v_{AMPK} \end{bmatrix} \cdot \left(1 - \dfrac{dAMP}{dATP}\right)^{-1}$
			with ATP + ADP + AMP = ANP
PCr	Phosphocreatine	30	$\dfrac{\partial PCr}{\partial t} = -v_{CK}$
			with PCr + Cr = 35 mM
AMPK*	AMP activated kinase (active form)	1e-4	$\dfrac{\partial AMPK^*}{\partial t} = v_{AMPK}$
			with AMPK + AMPK* = 0.1 mM

TABLE A2   Kinetic Equations for Metabolic Fluxes

Reaction	Kinetic Equation
GLC transport	$$v_{\text{GLUT}} = V_{\text{max,GLUT}} \cdot \left[\frac{GLC}{GLC + K_{T,\text{GLC}}} - \frac{GLCa}{GLCa + K_{T,\text{GLC}}} \right] \cdot \left[\frac{AMPK^*}{K_{\text{AMPK}}} \right]^{0.5}$$
Hexokinase	$$v_{\text{HK}} = V_{\text{max,HK}} \cdot \left[\frac{GLC}{GLC + K_{m,\text{GLC}}} \right] \cdot \left[\frac{ATP}{ATP + K_{m,\text{ATP}}} \right] \cdot \left[1 + \left(G6P \big/ K_{I,\text{G6P}} \right)^4 \right]^{-1}$$
Phosphoglucose isomerase	$$v_{\text{PGI}} = V_{\max f,\text{PGI}} \cdot \left[\frac{G6P}{G6P + K_{m,\text{G6P}}} \right] - V_{\max r,\text{PGI}} \cdot \left[\frac{F6P}{F6P + K_{m,\text{F6P,PGI}}} \right]$$
Phosphofructokinase	$$v_{\text{PFK}} = V_{\text{max,PFK}} \cdot \left[\frac{ATP}{ATP + K_{m,\text{ATP}}} \right] \cdot \left[\frac{F6P}{F6P + K_{m,\text{F6P,PFK}}} \right] \cdot AMP_{\text{act}} \cdot ATP_{\text{inh}}$$
Pyruvate kinase	$$v_{\text{PK}} = V_{\text{max,PK}} \cdot \left[\frac{GAP}{GAP + K_{m,\text{GAP}}} \right] \cdot \left[\frac{ADP}{ADP + K_{m,\text{ADP}}} \right] \cdot ATP_{\text{inh}}$$
Mitochondrial oxidation of pyruvate	$$v_{\text{mito}} = V_{\text{max,mito}} \cdot \left[\frac{PYR}{PYR + K_{m,\text{PYR}}} \right] \cdot \left[\frac{ADP}{ADP + K_{m,\text{ADP}}} \right] \cdot \left[\frac{O_2}{O_2 + K_{m,\text{O2}}} \right]$$ $$\cdot \left[\frac{1}{1 + 10 \cdot \dfrac{ATP}{ADP}} \right]$$
Lactate dehydrogenase	$$v_{\text{LDH}} = k_{\text{LDH},f} \cdot PYR - k_{\text{LDH},r} \cdot LAC$$
Lactate transport	$$v_{\text{MCT}} = V_{\text{max,MCT}} \cdot \left[\frac{LAC}{LAC + K_{T,\text{LAC}}} - \frac{LACa}{LACa + K_{T,\text{LAC}}} \right]$$
O_2 transport	$$v_{\text{O2}} = \frac{PS_{\text{cap}}}{V_c} \cdot \left[K_{\text{O2}} \cdot \left(\frac{Hb.OP}{O_2} - 1 \right)^{-1/nh} - O_2 \right]$$
Creatine kinase	$$v_{\text{CK}} = k_{\text{CK},f} \cdot PCr \cdot ADP - k_{\text{CK},r} \cdot Cr \cdot ATP$$
ATPase	$$v_{\text{ATPase}} = V_{\text{max,ATPase}} \cdot \left[\frac{ATP}{ATP + K_{m,\text{ATP}}} \right]$$
AMPK reaction	$$v_{\text{AMPK}} = k_{\text{AMPK},f} \cdot AMPK - k_{\text{AMPK},r} \cdot AMPK^*$$

(continued)

TABLE A2   Kinetic Equations for Metabolic Fluxes (Continued)

Reaction	Kinetic Equation
	Adenylate kinase equilibrium

$$ADP = \frac{ATP}{2} \cdot \left[-Q_{ADK} + \sqrt{u} \right]$$

with $u = Q_{ADK}^2 + 4 \cdot Q_{ADK} \cdot \left(\frac{ANP}{ATP} - 1 \right)$

$$\frac{dAMP}{dATP} = -1 + \frac{Q_{ADK}}{2} - 0.5 \cdot \sqrt{u}$$

$$AMP = ANP - ATP - ADP$$

$$+ Q_{ADK} \cdot \frac{ANP}{ATP \cdot \sqrt{u}}$$

Energy signaling (for v_{PFK} and v_{PK})

$$ATP_{inh} = \left[\frac{1 + n_{ATP} \cdot \frac{ATP}{K_{I,ATP}}}{1 + \frac{ATP}{K_{I,ATP}}} \right]^4 \qquad AMP_{act} = \left[\frac{1 + \frac{AMP}{K_{a,AMP}}}{1 + n_{AMP} \cdot \frac{AMP}{K_{a,AMP}}} \right]^4$$

TABLE A3   Model Parameters

Parameter Name	Value	Description	Units
$V_{max,GLUT}$	4.95	Reaction rate constant for glutamine synthase	mM·h^{-1}
$V_{max,HK}$	141.3	HK maximum reaction rate	mM·h^{-1}
$V_{maxf,PGI}$	82.2	PGI maximum forward reaction rate	mM·h^{-1}
$V_{maxr,PGI}$	86.4	PGI maximum reverse reaction rate	mM·h^{-1}
$V_{max,PFK}$	115	PFK maximum reaction rate	mM·h^{-1}
$V_{max,PK}$	2.2·105	PK maximum reaction rate	mM·h^{-1}
$V_{max,mito}$	175.1	Mitochondrial maximum reaction rate	mM·h^{-1}
$V_{max,ATPase}$	36.9	ATPase reaction rate	mM·h^{-1}
$K_{LDH,f}$	19	Forward reaction rate for LDH	h^{-1}
$K_{LDH,r}$	1.8	Reverse reaction rate for LDH	h^{-1}
$V_{max,MCT}$	0.4	Maximum LAC transport rate	mM·h^{-1}
$K_{T,GLC}$	0.5	Affinity constant for GLC transport	mM
$K_{m,GLC}$	0.105	Affinity constant for intracellular GLC	mM
$K_{m,ATP}$	0.05	Affinity constant for ATP	mM
$K_{m,ADP,PK}$	0.0013	Affinity constant of PK for ADP	mM
$K_{m,ADP,mito}$	0.005	Affinity constant of mitochondria for ADP	mM
$K_{m,G6P}$	0.5	Affinity constant for G6P	mM
$K_{I,G6P}$	0.6	Inhibition constant for G6P	mM
$K_{m,F6P,PGI}$	0.15	Affinity constant of PGI for F6P	mM
$K_{m,F6P,PFK}$	0.058	Affinity constant of PFK for F6P	mM
$K_{m,GAP}$	0.5	Affinity constant for GAP	mM
$K_{m,PYR}$	0.6	Affinity constant for PYR	mM
$K_{m,O2}$	0.01	Affinity constant of mitochondria for O$_2$	mM
K_{O2}	0.0897	Transport constant for O$_2$	mM·h^{-1}
$K_{T,LAC}$	1	Affinity constant for LAC transport	mM
$K_{CK,f}$	3.7·103	Forward reaction rate for CK	h^{-1}
$K_{CK,r}$	35.8	Reverse reaction rate for CK	h^{-1}

(*continued*)

TABLE A3   Model Parameters (Continued)

Parameter Name	Value	Description	Units
$K_{I,ATP}$	1	Inhibition constant for ATP	mM
$K_{a,AMP}$	0.0005	Activation constant for AMP	mM
n_{ATP}	0.25	Inhibition constant for ATP regulation	—
n_{AMP}	0.25	Activation constant for AMP regulation	—
$K_{f,AMPK}$	19	Forward reaction constant for AMPK activation	h^{-1}
$K_{r,AMPK}$	1.8	Reverse reaction constant for AMPK deactivation	h^{-1}
K_{AMPK}	$1 \cdot 10-4$	Activation constant for AMPK	mM
GLC_a	5	Arterial glucose	mM
O_{2a}	8.34	Arterial O_2	mM
LAC_a	0.3	Arterial LAC	mM
ANP	8.212	Total energy shuttles concentration	mM
V_c	0.7	Volumetric fraction of cells in tissue	—
nh	2.7	Reaction order constant for O_2 transport	—
PS_{cap}	0.11	Capillary transport constant	—
Hb.OP	8.6	Oxygenated hemoglobin	mM
Q_{ADK}	1.03	Adenylate kinase equilibrium constant	—

Kinetic equations, parameters, and experimental insights and steady-state concentrations for glycolysis were taken from Heinrich and Schuster (1996), Korzeniewski and Zoladz (2001), Lambeth and Kushmerick (2002), and Liguzinski and Korzeniewski (2006). Mitochondrial function (synthesized into one reaction) is based on works by Aubert and Costalat (2005) and experimental insights on the regulation by ADP/ATP ratio were taken from Heldt, Klingenberg, and Milovancev (1972). Adenylate kinase equilibrium was treated as described in Heinrich and Schuster (1996). Energy signaling and AMPK kinetics is based on works by Carling (2004) and insights reviewed in Hardie (2007).

CHAPTER 11

Cancer Gene Prediction Using a Network Approach

Xuebing Wu and Shao Li

CONTENTS

11.1 INTRODUCTION

Cancer is a genetic disease (Vogelstein and Kinzler 2004). Decades of research in molecular genetics have identified a number of important genes responsible for the genesis of various types of cancer (Futreal et al. 2004) and drugs targeting these mutated cancer genes have brought dramatic therapeutic advances and substantially improved and prolonged the lives of cancer patients (Huang and Harari 1999). However, cancer is extremely complex and heterogeneous. It has been suggested that 5% to 10% of the human genes probably contribute to oncogenesis (Strausberg, Simpson, and Wooster 2003), while current experimentally validated cancer genes only cover 1% of human genome (Futreal et al. 2004), suggesting that there are still hundreds or even thousands of cancer genes that remain to be identified. For example, in breast cancer, known susceptibility genes, including BRCA1 (Miki et al. 1994) and BRCA2 (Wooster et al. 1995), can only explain less than 5% of the total breast cancer incidence and less than 25% of the familial risk (Oldenburg et al. 2007). The same challenge is also faced by other types of cancer and other complex diseases, such as diabetes (Frayling 2007) and many brain diseases (Burmeister, McInnis, and Zollner 2008; Folstein and Rosen-Sheidley 2001). There is a long way to go from changes in genetic sequence to visible clinical phenotypes. The complex molecular interaction networks, together with environmental factors, further lower the penetrance of a single causal gene and complicate the relationship between genes and diseases. This high complexity and low penetrance might explain why so many disease genes remain unidentified.

Traditional gene mapping approaches, such as linkage analysis and association studies, have limited resolution to localize the causal genes in the genome, and the resultant region often contains hundreds of candidate genes (Altshuler, Daly, and Lander 2008). The functional testing and validation of causative genes are time consuming and laborious. The priority of candidate genes is usually determined by expert judgment based on the gene's known functions (Pharoah et al. 2007), which are often biased and limited by the scope of the expert. Alternatively, with the increasing availability of genome-wide sequence, genomics, proteomics, and epigenomics data, computational methods are exploited to predict and prioritize disease genes (Oti and Brunner 2007; Zhu and Zhao 2007), significantly reducing the number of candidate genes for further testing. Computational prediction and prioritization is complementary to genetic mapping, in terms of integrating existing knowledge on disease biology and relatively unbiased whole genome measurements.

More recently, large-scale molecular interaction network data have become available, and it turns out to be particularly powerful for disease gene prediction when used alone (Kohler et al. 2008; Oti et al. 2006) or combined with other data sources (Karni, Soreq, and Sharan 2009; Lage et al. 2007; Mani et al. 2008; Wu et al. 2008). Molecular interaction networks depict the basic skeleton of cellular processes, and network analysis has the ability to model the complex interactions among multiple genes and their higher-level organizations (Barabasi and Oltvai 2004; Han 2008; Zhu, Gerstein, and Snyder 2007). In this chapter, we will focus on network-based approaches for cancer gene prediction. Many of the methods discussed here are designed for general disease instead of cancer. Nonetheless, they can be applied to predict cancer genes as a special case, and most of these network-based methods have been demonstrated by applying them to various types of cancer.

11.2 MOLECULAR NETWORKS AND HUMAN DISEASES

Before going into the details of network-based gene prioritization methods, we will briefly introduce some basic concepts about molecular networks, the data sources and tools for building networks, and the working principles for network approaches in predicting disease genes.

Network is a simple but efficient abstraction of biological systems (Barabasi and Oltvai 2004). Nodes/vertices in a molecular network represent biomolecules, such as genes, proteins, and metabolites. Edges/links between nodes indicate physical or functional interactions, including transcriptional binding, protein-protein interaction, genetic interaction (such as synthetic lethal), biochemical reactions, and many others. An edge on a network (if it happens in the cell) shows that two molecules are functionally related to each other, and the distance on a network is correlated with functional similarity (Sharan, Ulitsky, and Shamir 2007). Network/graph theory provides multiple definitions and tools to measure the distance/proximity between two nodes on a network, which makes network analysis particularly suitable to the quantitative modeling of gene-gene and gene-disease relationships (see Box 11.1 for basic graph concepts).

BOX 11.1   BASIC GRAPH CONCEPTS

A **graph** is a pair G(V,E), where V is a set of nodes (or vertices) and E is a set of edges (or links, or interactions) connecting pairs of nodes. On molecular interaction networks, the nodes represent molecules such as genes or proteins, and the edges represent interactions such as protein-protein interaction, transcriptional binding between protein and DNA.

A graph can be represented by an **adjacent matrix** A, where $A_{ij} = 1$ if there is an edge between nodes i and j; otherwise $A_{ij} = 0$.

A **path** from node A to B is a sequence of nodes started with A and ended with B, such that from each of its nodes there is an edge to the next node in the sequence.

The **length** of a path is the number of edges in the path.

The **distance** of two nodes is usually defined as the length of the shortest path between the nodes. More complex definitions of graph distance are discussed in the main text.

The **kth-order** neighbor of a node is the node whose distance from it is k.

The **centrality** of a node measures how centrally a node is located in a given graph. Four commonly used centrality measures are degree, betweenness, closeness, and eigenvector centrality.

The **degree** of a node is the number of edges it is connected with.

The **eigenvector centrality** is a weighted version of the degree centrality, such that x_i of node i is proportional to the sum of the centralities of its neighbors:

$$x_i = \lambda^{-1} \sum_{j=1}^{n} A_{ij} x_j$$

Let the vector $x = (x_1, x_2, ..., x_n)$ be the centralities of the nodes; then we have

$$\lambda x = Ax$$

where x is an eigenvector of the adjacency matrix A with eigenvalue λ. Theoretical results show that there is only one eigenvector x with all centrality values non-negative and this is the unique eigenvector that corresponds to the largest eigenvalue λ. Eigenvector centrality assigns each node a centrality that not only depends on the quantity of its connections, but also on their qualities.

The **closeness** of a node measures the centrality of a node based on how close it is to other nodes in the network. It can be calculated by inverting the sum of the distances from it to other nodes in the network.

The **betweenness** of a node is the number of shortest paths between other nodes that run through the node of interest. Betweenness centrality characterizes the control of a node over the information flow of the network.

Until now, widely used large-scale human gene/protein networks have been generated mainly by four approaches: high throughput technology for large-scale screening of genetic interaction or protein-protein interaction, manual curation of high-quality interaction data from published small-scale experiment results, automatic text mining to extract gene interactions from the published literature, and computational prediction by integrating multiple genomics data. Generally, high-throughput technology such as yeast-2-hybrid (Fields and Song 1989; Fields and Sternglanz 1994) can yield relatively unbiased protein interaction data, but the false positive rate can reach 50% (Sprinzak, Sattath, and Margalit 2003; von Mering et al. 2002). In addition, though the interactomes (a full list of interactions) for species like yeast (Ito et al. 2001), worm (Li et al. 2004), and fly (Giot et al. 2003) have been extensively mapped using high-throughput technology, data generated in this way for human (Ghavidel, Cagney, and Emili 2005; Rual et al. 2005) composes only a small part of the known human interactome data. On the other hand, the most reliable experimental data comes from manual curation of interaction data reported by traditional small-scale experiments, and most of these data has been included in manually curated databases such as HPRD (Peri et al. 2003), BIND (Bader, Betel and Hogue 2003), and BioGRID (Breitkreutz et al. 2008). Occasionally traditional pathway-based databases are also used, including KEGG (Kanehisa and Goto 2000) and Reactome (Vastrik et al. 2007). Despite the intensive effort in mapping the human protein network, the current human interactome is far from complete (Hart, Ramani, and Marcotte 2006). Automatic literature mining techniques have also been developed to identify putative interacting relationships between human genes/proteins described in the published biomedical literature, such as the GENEWAYS system (Rzhetsky et al. 2004). Literature mining also has the advantage that is allows the construction of context-specific networks, such as the prostate cancer specific gene network (Ozgur et al. 2008) and angiogenesis network (Li, Wu, and Zhang 2006). In the LMMA (Li, Wu, and Zhang 2006) approach, we have also shown that the systematic integration of microarray data significantly refines the literature mined network and yields more biological insights. Finally, multiple computational approaches (Franke et al. 2006; Jansen et al. 2003; Lage et al. 2007; Rhodes et al. 2005; Xia, Dong, and Han 2006) have been developed to predict a comprehensive human interactome map, usually by integrating a number of unbiased genome-wide annotation data, such as sequence, expression, functional annotation, known interaction data, and many others. Among these datasets, homologous mapping is commonly used to transfer protein interactions from other organisms to human by sequence conservation. Typical high-quality interaction databases for other organisms include: BioGrid (Breitkreutz et al. 2008), BIND (Bader, Betel and Hogue 2003), MIPS (Mewes et al. 2004), DIP (Salwinski et al. 2004), MINT (Chatr-aryamontri et al. 2007), and IntAct (Kerrien et al. 2007). STRING (von Mering et al. 2005) and OPHID (Brown and Jurisica 2005) are two of the widely used databases hosting predicted interactions.

With all these network data available, studies on model organisms have shown that central positions on the network implicate important roles in cellular processes. For example, in yeast, the number of partners of a gene is positively correlated with lethal phenotypes (Jeong et al. 2001). With the increasing availability of human protein interaction data, network analysis has also shed light on human diseases. For example, consistent with the

observation from yeast, human disease genes tend to have higher network centrality, such
as higher degrees, compared to nonessential and nondisease genes (Feldman, Rzhetsky,
and Vitkup 2008; Goh et al. 2007; Xu and Li 2006), and cancer genes are found to be even
more central than other disease genes (Goh et al. 2007; Jonsson et al. 2006). Besides, consistent with the long-held assumption that genes that are closely related are more likely to
cause the same or similar diseases, network analysis shows that genes causing the same
or similar diseases are likely to interact directly or indirectly with each other (Lim et al.
2006; Oti et al. 2006; Oti and Brunner 2007; van Driel et al. 2006). For example, Lim et al.
(2006) show that many ataxia-causing proteins share interacting partners and form a small
tightly connected subnetwork. Recent genome-wide cancer mutation screen studies suggest that, though ~80 mutations can be found in a typical cancer, they tend to fall into a
few functional pathways (Wood et al. 2007). The functional relatedness of genes causing
similar diseases seems to be very general for human diseases, and network analysis provides powerful tools to fully exploit its potential in human disease study. Recently various
network-based approaches have emerged to predict disease genes based on the observations described above, generally achieving much better performance than traditional disease gene prediction approaches.

11.3 NETWORK APPROACH FOR CANCER GENE PREDICTION

For clarity we first give the typical settings for a network-based disease gene prediction
method (Figure 11.1). Given a list of N candidate genes which is assumed to contain at least
one disease gene, the goal is to pick out the true disease gene or to rank it at top Mi, where
M is much smaller than N. The candidate genes can be genes within a linkage interval
having been associated with the disease under study. Or, if there is no genetic mapping

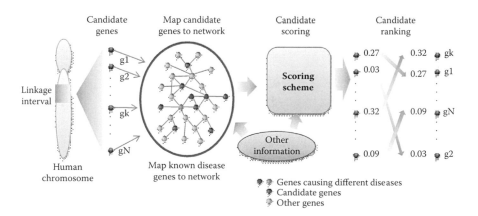

FIGURE 11.1 (See color insert following page 332.) Sketch map of network-based candidate gene prioritization and prediction. A list of candidate genes such as those in a linkage interval or all the human
genes are mapped onto a human gene/protein network, and if applicable, known disease genes and other
information (such as sequence characteristics and mRNA expression) are also mapped onto the network. A scoring scheme is used to score each candidate gene based on current data and outputs a rank
list of all candidate genes. Genes ranked above a certain position are predicted as disease causative.

information, one can simply use the entire human genome as the candidate list. Next, all candidate genes are mapped to a human gene/protein network, the construction of which is described in the previous section. If applicable, known disease genes and other information are mapped to the network too. After that, a scoring scheme scores each candidate gene according to its relative position on the network and additional information. The score is assumed to reflect the probability of the candidate gene to cause the disease under study, given the observed data sources. Finally, all candidate genes are ranked according to the score, and the top 1 or top M genes are predicted to be disease causing. The predictability of this score or the performance of the proposed approach is often assessed by cross-validation with known gene-disease relationships (the ability to rediscover known disease genes).

The scoring scheme is the key to a disease gene prediction method. In the following section, we will review different scoring functions used by different methods. To be clearer, we group these methods by the basic principles underlying their scoring schemes (Table 11.1).

11.3.1 Prioritize by Network Proximity

The common principle underlying all methods in this category is "guilt-by-proximity," that is, genes that lie closer to each other on the network are more likely to lead to the same disease. If some genes are already known to be related to the disease under study, then basically one can use the inverse of the distance (proximity) to these disease genes as the score. Otherwise, distance between candidate pairs is used. The methods described below differ in the way they define the distance measure and how the distance is combined with other information to rank candidate genes.

11.3.1.1 Proximity to Known Disease Genes of the Same Disease

Roughly about half of the diseases in the OMIM database (McKusick 2007) have at least one gene known to be involved in the particular disease. For these diseases, the most straightforward way to score and rank candidate genes is to use the proximity to known disease genes as the measure of the disease causing probability. If a candidate is more closely related to a known disease gene, it is more likely to be a disease gene too; therefore, it should get a higher score. If multiple disease genes are already known, then the final score will be the sum of scores across all known disease genes. This procedure can be viewed as a propagation of disease signal: known disease genes serve as the source of disease signal and this signal is propagated along paths on the network to other nodes, and the signal gradually damps as it travels to more distant nodes. Now the problem is how to define distance between two nodes in a network. Three types of distance measure can be found in disease gene finding approaches: direct neighbor, shortest path length, and global distance defined by diffusion kernel or random walk.

11.3.1.1.1 Direct Neighbor   In this type of measure, nodes that are directly connected have a distance of 1; otherwise they have a distance of infinity. Approaches employing this measure are actually doing neighbor counting: candidates with more neighbors causing the

TABLE 11.1   A Summary of Network-Based Disease Gene Prediction Methods

Method	Disease Tested	Network Data Sources
Proximity-based		
Direct neighbor		
Oti et al. 2006	General	HPRD, DIP
CPS (George et al. 2006)	General	OPHID
Aragues et al. 2008	Cancer	HPRD, DIP, MIPS, MINT, BioGrid, IntAct
Furney et al. 2008a	Cancer	DIP, MIPS
ENDEAVOUR (Aerts et al. 2006)	General	BIND
Shortest path		
Krauthammer et al. 2004	Alzheimer's disease	Literature mining by GENEWAYS
Liu et al. 2006	Alzheimer's disease	Inferred from multiple dataset
Radivojac et al. 2008	General	HPRD, OPHID
Prioritizer (Franke et al. 2006)	General	Inferred from multiple dataset
Diffusion kernel		
Kohler et al. 2008	General	HPRD, BIND, BioGrid, STRING, DIP, IntAct
Chen et al. 2009	General	HPRD, BIND, BioGrid
Similarity-assisted		
Ala et al. 2008	General	Coexpression
Miozzi et al. 2008	General	Coexpression
Lage et al. 2007	General	MINT, BIND, IntAct, KEGG, Reactome
CIPHER (Wu et al. 2008)	General	HPRD, OPHID
AlignPI (Wu et al. 2009)	General	HPRD
Centrality-based		
Ozgur et al. 2008	Prostate cancer	Literature mining by GIN (Ozgur et al. 2008)
Ortutay and Vihinen 2009	Immunodeficiency	HPRD
Gudivada et al. 2008	Cardiovascular disease	Genomic-phenomic Semantic web
Others		
Mani et al. 2008	Cancer	B-cell interactome, Co-expression
Karni et al. 2009	General	HPRD

disease are more likely to be related to the disease. For example, Oti et al. (2006) predict candidate genes as those that directly interact with known causative genes of the same disease, and they validate this method against 289 diseases with at least two known disease genes in OMIM. Though the performances vary for different protein network datasets, all are much better than random selection. By applying this method to diseases with both known genes and uncharacterized loci, they are able to predict 300 novel disease candidate genes, of which 10% are confirmed by literature evidence outside OMIM. The same strategy is used in the CPS method in the study of George et al. (2006). When benchmarking with protein interaction data from OPHID, the method has a sensitivity of 0.42 and a specificity of 1.0. In another study on cancer gene prediction, Aragues, Sander, and Oliva (2008) define the cancer linker degree (CLD) of a gene as the number of its neighbors that are known to be involved in cancer. They find that CLD of a gene is a good indicator of the probability of being a cancer gene.

Similar results are obtained by Furney et al. (2008b). By integrating protein interaction data with protein sequence conservation, protein domain, gene structure, and regulatory data, Furney et al. train Bayesian classifiers to prioritize proto-oncogenes and tumor suppressor genes. For protein interaction data, they use the number of interactions and the number of interactions with cancer genes, assuming that cancer genes have a higher degree and are more likely to interact with other cancer genes. The study by Furney et al. is a typical data integration strategy for gene prioritization. First, a number of data sources/evidences are collected for each candidate gene, and then some machine learning algorithms are used to integrate these features and generate ranking scores. Often data sources are explored in a relatively simple fashion. Another example is provided by Aerts et al. (2006). In this study, up to 12 data sources, including protein interaction data in the database BIND (Bader, Betel and Hogue 2003), are used separately to calculate the similarity between training genes (known disease genes) and candidate genes, yielding 12 ranking lists. A rank aggregation algorithm based on order statistics is used to combine these rank lists into a single rank. Again, only direct neighbors are considered for protein interaction data, but instead of neighbor counting, Aerts et al. use the number of common neighbors as the similarity score between known disease genes and candidate genes.

11.3.1.1.2 Shortest Path Length   The direct neighbor strategy has some limitations. It is quite possible that two functionally related genes do not interact directly with each other. For example, they may function in different steps of a signaling cascade, yet still lead to the same disease (Brunner and van Driel 2004; Wood et al. 2007). The direct neighbor strategy is more likely to be true for cases where two genes function in the same protein complex (Lage et al. 2007), instead of a pathway. To make use of indirect interactions, one can take higher-order neighborhoods into consideration. The shortest path length measure of distance considers the influence between nodes that are reachable. The length of the shortest path between two biomolecules in molecular interaction networks is assumed to be related to the speed of information communication and/or the strength of the functional association between the two molecules. Thus, the shortest path length is a good measure of functional relatedness, as demonstrated by its correlation with functionally similarity (based on Gene Ontology) (Sharan, Ulitsky, and Shamir 2007). One of the pioneering works to apply shortest path analysis to gene prioritization is from the Rzhetsky group, with a method called Molecular Triangulation (Krauthammer et al. 2004). They use an automatic literature mining system to construct a network around four Alzheimer's disease (AD) genes, and then calculate the shortest path length between all other nodes to these four seed genes. The statistical significance of the distance serves as the final score. The method performs well in predicting additional AD gene candidates identified manually by an expert. This approach was later extended by Liu et al. (2006) by applying shortest path length scoring on a brain-specific gene network, and based on the same four AD seed genes, they were able to rank 37 AD associated genes within the top 46 high-scoring genes.

Like the direct neighbor approach, shortest path analysis has also been used in data integration methods to transform protein interaction data into feature sets. Radivojac et al.

(2008) integrate human protein interaction network, protein sequence, function, physico-chemical and structural properties to train Support Vector Machines (SVM) that are able to predict gene-disease associations with relatively high accuracy. Protein network data are used to calculate the distance between candidate proteins and disease causing proteins, which serves as one important feature for the classifier. A case study for leukemia is given in this study. The training set contains 80 genes associated with leukemia, which are manually curated from OMIM, Swiss-Prot (Boeckmann et al. 2003), and HPRD. Cross-validation shows an accuracy of 77.5% and 15 novel genes are predicted to be associated with leukemia. The authors are able to find from the published literature strong association for 8 of the 15 predictions. One limitation of this approach is that the SVM requires at least 10 known disease-related genes to train the model and to predict novel disease genes.

11.3.1.1.3 *Global Distance Measure*   The problem with shortest path length is that it considers only one of the shortest paths, ignoring the contribution of other shortest paths and other paths with longer length. Most of the time there will be more than one path and even more than one shortest path between two nodes, and the existence of these paths shows additional relatedness between two genes. Another defect is that the shortest path length lacks resolution: the lengths are integers and the longest path in a biological network is typically very small, due to the small world property of biological networks (Jeong et al. 2000; Watts and Strogatz 1998). The so-called global distance measure, mainly diffusion-type distance measure, overcomes these drawbacks by considering the topology of the entire network (see illustrations in Kohler et al. 2008). The diffusion kernel K of a graph G is defined as $K = e^{-\beta L}$, where β controls the magnitude of the diffusion. The matrix L is the Laplacian of the graph, defined as $D - A$, where A is the adjacency matrix of the interaction graph and D is a diagonal matrix containing the nodes' degrees. The inverse Laplacian takes into account all powers of diffusion and thus incorporates all paths along the network. Kohler et al. (2008) propose using the following scoring function to quantify the association between a candidate gene j and a disease:

$$S_j = \sum_i K_{ij}$$

where i represents known disease genes. By applying this approach and another similar random walk approach to an assembled human protein-protein interaction network, they show that methods based on global distance measure significantly outperform those based on local distance measure and non-network approaches. This result is consistently observed for monogenic disorders, polygenic disorders, and cancer. Similar random walk algorithms have been widely used in social- and Web-network analysis to find important nodes (persons or web pages) on the network, such as the PageRank algorithm (Brin and Page 1998) used by Google to rank web pages. By fixing known disease genes as root nodes, some of these algorithms have recently been exploited to prioritize disease genes based on protein network (Chen, Aronow, and Jegga 2009).

11.3.1.2 Proximity of Candidate Gene Pairs: Enabling de Novo Discovery

All the approaches discussed above require at least one disease gene known to cause the disease under study, which covers only about half of human diseases. For genetically unrecognized diseases, these methods do not work. We call methods that do not rely on known disease genes of the same disease *de novo* methods. To enable *de novo* prediction, one has to add some other disease-specific information, such as disease similarity, to use genes causing a similar disease as a surrogate. We will discuss this type of information later. Here we introduce another method, called Prioritizer (Franke et al. 2006), which does not rely on such phenotype information. Prioritizer assumes the disease-specific information is provided when the candidate genes are available, for example, from a linkage locus associated with the disease. Prioritizer takes at least two genomic regions as input, each containing many candidate genes. Each of the regions is supposed to contain at least one gene causing the disease under study. Assuming the two disease genes should be close to each other on the network, the scoring scheme is designed such that a candidate gene has a higher score if it has a smaller distance to genes in another region. A permutation test is introduced to correct the topology differences and yield a p-value based on which all candidate genes are prioritized. Theoretically Prioritizer can be used in *de novo* discovery of disease genes when multiple genetic regions are given, and this is demonstrated by a case study on breast cancer. Ten 100-gene artificial loci are constructed around 10 known breast cancer genes, and Prioritizer is able to rank 2 to 4 of the 10 breast cancer genes in the top 10 of each locus, when using different gene networks. When the candidate genes in a region are fixed to some known disease genes, this method is essentially the shortest path analysis discussed in the above section. Another method employing this principle is CPS (George et al. 2006), which predicts genes directly interacting with genes from another locus as disease genes.

11.3.2 Phenotype Similarity-Assisted Methods

A natural generalization of the "guilt-by-proximity" principle is that genes causing similar (instead of the same) diseases are likely to be closely related. The additional information provided by similar diseases enables *de novo* prediction of causative genes for diseases without known causative genes, and will also improve the performance for those with known causative genes. Then two questions remain to be addressed: (1) how to define and compute the similarity between diseases, and (2) how to incorporate disease similarity into disease gene prediction approaches.

11.3.2.1 Calculating and Validating Phenotypic Similarity

A disease can be represented by a set of terms describing its clinical symptoms, namely, phenotypes. The phenotypic similarity between two diseases quantifies the overlap or semantic similarity between two sets of terms (Brunner and van Driel 2004; Oti and Brunner 2007). Four different approaches (Care et al. 2009; Lage et al. 2007; Robinson et al. 2008; van Driel et al. 2006) have been proposed to calculate the phenotypic similarity for diseases in OMIM.

van Driel et al. (2006) use a text mining technique to map OMIM disease records to a set of standardized terms, that is, terms defined in MeSH (Medical Subject Headings;

Lowe and Barnett 1994), and then a vector is created for each disease, with each element in the vector representing the number of times the term occurs in the disease record. After adjusting for the hierarchical relationship between different terms, the relative frequency of terms, and size of each record, each disease is represented by a high dimension feature vector of weighted MeSH terms. The similarity between any two diseases is then calculated as the cosine of the angle between the two vectors. Essentially, the method amounts to detecting standardized terms that are (1) common to the description of both diseases, and (2) do not occur too frequently among all diseases such that they are informative about the disease under study. Lage et al. (2007) propose a similar method, and the major difference is, instead of using MeSH as the standard vocabulary, they use UMLS (the unified medical language system; Bodenreider 2004), a more general system containing MeSH and several other vocabularies.

Different approaches have been proposed to evaluate the quality of the calculated disease similarity data. Instead of directly assessing the quality of disease similarity, van Driel et al. (2006) correlated the similarity with the functional relatedness between disease genes. They find that the genes that lead to more similar diseases are more likely to have similar protein sequences, more likely to interact with each other, and more likely to share Pfam domains and Gene Ontology annotations, thus demonstrating that the phenotypic similarity reflects real biological knowledge. Lage et al. (2007) directly evaluate the phenotypic similarity score by comparing it with a putative golden positive set of ~7000 disease pairs. A disease pair is included in this set if one disease is referred to in the text record of another disease. One hundred disease pairs randomly selected from the putative golden positive set are subject to expert OMIM curators' evaluation, and over 90% of them are judged to have a high degree of phenotypic overlap. The phenotypic similarity is then evaluated by calculating the percentage of disease pairs attaining at least a given similarity threshold present in the putative golden positive set. Recently, Care et al. (2009) proposed to use a more stringent golden putative set by only accepting disease pairs with reciprocal references, resulting in a set of about 4000 disease pairs. However, this set has not been evaluated by expert OMIM curators. Interestingly, based on this stringent disease pair set, Care et al. find that the mapping of free text to standard vocabulary is not necessary, as a simple word counting method outperforms the UMLS-based method. However, if the disease ID is also counted as terms, the evaluation procedure will prefer the word counting method; thus the comparison is biased. Further studies are needed to exclude this bias and show whether simple word counting is also more powerful than MeSH and other standard vocabularies. All these three phenotypic similarity datasets (Care et al. 2009; Lage et al. 2007; van Driel et al. 2006) have been used in disease gene prioritization, and all show significant improvement compared to methods that do not employ phenotypic similarity data.

More recently, the Human Phenotype Ontology (HPO) was created to standardize the annotation of OMIM diseases (Robinson et al. 2008). Ontology is a special type of standard vocabulary that is particularly suited for knowledge representation and computation, and the usefulness of ontology in biology is evidenced by the great success of Gene Ontology (GO) (Ashburner et al. 2000). GO annotation is now widely accepted as the representation of gene functions, and various methods have been developed to calculate the functional similarity between genes using GO annotations (Xu, Du, and Zhou 2008). Most of these methods can be applied to calculating disease phenotypic similarity using HPO, because HPO is designed to have the same structure as GO (Robinson et al. 2008). In fact, along with the publication of HPO, Robinson et al. (2008) also applied a classic approach (Lord et al. 2003) for calculating gene functional similarity to generate an HPO-based phenotypic similarity for 727 OMIM diseases, which have been classified into one of 21 physiological disorder classes (Goh et al. 2007). Though this similarity data has not been evaluated using the same methods as the UMLS-based disease similarity data, the HPO-based disease similarity network shows a pattern consistent with the physiological disorder classes. Compared to MeSH and UMLS, HPO has several potential advantages for computational phenotype analysis. First, HPO is specifically designed for the needs of describing human hereditary diseases and their phenotypes, and second, as demonstrated by GO, the ontology framework may be more powerful in knowledge representation and computation. Finally, instead of annotating diseases using automatic text mining, HPO experts have manually annotated almost all the OMIM diseases. It is expected that the phenotypic similarity calculated based on HPO will provide more strong support for disease gene prioritization, though so far no such study is available.

11.3.2.2 Modeling with Molecular Network and Phenotype Similarity

The hypothesis underlying most if not all similarity-assisted methods is that similar diseases are caused by functionally related genes. Methods of this type differ in the way to model such correlation and how they incorporate phenotypic similarity information into the model.

11.3.2.2.1 Group Diseases by Similarity   The simplest way to exploit phenotypic similarity perhaps is to treat diseases showing a certain level of similarity as the same disease; thus more known disease genes are available for model training or seed propagation. For example, van Driel et al. (2006) have shown that, for the MeSH-based similarity score, biologically meaningful relationships were mostly detected in disease pairs with a similarity score equal to or greater than 0.4. Ala et al. (2008) use this phenotype similarity data, and group diseases according to this threshold. They then employ essentially a neighbor counting strategy, together with a human-mouse conserved coexpression network, to predict disease genes. A similar procedure has been applied to a different dataset (Miozzi et al. 2008).

11.3.2.2.2 Weighted Neighbor Counting   Lage et al. (2007) propose a Bayesian model to systematically integrate the UMLS-based similarity score with a weighted human protein-protein interaction network. Basically, for each candidate gene, all the direct neighbors are annotated with, if any, diseases associated with them, and weighted by the similarity to the disease under study. At the same time, all the edges are weighted with a confidence score. Based on these observed data and a uniform *priori*, the posterior probability of the candidate gene to be associated with the disease under study is obtained via the Bayesian formula. This is essentially a weighted version of neighbor counting: the neighbors of the gene under consideration are weighted by the confidence of the edges (protein-protein interactions), and the similarity between the disease they lead to and the disease under study. The more reliably a gene is connected to neighbors associated with diseases similar to the disease under study, the more likely the gene is involved in the disease. When applying this approach to 669 genetic loci with known disease genes, they are able to rank the disease gene as the top candidate in 298 loci, significantly outperforming all other methods compared in this study. As the first study to incorporate phenome-wide disease similarity information into disease gene prioritization, it clearly demonstrates the benefits of phenotype data. They then apply the method to 870 genetic loci without the known causative genes and predict a list of 113 candidates for 91 loci, 24 of which are likely to be true predictions according to the recently published literature.

11.3.2.2.3 Prioritize by Interactome-Phenome Correlation   Using the same type of data (phenotypic similarity and protein networks), we have proposed a novel method, CIPHER (Correlating protein Interaction network and PHEnotype network to pRedict disease genes), with drastically different formulation (Wu et al. 2008). We choose to directly model the correlation between disease phenotypic similarity and gene functional relatedness, and use the correlation to prioritize candidate genes. We hypothesize that the phenotypic similarity between any two diseases p and p' can be explained by the proximity of their disease genes on the network:

$$S_{p,p'} = C_p + \sum_i \beta_{p,i} \sum_{i'} \exp\left(-L_{i,i'}^2\right)$$

or

$$S_{p,p'} = C_p + \sum_i \beta_{p,i} \Phi_{i,p'}$$

where C_p and $\beta_{p,i}$ are constants for a fixed disease p, and i and i' indicate disease genes of p and p', respectively. $L_{i,i'}$ is the graph distance between gene i and i', which is transformed into proximity by a Gaussian kernel function. The distance measure can be any of the direct neighbor (CIPHER-DN), shortest path (CIPHER-SP), or diffusion kernels.

$$\Phi_{i,p'} = \sum_{i'} \exp\left(-L_{i,i'}^2\right)$$

is defined as the proximity between gene i and disease p' by summing the gene proximity over all known disease genes of p'. This is the classical measure used in shortest path analysis to prioritize candidate genes (Franke et al. 2006; Krauthammer 2004), which do not rely on the phenotypic similarity information. Instead, we choose to evaluate the ability of gene-disease proximity in explaining the disease similarity for a pair of genes and diseases (i, p). We create a phenome-wide vector for each gene i: $\Phi_i = (\Phi_{i,p'})$, and each disease p: $S_p = (S_{p,p'})$, with p' varying for all human diseases. Then we use the correlation

TABLE 11.2 The Ranks and Percentages of Known Breast Cancer Susceptibility Genes in Genome-Wide *de Novo* Prioritization

Known Breast Cancer Gene	Rank by CIPHER-SP	Rank by CIPHER-DN
BRCA1	1	2
AR	3	3
ATM	19	4
CHEK2	66	19
BRCA2	139	49
STK11	150	21
RAD51	174	36
PTEN	188	24
BARD1	196	41
TP53	287	45
RB1CC1	798	6360
NCOA3	973	343
PIK3CA	1644	367
PPM1D	1946	7318
CASP8	4978	2397
TGF1	7116	3502

between these two vectors as the final score for association between gene i and disease p. The usefulness of this score is validated by both systematic large-scale cross-validation, and a case study for breast cancer. We have shown that the proposed CIPHER approach can accurately pinpoint the true disease genes from linkage loci or from the whole genome. Further analysis shows that CIPHER is robust to noise in the phenotype similarity data and the protein network data. Without any modification, CIPHER can be applied to *de novo* discovery, that is, to diseases without known disease genes (without mapped locus or with mapped but uncharacterized loci).

A case study for breast cancer is presented to demonstrate CIPHER's ability in *de novo* discovery of breast cancer genes. Sixteen known breast cancer genes are treated as non-breast cancer genes and then the whole human genome is prioritized by CIPHER. When using a shortest path length measure of distance (CIPHER-SP), the well-characterized breast cancer gene BRCA1 is ranked at the top, and 10 of the 16 genes are ranked in the top 300, roughly the top 1% of the human genome (Table 11.2). In addition, among the top 10% of the prioritized human genome the same *de novo* prioritization identifies 15 genes that have been suggested recently to be novel breast cancer genes, including AKT1, ranked at 27, a novel oncogene, and recently a transforming mutation was identified in human breast, colorectal, and ovarian cancers (Carpten et al. 2007). The case study also shows that, though direct neighbor distance measure (CIPHER-DN) works better on ranking known breast cancer genes than CIPHER-SP, it fails to assign ranks to many of the novel susceptibility genes.

All the advantages of CIPHER enable us to perform genome-wide candidate gene prioritization for almost all human diseases, leading to a comprehensive genetic landscape of human diseases. Automatic clustering and enrichment analysis of the landscape reveal the modularity of human disease and gene relationships (Wu et al. 2008).

11.3.2.2.4 Network Alignment   To fully explore the modularity of the human disease genetic landscape, Wu, Liu, and Jiang (2009) borrow ideas from the study of conservation in protein networks (Sharan et al. 2005), or network alignment. Sharan et al. propose a local alignment technique to identify conserved modules between two or more protein interaction networks. To apply this technique, Wu, Liu, and Jiang (2009) created a human disease network by linking diseases with a phenotypic similarity score larger than a given threshold, resulting in a human disease similarity network. Then they used the network alignment technique to compare the human disease network and human protein network, and identify 39 disease modules together with corresponding gene modules, or bimodules. Examining the functions of genes and categories of diseases, they show that these bimodules represent disease families and their common pathways. After validating the bimodule identification method, they propose to use it for disease gene prediction. Essentially, they predict a candidate gene to cause a disease if it is linked to the disease in a bimodule. This approach, named AlignPI (Wu, Liu, and Jiang 2009), attains similar performance with CIPHER.

11.3.3 Prioritize by Network Centrality

The working principle for methods in this category is totally different from those discussed above. Here we assume that genes with higher centrality on a network are more likely to cause disease. To be more informative, the network is often specially designed.

11.3.3.1 Centrality in a Context-Specific Gene Network

Ozgur et al. (2008) introduce a sophisticated automatic literature mining approach to construct a disease-specific gene interaction network, in their example, a prostate cancer network. Hypothesizing that genes with high centrality in a disease-specific network are likely to be related to the disease, they used several network centrality measures to rank genes in the prostate cancer network, and found that two measures, degree and eigenvector, were highly informative of known prostate cancer genes. Specifically, 19 of the top 20 genes returned by the approach have supportive evidence from either OMIM or PGDB (Prostate Gene DataBase; Li et al. 2003), a curated database of genes related to prostate cancer. One limitation of the approach is that, similar to the Molecular Triangulation approach (Krauthammer et al. 2004), it relies on a list of seed genes (genes known to be involved in the disease) to construct the network, yet to what extent the choice of seed genes influences the results is not discussed. In a second study, Ortutay and Vihinen (2009) create a human immune gene interaction network by linking curated immune genes with interaction data from HPRD, and use multiple centralities, including degree and closeness, to prioritize candidate genes for immunodeficiencies.

11.3.3.2 Centrality in a Genomic-Phenomic Network

So far we have focused on networks whose nodes are genes or proteins. There are also other network approaches using more complicated networks. For example, Gudivada et al. (2008)

create a network of various concepts, with edges representing the association between genes and Gene Ontology annotations, pathways, mouse phenotypes, and human clinical features, therefore establishing a semantic web of integrated genomic and phenomic knowledge. Assuming that disease-causing genes tend to play functionally important roles and share similar biochemical characteristics with genes causing diseases with similar clinical features, the authors use a Google-like search and ranking algorithm (Mukherjea 2005) to prioritize candidate genes. The efficiency of the proposed approach is tested in prioritizing candidate genes for cardiovascular diseases.

11.3.4 Other Methods

Here we discuss several interesting and promising approaches that do not fall into the above categories. These methods are interesting because they do not rely on known disease gene or disease similarity, yet still are able to find the causal gene based on the genome-wide secondary response.

Mani et al. (2008) propose a method called Interactome Dysregulation Enrichment Analysis (IDEA) to predict oncogenes. Using interactome and microarray data, they first identify dysregulated interactions, that is, gene pairs with annotated interaction but significantly changed correlation according to gene expression profiling of normal and tumor samples. Then genes with an unusually high number of dysregulated interactions in their neighborhood are predicted as oncogenes. The assumption is that genes implicated in cancer initiation and progression will show dysregulated interactions with their molecular partners. In three B-cell tumor phenotypes, the method correctly identifies the known genes in the top 20 candidates out of about 8000 genes. The IDEA method exploits direct neighbors only. As demonstrated by other examples discussed in previous sections, shortest path-based analysis might yield higher coverage and more novel predictions that are not so obvious from protein interaction data.

A more sophisticated network-based approach has been proposed to solve a problem with similar settings. With the protein interaction network available, Karni, Soreq, and Sharan (2009) attempted to predict the causal gene from expression profile data assumed to be perturbed by the gene. They first identified a set of disease-related genes whose expression is changed in the disease state, then, based on a parsimonious assumption, an algorithm sought the smallest set of genes that could best explain the expression changes of the disease-related genes in terms of probable pathways leading from the causal to the affected genes in the network. Experiments with both simulated and real knock-out data show that the proposed approach attains very high accuracy. Further validations on expression data from different types of cancer show high accuracy in pinpointing known oncogenes. For example, using expression profiles for a subset of acute leukemias involving chromosomal translocation of the mixed leukemia gene (MLL), the algorithm correctly assigns MLL an average rank of 1.5, out of 168 genes in the neighboring region. When applying the algorithm to four breast cancer datasets, the major causal genes BRCA1 and BRCA2 are ranked very high. They are also able to show that the algorithm outperforms a naive algorithm that ranks disease-associated genes according to their shortest path length in the network to the directly affected genes.

11.4 DISCUSSION

Five years after the first network-based candidate gene prioritization method (Krauthammer et al. 2004), now there are more than 20 available in the published literature (Table 11.1), calling for a comprehensive comparison among them. Unfortunately, a systematic and rigorous direct comparison is very difficult and seldom occurs in the literature, mostly because different methods use different types of data sources, and are trained and tested on customized datasets which are often unavailable to others. For methods running with the same type of data sources, one can re-implement different methods proposed by other groups, and compare them using one dataset that is probably not the original dataset on which most methods were tested. Such a comparing scheme is only feasible for comparing methods that are easy to implement. For example, Kohler et al. (2008) implemented four algorithms that are purely network based and compared their performance, showing the superiority of global distance measures. For situations where methods are not easy to implement, researchers often compare self-reported performances along with the original publications. Self-reported performances are often transformed into so-called (average) fold enrichment, that is, the average fold of enriching the true disease genes among a short top list, compared to random selection (Lage et al. 2007; Wu et al. 2008). According to this criterion, disease similarity-assisted methods significantly outperform previous methods, and we are able to show that CIPHER works even better, especially for higher recall. The problem with the fold enrichment criterion is that it is influenced by the total number of candidate genes and the size of the top list, while these numbers often vary across different methods. For comprehensive comparison, a community-wide effort is needed, to establish a publicly available data platform, including widely used different data sources, a training dataset of known gene-disease associations, and a blinded test dataset. Such efforts have recently been performed in a related field, the mouse gene function prediction (Pena-Castillo et al. 2008).

Most of the methods discussed here are not designed particularly for cancer, though they can be applied to cancer without any modification. Here we discuss some cancer-specific issues. Though these issues are not particularly related to network-based approaches, it will be important for us to realize their impact. First, prediction methods generally do not differentiate two types of cancer genes that are different in many aspects and thus fail to generate more testable hypotheses that could guide further experimental validation. Genes that can initialize tumorigenesis are traditionally divided into oncogene and tumor suppressor gene, though more recently stability gene has been proposed to be a further type of cancer gene (Vogelstein and Kinzler 2004). Study has shown that a classifier using protein conservation, gene sequence, protein domains, protein interactions, and regulatory data is able to differentiate oncogenes from tumor suppressor genes (Furney et al. 2008a). Specifically, they show that tumor suppressor genes have higher degree than oncogenes, while oncogene evolution appears to be more highly constrained (Furney et al. 2008a). Together, these results imply that oncogenes and tumor suppressor genes may be inherently different. Taking the difference into consideration may further improve the prediction of cancer genes. In addition, the experimental procedures to verify oncogenes and tumor suppressor genes are different; computational prediction will facilitate the verification if it can tell an oncogene from a tumor suppressor gene. Another special issue for cancer gene prediction is that there are several cancer-specific genome-wide data sources which may greatly advance the prediction of cancer genes. For example, large-scale sequencing of the human cancer genome has identified thousands of genes carrying nonsilent mutations in breast or colon cancer samples (Sjoblom et al. 2006; Wood et al. 2007), while array-based techniques, such as array comparative genomic hybridization (aCGH; Pinkel et al. 1998) and representational oligonucleotide microarray analysis (ROMA; Lucito et al. 2003), have been developed to localize genes with altered copy numbers (amplified or deleted) in cancer samples. Combining candidate genes identified from the above large-scale screen and computational cancer gene prioritization methods will greatly facilitate the discovery of human cancer-causing genes.

With the development of high-throughput techniques in exploring the human cancer genome, and the increasing quality in large-scale detection of protein interactions, network-based cancer gene discovery will remain promising and continue to be an active research area. Progress in this area will also benefit from other network-based research, such as the network-based prediction of protein functions (Sharan, Ulitsky, and Shamir 2007), especially functions of cancer genes (Hu et al. 2007), and the discovery of novel drug targets for cancer (Campillos et al. 2008; Huang and Harari 1999), since the formulations are similar; thus novel methods developed for one problem may also apply to the other. We expect that network analysis will provide both systems thinking and methodology advantages on our way to understanding the complexity of life.

ACKNOWLEDGMENTS

This work is supported by the 863 project of China (No. 2006AA02Z311), the NSFC (Nos. 6903 and 60721003), and NCET-07-0486.

REFERENCES

Aerts, S., Lambrechts, D., Maity, S. et al. 2006. Gene prioritization through genomic data fusion. *Nature Biotechnol* 24: 537–544.

Ala, U., Piro, R. M., Grassi, E. et al. 2008. Prediction of human disease genes by human-mouse conserved coexpression analysis. *PLoS Comput Biol* 4: e1000043.

Altshuler, D., Daly, M. J., and Lander, E. S. 2008. Genetic mapping in human disease. *Science* 322: 881–888.

Aragues, R., Sander, C., and Oliva, B. 2008. Predicting cancer involvement of genes from heterogeneous data. *BMC Bioinformatics* 9: 172.

Ashburner, M., Ball, C. A., Blake, J. A. et al. 2000. Gene ontology: tool for the unification of biology. The Gene Ontology Consortium. *Nature Genet* 25: 25–29.

Bader, G. D., Betel, D., Hogue, C. W. 2003. BIND: the biomolecular interaction network database. *Nucl. Acids Res.* 31: 248–250.

Barabasi, A. L. and Oltvai, Z. N. 2004. Network biology: understanding the cell's functional organization. *Nature Rev Genet* 5: 101–113.

Bodenreider, O. 2004. The Unified Medical Language System (UMLS): integrating biomedical terminology. *Nucleic Acids Res* 32: D267–D270.

Boeckmann, B., Bairoch, A., Apweiler, R. et al. 2003. The SWISS-PROT protein knowledgebase and its supplement TrEMBL in 2003. *Nucleic Acids Res* 31: 365–370.

Breitkreutz, B. J., Stark, C., Reguly, T. et al. 2008. The BioGRID Interaction Database: 2008 update. *Nucleic Acids Res* 36: D637–D640.

Brin, S. and Page, L. 1998. The anatomy of a large-scale hypertextual Web search engine. In: *7th International World Wide Web Conference: April 14–18 1998.* Brisbane, Australia: Elsevier Science, pp. 107–117.

Brown, K. R. and Jurisica, I. 2005. Online predicted human interaction database. *Bioinformatics* 21: 2076–2082.

Brunner, H. G. and van Driel, M. A. 2004. From syndrome families to functional genomics. *Nature Rev Genet* 5: 545–551.

Burmeister, M., McInnis, M. G., and Zollner, S. 2008. Psychiatric genetics: progress amid controversy. *Nature Rev Genet* 9: 527–540.

Campillos, M., Kuhn, M., Gavin, A. C., Jensen, L. J., and Bork, P. 2008. Drug target identification using side-effect similarity. *Science* 321: 263–266.

Care, M. A., Bradford, J. R., Needham, C. J., Bulpitt, A. J., and Westhead, D. R. 2009. Combining the interactome and deleterious SNP predictions to improve disease gene identification. *Hum Mutat* 30: 485–492.

Carpten, J. D., Faber, A. L., Horn, C. et al. 2007. A transforming mutation in the pleckstrin homology domain of AKT1 in cancer. *Nature* 448: 439–444.

Chatr-aryamontri, A., Ceol, A., Palazzi, L. M. et al. 2007. MINT: the Molecular INTeraction database. *Nucleic Acids Res* 35: D572–D574.

Chen, J., Aronow, B. J., and Jegga, A. G. 2009. Disease candidate gene identification and prioritization using protein interaction networks. *BMC Bioinformatics* 10: 73.

Feldman, I., Rzhetsky, A., and Vitkup, D. 2008. Network properties of genes harboring inherited disease mutations. *Proc Natl Acad Sci USA* 105: 4323–4328.

Fields, S. and Song, O. 1989. A novel genetic system to detect protein-protein interactions. *Nature* 340: 245–246.

Fields, S. and Sternglanz, R. 1994. The two-hybrid system: an assay for protein-protein interactions. *Trends Genet* 10: 286–292.

Folstein, S. E. and Rosen-Sheidley, B. 2001. Genetics of autism: complex aetiology for a heterogeneous disorder. *Nature Rev Genet* 2: 943–955.

Franke, L., van Bakel H., Fokkens, L. et al. 2006. Reconstruction of a functional human gene network, with an application for prioritizing positional candidate genes. *Am. J. Hum. Genet.* 78: 1011–1025.

Frayling, T. M. 2007. Genome-wide association studies provide new insights into type 2 diabetes aetiology. *Nature Rev Genet* 8: 657–662.

Furney, S. J., Calvo, B., Larranaga, P., Lozano, J. A., and Lopez-Bigas, N. 2008a. Prioritization of candidate cancer genes: an aid to oncogenomic studies. *Nucleic Acids Res* 36: e115.

Furney, S. J., Madden, S. F., Kisiel, T. A., Higgins, D. G., and Lopez-Bigas, N. 2008b. Distinct patterns in the regulation and evolution of human cancer genes. *In Silico Biol* 8: 33–46.

Futreal, P. A., Coin, L., Marshall, M. et al. 2004. A census of human cancer genes. *Nature Rev Cancer* 4: 177–183.

George, R. A., Liu, J. Y., Feng, L. L. et al. 2006. Analysis of protein sequence and interaction data for candidate disease gene prediction. *Nucleic Acids Res* 34: e130.

Ghavidel, A., Cagney, G., and Emili, A. 2005. A skeleton of the human protein interactome. *Cell* 122: 830–832.

Giot, L., Bader, J. S., Brouwer, C. et al. 2003. A protein interaction map of *Drosophila melanogaster.* *Science* 302: 1727–1736.

Goh, K. I., Cusick, M. E., Valle, D. et al. 2007. The human disease network. *Proc Natl Acad Sci USA* 104: 8685–8690.

Gudivada, R. C., Qu, X. A., Chen, J. et al. 2008. Identifying disease-causal genes using Semantic Web-based representation of integrated genomic and phenomic knowledge. *J. Biomed. Inform.* 41: 717–729.

Han, J. D. 2008. Understanding biological functions through molecular networks. *Cell Res* 18: 224–237.

Hart, G. T., Ramani, A. K., and Marcotte, E. M. 2006. How complete are current yeast and human protein-interaction networks? *Genome Biol* 7: 120.

Hu, P., Bader, G., Wigle, D. A., and Emili, A. 2007. Computational prediction of cancer-gene function. *Nature Rev Cancer* 7: 23–34.

Huang, S. M. and Harari, P. M. 1999. Epidermal growth factor receptor inhibition in cancer therapy: biology, rationale and preliminary clinical results. *Invest New Drugs* 17: 259–269.

Ito, T., Chiba, T., Ozawa, R. et al. 2001. A comprehensive two-hybrid analysis to explore the yeast protein interactome. *Proc Natl Acad Sci USA* 98: 4569–4574.

Jansen, R., Yu, H., Greenbaum, D. et al. 2003. A Bayesian networks approach for predicting protein-protein interactions from genomic data. *Science* 302: 449–453.

Jeong, H., Mason, S. P., Barabasi, A. L., and Oltvai, Z. N. 2001. Lethality and centrality in protein networks. *Nature* 411: 41–42.

Jeong, H., Tombor, B., Albert, R., Oltvai, Z. N., and Barabasi, A. L. 2000. The large-scale organization of metabolic networks. *Nature* 407: 651–654.

Jonsson, P. F., Bates, P. A. 2006. Global topological features of cancer proteins in the human interactions. *Bioinformatics* 22: 2291–2297.

Kanehisa, M. and Goto, S. 2000. KEGG: Kyoto encyclopedia of genes and genomes. *Nucleic Acids Res* 28: 27–30.

Karni, S., Soreq, H., and Sharan, R. 2009. A network-based method for predicting disease-causing genes. *J. Comput. Biol.* 16: 181–189.

Kerrien, S., Alam-Faruque, Y., Aranda, B. et al. 2007. IntAct: open source resource for molecular interaction data. *Nucleic Acids Res* 35: D561–D565.

Kohler, S., Bauer, S., Horn, D., and Robinson, P. N. 2008. Walking the interactome for prioritization of candidate disease genes. *Am. J. Hum. Genet.* 82: 949–958.

Krauthammer, M., Kaufmann, C. A., Gilliam, T. C., and Rzhetsky, A. 2004. Molecular triangulation: bridging linkage and molecular-network information for identifying candidate genes in Alzheimer's disease. *Proc Natl Acad Sci USA* 101: 15148–15153.

Lage, K., Karlberg, E. O., Storling, Z. M. et al. 2007. A human phenome-interactome network of protein complexes implicated in genetic disorders. *Nature Biotechnol* 25: 309–316.

Li, L. C., Zhao, H., Shiina, H., Kane, C. J., and Dahiya, R. 2003. PGDB: a curated and integrated database of genes related to the prostate. *Nucleic Acids Res* 31: 291–293.

Li, S., Armstrong, C. M., Bertin, N. et al. 2004. A map of the interactome network of the metazoan *C. elegans. Science* 303: 540–543.

Li, S., Wu, L., and Zhang, Z. 2006. Constructing biological networks through combined literature mining and microarray analysis: a LMMA approach. *Bioinformatics* 22: 2143–2150.

Lim, J., Hao, T., Shaw, C. et al. 2006. A protein-protein interaction network for human inherited ataxias and disorders of Purkinje cell degeneration. *Cell* 125: 801–814.

Liu, B., Jiang, T., Ma, S. et al. 2006. Exploring candidate genes for human brain diseases from a brain-specific gene network. *Biochem Biophys Res Commun* 349: 1308–1314.

Lord, P. W., Stevens, R. D., Brass, A., and Goble, C. A. 2003. Investigating semantic similarity measures across the Gene Ontology: the relationship between sequence and annotation. *Bioinformatics* 19: 1275–1283.

Lowe, H. J. and Barnett, G. O. 1994. Understanding and using the medical subject headings (MeSH) vocabulary to perform literature searches. *JAMA* 271: 1103–1108.

Lucito, R., Healy, J., Alexander, J. et al. 2003. Representational oligonucleotide microarray analysis: a high-resolution method to detect genome copy number variation. *Genome Res* 13: 2291–2305.

Mani, K. M., Lefebvre, C., Wang, K. et al. 2008. A systems biology approach to prediction of oncogenes and molecular perturbation targets in B-cell lymphomas. *Mol Syst Biol* 4: 169.

McKusick, V. A. 2007. Mendelian inheritance in man and its online version, OMIM. *Am. J. Hum. Genet.* 80: 588–604.

Mewes, H. W., Amid, C., Arnold, R. et al. 2004. MIPS: analysis and annotation of proteins from whole genomes. *Nucleic Acids Res* 32: D41–D44.

Miki, Y., Swensen, J., Shattuck-Eidens, D. et al. 1994. A strong candidate for the breast and ovarian cancer susceptibility gene BRCA1. *Science* 266: 66–71.

Miozzi, L., Piro, R. M., Rosa, F. et al. 2008. Functional annotation and identification of candidate disease genes by computational analysis of normal tissue gene expression data. *PLoS One* 3: e2439.

Mukherjea, S. 2005. Information retrieval and knowledge discovery utilising a biomedical Semantic Web. *Brief Bioinform* 6: 252–262.

Oldenburg, R. A., Meijers-Heijboer, H., Cornelisse, C. J., and Devilee, P. 2007. Genetic susceptibility for breast cancer: how many more genes to be found? *Crit Rev Oncol Hematol* 63: 125–149.

Ortutay, C. and Vihinen, M. 2009. Identification of candidate disease genes by integrating Gene Ontologies and protein-interaction networks: case study of primary immunodeficiencies. *Nucleic Acids Res* 37: 622–628.

Oti, M. and Brunner, H. G. 2007. The modular nature of genetic diseases. *Clin Genet* 71: 1–11.

Oti, M., Snel, B., Huynen, M. A., and Brunner, H. G. 2006. Predicting disease genes using protein-protein interactions. *J. Med. Genet.* 43: 691–698.

Ozgur, A., Vu, T., Erkan, G., and Radev, D. R. 2008. Identifying gene-disease associations using centrality on a literature mined gene-interaction network. *Bioinformatics* 24: i277–i285.

Pena-Castillo, L., Tasan, M., Myers, C. L. et al. 2008. A critical assessment of *Mus musculus* gene function prediction using integrated genomic evidence. *Genome Biol* 9 (Suppl 1): S2.

Peri, S., Navarro, J. D., Amanchy, R. et al. 2003. Development of human protein reference database as an initial platform for approaching systems biology in humans. *Genome Res* 13: 2363–2371.

Pharoah, P. D., Tyrer, J., Dunning, A. M., Easton, D. F., and Ponder, B. A. 2007. Association between common variation in 120 candidate genes and breast cancer risk. *PLoS Genet* 3: e42.

Pinkel, D., Segraves, R., Sudar, D. et al. 1998. High resolution analysis of DNA copy number variation using comparative genomic hybridization to microarrays. *Nature Genet* 20: 207–211.

Radivojac, P., Peng, K., Clark, W. T. et al. 2008. An integrated approach to inferring gene-disease associations in humans. *Proteins* 72: 1030–1037.

Rhodes, D. R., Tomlins, S. A., Varambally, S. et al. 2005. Probabilistic model of the human protein-protein interaction network. *Nature Biotechnol* 23: 951–959.

Robinson, P. N., Kohler, S., Bauer, S. et al. 2008. The Human Phenotype Ontology: a tool for annotating and analyzing human hereditary disease. *Am. J. Hum. Genet.* 83: 610–615.

Rual, J. F., Venkatesan, K., Hao, T. et al. 2005. Towards a proteome-scale map of the human protein-protein interaction network. *Nature* 437: 1173–1178.

Rzhetsky, A., Iossifov, I., Koike, T. et al. 2004. GeneWays: a system for extracting, analyzing, visualizing, and integrating molecular pathway data. *J. Biomed. Inform.* 37: 43–53.

Salwinski, L., Miller, C. S., Smith, A. J. et al. 2004. The Database of Interacting Proteins: 2004 update. *Nucleic Acids Res* 32: D449–D451.

Sharan, R., Suthram, S., Kelley, R. M. et al. 2005. Conserved patterns of protein interaction in multiple species. *Proc Natl Acad Sci USA* 102: 1974–1979.

Sharan, R., Ulitsky, I., and Shamir, R. 2007. Network-based prediction of protein function. *Mol Syst Biol* 3: 88.

Sjoblom, T., Jones, S., Wood, L. D. et al. 2006. The consensus coding sequences of human breast and colorectal cancers. *Science* 314: 268–274.

Sprinzak, E., Sattath, S., and Margalit, H. 2003. How reliable are experimental protein-protein interaction data? *J. Mol. Biol.* 327: 919–923.

Strausberg, R. L., Simpson, A. J., and Wooster, R. 2003. Sequence-based cancer genomics: progress, lessons and opportunities. *Nature Rev Genet* 4: 409–418.

van Driel, M. A., Bruggeman, J., Vriend, G., Brunner, H. G., and Leunissen, J. A. 2006. A text-mining analysis of the human phenome. *Eur J. Hum. Genet.* 14: 535–542.

Vastrik, I., D'Eustachio, P., Schmidt, E. et al. 2007. Reactome: a knowledge base of biologic pathways and processes. *Genome Biol* 8: R39.

Vogelstein, B. and Kinzler, K. W. 2004. Cancer genes and the pathways they control. *Nature Med* 10: 789–799.

von Mering, C., Jensen, L. J., Snel, B. et al. 2005. STRING: known and predicted protein-protein associations, integrated and transferred across organisms. *Nucleic Acids Res* 33: D433–D437.

von Mering, C., Krause, R., Snel, B. et al. 2002. Comparative assessment of large-scale data sets of protein-protein interactions. *Nature* 417: 399–403.

Watts, D. J. and Strogatz, S. H. 1998. Collective dynamics of "small-world" networks. *Nature* 393: 440–442.

Wood, L. D., Parsons, D. W., Jones, S. et al. 2007. The genomic landscapes of human breast and colorectal cancers. *Science* 318: 1108–1113.

Wooster, R., Bignell, G., Lancaster, J. et al. 1995. Identification of the breast cancer susceptibility gene BRCA2. *Nature* 378: 789–792.

Wu, X., Jiang, R., Zhang, M. Q., and Li, S. 2008. Network-based global inference of human disease genes. *Mol Syst Biol* 4: 189.

Wu, X., Liu, Q., and Jiang, R. 2009. Align human interactome with phenome to identify causative genes and networks underlying disease families. *Bioinformatics* 25: 98–104.

Xia, K., Dong, D., and Han, J. D. 2006. IntNetDB v1.0: an integrated protein-protein interaction network database generated by a probabilistic model. *BMC Bioinformatics* 7: 508.

Xu, J. and Li, Y. 2006. Discovering disease-genes by topological features in human protein-protein interaction network. *Bioinformatics* 22: 2800–2805.

Xu, T., Du, L., and Zhou, Y. 2008. Evaluation of GO-based functional similarity measures using *S. cerevisiae* protein interaction and expression profile data. *BMC Bioinformatics* 9: 472.

Zhu, M., Zhao, S. 2007. Candidate gene indentification approach progress and challenges. *Int. J. Biol. Sci.* 3: 420–427.

Zhu, X., Gerstein, M., and Snyder, M. 2007. Getting connected: analysis and principles of biological networks. *Genes Dev* 21: 1010–1024.

II

Cancer Biology: Basic Concepts and Cutting-Edge Topics

CHAPTER 12

Cancer Genomics to Cancer Biology

Maria Luz Jaramillo and Chabane Tibiche

CONTENTS

12.1 EVOLUTIONARY THEORY OF CANCER

Development of cancer consists of a step-wise accumulation of DNA mutations and other genetic alterations resulting in a dysregulation of cell growth. As a consequence, cells become progressively more abnormal as more genes become altered, thus leading to the transformation of these cells into the neoplastic or malignant state (Nowell 1976). The abnormal behaviors demonstrated by these "transformed" cancer cells are largely the result of a series of mutations in key regulatory genes: oncogenes, which positively regulate cell growth, and tumor suppressor genes, which negatively regulate cell growth. In addition, the genes that are in control of DNA repair often become damaged themselves, resulting in cells that are capable of generating increasing levels of genetic chaos to which they can become even more susceptible, fueling a vicious cycle.

Most cancers are considered to descend from a single mutant precursor cell. As that cell divides, the resulting generation of cells may acquire genetic and nongenetic modifications possibly resulting in different biological behaviors over a period of time. If alterations result in an advantage with respect to increased cell division and/or resistance to cell death, these altered cells will overgrow the population. As a consequence of these alterations, tumor cells acquire a wide range of behaviors or phenotypes that are not characteristic of their normal cell counterparts. These phenotypic changes will be outlined further in the section on cancer cell biology (Section 12.4).

12.2 CANCER GENOMICS

DNA can be altered in several ways. Mutations can consist of a single nucleotide change to the protein coding sequence resulting in a missense mutation (causing a single amino acid change) or a frameshift or nonsense mutation (causing premature truncation or a nonfunctional protein). In cancer, somatic mutations that frequently occur in the coding regions of protooncogenes, such as Ras and Myc, often result in activation (gain of function) and increase their ability to stimulate cell growth or inhibit cell death. On the other hand, mutations in tumor suppressor genes generally result in a loss of function, as in the case of mutated p53 or APC genes. Mapping of tumor mutations is important in the analysis of the genetic changes that lead to cancer (as discussed further below). It is also possible to analyze germ-line mutations in order to identify dysfunctional genes leading to the development of a familial form of cancer.

Larger regions of DNA alterations encompassing many genes, as seen in translocations, gene amplification, inversions, duplications, and deletions or even entire chromosomes, are often generated during tumor progression. All of these genetic mechanisms can contribute to the programming of the neoplastic state. Whole genome profiling (such as comparative genome hybridization or high density single nucleotide polymorphism arrays) has shown that a variety of genomic changes generally occur during tumor initiation and progression. Examples of physical loss or gain of chromosomal regions which occur in epithelial cancers include human epidermal growth factor receptor 2 (HER2) amplification in breast cancer and loss of PTEN (tumor suppressor) in neuroblastoma.

In addition, gene expression can also be altered by epigenetics, which involve nongenetic changes at the level of the DNA (such as promoter methylation) or chromatin (such as histone acetylation). Although these types of alterations were originally linked to development, they are also clearly associated with tumorigenesis as evidenced by the hypermethylation induced silencing of the promoter regions of certain tumor suppressors such as PTEN and p16 INK4 (Chang, Huang, and Wang 2008; Lopez et al. 2009; Ohm et al. 2007). Efforts have now begun to analyze epigenetic changes at a global level (He et al. 2008).

In summary, every step of the gene expression pathway is subject to modification, resulting in dysregulation (Figure 12.1). The mapping of such changes which translate into altered levels or functions of proteins involved in cellular homeostasis forms the basis of cancer genomics (Carr et al. 2004). By linking a particular genotype (genetic constitution) to the cancer phenotype, or biological manifestation of the disease, this type of profiling can lead to "signatures" that can be

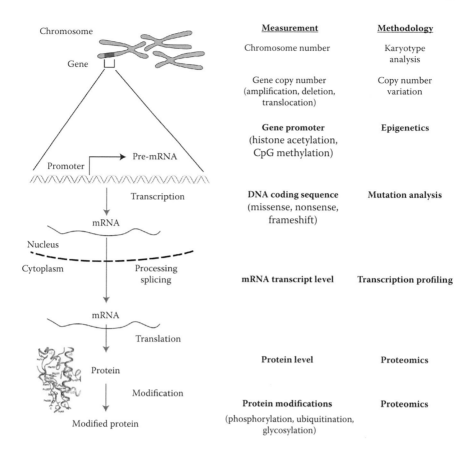

FIGURE 12.1 Gene expression pathway and cancer genomics. Cancer-associated modifications that are associated with various processes in gene expression are indicated.

used as predictive diagnostic (identification of disease), prognostic (clinical outcome), and theranostic (ability to predict therapeutic response) markers (Jain 2007; van't Veer and Bernards 2008). For example, in the case of non-small cell lung cancer (NSCLC), activating mutations in ErbB1/EGFR better predict the ability of patients to respond to EGFR targeted inhibitors rather than EGFR expression level alone (Kancha et al. 2009; Lynch et al. 2004). Along similar lines, the presence of an activating Ras mutation, which lies downstream of EGFR, is predictive of nonresponsiveness (Massarelli et al. 2007). In other cases, overexpression of family member ErbB3 is linked to resistance toward EGFR targeted inhibitors (Fujimoto et al. 2005).

Initially, genomic studies concentrated on global mRNA expression-based profiling due to the rapid development of microarray technology and its relative ease of use. These studies have focused on transcriptional changes that occur in large numbers of tumor types for the purpose of classifying tumors and predicting clinical outcome. As an example, in addition to standard clinical and histopathological markers, a 70-gene signature now forms the basis of a diagnostic test approved by the U.S. Food and Drug Administration (FDA) to help assess risk and optimize treatment strategies in certain breast cancer patients (Glas et al. 2006). However, among the assortment of known breast cancer gene expression signatures linked to poor prognosis to date, there exists little overlap. The existence of a variety of gene signatures serves to illustrate the diversity of gene expression patterns associated with the cancer phenotype. In general, the heterogeneity of cancer, both genetic and nongenetic in origin, can severely complicate diagnosis and treatment.

The sequencing of the human genome has led to several initiatives that aim to comprehensively sequence tumor genomes from a wide variety of tissue types. Notably, not all mutations are necessarily associated with cancer progression and cancer "driver" mutations are operationally distinguished from cancer "passenger" gene mutations by their ability to be functionally selected for during evolution of the tumor (Haber and Settleman 2007). Recently, global gene mutation studies have started to reveal some interesting results (Greenman et al. 2007). Large-scale comprehensive sequencing of the 30,000 or so genes found in human genomes thus far have suggested that there exists a large heterogeneity of gene mutation among tumor samples, with an average of 14 to 15 driver mutations per patient and very little overlap (Sjoblom et al. 2006). Whereas some gene mutations have a high frequency, the majority of them have a low mutation frequency. In addition to transcriptomics and genomics, proteomic approaches have been utilized to try to better generate molecular descriptors of the transformed state of the cell. Proteomics offers the possibility to globally study protein level, posttranslation modification, including phosphorylation, and enzyme activity (Latterich, Abramovitz, and Leyland-Jones 2008; Preisinger et al. 2008).

Taken together, these unbiased approaches can lead to a descriptive understanding of the differences that occur between normal and tumor cells. However, difficulties arise due to the inherent complexity associated with cancer and its heterogeneity, genetic or otherwise. This all points to a need for a systems biology approach to reveal the underlying logic focusing on the involvement of a limited number of signaling pathways. Using an innovative approach to tackle this problem, Cui et al. (2007) mapped all cancer driver-mutating genes on a human signaling network. This important study demonstrated the utility of such an approach and revealed the presence of 12 major clusters of signaling interactions which are implicated to various extents in different kinds of cancer.

To overcome problems in determining changes specifically associated with the neoplastic state, large-scale functional studies are beginning to be used to isolate and determine the functional relevance of cancer-specific changes (Kim and Hahn 2007; Liu 2008). Typical functional genomics studies include gain-of-function screens by expressing cDNA libraries or loss-of-function screens using siRNA-mediated silencing in a variety of functional assays that are believed to be relevant to the cancer phenotype. Examples using these approaches have recently been exploited and have led to the identification of a variety of genes that are causally implicated in the growth of several cancer cell lines (Luo et al. 2008).

12.3 SIGNALING PATHWAYS

In biology, cell signaling is based on the ability of a cell to give a proper response to its environment. The resulting activation of a series of signal transduction pathways involves an ordered sequence of biochemical reactions inside the cell, which are carried out by enzymes, activated by second messengers, and which result in a biological response or phenotype. These signals are essential to many cellular processes, such as control of cell growth, cell survival, differentiation, metabolism, and migration. Although often depicted

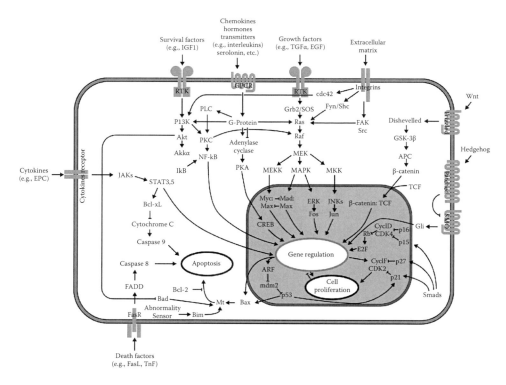

FIGURE 12.2   Typical signal transduction pathway (obtained from http://en.wikipedia.org/wiki/File:Signal_transduction_v1.png).

as separate and linear pathways, a lot of cross-talk occurs among the signaling cascades, particularly at the level of cytoplasmic effectors such as kinases and adaptors as well as nuclear signaling components, mainly transcription factors.

Most signal transduction involves the binding of extracellular signaling molecules (or ligands) to cell surface receptors that are localized to the plasma membrane (Figure 12.2). Intracellular receptors also exist, including steroid receptors which bind lipophillic hormones that can cross the plasma membrane. In addition to binding soluble peptide or protein ligands, signaling cascades can also be triggered through cell-substratum interactions, as is the case for the integrins, cell surface integrin receptors which bind constituents of the extracellular matrix (ECM).

Several classifications of cell surface receptors exist: multipass transmembrane proteins, such as ion channel-lined receptors or G-protein coupled receptors (GPCR) or single transmembrane protein such as growth factor receptors. In the case of GPCR, signaling is linked to heterotrimeric G proteins. In contrast, single transmembrane receptors exemplified by receptor tyrosine kinases (such as EGFR) and Ser-Thr receptor kinases (such as the transforming growth factor-β receptor family) have integral kinase activity. Upon binding of ligand, the kinase domain of these receptors is activated, initiating the phosphorylation of downstream cytoplasmic effector molecules, which generally results in the activation of transcription.

To illustrate the complexity of these pathways and the signaling connections involved, we have outlined the EGFR/erbB family signal transduction in Figure 12.3. The epidermal

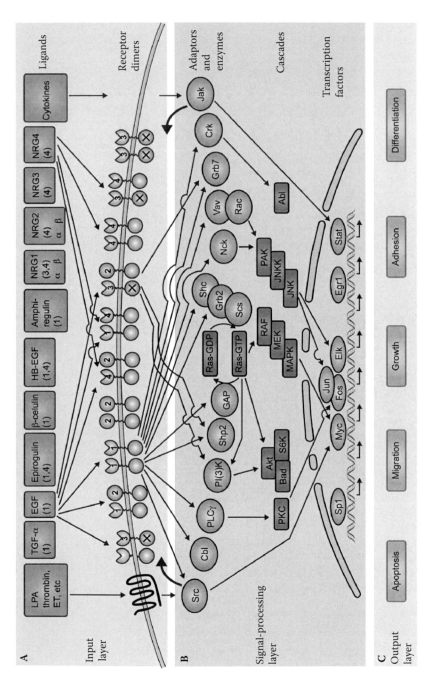

FIGURE 12.3 EGFR/erbB signal transduction pathway. (Yarden Y. and Sliwkowski M.X. 2001. *Nature Rev Mol Cell Biol* 2:127–137. Reprinted with permission.)

growth factor receptor (EGFR also known as ErbB or human EGF receptor HER) family consists of four tyrosine kinase receptors, designated ErbB1 (EGFR, HER1), ErbB2 (HER2), ErbB3 (HER3), and ErbB4 (HER4), all of which bind a specific set of ligands including epidermal growth factor (EGF), transforming growth factor (TGF)-α, epiregulin, betacellulin, amphiregulin (Areg), heparin-binding EGF-like growth factor (HB-EGF), heregulin, and the neuregulins. The nature and strength of the signal transduction pathway depend on the availability of ligands (input layer) and/or receptor molecules (Yarden and Sliwkowski 2001). Upon ligand binding, homo- or heterodimerization of these receptor pairs results in kinase activation and tyrosine phosphorylation of various downstream substrates, such as PLCγ or adaptor proteins such as Grb7. Consequently, these effector molecules transmit the signal through the cytoplasm, which results in the translocation or activation of transcription factors in the nucleus, which ultimately concludes in a coordinated biological response, such as growth or differentiation (output layer). Notably, the wiring and biological consequence of most signaling pathways are plastic in nature and may be alternatively configured in distinct cell types and under different conditions such as during development or tumor progression. Moreover, transactivation of EGFR can be triggered by other receptor systems, rendering the EGF receptor, an important intersection between various signaling systems (Hynes and MacDonald 2009; Yarden and Sliwkowski 2001; Yen et al. 2002).

12.4 CANCER CELL BIOLOGY

As detailed in an excellent paper "The Hallmarks of Cancer" by Hanahan and Weinberg (2000), a minimum of six biological traits or phenotypes is necessary in order for a tumor to undergo malignant growth. In this model of cancer growth and spread, several biological changes must occur which act to overcome the normal anticancer defense mechanism that is built into normal cells. This is in accordance with the generally accepted theory that tumor cells evolve progressively via a series of premalignant states into invasive cancers (Foulds 1954). Experimental *in vitro* models of immortalization and malignant transformation indicate that the disruption of a limited number of cellular pathways is sufficient to transform a normal cell into a malignant one (Hahn 2002). Many tumor studies in mice also support the premise that tumorigenesis involves multiple rate-limiting steps. Although most of these characteristics are shared by most cancer types, this analysis will focus mainly on the acquired capabilities involved in the development of carcinomas, a tumor derived from epithelial cells. These cancer types encompass those originating from cells that line the surface of our skin and organs such as lung, breast, and colon, and represent about 80% to 90% of all cancer cases reported in North America (American Cancer Society 2008).

12.4.1 Immortalization: The Ability to Undergo an Unlimited Number of Cell Divisions

The fact that normal cells can only divide a finite number of times before entering a state of permanent growth arrest, termed replicative senescence, is a strong barrier to cellular transformation. This cellular "aging" occurs largely due to the fact that each cell division results in a shortening of the ends of their chromosomal DNA, called telomeres (Braig and Schmitt 2006). In contrast to normal cells, the majority of immortalized cells express telomerase activity that functions to maintain the integrity of the chromosomes during cell division. One form of irreversible growth arrest, also referred to as cellular senescence, can be induced following either oncogene activation, loss of tumor suppressor signaling, or in response to DNA damage. Oncogene-induced senescence has recently been recognized as a tumor-suppressive mechanism *in vivo*, both in human lesions and in several mouse tumor models (Prieur and Peeper 2008). In all these cases inactivation of other pathways involving the p53 and retinoblastoma (Rb) proteins, which act at the level of cell cycle checkpoints, is necessary to overcome senescence and achieve immortalization.

It should be noted that although immortalization is necessary, it is not sufficient by itself to result in transformation into a tumorigenic state (Hahn and Meyerson 2001). This is supported by the fact that although many tumor-derived cell lines appear to be immortalized (i.e., capable of being propagated in perpetuity in tissue culture) they do not necessarily form tumors in animals. Notably, the requirements for human versus mouse cancer transformation differ significantly, in that the induction of tumors in mouse models of cancer requires fewer genetic changes in mouse versus human models of cancer (Hahn 2002).

12.4.2 Growth Signal Self-Sufficiency

Most cells require external cues in order to undergo cell proliferation. Normally, cells will not divide unless they receive extracellular signals, such as those from growth factors and hormones, which cause the cells to enter into the cell cycle. Another form of growth signal dependence arises from the interaction of integrin receptors on the cells with components of the extracellular matrix (Walker and Assoian 2005). In contrast, tumor cells invariably show reduced dependence on exogenous growth stimulation.

This growth factor independence often occurs as a result of constitutive activation of a growth factor receptor signaling pathway, such as EGFR. In some cases, the ligands of these pathways are overexpressed in the tumor cells, resulting in autocrine stimulation (Rusch, Mendelsohn, and Dmitrovsky 1996) . In some cases, growth factor receptors are vastly overexpressed, such as HER2 in breast cancer (Natali et al. 1990), or mutated, such as EGFR in glioblastoma (Gan, Kaye, and Luwor 2009), rendering them ligand-independent in their signaling. Commonly, critical components of the signaling pathways are dysregulated, such as the case of Ras, which is activated by genetic mutations in more than 25% of human cancers.

12.4.3 Insensitivity to Antigrowth Signals

The failure of cancer cells to respond to "stop" signals, including those from soluble factors coming from both the tumor cells and from neighboring cells in their microenvironment, is an important factor in their ability to become tumorigenic. Customarily, cancer cells also fail to undergo contact inhibition, the growth inhibition that results from physical contact with adjacent cells. Contact inhibition was one of the earliest cell-based mechanisms identified during transformation (Nelson and Daniel 2002) and forms the basis of the transformation-tumorigenicity focus forming assay which was used to identify many oncogenes, such as Ras (Pulciani et al. 1982). Transforming growth factor-β (TGF-β) is one of the best known soluble signaling molecules involved in growth inhibition. Secreted by many cell types, TGF-β acts by preventing the phosphorylation of Rb, through the suppression of Myc expression or the synthesis of cyclin/CDK blockers p15/Ink 4B or p21Cip/WAF1, thus blocking advance through the G1 phase of the cell cycle (Moses, Yang, and Pietenpol 1990). In some cases, resistance to antigrowth signals may involve downregulation or mutations in TGF-β receptors. Alternatively, components of the TGF-β signaling pathways, such as Smad4, or the cell cycle machinery, such as p15/Ink 4B or CDK4, may lose expression or become unresponsive through mutations (Massague and Gomis 2006). Finally, Rb itself may be the ultimate target, either by mutation of the gene or functional inactivation through sequestration by viral oncoproteins, such as the E7 protein expressed by the human papilloma virus.

In addition to the acquisition of a quiescent state from which the cell can reemerge at a later point in time, an alternative means to permanently stop cell division is by inducing differentiation, which is associated with the post-mitotic state. Although its links to the cell cycle regulation are not well worked out, the differentiation process involves the coordinated regulation of gene expression that results in differential morphological and biological properties for the cells, often resulting in cells that have very limited potential for cell division.

12.4.4 Evasion of Apoptosis

The ability of a tumor cell population to expand in number depends on both its ability to divide as well as its ability to undergo cell death. Programmed cell death, otherwise known as apoptosis, represents a major mechanism by which cell attrition is achieved. It is involved in a wide diversity of tissue-homeostatic, developmental, and oncogenic processes. Present in virtually all cell types, the apoptotic regulatory machinery senses both soluble and ECM- or membrane-bound signals which influence whether a cell should live or die. The ability of epithelial cells to undergo apoptosis in response to detachment from their surroundings as well as cell-cell adherence-based survival signals, known as anoikis, contributes to the ability of cells to maintain their bona fide architecture and location *in situ*. Notably, the ability of cells to survive (and grow) when detached from other cells and their extracellular matrix underlies another well-known transformation assay, namely, anchorage-independent growth (Chiarugi and Giannoni 2008). This assay, in which cell growth is assessed following suspension into a semisolid medium, correlates well with the tumorigenicity of tumor cells in animal xenograft transplantation models (Shin et al. 1975).

The process of apoptosis is controlled by a diverse range of cell signals, which may originate either extracellularly (via extrinsic inducers) or intracellularly (via intrinsic inducers). Examples of the "sensor" components which respond to *extrinsic* inducers include receptors that bind to survival ligands such as IGF-1 or IGF-2 or IL-3 or death-inducing ligands such as FAS ligand or TNF-α (Ashkenazi and Dixit 1998). Intracellular sensors also exist which transduce abnormal signals from DNA damage, oncogene activation, or hypoxia into cell death. The apoptotic machinery comprises the bcl2 family of proteins, which can be either pro-apoptotic or anti-apoptotic, acting on an array of intracellular proteases named caspases, which execute the death program through the destruction of cellular structures and the genome (Sanfilippo and Blaho 2003).

Resistance to apoptosis can occur through p53 tumor suppressor mutation, which occurs in more than 50% of human cancers, resulting in resistance to apoptosis in response to DNA damage as well as other intracellular signals (Joerger and Fersht 2008). Resistance to signals emanating from extracellular inducers involves the use of nonsignaling death decoy FAS receptors (e.g., TRAIL) as well as alterations in the prosurvival PI3 kinase/Akt/PTEN pathways which lie downstream of IGF-1/2 or IL-3 receptors (Fulda 2009).

12.4.5 Productive Interaction with the Microenvironment, Including Vascular Remodeling

While tumor cells play a prominent role in tumorigenesis, the role of supporting stromal cells within the microenvironment should not be overlooked. During carcinoma formation, cancer cells release various cytokines and growth factors into their surroundings and recruit and reprogram fibroblasts, endothelial cells, and infiltrating inflammatory cells that make up the stromal cell compartment of the tumor mass in order to establish a tumor microenvironment. Heterotypic interactions between various cell types are likely to operate in the majority of solid tumors, considering the instructive role of paracrine signaling in the regulation of tumorigenicity and metastatic potential of immortalized cancer cells. Furthermore these tumor-associated cells may co-evolve with the neighboring transformed epithelial cells in order to sustain tumor growth (Polyak, Haviv, and Campbell 2009). Recent studies have identified new roles for cancer-associated fibroblasts in promoting tumor progression, through stimulation of inflammatory pathways and induction of extracellular matrix-remodeling proteases (Radisky and Radisky 2007). The interactions that take place between immune and cancer cells are rather complex, involving multiple cascades of cytokines, chemokines, and/or growth factors (Sheu et al. 2008). The link between inflammation and cancer is nonetheless well documented (Coussens and Werb 2002) and the presence of immune cells, particularly macrophages, has been shown to stimulate tumor cell motility and metastasis as well as angiogenesis through the release of inflammatory cytokines by immune cells, in response to the presence of tumor cells (Sica, Allavena, and Mantovani 2008).

All cells, particularly fast growing tumor cells, need a constant source of oxygen and nutrients, such as glucose, from their blood supply. In addition, successful tumors need to be able to recruit endothelial cells to induce the formation of blood vessels in a process termed angiogenesis (Kerbel 2008). Angiogenesis is carefully controlled by counterbalancing positive and negative signals resulting in an angiogenic switch occurring in early to mid-stage tumor development (Persano, Crescenzi, and Indraccolo 2007), just before the onset of rapid clonal expansion that can be seen in macroscopic tumors. Angiogenesis promoting signals are best exemplified by vascular endothelial growth factor (VEGF) which acts through the tyrosine kinase receptor VEGFR1 or 2. Currently, strategies to inhibit VEGFR or VEGF, through neutralizing antibodies such as bevacizumab, are proving to be quite successful in a variety of clinical trials, highlighting the importance of the neovascularization step during tumor progression.

12.4.6 Tumor Metastasis

Tumors that are capable of invading into adjacent tissues and spreading to other sites to form metastases account for over 90% of deaths due to cancer (Sporn 1996). In this process, tumor cells must leave the initial or primary tumor and travel through the blood or lymphatic system to reach and eventually colonize a distant location. Normally, cells sense their role, both in relation to their spatial location and place in time within the body such that their presence, architecture, and function remain beneficial to the organism as a whole. The invasion-metastasis cascade that occurs during metastasis breaches this barrier in a complex process that is not genetically or biochemically completely understood.

The ability of cells to undergo an epithelial-to-mesenchymal transition (EMT) appears to be fundamental to the ability of epithelial cells to detach from the primary tumor and invade adjacent tissue. This process is similar to the differentiation process, which plays a key role in many steps during embryogenesis such as mesoderm formation and neural tube formation (Baum, Settleman, and Quinlan 2008). In this process, epithelial cells lose some of their markers, such as the cell adhesion molecule E-cadherin, and gain characteristics of mesenchymal cells such as the ability to migrate and invade surrounding tissue. Induction of EMT occurs via a diverse set of stimuli, including growth factor signaling, tumor-stromal cell interactions and hypoxia and several pathways involving EGF, TGF-β, Src, Ras, Ets, integrins, Wnt/beta-catenin and Notch, culminating in the expression of specific transcription factors, including Snail1, Snail2/Slug, and Twist (Moustakas and Heldin 2007). Originally discovered through the study of developmental genetics, the expression of many of the transcription factors, such as Twist, is often correlated with invasiveness of tumors (Yang, Mani, and Weinberg 2006).

In addition to the loss of E-cadherin, which acts as a metastasis suppressor by inhibiting epithelial cell to cell adhesion, the appearance of certain other cell adhesion molecules such as N-cadherin appears to play a functional role in motility and invasion (Hazan et al. 2004). As well as changes in cell-cell adhesion molecule composition, changes in integrin expression often occur which correlate with altered binding to the tumor-associated ECM environment. Proteases also serve a vital function in metastasis by degrading extracellular matrix proteins, thus facilitating invasion across basement membranes, into nearby stroma and across endothelial cells (VanSaun and Matrisian 2006). A more detailed overview of the signaling pathways involved in EMT is not presented here, but is the subject of another chapter in this book (Chapter 13).

12.4.7 Genomic Instability, an Accelerator for Tumor Evolution

Taking into consideration that many of the changes that are required in the evolution of a tumor involve mutations and other forms of stable genetic alteration, there needs to be an accelerated means to produce the mutations that are selected for in this selection process. Under normal circumstances, DNA damage is sensed, resulting in either cell cycle arrest to allow for repair or apoptosis if the damage is extensive. To produce the genomic instability and variability needed for the selective advantage, the machinery of the DNA repair system, including the sensors of damaged DNA, such as the p53 system, is commonly affected in tumors (Jeggo 2005).

12.4.8 Cell-Based Assays to Study Tumor Progression

As summarized above, various biological hurdles must be overcome in order for a cell to become cancerous. Traditionally, two-dimensional monolayer cell growth assays have been relied upon in preclinical studies designed to assess the efficacy of chemotherapeutics and other molecularly targeted inhibitors. These assays generally measure relatively short-term growth of carcinoma cell lines, usually through the evaluation of metabolic activity using reducing dyes such as MTT. These monolayer growth assays fail to measure anchorage-independent survival and growth, one of the distinguishing features of transformed cells (Walker, Fournier, and Assoian 2005). On the other hand, assays performed under anchorage-independent conditions, such as colony-forming or clonogenic assays, tend to measure more accurately the *in vivo* tumorigenic potential of tumor cell lines (Shin et al. 1975).

An additional limitation of both monolayer and anchorage-independent tumor cell growth assays is their failure to take into account the effect of inhibitors on processes related to metastasis, such as motility and invasion. It has been proposed that increased cell motility and invasion may impact not only metastasis, but also growth at the primary tumor site. This "self-seeding" hypothesis postulates that cell motility may contribute to the high cell density, rapid growth rate, and large primary tumor size of more aggressive tumors (Norton and Massague 2006). Accordingly, effects of inhibitors on primary tumor growth may also be underestimated if effects on motility and invasion are not assessed. In support of this more comprehensive approach to cell-based assay testing, we utilized anchorage-independent cell growth as well as motility assays and demonstrated that these tests detect the effects of two epidermal growth factor receptor (EGFR) inhibitors, the small molecule inhibitor AG1478 and the ligand-blocking antibody 225 mAb, on A549 non-small cell lung cancer cells more sensitively than monolayer growth assays (Jaramillo et al. 2008). These results explain, in part, why these cells are sensitive to EGFR inhibitors in *in vivo* xenograft animal models, but are relatively resistant in conventional *in vitro* monolayer growth assays.

To take into account the role of the tumor microenvironment, heterotypic assays involving two or more cell types in three-dimensional (3D) tissue extracellular matrix microenvironments are beginning to be developed (Hu et al. 2009; Karnoub et al. 2007; Zahir and Weaver 2004). These types of approaches are leading to some interesting microenvironment-based therapeutic targets, such as COX-2 and CCL-5 (Hu et al. 2009; Karnoub et al. 2007).

12.5 A BIOINFORMATICS APPROACH TO CANCER BIOLOGY

To connect the plethora of signaling pathways that are being characterized to the roles that they play in the development of cancer, we attempted to link a number of gene ontology (GO) terms to a large selection of biological pathways associated with cancer. The GO terms were selected from Amigo, a Gene Ontology database, which consists of a collection of three structured controlled vocabularies of terms that describe gene products in terms of their associated biological processes, cellular components, and molecular functions. Eight cancer-associated GO terms were selected from the AMIGO database to reflect the biological phenotypes first described by Hanahan and Weinberg (2000) as outlined above. Table 12.1 lists the GO terms chosen for the

TABLE 12.1   Annotated Genes in GO Terms Associated with Cancer Hallmarks

Cancer Hallmarks	GO Term Definition	GO Term	Number of Annotated Genes with This GO Term (Whole Genome)	Annotated Genes in Selected Cancer Pathways (64)
Limitless replicative potential	Cell aging	GO:0007569	29	5
Insensitivity to growth inhibitors	Positive regulation of cell proliferation	GO:0008284	314	98
Growth signal self-sufficiency	Negative regulation of cell proliferation	GO:0008285	286	54
Sustained angiogenesis	Angiogenesis	GO:0001525	164	40
Tissue invasion and metastasis	Epithelial to mesenchymal transition	GO:0001837	334	70
	Cell motility	GO:0048870		
Evasion of apoptosis	Anti-apoptosis	GO:0006916	192	67

analysis as well as the total number of genes associated with each GO term at the time of analysis. It should be noted, however, that the AMIGO gene ontology database is constantly being updated in regard to both GO terms and the gene products associated with them.

We used the Ambion database of signaling pathways (http://www.ambion.com/tools/pathway/) as a source for signaling pathways that are involved in a variety of biological and biochemical processes. To date, more than 3700 gene products are involved in the 387 signaling pathways listed. Among those, we selected a subset of 64 pathways that are known to be related to cancer, containing 777 genes, of which 214 have associated GO terms, thereby representing 27.5% of pathway genes. We then performed a search for the particular GO terms associated with each of the biological phenotypes under investigation among the genes of these cancer selected pathways in order to gauge their involvement in a particular phenotype associated with cancer (such as cell proliferation or anti-apoptosis). In the final overview, after conducting statistical analysis in which the degree of enrichment of each pathway in a given phenotype is compared to that which would be expected randomly, 64 pathways were identified whose genes were found to be associated with one or more of the six cancer biological phenotypes. Of those, we eliminated pathways that were redundant or only peripherally related to cancer, resulting in a net total of 37 pathways.

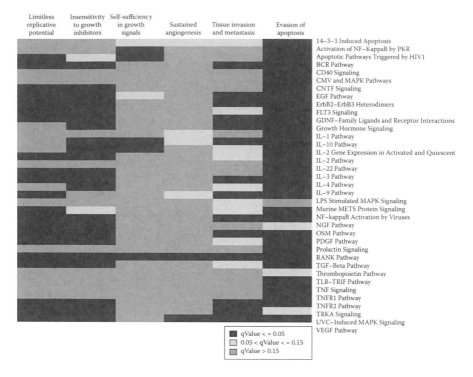

FIGURE 12.4 (See color insert following page 332.) Heatmap demonstrating associations of cancer-linked GO terms (linked to cancer hallmarks) in various signaling pathways. Significance value q < 0.05 (red in color) implies more significant association whereas q > 0.15 (green in color) implies less significant association.

This novel approach is shown schematically in a heat map in which the signaling pathways are listed horizontally, the biological phenotypes listed vertically, and the significance of the association, q value, indicated by color (as indicated, Figure 12.4). This global approach was designed to provide a global overview of the relative contributions of many pathways to the cancer state. Undoubtedly, such an approach will undoubtedly improve over time as the boundaries of pathways become better defined and gene ontology of many of the involved genes becomes better annotated.

When we performed our analysis, we found that most pathways are involved in more than one phenotype, suggesting that the majority of these phenotypes are associated with each other, to a certain degree. The EGFR/ErbB cascade is an example of a pathway that ties into several phenotypes such as growth, motility, and differentiation, which act together to generate a complex cancer phenotype. EGFR/ErbB signaling is generally linked with the development of the majority of solid cancers through its effects on both the MAPK pathway, leading to cell cycle progression, and the PI3K pathway, causing cell survival. As a consequence, many EGFR antagonists (such as cetuximab and Iressa) have been developed for cancer treatment. Similarly, the TGF-β pathway appears to be quite pleotropic in nature, acting through a variety of effector molecules and on a variety of cell types, including epithelial cells, fibroblasts, and immune cells. Another interesting feature of this type of analysis is the participation of many cytokine pathways, such as those involving interleukins, in several of the cancer phenotypes. Produced by a variety of cells and originally identified by their ability to select, attract, and activate leukocytes, interleukins play an important role in a number of tumor-related processes, thereby highlighting the role of the tumor microenvironment in many of these processes.

Our analysis also links the tumor metastasis phenotype with a variety of other cancer cell phenotypes. In addition to the self-seeding hypothesis which states that cell motility in cancer plays an important role in the ability of a primary tumor to repopulate both itself and distant organs, several recent studies link the ability of tumor cells to undergo EMT with other biological processes. Traditionally associated with metastasis, a recent study indicates that EMT may also participate in the early stages of tumor development by preventing oncogene-induced cellular senescence during the initial cell transformation. In particular, transcription factors capable of inducing EMT, for example, Twist-1 and Twist-2, were found to be capable of overriding Ras oncogene-induced premature senescence by abrogating key regulators of the p53- and Rb-dependent pathways (Ansieau et al. 2008). Cells that have undergone an EMT are also more resistant to anoikis, which is consistent with the ability of disseminated cells to survive in the bloodstream during the metastatic process. Several of the features of cells undergoing an EMT, including resistance to chemotherapeutics and anoikis, are similar to the characteristics of cancer stem cells (Polyak, Haviv, and Campbell 2009). This pluripotent subpopulation of tumor cells generates tumors through the processes of self-renewal and differentiation into multiple cell types and is linked to the ability of a tumor to relapse and metastasize. Several lines of evidence are beginning to emerge regarding the relevance and importance of the signaling pathways that converge on the EMT phenotype during all stages of tumorigenesis.

12.6 CONCLUDING REMARKS

From cancer genomics to signaling pathways to cancer biology, the increase in complexity is apparent as one increases the scale of view. However, it is necessary to take both the heterogeneity and complexity of cancer signaling into account in order to gain a better understanding of the overall process and allow for the design of better and more specific cancer therapeutics. Bioinformatics is an absolute necessity in order to deal with these large volumes of information, and will likely lead the way to the next generation of cancer-based studies and drug development.

REFERENCES

American Cancer Society. 2008. Cancer statistics.

Ansieau S., Bastid J., Doreau A. et al. 2008. Induction of EMT by twist proteins as a collateral effect of tumor-promoting inactivation of premature senescence. *Cancer cell* 14:79–89.

Ashkenazi A. and Dixit V. M. 1998. Death receptors: signaling and modulation. *Science* 281:1305–1308.

Baum B., Settleman J., and Quinlan M. P. 2008. Transitions between epithelial and mesenchymal states in development and disease. *Sem Cell Develop Biol* 19:294–308.

Braig M. and Schmitt C. A. 2006. Oncogene-induced senescence: putting the brakes on tumor development. *Cancer Res* 66:2881–2884.

Carr K. M., Rosenblatt K., Petricoin E. F., and Liotta L.A. 2004. Genomic and proteomic approaches for studying human cancer: prospects for true patient-tailored therapy. *Human Genomics* 1:134–140.

Chang J. W., Huang T. H., and Wang Y. C. 2008. Emerging methods for analysis of the cancer methylome. *Pharmacogenomics* 9:1869–1878.

Chiarugi P. and Giannoni E. 2008. Anoikis: a necessary death program for anchorage-dependent cells. *Biochem Pharmacol* 76:1352–1364.

Coussens L. M. and Werb Z. 2002. Inflammation and cancer. *Nature* 420:860–867.

Cui Q., Ma Y., Jaramillo M. et al. 2007. A map of human cancer signaling. *Mol Syst Biol* 3:152.

Foulds L. 1954. The experimental study of tumor progression: a review. *Cancer Res* 14:327–339.

Fujimoto N., Wislez M., Zhang J. et al. 2005. High expression of ErbB family members and their ligands in lung adenocarcinomas that are sensitive to inhibition of epidermal growth factor receptor. *Cancer Res* 65:11478–11485.

Fulda S. 2009. Tumor resistance to apoptosis. *Int J Cancer* 124:511–515.

Gan H. K., Kaye A. H., and Luwor R. B. 2009. The EGFRvIII variant in glioblastoma multiforme. *J Clin Neurosci* 16:748–754.

Glas A. M., Floore A., Delahaye L. J. et al. 2006. Converting a breast cancer microarray signature into a high-throughput diagnostic test. *BMC Genomics* 7:278.

Greenman C., Stephens P., Smith R. et al. 2007. Patterns of somatic mutation in human cancer genomes. *Nature* 446:153–158.

Haber D. A. and Settleman J. 2007. Cancer: drivers and passengers. *Nature* 446:145–146.

Hahn W. C. 2002. Immortalization and transformation of human cells. *Mol Cell* 13:351–361.

Hahn W. C. and Meyerson M. 2001. Telomerase activation, cellular immortalization and cancer. *Ann Med* 33:123–129.

Hanahan D. and Weinberg R. A. 2000. The hallmarks of cancer. *Cell* 100:57–70.

Hazan R. B., Qiao R., Keren R., Badano I., and Suyama K. 2004. Cadherin switch in tumor progression. *Ann NY Acad Sci* 1014:155–163.

He X., Chang S., Zhang J. et al. 2008. MethyCancer: the database of human DNA methylation and cancer. *Nucleic Acids Res* 36:D836–D841.

Hu M., Peluffo G., Chen H. et al. 2009. Role of COX-2 in epithelial-stromal cell interactions and progression of ductal carcinoma *in situ* of the breast. *Proc Natl Acad Sci USA* 106:3372–3377.

Hynes N. E. and MacDonald G. 2009. ErbB receptors and signaling pathways in cancer. *Current Opin Cell Biol* 21:177–184.

Jain K. K. 2007. Cancer biomarkers: current issues and future directions. *Curr Opin Mol Ther* 9:563–571.

Jaramillo M. L., Banville M., Collins C. et al. 2008. Differential sensitivity of A549 non-small lung carcinoma cell responses to epidermal growth factor receptor pathway inhibitors. *Cancer Biol Ther* 7:557–568.

Jeggo P. A. 2005. Genomic instability in cancer development. *Adv Exp Med Biol* 570:175–197.

Joerger A. C. and Fersht A. R. 2008. Structural biology of the tumor suppressor p53. *Annu Rev Biochem* 77:557–582.

Kancha R. K., von Bubnoff N., Peschel C., and Duyster J. 2009. Functional analysis of epidermal growth factor receptor (EGFR) mutations and potential implications for EGFR targeted therapy. *Clin Cancer Res* 15:460–467.

Karnoub A. E., Dash A. B., Vo A. P. et al. 2007. Mesenchymal stem cells within tumour stroma promote breast cancer metastasis. *Nature* 449:557–563.

Kerbel R. S. 2008. Tumor angiogenesis. *New Engl J Med* 358:2039–2049.

Kim S. Y. and Hahn W. C. 2007. Cancer genomics: integrating form and function. *Carcinogenesis* 28:1387–1392.

Latterich M., Abramovitz M., and Leyland-Jones B. 2008. Proteomics: new technologies and clinical applications. *Eur J Cancer* 44:2737–2741.

Liu E. T. 2008. Functional genomics of cancer. *Curr Opin Genet Dev* 18:251–256.

Lopez J., Percharde M., Coley H. M., Webb A., and Crook T. 2009. The context and potential of epigenetics in oncology. *Br J Cancer* 100:571–577.

Luo B., Cheung H. W., Subramanian A. et al. 2008. Highly parallel identification of essential genes in cancer cells. *Proc Natl Acad Sci USA* 105:20380–20385.

Lynch T. J., Bell D. W., Sordella R. et al. 2004. Activating mutations in the epidermal growth factor receptor underlying responsiveness of non-small-cell lung cancer to gefitinib. *N Engl J Med* 350:2129–2139.

Massague J. and Gomis R. R. 2006. The logic of TGFbeta signaling. *FEBS Lett* 580:2811–2820.

Massarelli E., Varella-Garcia M., Tang X. et al. 2007. KRAS mutation is an important predictor of resistance to therapy with epidermal growth factor receptor tyrosine kinase inhibitors in non-small-cell lung cancer. *Clin Cancer Res* 13:2890–2896.

Moses H. L., Yang E. Y., and Pietenpol J. A. 1990. TGF-beta stimulation and inhibition of cell proliferation: new mechanistic insights. *Cell* 63:245–247.

Moustakas A. and Heldin C. H. 2007. Signaling networks guiding epithelial-mesenchymal transitions during embryogenesis and cancer progression. *Cancer Sci* 98:1512–1520.

Natali P. G., Nicotra M. R., Bigotti A. et al. 1990. Expression of the p185 encoded by HER2 oncogene in normal and transformed human tissues. *Int J Cancer* 45:457–461.

Nelson P. J. and Daniel T. O. 2002. Emerging targets: molecular mechanisms of cell contact-mediated growth control. *Kidney Int* 61:S99–S105.

Norton L. and Massague J. 2006. Is cancer a disease of self-seeding? *Nature Med* 12:875–878.

Nowell P. C. 1976. The clonal evolution of tumor cell populations. *Science* 194:23–28.

Ohm J. E., McGarvey K. M., Yu X. et al. 2007. A stem cell-like chromatin pattern may predispose tumor suppressor genes to DNA hypermethylation and heritable silencing. *Nature Genet* 39:237–242.

Persano L., Crescenzi M., and Indraccolo S. 2007. Anti-angiogenic gene therapy of cancer: current status and future prospects. *Mol Aspects Med* 28:87–114.

Polyak K., Haviv I. and Campbell I. G. 2009. Co-evolution of tumor cells and their microenvironment. *Trends Genet* 25:30–38.

Polyak K. and Weinberg R. A. 2009. Transitions between epithelial and mesenchymal states: acquisition of malignant and stem cell traits. *Nature Rev* 9:265–273.

Preisinger C., von Kriegsheim A., Matallanas D. and Kolch W. 2008. Proteomics and phosphoproteomics for the mapping of cellular signalling networks. *Proteomics* 8:4402–4415.

Prieur A. and Peeper D.S. 2008. Cellular senescence *in vivo*: a barrier to tumorigenesis. *Curr Opin Cell Biol* 20:150–155.

Pulciani S., Santos E., Lauver A. V. et al. 1982. Oncogenes in human tumor cell lines: molecular cloning of a transforming gene from human bladder carcinoma cells. *Proc Natl Acad Sci USA* 79:2845–2849.

Radisky E. S. and Radisky D. C. 2007. Stromal induction of breast cancer: inflammation and invasion. *Rev Endocr Metab Disorders* 8:279–287.

Rusch V., Mendelsohn J., and Dmitrovsky E. 1996. The epidermal growth factor receptor and its ligands as therapeutic targets in human tumors. *Cytokine Growth Factor Rev* 7:133–141.

Sanfilippo C. M. and Blaho J. A. 2003. The facts of death. *Int Rev Immunol* 22:327–340.

Sheu B. C., Chang W. C., Cheng C. Y. et al. 2008. Cytokine regulation networks in the cancer microenvironment. *Front Biosci* 13:6255–6268.

Shin S. I., Freedman V. H., Risser R., and Pollack R. 1975. Tumorigenicity of virus-transformed cells in nude mice is correlated specifically with anchorage independent growth *in vitro*. *Proc Natl Acad Sci USA* 72:4435–4439.

Sica A., Allavena P., and Mantovani A. 2008. Cancer related inflammation: the macrophage connection. *Cancer Lett* 267:204–215.

Sjoblom T., Jones S., and Wood L. D. et al. 2006. The consensus coding sequences of human breast and colorectal cancers. *Science* 314:268–274.

Sporn M. B. 1996. The war on cancer. *Lancet* 347:1377–1381.

van't Veer L. J. and Bernards R. 2008. Enabling personalized cancer medicine through analysis of gene-expression patterns. *Nature* 452:564–570.

VanSaun M. N. and Matrisian L. M. 2006. Matrix metalloproteinases and cellular motility in development and disease. *Birth Defects Res C Embryo Today* 78:69–79.

Walker J. L. and Assoian R. K. 2005. Integrin-dependent signal transduction regulating cyclin D1 expression and G1 phase cell cycle progression. *Cancer Metastasis Rev* 24:383–393.

Walker J. L., Fournier A. K., and Assoian R. K. 2005. Regulation of growth factor signaling and cell cycle progression by cell adhesion and adhesion-dependent changes in cellular tension. *Cytokine Growth Factor Rev* 16:395–405.

Yang J., Mani S. A., and Weinberg R. A. 2006. Exploring a new twist on tumor metastasis. *Cancer Res* 66:4549–4552.

Yarden Y. and Sliwkowski M. X. 2001. Untangling the ErbB signalling network. *Nature Rev Mol Cell Biol* 2:127–137.

Yen L., Benlimame N., Nie Z. R. et al. 2002. Differential regulation of tumor angiogenesis by distinct ErbB homo- and heterodimers. *Mol Biol Cell* 13:4029–4044.

Zahir N. and Weaver V. M. 2004. Death in the third dimension: apoptosis regulation and tissue architecture. *Curr Opin Genet Dev* 14:71–80.

CHAPTER 13

Epithelial-to-Mesenchymal Transition (EMT)

The Good, the Bad, and the Ugly

Anne E.G. Lenferink

CONTENTS

13.1 INTRODUCTION

Epithelial cells usually form tightly connected sheets of monolayers, which occurs as a result of the subcellular distribution of cadherins, specific integrins, cell-cell junctions (desmosomes, tight junctions, adherens junctions, and gap junctions), as well as the polarized organization of the cytoskeleton. This organization establishes an apical-basal polarity through their association with a basement membrane, which, under normal circumstances, allows cells to move laterally within the epithelial layer (Birchmeier, Birchmeier, and Brand-Saberi 1996) but prevents their entrance into the underlying extracellular matrix (ECM).

Epithelial-to-mesenchymal transition (EMT) has long been known to play an important role in the generation of mesenchymal cells from the primitive epithelium during metazoan embryogenesis. In contrast to epithelial cells, mesenchymal cells do not form organized layers, are not associated with a basement membrane, and maintain only focal contacts with neighboring mesenchymal cells. In addition, mesenchymal cells in culture tend to be highly motile, either as chains or as individual cells moving around using the mechanism of "extension-adhesion-retraction" or amoeboid crawling (Thiery and Chopin 1999). The concept of EMT was proposed in the 1960s in studies using chicken embryos (Krug, Mjaatvedt, and Markwald 1987; Trelstad, Hay, and Revel 1967; Trelstad, Revel, and Hay 1966). However, it was not until the mid 1980s that Krug and co-workers (1987) used the word combination "epithelial-to-mesenchymal transition" in the context of a cellular change induced by the ECM and that the EMT process was recognized as a discrete cellular program (Greenburg and Hay 1982, 1986, 1988). Currently EMT is defined, although not strictly, by the occurrence of three phenotypic changes in the epithelial cell: (1) acquisition of a spindle-shaped morphology; (2) repression of epithelial markers [such as E-cadherin, occludin, e.g., Zona Occludens 1 (ZO-1), and cytokeratins] and gain of mesenchymal markers (e.g., vimentin, N-cadherin, and fibronectin); and (3) increased motility and the capacity to invade the ECM, which is considered to be a functional hallmark of EMT (Cui et al. 2007).

For EMT to occur the epithelial cell requires a series of complex changes in its architecture and behavior. These changes are driven by the extracellular signals the cell receives and then processes into a well-orchestrated integrated response (Figure 13.1). The EMT process is in many cases reversible and cells can undergo a so-called mesenchymal-to-epithelial transition (MET). This reciprocal process, which provides important cellular flexibility that is necessary during tissue construction in normal development (Chaffer, Thompson, and Williams 2007), has recently also been implicated in pathological processes such as

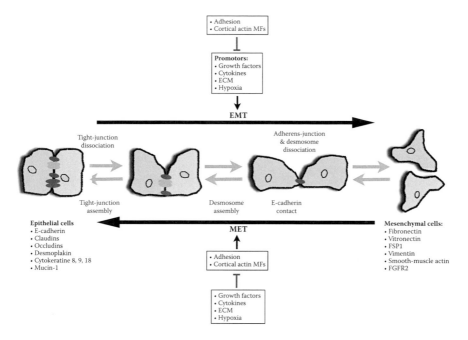

FIGURE 13.1 (See color insert following page 332.) Simplified representation of the balance between the mechanisms driving epithelial-to-mesenchymal transition (EMT) and mesenchymal-to-epithelial transition (MET).

cancer (Boyer and Thiery 1993; Davies 1996). In addition to the idea that EMT is imperative for the escape of tumor cells from the primary tumor, these observations support the notion that the reverse MET process allows these tumor cells to form secondary tumors (Chaffer et al. 2005, 2006).

13.2 EMT AND MET IN DEVELOPMENT

As previously mentioned, EMT and MET are two important events which orchestrate the interconversion between epithelial and mesenchymal cell morphology, thus allowing for the reorganization of germ layers and tissues during embryonic development (Shook and Keller 2003). The first EMT occurs in the very early stages of development during the formation of mesoderm from the primitive ectoderm, whereas the first MET takes place even earlier during the pre-implantation stage when the trophectoderm is formed. At later stages, additional EMT conversions occur during the establishment of (1) the neural crest (Kuriyama and Mayor 2008; Snarr, Kern, and Wessels 2008), (2) the heart valves (Snarr, Kern, and Wessels 2008), (3) the sclerotome (Hay 2005), (4) coronary vessel progenitor cells (Wessels and Perez-Pomares 2004), (5) the secondary palate (Kang and Svoboda 2005), and (6) during male Müllerian duct regression (Arias 2001; Klattig and Englert 2007). It is interesting to note that, even though the mechanisms of EMT underlying these processes are quite well understood through the use of cultured cell models (Arias 2001; Savagner 2001), the entire development process has yet to be characterized within a single species. More details regarding the mechanisms, mechanics, and function of EMTs occurring in early development can be found in an excellent review that was published in 2003 by Shook and Keller. However, to illustrate the role of EMT in embryonic development we will further outline two examples: (1) the mesoderm formation in early development, and (2) the formation of the heart valves at a later stage of development.

13.2.1 EMT during Mesoderm Formation during Gastrulation

Most of our current knowledge of the EMT program guiding the gastrulation process was generated using fruit fly, fish, amphibian, and bird model systems (Kimelman 2006). Nonetheless, most of the principles described for these model organisms also apply to the embryonic development of mammals (Viebahn 1995). Prior to implantation, the embryonic blastocyst consists of the mesenchymal inner cell mass (embryoblast) which is surrounded by the trophectoderm, and which is considered to be the first epithelial structure that is formed in a mammalian embryo (Nishioka et al. 2008). Around 16 days post-fertilization, ectoderm cells invade the embryoblast, causing the inner cell mass to be divided into the epiblast and hypoblast lineages. The epithelial-shaped epiblast cells then undergo EMT, invade the basal membrane, and migrate along the narrow space underneath the ectoderm, thereby forming the mesoderm (Arnold and Robertson 2009; Nakaya and Sheng 2008; Tam and Beddington 1987; Tam, Williams, and Chan 1993). As outlined above, the mammalian gastrulation process results in the formation of three primitive embryonic germ layers, the endoderm, ectoderm, and mesoderm. Multiple signaling pathways tightly regulate this process. For example, the bone morphogenetic proteins (BMPs) play an important role in early embryonic patterning, a process that is responsible for establishing three-dimensional polarities (Kishigami and Mishina 2005). In addition, fibroblast growth factor receptor 1 (FGFR1) was shown to orchestrate the epiblast EMT and morphogenesis of the mesoderm by controlling the expression of the transcription factor Snail1 and the adherens junction protein E-cadherin (Ciruna and Rossant 2001). The importance of Snail1 in this process was confirmed by the observation that mouse embryos deficient in this transcription factor display a morphologically abnormal mesoderm (Carver et al. 2001). Furthermore, expression of a glycogen synthase kinase (GSK)-3β mediated proteasome degradation resistant form of β-catenin in oocytes caused a premature EMT of the epiblast, thereby generating mesenchymal cells lacking E-cadherin (Kemler et al. 2004).

13.2.2 EMT during Heart Valve Development

The embryonic development of mammalian heart valves requires the formation of endocardial cushions. These structures are formed in the atrioventricular (AV) canal and outflow tract (OFT) regions of the heart as a result of the secretion of a hyaluronan- and chondroitin sulfate proteoglycan-rich ECM by the myocardial cells. During embryonic development, and as the result of a signal that is produced by the myocardium of the AV and OFT regions, endocardial cells undergo EMT (Bolender and Markwald 1979; Markwald, Fitzharris, and Manasek 1977), causing these cells to invade the proteoglycan-rich ECM, also known as cardiac jelly (Markwald, Fitzharris, and Manasek 1977), and to fill the cushion space, thereby establishing the rudimental onset for the formation of the cardiac septa and valves (Eisenberg and Markwald 1995; Person, Klewer, and Runyan

2005). Experiments have shown that several of the transforming growth factor (TGF)-βs and BMPs (Nakajima et al. 2000; Okagawa, Markwald, and Sugi 2007; Stevens et al. 2008; Townsend et al. 2008; van Wijk, Moorman, and van den Hoff 2007; Yamagishi, Ando, and Nakamura 2009), and Notch receptor signaling (Bailey, Singh, and Hollingsworth 2007; Krebs et al. 2000) play critical roles in the regulation of this EMT during cardiac development.

13.2.3 MET during Development

Processes such as embryonic development and organogenesis are very dynamic and the cells that are involved therefore have a need to be able to rapidly switch between a sessile epithelial and a motile mesenchymal cell state. While EMT induces epithelial cells to adopt a mesenchymal phenotype, MET allows cells to revert back from a mesenchymal to an epithelial cell state. MET occurs on several occasions during embryonic development, and is responsible for generating the first embryonic epithelium (Larue and Bellacosa 2005). Also, later in development, processes such as kidney organogenesis (Ekblom 1989) and somitogenesis (Christ and Ordahl 1995), rely heavily on MET. The MET process appears to be active during adult life as well. For example, mesenchymal stem cells (MSCs) are frequently found in adult tissues, and the isolation and subsequent *in vitro* culturing of these cells results in the generation of a variety of mature cell types. This demonstrates the high degree of plasticity these cells display and raises the question as to whether the assumption that EMT/MET transitions are absent in the adult organism is incorrect (Zipori 2004, 2006). The molecular mechanism by which these MSCs regulate their capacity to transdifferentiate remains, however, largely unsolved, although several cell adhesion molecules, growth factors, signaling pathways, and transcription factors have been implicated in the MET process (Chaffer, Thompson, and Williams 2007).

13.3 EMT IN HOMEOSTASIS

We have seen that, even though EMT is crucial for and very common in normal embryonic development, there are indications that the process continues to play a role during adult life. For example, the replacement of cells necessary for the proper functioning of tissues and organs relies strongly on EMT-like events. The cells involved in this maintenance function are made up of a small population of immortal stem cells, which, under specific culture conditions, display a long-term capacity to self-renew as well as differentiate. The embryonic stem cell (ESC), which originates from the epiblast, is probably the most and best-studied mammalian stem cell type (Thomson et al. 1998). As mentioned earlier, epiblast cells are highly epithelial, are not motile, and are engaged in E-cadherin-mediated cell-cell interactions (Vestweber et al. 1987). Nonetheless, isolation and the subsequent maintenance of these ESCs under feeder-free culturing conditions results in the appearance of mesenchymal-like cells at the periphery of the established colonies. These cells do not express E-cadherin, but do express β-catenin, N-cadherin, and vimentin. In addition, they are motile and have upregulated the expression of EMT-related transcription factors Snail1 and Snail2/Slug (Ullmann et al. 2007), suggesting that the highly epithelial epiblast cells have undergone EMT.

13.4 EMT AND MET IN DISEASE

It is widely accepted that the strategies that cells employ during embryonic development, such as the induction of EMT, are also exploited in a variety of disease states. For example, emerging evidence suggests that EMT is an important event in fibrosis (Wynn 2008), as illustrated by its role in fibrotic disease of the kidney (Bedi, Vidyasagar, and Djamali 2008). EMT is important also in the progression of cancer, especially in the case of carcinomas, which make up the vast majority of solid human tumors and which share an epithelial origin. To illustrate the role of EMT in disease we will further outline the process in the context of kidney fibrosis and cancer progression. In addition, a section is dedicated to the recent discussion of the role of MET in tumor development.

13.4.1 EMT in Kidney Fibrosis

Fibrosis can be defined by the overgrowth, hardening, and/or scarring of various tissues as a result of excess deposition of ECM components, and is the end result of chronic inflammatory reactions. These can be induced by many different stimuli, such as autoimmune and allergic responses, persistent infections, and tissue injury (Wynn 2008). During the fibrosis process it is the myofibroblasts that play a central role. For example, renal damage is characterized by tubular atrophy and peritubular capillar loss. It has long been known that EMT is an important event in kidney development (Hay and Zuk 1995), but recently the process has also been attributed a role in interstitial fibrosis (Burns, Kantharidis, and Thomas 2007), during which the accumulation of myofibroblasts results in the overproduction of interstitial ECM. The accumulation of these interstitial myofibroblasts is still somewhat controversial, in that these cells can originate from different sources: (1) through activation and proliferation of resident renal fibroblasts, (2) from circulating bone marrow precursors, or (3) from tubulo-epithelial cells that have undergone EMT and transformed into myofibroblasts. The latter process can be induced by a range of growth factors and cytokines, such as, for example, via TGF-β, and the subsequent induction of integrin-linked kinase (ILK) and the Smads. In addition, connective tissue growth factor (CTGF; Burns et al. 2006), angiotensin II (Carvajal et al. 2008; Mezzano, Ruiz-Ortega, and Egido 2001), and various interleukins (Pesce et al. 2006; Reiman, Mauldin, and Neal Mauldin 2008; Wynn et al. 1995) have all been shown to play their part in these EMT processes (Li and Liu 2007; Ziyadeh 2008). It is interesting to note that TGF-β is pivotal in the induction of fibrosis, in that it is also involved in the development of pulmonary (Willis and Borok 2007) and hepatic fibrosis (Gressner et al. 2002).

13.4.2 EMT in Cancer

The transformation of a normal epithelial cell into a tumor cell is a highly stochastic process. Benign noninvasive and nonmetastatic tumor cells acquire attributes that allow these cells to penetrate the surrounding tissues and eventually metastasize, transit in the blood or lymphatic vessels, extravasate, and finally proliferate in a distant organ (Chambers, Groom, and MacDonald 2002). Tumor cells can use different strategies to invade their surroundings. One strategy is by adopting a mesenchymal morphology through the EMT process. The elongated mesenchymal-like cells produce increased amounts of matrix metalloproteinases (MMPs) which degrade the surrounding ECM thereby creating a "trail" through which the tumor cells can migrate, likely through a mechanism that involves activation of the Rho family of GTPase 1 (Rac1) and inhibition of Ras homolog gene family member A (RhoA) GTPase at the cell's leading edge (Kurisu et al. 2005; Yamazaki, Kurisu, and Takenawa 2009). Cell motility and invasion mechanisms have mostly been studied using *in vitro* cell culture systems. Accordingly, it has therefore been difficult to translate the information that is generated to *in vivo* situations. Nonetheless the most convincing evidence is the fact that several known *in vitro* EMT regulators have also been found to enhance tumor formation and/or infiltration and metastasis *in vivo* (Hugo et al. 2007; Thiery 2002). For example, cells at the invasive edge of colon carcinomas, but not those in the center, display a strong nuclear β-catenin staining (Brabletz et al. 1998; Hlubek et al. 2007) and express high levels of the EMT related transcription factor Snail1 (Franci et al. 2006). Snail1 can also be detected in a subset of human clinical tumor samples (Becker et al. 2007) and can be linked to decreased survival (Waldmann et al. 2008). Despite these observations the involvement of EMT in the dissemination of human cancers is still being critically debated (Tarin, Thompson, and Newgreen 2005; Thompson, Newgreen, and Tarin 2005). Nevertheless, the presence of EMT markers specifically at the tumor-host interface is strong evidence that EMT is involved in the regulation of tumor cell invasiveness.

13.5 MET IN CANCER

In addition to its role in development, MET also appears to be critical for a metastatic tumor cell to be successful in its endeavor to repopulate distant organs (Hugo et al. 2007). This idea is best illustrated by the fact that the epithelial gatekeeper E-cadherin can be found in many carcinoma metastases. Also, the fact that implantation in the mouse mammary fat pad of the aggressive E-cadherin negative MDA-MB231 mammary tumor cell line leads to the formation of E-cadherin positive lung metastasis (reported as unpublished data in Vincan et al. 2007; Vincan, Whitehead, and Faux 2008; Wells, Yates, and Shepard 2008). Similar observations were made following the implantation of the human LIM1863-Mph colon (Vincan et al. 2007; Vincan, Whitehead, and Faux 2008) and a panel of TSU-Pr1 bladder carcinoma cell lines (Chaffer et al. 2005, 2006) into nude mice, as well as when the human DU145 prostate cell line was co-cultured with hepatocytes (Yates et al. 2007). These results imply that disseminated tumor cells must also undergo MET, hence the re-expression of E-cadherin, in order to successfully initiate tumor growth at the site of metastasis.

Another hypothesis explaining these results is that the E-cadherin positive cells found at the site of metastasis have originated from cells that have simply escaped or were shed by the primary tumor (Bockhorn, Jain, and Munn 2007), and have successfully formed secondary tumors (Paget 1989; Ribatti, Mangialardi, and Vacca 2006). One has to wonder, though, how these cells would be able to exit the blood or lymphatic vessels. Cells could squeeze through pores, adopting an amoeboid phenotype, which is an extremely rapid process that does not require protease activity for degradation of the ECM, or other processes that are required for EMT. On the other hand, the cells may not have to do any of this. Some studies suggest that tumor cells simply get stuck in the capillaries and this allows them to proliferate in an organ. Although these are intriguing possibilities, there is currently enough evidence to suggest that the loss of E-cadherin correlates strongly to tumor cell dissemination (Jeanes, Gottardi, and Yap 2008; Kopfstein and Christofori 2006), and that MET is required for the successful development and growth of metastases (Sabbah et al. 2008; Wells, Yates, and Shepard 2008).

13.6 SIGNALING CASCADES INVOLVED IN EMT

The canonical signaling pathways and cross-talk between these signaling pathways regulating EMT in cancer have mostly been defined by what is known from developmental studies and the use of *in vitro* cell culture models. Sabbah et al. (2008) published an excellent overview of the molecular signatures that are involved in EMT. A simplified schematic of the most important signaling cascades inducing EMT that will be further discussed in detail below is shown in Figure 13.2. It should be noted that this overview will not discuss the signaling cascades involved in tumorigenesis, since these are the subject of another chapter in this book (Chapter 12).

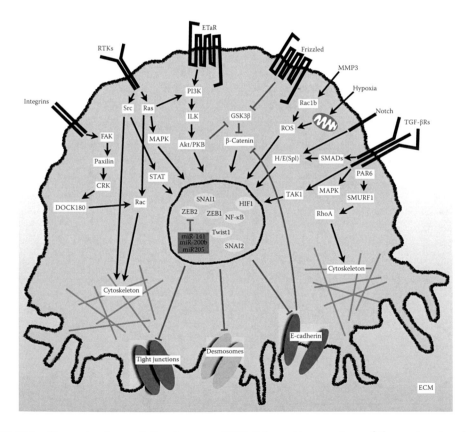

FIGURE 13.2   (See color insert following page 332.) Schematic overview of the major signal transduction pathways that induce epithelial-to-mesenchymal transition (EMT).

13.6.1 TGF-*β*

Many publications have shown that TGF-β is a key inducer of EMT during cancer development (Cui et al. 1996; Derynck, Akhurst, and Balmain 2001). The ability of TGF-β1 to induce an EMT was first described by Miettinen et al. (1994), whereas similar observations were later made for the other TGF-βs, TGF-β2 (Piek et al. 1999) and TGF-β3 (Valcourt et al. 2005). TGF-β transmits its signal over the cell membrane through a heteromeric complex of the transmembrane serine-threonine kinase TGF-β Type I (TβRI) and Type II (TβRII) receptors. Receptor dimerization leads to the recruitment and phosphorylation of receptor Smad2 and Smad3 (Wrighton, Lin, and Feng 2009), which trimerize with common Smad4 and this complex then translocates to the nucleus where the interaction with transcription factors ultimately regulates gene transcription (Figure 13.2; see also Feng and Derynck 2005). Transcription factors such as Snail1 and Snail2/Slug, ZEB1/δEF1 and ZEB2/SIP1, those that belong to the high mobility group A2 (HMGA2) family, and members of the basic helix-loop-helix (bHLH) family such as E12, E47, Twist1, and Id1-3, are all involved in the regulation of the EMT process, either in a Smad-dependent or -independent manner (Peinado, Olmeda, and Cano 2007). Smad activation of these transcription factors can ultimately result in the loss of desmosomes, tight and adherens junctions, and the gain of mesenchymal attributes such as N-cadherin, β5-integrin, and fibronectin. Even though Smad signaling is crucial, a complete TGF-β-induced EMT response also depends on the activation of non-Smad signals. These signals can either be directly or indirectly induced by the TβRs and provide control over additional processes such as survival, migration, and proliferation. An extensive overview of the signaling cascades activated during the TGF-β-induced EMT can be found in an excellent review recently published by Xu, Lamouille, and Derynck (2009). For example, the TβRs trigger migration signals by directly activating Rho GTPases and phosphoinositide 3-kinase (PI3K), whereas the inactivation (degradation) of Rho GTPases by the E3 ubiquitin ligase Smurf1 relies on the activation of Par6. In addition, survival signals such as those generated by the p38- and c-Jun N-terminal kinases (JNK) are induced by TRAF and TAK1, whereas proliferation can be controlled through Shc-induced activation of Erk mitogen-activated protein kinase (MAPK).

In this context it is noteworthy to mention that TGF-β signaling is actively involved in the regulation of the balance between a cancer cell and its microenvironment (Naber, ten Dijke, and Pardali 2008).

13.6.2 Jagged and Delta-Like

Notch signaling is involved in the maintenance of the balance between cell proliferation and apoptosis, but also has a role during development and in the function of many organs. There are four mammalian Notch receptors (Notch1-4), integral membrane proteins that can be activated through the ligands Jagged (Jag1 and Jag2), and Delta-like (Dll-1, Dll-3, and Dll-4) that are expressed on the cell surface of neighboring cells. Upon ligand binding, Notch is proteolytically cleaved within its transmembrane domain by Presenilin-1, the enzymatic component of the γ-secretase complex. This cleavage gives rise to the so-called Notch intracellular domain (NIC) which enters the nucleus, binds transcription factor

RBPJK/CBF1/Su(H), and finally activates the hairy and enhancer of split (Hes), and the Hes-related repressor (Hesr, also known as Hey, Herp, Hrt, CHF, or Gridlock) gene families. A review describing the mechanistic aspects of Notch signaling was recently published by Kopan et al. (Kopan and Ilagan 2009).

Notch signaling is known to play an important role in embryonic development and in disease through the induction of EMT. For example, activation of Notch signaling during heart valve development leads to TGF-β2 expression which, via the induction of Snail1, causes the repression of E-cadherin and consequently leads to EMT (Person, Klewer, and Runyan 2005; Zavadil et al. 2004). Inversely, TGF-β can also induce Jag1, which itself in turn can induce EMT through activation of Notch (Niimi et al. 2007; Zavadil et al. 2004).

Notch is also involved in cancer progression (Balint et al. 2005; Postovit et al. 2008; Wang et al. 2008; Zavadil et al. 2004). In epithelial and carcinoma cells, Notch signaling can be activated by oncogenic Ras, whose activity leads to an increase in NIC levels and the upregulation of Dll-1. This upregulation and the consequent activation of Notch, in cooperation with TGF-β signaling, then result in the downregulation of E-cadherin and the induction of EMT (Grego-Bessa et al. 2004).

13.6.3 Wnt

Mutations in the adenomatous polyposis coli (APC) tumor suppressor gene are the cause of the familial adenomatous polyposis (FAP) syndrome, and are an early causative event in the development of sporadic cancer. It has been shown that not only a mutated (dysfunctional) APC gene but also mutations in the axin-1 and -2, and β-catenin genes result in a genetic predisposition to cancer (Fodde, Smits, and Clevers 2001; Giles, van Es, and Clevers 2003; Reya and Clevers 2005). Wnt ligands bind initially to the LDL receptor-related proteins (LRP) 5 and 6. This ternary complex then interacts with the G-protein-coupled receptors Frizzled and axin. The following recruitment and activation of the phospho-protein Dishevelled (Dsh) results in the inhibition of the axin/GSK-3β/APC complex (Nathke 2004), thereby causing the pool of cytoplasmic β-catenin to stabilize and translocate to the nucleus where its interaction with TCF/LEF family transcription factors promotes the gene expression of a set of genes implicated in EMT, including c-Myc, cyclin D1, fibronectin, MMP7, Id2, CD44, axin-2, and TCF-1 (Huelsken and Behrens 2002). Free β-catenin is usually rapidly phosphorylated by GSK-3β and subsequently degraded by the ubiquitin-proteasome pathway. However, during tumorigenesis in the presence of a nonfunctional APC, this degradation does not take place, causing β-catenin to accumulate, resulting in its constitutive translocation to the nucleus and activation of its target genes (Brabletz, Schmalhofer, and Brabletz 2009; Guaita et al. 2002; Huang et al. 2008; Jamora and Fuchs 2002; Vincan and Barker 2008). Wnt signaling also inhibits E-cadherin-mediated cell adhesion through the induction of Snail1. Furthermore, Snail1 in turn can also induce the expression of TCF causing a repression of E-cadherin through the β-catenin/TCF complex (Guaita et al. 2002; Jamora and Fuchs 2002).

13.6.4 Endothelins

A role for endothelin 1 (ET-1) in the induction of EMT was demonstrated in ovarian carcinoma cells. Activation of endothelin A receptor (ET$_A$R) in HEY and OVCA 433 ovarian carcinoma cells activated an ILK-mediated signaling cascade causing the downregulation of E-cadherin and β-catenin and the upregulation of N-cadherin, and the inhibition of GSK-3β. The latter consequently resulted in the stabilization of both Snail1 and β-catenin, events that culminated in the regulation of transcriptional programs that control EMT and caused these cells to acquire a fibroblastoid and invasive phenotype (Rosano et al. 2005, 2006).

It has also been shown that ET-1 activation of ET_AR increases TGF-β1 production, which ultimately drives EMT during pulmonary fibrosis (Jain et al. 2007).

13.6.5 ErbB Receptors

The epidermal growth factor receptor (EGFR) family consists of four different tyrosine kinase receptors, designated ErbB1 (EGFR, HER1), ErbB2 (HER2), ErbB3 (HER3), and ErbB4 (HER4), all of which bind a specific set of ligands, including epidermal growth factor (EGF), amphiregulin (Areg), heparin-binding EGF-like growth factor (HB-EGF), transforming growth factor (TGF)-α, betacellulin (BTC), epigen, epiregulin (EPR), and neuregulin 1-3 (NRG1-3). Binding of these ligands induces either the homo- or heterodimerization of the ErbB receptors and leads to the activation of specific signaling pathways (Lemmon 2009; Sorkin and Goh 2009) that regulate a plethora of cellular processes, including proliferation, apoptosis, cell polarity, migration, and invasion.

The ErbB receptors have been shown to be involved in the initiation and progression of a number of human cancers (Feigin and Muthuswamy 2009). For example, overexpression of EGFR is linked to poor prognosis in breast cancer patients, whereas chronic exposure to EGF and TGF-α is correlated to EMT (Barr et al. 2008; Cattan et al. 2001; Shrader et al. 2007; Sok et al. 2006) either through activation of the Janus-activated kinase and activator of transcription 3 (Jak/Stat3) cascade and the upregulation of the transcription factor Twist (Lo et al. 2007). In addition, expression of the truncated variant of the EGF receptor, EGFRvIII has been correlated with the invasive potential of the tumor and was identified in patients with non-small lung cancer carcinoma (NSCLC), advanced head and neck squamous cell carcinoma (HNSCC), and glioblastoma (Shrader et al. 2007; Sok et al. 2006).

Overexpression of another member of the ErbB receptor family, ErbB2, can be observed in approximately 18% to 20% of breast cancers. Although more frequently observed in estrogen and progesterone receptor negative tumors, ErbB2 overexpression is associated with a higher rate of recurrence and mortality in patients with newly diagnosed breast cancer who do not receive adjuvant therapy. Signaling induced by the oncogenic ErbB2 in combination with upregulation of transcription factor Twist can induce a complete EMT, allowing tumor cells to disseminate and avoid apoptosis (Ansieau et al. 2008).

13.6.6 Fibroblast Growth Factor

Fibroblast growth factor (FGF) signaling is initiated by its association with heparan sulfate proteoglycan which mediates the subsequent interaction with the FGF receptors (FGFR). As a result, the receptor undergoes dimerization, its intrinsic tyrosine kinase is activated, and signaling cascades involving MAPK, phospholipase Cγ (PLCγ), and PI3K/Akt are activated (Dailey et al. 2005; Kouhara et al. 1997). FGF signaling is implicated in the control of cellular proliferation, differentiation, migration, and survival. Through its crosstalk with the Wnt cascade FGF is involved in, for example, human colorectal carcinogenesis and mouse mammary tumor virus (MMTV)-induced carcinogenesis, but also early embryogenesis and body-axis formation (Katoh and Katoh 2006). Moreover, FGFR1 has been shown to be over-expressed in ~40% of poorly differentiated prostate adenocarcinomas (Ozen et al. 2001), while the prolonged activation of FGFR1 can be linked to EMT in an inducible mouse model of prostate cancer (Acevedo et al. 2007; Freeman et al. 2003). Recently, Lehembre et al. (2008) published a paper showing that the FGFR can cause cells to remain sessile or become motile. On one hand, activation of FGFR by FGF results in protein kinase (PK)Cα-mediated activation of PLCγ and triggers a short burst of Ras/MAPK activity, which results in cell adhesion and proliferation. On the other hand, complex formation of FGFR with neural cell adhesion molecule (NCAM) and PLCγ causes activation of the Raf-kinase PKCβII and consequently a sustained MAPK activation. Upregulation of NCAM, as a result of E-cadherin loss during EMT, causes an NCAM subpopulation to move into lipid rafts where they interact with p59[Fyn] resulting in the phosphorylation of the focal adhesion kinase (FAK), the assembly of focal adhesions, and cell migration.

13.6.7 Platelet-Derived Growth Factor

There are five distinct dimeric platelet-derived growth factor (PDGF) isoforms: PDGF-AA, -AB, -BB, -CC, and -DD (Heldin, Eriksson, and Ostman 2002) which, by binding to the PDGF α- and β-receptors, cause receptor dimerization and tyrosine autophosphorylation. This then leads to the activation of signaling molecules such as c-Src, PLC-γ, PI3K, and the growth factor receptor-bound protein 2 (Grb2)/son of Sevenless (Sos) complex (Heldin, Ostman, and Ronnstrand 1998).

PDGF is a strong chemoattractant for mesenchymal cells (Monypenny et al. 2009), and has been shown to increase cell migration and prevent apoptosis (Schneller 2001). In addition, PDGF ligands and receptors play an important role in the development of the cranial and cardiac neural crest, blood vessel development, and hematopoesis (Andrae, Gallini, and Betsholtz 2008; Battegay et al. 1994). In a pathological setting PDGF signaling has been implicated in the development of certain gliomas, sarcomas, and leukemias, whereas its role in the progression of epithelial cancers appears to be at the level of angiogenesis and tumor pericyte recruitment (Ostman 2004). Its role as an EMT inducer is still somewhat uncertain. Expression of PDGF has been correlated to bad prognosis in breast carcinoma (Seymour, Dajee, and Bezwoda 1993), whereas the increased expression of PDGF specific genes has been correlated to EMT (Jechlinger et al. 2003). Also, PDGF was shown to maintain EMT and promote metastasis in mouse mammary carcinomas possibly through activation of Stat1 (Jechlinger et al. 2006). Recently, it was also reported that TGF-β causes the upregulation of the PDGF-A ligand and the PDGFα and β receptors (Gotzmann et al. 2006), and the nuclear accumulation of β-catenin in malignant fibroblastoid hepatocytes, thereby conferring stem cell characteristics to these cells (Fischer et al. 2007). Lastly, over-expression of PDGF-D was shown to induce changes in cellular morphology and motility in PC3 human prostate cancer cells, and includes the loss of E-cadherin and ZO-1 and the gain of vimentin, all of which are considered to be EMT-related events (Kong et al. 2008).

13.6.8 Vascular Endothelial Growth Factor

The most important member of the vascular endothelial growth factor (VEGF) family is VEGF-A, also known as just VEGF. This ligand binds the VEGF receptor 1 (VEGFR/Flt-1), a tyrosine kinase receptor, thereby activating the MAPK and PI3K/Akt cascades. VEGFR-1 signaling is a critical mediator of both developmental and physiologic angiogenesis (Fong et al. 1995) and was recently also described as a functional mediator of the increased invasion and migration of tumor cells (Fan et al. 2005; Wey et al. 2005). Similar to FGF, long-term VEGF-A and VEGF-B exposure of L3.6pl human pancreatic cancer cells resulted in an EMT phenotype that was characterized by, for example, the acquisition of a spindle-shaped morphology, the loss of ZO-1, and the relocalization of membrane-bound E-cadherin and β-catenin to the cytoplasm and nucleus (Yang et al. 2006). In addition, EMT characteristics could also be observed when transfecting the well-differentiated MCF-7 and T47D human breast tumor cells with a VEGF-A construct. These cells, which normally grow in islands, gave rise to single cells with membrane ruffles that were budding off the islands of clustered cells, and increased the levels of Snail1, Snail2/Slug, and Twist, whereas a reduction of E-cadherin could be detected (Wanami et al. 2008).

13.6.9 Insulin-Like Growth Factors

Insulin-like growth factors (IGF-1/somatomedin C and IGF-2) have a role in embryonic as well as cancer development (Baserga 2009a, 2009b). For example, *in vivo* overexpression of insulin-like growth factor receptor 1 (IGFR1) in the transgenic Rip1Tag2 mouse model for pancreatic cancer resulted in increased metastasis to various organs (Lopez and Hanahan 2002). This is similar to observations in human primary tumors in which the overexpression of IGFR1 can be correlated to an increased incidence of metastasis (Gydee et al. 2004; Jiang et al. 2004). IGF is secreted not only by cancer cells but also by stromal cells (Kawada et al. 2006), soft tissue and bone (Stearns et al. 2005), causing auto- and paracrine activation of the IGFR1. The interaction of IGF with IGFR1 activates not only the PI3K/Akt and Ras/MAPK cascades, which are responsible for the anti-apoptotic and mitogenic effects, but also FAK. FAK activation results in changes in cell adhesion and migration, possibly through the modulation of ZO-1, and RhoA GTPase, Rac1 and cell division cycle 42 (Cdc42; Mauro et al. 2003). Recently it was also shown that IGF1 stimulation of the MCF-10A breast cancer cell line transfected with IGFR1 causes EMT through the specific activation of Akt1 but not Akt2 (Graham et al. 2008; Irie et al. 2005). Other evidence supporting the role of IGF in EMT comes from studies involving the androgen refractory prostate cell line (ARCaP) in which exposure to IGF causes the upregulation of transcription factor ZEB1 and the concomitant loss of E-cadherin and gain of N-cadherin and fibronectin (Graham et al. 2008).

13.6.10 Hepatocyte Growth Factor

Hepatocyte growth factor (HGF), also known as scatter factor (SCF), binds to the dimeric c-MET tyrosine kinase receptor resulting in receptor activation and downstream signaling of intracellular transducers comprising Ras/MAPK, PI3K, PLC-γ, Src-related tyrosine kinases, and growth factor receptor-bound protein-2 (Grb-2)-associated binder 1 (Gab-1).

The induction of EMT-related properties, in particular the dissolution of adherens junctions and scattering leading to metastasis, is regulated through the Rac-independent activation of the PI3K/Akt cascade in a manner, provided that there is basal MAPK activity (Bardelli et al. 1999; Khwaja et al. 1998), which leads to the upregulation of Snail1 (Grotegut et al. 2006). Recently it was also reported that activation of RON, a close relative of the c-MET receptor, causes a decrease in cell surface E-cadherin and the nuclear translocation of β-catenin in pancreatic L3.6pl tumor cells exposed to macrophage stimulating protein (MSP). The use of blocking RON mAbs prevented these EMT features *in vitro* as well as a more than 50% reduction in the subcutaneous and orthotopic tumor growth of these cells in mice (Camp et al. 2007).

13.6.11 Integrins

The integrin glycoproteins are transmembrane heterodimeric combinations of transmembrane α- and β-subunits, which bind to the ECM via a globular extracellular domain. Their cytoplasmic portion associates with several intracellular signaling molecules and to the cytoskeleton in a manner similar to E-cadherin. Ligand-induced clustering of integrins leads to the recruitment of adaptor proteins such as Shc and intracellular kinases such as FAK, Src, and ILK. Once activated, these proteins then trigger the Rho family of GTPases, and the Ras/MAPK and PI3K/Akt cascades (Giancotti and Ruoslahti 1999; Valles, Beuvin, and Boyer 2004). In addition, activated Src can phosphorylate paxillin which, with adaptor protein Crk, forms a paxillin/Crk/DOCK180 complex that is responsible for activation of Rac1 and which controls cell migration and EMT (Valles, Beuvin, and Boyer 2004). β-integrins can also induce EMT via the activation of ILK either through the engagement of the integrins with the ECM, or as a downstream effector of growth factor receptors (Streuli and Akhtar 2009), such as TGF-β (Bhowmick et al. 2001; Galliher and Schiemann 2006; Kim et al. 2009). ILK phosphorylates targets such as Akt and downregulates GSK-3β which causes activation of the Wnt/β-catenin cascade leading to an increase of Snail1 and the induction of EMT (Novak et al. 1998; Somasiri et al. 2000).

13.7 NEW PLAYERS IN THE EMT FIELD

In the past few years, mostly through the use of global analysis approaches such as gene expression microarrays and proteomics, a number of novel EMT mediators have entered the EMT arena. Several of these new players will be discussed briefly.

13.7.1 Non-Coding RNAs

Micro (mi)RNAs consist of a large family of small 21 to 23 nucleotide (nt) RNAs (Wienholds et al. 2005) which are synthesized by polymerase II, and then processed into ~70-nt stem-loop pre-miRNAs by Drosha RNase III endonuclease (Lee et al. 2003). These pre-miRNAs are transported out of the nucleus by exportin 5 (Lin et al. 2007), and further processed in the cytosol to the final ~22-nt mature miRNAs by Dicer (Hutvagner et al. 2001). The (near) perfect complementarity with which these miRNAs target mRNAs leads to mRNA degradation, whereas in the case of imperfect complementarity, translation is repressed. MiRNAs have been implicated in embryonic development (Wienholds et al.

2005; Wienholds and Plasterk 2005), but have also been identified as tumor promotors or suppressors (Asangani et al. 2008; Huang et al. 2008; Ma and Weinberg 2008; Verghese et al. 2008), and recognized as initiators of EMT. For example, the miR-200 (miR-200a, miR-200b, miR-200c, miR-141, and miR-429) and miR-205 family of miRNAs (Bracken et al. 2009; Gregory et al. 2008a, 2008b; Park and Peter 2008; Peter 2009) have been correlated to E-cadherin suppression during EMT, whereas a natural antisense transcript which regulates ZEB2 has been described to be involved in EMT (Beltran et al. 2008). Several other miRs are involved in the regulation of EMT and metastasis. For example, overexpression of miR-29a, which suppresses the expression of the tristetraprolin (TTP) protein, in cooperation with oncogenic Ras signaling, causes EMT and metastasis, whereas the Twist-induced expression of miR-10b can be found in breast carcinoma patients with metastasis (Negrini and Calin 2008).

13.7.2 Interleukin-Like EMT Inducer

It was shown that EpH4 cells (Grunert, Jechlinger, and Beug 2003; Janda et al. 2002a) transfected with oncogenic ras (EpRas) are tumorigenic, can undergo EMT in response to TGF-β (Jechlinger et al. 2006; Oft et al. 1996), but fail to metastasize from primary tumors established in immunocompromised mice (Jechlinger et al. 2006). Generation of several EpH4 cell lines expressing various Ras effector mutants showed that EMT and metastasis required Ras-induced hyperactivation of MAPK and not PI3K (Janda et al. 2002b). Expression profiling of these cell lines, using polysome-bound mRNA, identified interleukin-like EMT inducer (ILEI/FAM3C), to be upregulated. Functional studies in which ILEI expression levels were downregulated in the EpRas cells using RNAi before and after EMT prevented and reverted the TGF-β-dependent EMT, whereas stable overexpression of ILEI in the EpH4 and EpRas cell lines caused EMT, tumor growth, and metastasis independent of TGF-β. Furthermore, human metastatic breast cancer, as well as the tumor-host borders of invasive colon carcinomas, displayed a strongly enhanced cytoplasmic ILEI expression pattern (Waerner et al. 2006).

Recently it was shown that ILEI cooperates with Ras to govern a hepatocellular EMT, which is TGF-β independent but involves PDGFR/β-catenin and PDGFR/Stat3 signaling. Also, evaluation of clinical human hepatocellular cancer samples suggests that a strong cytoplasmic ILEI staining could serve as a predictor of poor differentiation and prognosis in these patients (Lahsnig et al. 2009).

13.7.3 Clusterin

Clusterin, also known as apolipoprotein J, testosterone-repressed prostate message-2, or sulfated glycoprotein-2, is a secreted glycoprotein that is expressed in virtually all tissues, and can be found in all human fluids (Gleave and Miyake 2005; Pucci et al. 2004; Trougakos and Gonos 2002; Trougakos et al. 2004). Clusterin is involved in many different but important, albeit sometimes opposite, normal physiological and cellular processes (Shannan et al. 2006, 2007), but has also been reported to play a role in malignant situations such as tumor formation and metastasis (Kang et al. 2004; Lau et al. 2006; Miyake et al. 2000a, 2000b; Redondo et al. 2000, 2006; Shannan et al. 2006a, Shannan et al. 2006c; Xie et al. 2005a, 2005b).

Even though clusterin mRNA levels are upregulated by the EMT inducing growth factor TGF-β (Jin and Howe 1997, 1999), there is no direct evidence to date that shows that clusterin is directly involved in EMT induction, but three very recent reports, however, hint at this possibility. First, Chou et al. (2009) presented data showing that silencing of clusterin in mesenchymal human CL1.0 lung adenocarcinoma cells caused a reduction in migration and invasion *in vitro*, and *in vivo* using an experimental metastasis mouse model. It was proposed that clusterin regulates EMT by modulating the transcription factor Snail2/Slug, which has been shown in many studies to play a seminal role in EMT. Second, we showed that a TGF-β induced EMT can be blocked by anti-clusterin antibodies (both *in vitro* and *in vivo*), implying a tumor-promoting role for secreted clusterin (Lenferinh, et al., 2009). Third, Mathias et al. (2009) found that normal Madin-Darby canine kidney (MDCK) cells transformed with oncogenic Ras undergo EMT. Secretome-based proteomic profiling of these cells revealed that EMT was accompanied by a significant downregulation of clusterin. The discrepancy between this observation and the upregulation of clusterin in tumors and tumor cell lines reported by many others can possibly be explained by the fact that the MDCK cell line is not transformed. This also implies that clusterin's expression and associated function may be very different under pathological conditions such as tumor growth and metastasis.

13.7.4 Hypoxia

The growth of tumors requires an increase in the local vasculature to assure that sufficient nutrients and oxygen are available. In situations where this is not the case, tumor cells can adjust to a low nutrient and oxygen-poor milieu by switching on specific pathways that allow these tumor cells to survive and to even grow under these unfavorable circumstances (Harris 2002). One way to respond to low oxygen levels is through the hypoxia-inducible helix-loop-helix transcription factor 1 (HIF-1), a heterodimer consisting of the hypoxic response factor HIF-1α and the constitutively expressed aryl hydrocarbon receptor nuclear translocator (ARNT/HIF-1β). Under low oxygen conditions HIF-1 binds to hypoxia-response elements (HREs), and activates the expression of hypoxia-response genes, such as pro-angiogenic VEGF. Induction of HIF signaling under these conditions has also been shown to induce EMT (Higgins et al. 2007). For example, human HK-2 and HKC tubular epithelial cell lines grown under hypoxic conditions showed reduced expression of E-cadherin and ZO-1, and an increase in mesenchymal markers such as vimentin and α-smooth muscle actin (αSMA), as a result of the HIF-1α-mediated overexpression of the EMT inducing transcription factor Twist. *In vivo*, using the rat remnant kidney, both Twist and HIF-1α were found to be overexpressed in tubular epithelial cells showing EMT (Sun et al. 2009). There is also evidence that in the context of EMT, HIF cooperates with other EMT inducing pathways such as TGF-β (Sanchez-Elsner et al. 2001; Zhang et al. 2003) and Notch (Gustafsson et al. 2005; Sahlgren et al. 2008), and can also be mediated by the increased phosphorylation and activation of PI3K/Akt (Yan et al. 2009).

13.7.5 Matrix Metalloproteinases

The identification of the matrix metalloproteinase (MMP) family is based on the ability of the protein to specifically cleave certain components of the ECM. The expression of the MMPs is under normal conditions very low and tightly regulated, but this can change rapidly during tissue remodeling or cancer development. Certain MMPs, including MMP1, -2, -3, -7, -9, -11, and -14, have frequently been found to be overexpressed in cancer, possibly facilitating the ability of these tumor cells to invade and metastasize. More recently, specific MMPs, such as MMP3 (stromelysin; Lochter et al. 1998; Radisky et al. 2005), MMP7 (matrilysin; McGuire, Li, and Parks 2003), MMP9 (gelatinase B; Illman et al. 2006; Orlichenko and Radisky 2008), and MMP28 (epilysin; Illman et al. 2006) have also been shown to induce EMT. MMP3, for example, can induce the expression of an alternatively spliced Rac1 (Rac1b), which increases the cellular levels of ROS, causing an upregulation of Snail and induction of EMT (Jorda et al. 2005; Radisky et al. 2005), whereas overexpession of MMP9 in MDCK cells was held, at least in part, responsible for the increased invasive properties of these cells (Jorda et al. 2005).

13.8 EMT AND STEM CELLS

Several studies have revealed that tumors contain a small minority of so-called cancer stem cells (CSCs), stem-like cells that are capable of self-renewal and that are able to initiate and drive tumor growth (Reya et al. 2001). CSCs were originally discovered in hematopoietic cell populations (Bonnet and Dick 1997), but have since then also been identified in solid tumors (O'Brien, Kreso, and Dick 2009). Disseminated tumor cells may need the ability to self-renew in order to spawn micro-metastases, which has led to the hypothesis that EMT can generate mesenchymal-like cells that share properties in common with the less differentiated cancer stem cells. Indeed, Mani et al. (2008) showed that ectopic expression of the transcription factors Twist or Snail1 leads to the induction of EMT in immortalized human mammary epithelial cells (HMLEs). These cells acquired a fibroblast-like morphology with mesenchymal characteristics such as downregulated E-cadherin and upregulated N-cadherin, vimentin, and fibronectin. Additional flow cytometry experiments were conducted in which cells expressing the stem cell markers CD44 and CD24 were selected. The results of these experiments showed that the majority of the mesenchymal cells generated through EMT expressed high levels of CD44 and low levels of CD24, a characteristic that is associated with human normal and tumor breast stem cells (Al-Hajj and Clarke 2004; Al-Hajj et al. 2003, 2004; Sleeman et al. 2006). Similar results were obtained by the sequential introduction of the telomerase catalytic subunit (hTERT), SV40 large T and small t (SV40T/t), and the oncogenic allele of H-Ras, H-RasV12, into primary human mammary epithelial cells (Morel et al. 2008).

13.9 CONCLUSIONS AND FUTURE PERSPECTIVES

Epithelia are typically organized as single or multilayered sheets of cuboidal cells that are connected by cell-cell adhesive junctions. This type of organization ensures the mechanical integrity of adjacent tissues and keeps organs and tissues separated. It also prevents epithelial cells from autonomously changing their shape or from wandering off. Nonetheless, the process is active at specific stages and circumstances in the life of metazoa. EMT makes its first appearance in the epithelial blastoderm, where it allows, albeit in a strictly controlled and timely manner, the sessile epithelial cells to change shape, to become motile and initiate the morphogenesis process. At a later point during embryonic development, EMT is responsible for the formation of germ layers, the gastrulation process, and the development of tissues such as muscle and bone. Even in adult life, EMT has a role in that it provides plasticity to epithelial cells during regenerative processes such as wound healing. These are the good sides of EMT.

It is during cancer development that things turn bad. The majority of solid tumors (carcinomas) have an epithelial origin. There is accumulating evidence that due to their long-term residence in the epithelial tissues, the cancer stem cells over time accumulate oncogenic lesions that allow these cells to revisit and reactivate the EMT program. This then provides carcinoma cells with the plasticity that allows them to escape the primary tumor and set out on a journey in search of new grounds to populate. Even though doubt has been cast upon the idea, many examples in the recent literature suggest a role for EMT in clinical cancer. The reports that cancer stem cells contain EMT features, that cancer cells that undergo EMT acquire stem-like characteristics (Mani et al. 2008, Polyak and Weinberg 2009), the evidence that tumor cells may already be present in the circulation even before a primary tumor is discovered (Husemann et al. 2008), and the overwhelming complexity of the EMT process pose a significant barricade for the successful development of EMT specific inhibitors with low toxicity. And this is what makes it all very ugly. Nonetheless there is hope. If the "self-seeding" hypothesis, which assumes that a tumor cell that escapes the primary tumor to "re-seed" itself in the vicinity of its primary tumor (Norton and Massague 2006), proves to be true, the development of therapeutics that will prevent further dissemination will be very beneficial. In addition, the use of a bioinformatics approach to study both embryonic and tumor development will reveal the molecular networks that coordinate the EMT process, thus holding a promise for the development of new and specific cancer therapeutics.

13.10 ACKNOWLEDGMENTS

The author would like to thank Dr. Maria Jaramillo for the critical reading of the manuscript, and would like to apologize to all researchers whose work was not mentioned due to space restrictions.

REFERENCES

Acevedo V. D., Gangula R. D., Freeman K. W. et al. 2007. Inducible FGFR-1 activation leads to irreversible prostate adenocarcinoma and an epithelial-to-mesenchymal transition. *Cancer Cell* 12:559–571.

Al-Hajj M., Becker M. W., Wicha M., Weissman I., and Clarke M. F. 2004. Therapeutic implications of cancer stem cells. *Curr Opin Genet Dev* 14:43–47.

Al-Hajj M. and Clarke M. F. 2004. Self-renewal and solid tumor stem cells. *Oncogene* 23:7274–7282.

Al-Hajj M., Wicha M. S., Benito-Hernandez A., Morrison S. J., and Clarke M. F. 2003. Prospective identification of tumorigenic breast cancer cells. *Proc Natl Acad Sci USA* 100:3983–3988.

Andrae J., Gallini R., and Betsholtz C. 2008. Role of platelet-derived growth factors in physiology and medicine. *Genes Dev* 22:1276–1312.

Ansieau S., Bastid J., Doreau A. et al. 2008. Induction of EMT by twist proteins as a collateral effect of tumor-promoting inactivation of premature senescence. *Cancer Cell* 14:79–89.

Arias A. M. 2001. Epithelial mesenchymal interactions in cancer and development. *Cell* 105:425–431.

Arnold S. J. and Robertson E. J. 2009. Making a commitment: cell lineage allocation and axis patterning in the early mouse embryo. *Nature Rev Mol Cell Biol* 10:91–103.

Asangani I. A., Rasheed S. A., Nikolova D. A. et al. 2008. MicroRNA-21 (miR-21) post-transcriptionally downregulates tumor suppressor Pdcd4 and stimulates invasion, intravasation and metastasis in colorectal cancer. *Oncogene* 27:2128–2136.

Bailey J. M., Singh P. K., and Hollingsworth M. A. 2007. Cancer metastasis facilitated by developmental pathways: Sonic hedgehog, Notch, and bone morphogenic proteins. *J. Cell. Biochem.* 102:829–839.

Balint K., Xiao M., Pinnix C. C. et al. 2005. Activation of Notch1 signaling is required for beta-catenin-mediated human primary melanoma progression. *J. Clin. Invest.* 115:3166–3176.

Bardelli A., Basile M. L., Audero E. et al. 1999. Concomitant activation of pathways downstream of Grb2 and PI 3-kinase is required for MET-mediated metastasis. *Oncogene* 18:1139–1146.

Barr S., Thomson S., Buck E. et al. 2008. Bypassing cellular EGF receptor dependence through epithelial-to-mesenchymal-like transitions. *Clin Exp Metastasis* 25:685–693.

Baserga R. 2009a. Customizing the targeting of IGF-1 receptor. *Future Oncol* 5:43–50.

Baserga R. 2009b. The insulin receptor substrate-1: a biomarker for cancer? *Exp Cell Res* 315:727–732.

Battegay E. J., Rupp J., Iruela-Arispe L., Sage E. H., and Pech M. 1994. PDGF-BB modulates endothelial proliferation and angiogenesis *in vitro* via PDGF beta-receptors. *J. Cell. Biol.* 125:917–928.

Becker K. F., Rosivatz E., Blechschmidt K. et al. 2007. Analysis of the E-cadherin repressor Snail in primary human cancers. *Cells Tissues Organs* 185:204–212.

Bedi S., Vidyasagar A., and Djamali A. 2008. Epithelial-to-mesenchymal transition and chronic allograft tubulointerstitial fibrosis. *Transplant Rev (Orlando)* 22:1–5.

Beltran M., Puig I., Pena C. et al. 2008. A natural antisense transcript regulates Zeb2/Sip1 gene expression during Snail1-induced epithelial-mesenchymal transition. *Genes Dev* 22:756–769.

Bhowmick N. A., Zent R., Ghiassi M., McDonnell M., and Moses H. L. 2001. Integrin beta 1 signaling is necessary for transforming growth factor-beta activation of p38MAPK and epithelial plasticity. *J. Biol. Chem.* 276:46707–46713.

Birchmeier C., Birchmeier W., and Brand-Saberi B. 1996. Epithelial-mesenchymal transitions in cancer progression. *Acta Anat (Basel)* 156:217–226.

Bockhorn M., Jain R. K., and Munn L. L. 2007. Active versus passive mechanisms in metastasis: do cancer cells crawl into vessels, or are they pushed? *Lancet Oncol* 8:444–448.

Bolender D. L. and Markwald R. R. 1979. Epithelial-mesenchymal transformation in chick atrioventricular cushion morphogenesis. *Scan Electron Microsc* 3:313–321.

Bonnet D. and Dick J. E. 1997. Human acute myeloid leukemia is organized as a hierarchy that originates from a primitive hematopoietic cell. *Nature Med* 3:730–737.

Boyer B. and Thiery J. P. 1993. Epithelium-mesenchyme interconversion as example of epithelial plasticity. *Apmis* 101:257–268.

Brabletz T., Jung A., Hermann K. et al. 1998. Nuclear overexpression of the oncoprotein beta-catenin in colorectal cancer is localized predominantly at the invasion front. *Pathol Res Pract* 194:701–704.

Brabletz S., Schmalhofer O., and Brabletz T. 2009. Gastrointestinal stem cells in development and cancer. *J. Pathol.* 217:307–317.

Bracken C. P., Gregory P. A., Khew-Goodall Y., and Goodall G. J. 2009. The role of microRNAs in metastasis and epithelial-mesenchymal transition. *Cell Mol Life Sci* 66:1682–1699.

Burns W. C., Kantharidis P., and Thomas M. C. 2007. The role of tubular epithelial-mesenchymal transition in progressive kidney disease. *Cells Tissues Organs* 185:222–231.

Burns, W. C., Twiqq, S. M., Forbes, J. M., Pete, J., Tihellis, C., Thallas-Bonke, V., Thomas, M. C., Cooper, M. E., and Kantharidis, P. (2006). Connective tissue growth factor plays and important role in advanced glycation end product-induced tubular epthelial-to mesenchynal transition: implications for diabetic renal disease. *J. Am. Soc. Nephrol.* 2006 Sep: 17(9):2484–94.

Camp E. R., Yang A., Gray M. J. et al. 2007. Tyrosine kinase receptor RON in human pancreatic cancer: expression, function, and validation as a target. *Cancer* 109:1030–1039.

Carvajal G., Rodriguez-Vita J., Rodrigues-Diez R. et al. 2008. Angiotensin II activates the Smad pathway during epithelial mesenchymal transdifferentiation. *Kidney Int* 74:585–595.

Carver E. A., Jiang R., Lan Y., Oram K. F., and Gridley T. 2001. The mouse snail gene encodes a key regulator of the epithelial-mesenchymal transition. *Mol Cell Biol* 21:8184–8188.

Cattan N., Rochet N., Mazeau C. et al. 2001. Establishment of two new human bladder carcinoma cell lines, CAL 29 and CAL 185: comparative study of cell scattering and epithelial to mesenchyme transition induced by growth factors. *Br. J. Cancer.* 85:1412–1417.

Chaffer C. L., Brennan J. P., Slavin J. L. et al. 2006. Mesenchymal-to-epithelial transition facilitates bladder cancer metastasis: role of fibroblast growth factor receptor-2. *Cancer Res* 66:11271–11278.

Chaffer C. L., Dopheide B., McCulloch D. R. et al. 2005. Upregulated MT1-MMP/TIMP-2 axis in the TSU-Pr1-B1/B2 model of metastatic progression in transitional cell carcinoma of the bladder. *Clin Exp Metastasis* 22:115–125.

Chaffer C. L., Thompson E. W., and Williams E. D. 2007. Mesenchymal to epithelial transition in development and disease. *Cells Tissues Organs* 185:7–19.

Chambers A. F., Groom A. C., and MacDonald I. C. 2002. Dissemination and growth of cancer cells in metastatic sites. *Nature Rev Cancer* 2:563–572.

Chou T. Y., Chen W. C., Lee A. C. et al. 2009. Clusterin silencing in human lung adenocarcinoma cells induces a mesenchymal-to-epithelial transition through modulating the ERK/Slug pathway. *Cell Signal* 21:704–711.

Christ B. and Ordahl C. P. 1995. Early stages of chick somite development. *Anat Embryol (Berl)* 191:381–396.

Ciruna B. and Rossant J. 2001. FGF signaling regulates mesoderm cell fate specification and morphogenetic movement at the primitive streak. *Dev Cell* 1:37–49.

Cui W., Fowlis D. J., Bryson S. et al. 1996. TGFbeta1 inhibits the formation of benign skin tumors, but enhances progression to invasive spindle carcinomas in transgenic mice. *Cell* 86:531–542.

Cui Q., Ma Y., Jaramillo M. et al. 2007. A map of human cancer signaling. *Mol Syst Biol* 3:152.

Dailey L., Ambrosetti D., Mansukhani A. and Basilico C. 2005. Mechanisms underlying differential responses to FGF signaling. *Cytokine Growth Factor Rev* 16:233–247.

Davies J. A. 1996. Mesenchyme to epithelium transition during development of the mammalian kidney tubule. *Acta Anat (Basel)* 156:187–201.

Derynck R., Akhurst R. J., and Balmain A. 2001. TGF-beta signaling in tumor suppression and cancer progression. *Nature Genet* 29:117–129.

Eisenberg L. M. and Markwald R. R. 1995. Molecular regulation of atrioventricular valvuloseptal morphogenesis. *Circ Res* 77:1–6.

Ekblom P. 1989. Developmentally regulated conversion of mesenchyme to epithelium. *FASEB J* 3:2141–2150.

Fan F., Wey J. S., McCarty M. F. et al. 2005. Expression and function of vascular endothelial growth factor receptor-1 on human colorectal cancer cells. *Oncogene* 24:2647–2653.

Feigin M. E. and Muthuswamy S. K. 2009. ErbB receptors and cell polarity: new pathways and paradigms for understanding cell migration and invasion. *Exp Cell Res* 315:707–716.

Feng X. H. and Derynck R. 2005. Specificity and versatility in TGF-beta signaling through Smads. *Annu Rev Cell Dev Biol* 21:659–693.

Fischer A. N., Fuchs E., Mikula M. et al. 2007. PDGF essentially links TGF-beta signaling to nuclear beta-catenin accumulation in hepatocellular carcinoma progression. *Oncogene* 26:3395–3405.

Fodde R., Smits R., and Clevers H. 2001. APC, signal transduction and genetic instability in colorectal cancer. *Nature Rev Cancer* 1:55–67.

Fong G. H., Rossant J., Gertsenstein M., and Breitman M.L. 1995. Role of the Flt-1 receptor tyrosine kinase in regulating the assembly of vascular endothelium. *Nature* 376:66–70.

Franci C., Takkunen M., Dave N. et al. 2006. Expression of Snail protein in tumor-stroma interface. *Oncogene* 25:5134–5144.

Freeman K. W., Welm B. E., Gangula R.D. et al. 2003. Inducible prostate intraepithelial neoplasia with reversible hyperplasia in conditional FGFR1-expressing mice. *Cancer Res* 63:8256–8263.

Galliher A. J. and Schiemann W. P. 2006. Beta3 integrin and Src facilitate transforming growth factor-beta mediated induction of epithelial-mesenchymal transition in mammary epithelial cells. *Breast Cancer Res* 8:R42.

Giancotti F. G. and Ruoslahti E. 1999. Integrin signaling. *Science* 285:1028–1032.

Giles R. H., van Es J. H., and Clevers H. 2003. Caught up in a Wnt storm: Wnt signaling in cancer. *Biochim Biophys Acta* 1653:1–24.

Gleave M. and Miyake H. 2005. Use of antisense oligonucleotides targeting the cytoprotective gene, clusterin, to enhance androgen- and chemo-sensitivity in prostate cancer. *World. J. Urol.* 23:38–46.

Gotzmann J., Fischer A. N., Zojer M. et al. 2006. A crucial function of PDGF in TGF-beta-mediated cancer progression of hepatocytes. *Oncogene* 25:3170–3185.

Graham T. R., Zhau H. E., Odero-Marah V. A. et al. 2008. Insulin-like growth factor-I-dependent up-regulation of ZEB1 drives epithelial-to-mesenchymal transition in human prostate cancer cells. *Cancer Res* 68:2479–2488.

Greenburg G. and Hay E. D. 1982. Epithelia suspended in collagen gels can lose polarity and express characteristics of migrating mesenchymal cells. *J. Cell. Biol.* 95:333–339.

Greenburg G. and Hay E. D. 1986. Cytodifferentiation and tissue phenotype change during transformation of embryonic lens epithelium to mesenchyme-like cells *in vitro*. *Dev Biol* 115:363–379.

Greenburg G. and Hay E. D. 1988. Cytoskeleton and thyroglobulin expression change during transformation of thyroid epithelium to mesenchyme-like cells. *Development* 102:605–622.

Grego-Bessa J., Diez J., Timmerman L., and de la Pompa J.L. 2004. Notch and epithelial-mesenchyme transition in development and tumor progression: another turn of the screw. *Cell Cycle* 3:718–721.

Gregory P. A., Bert A. G., Paterson E. L. et al. 2008a. The miR-200 family and miR-205 regulate epithelial to mesenchymal transition by targeting ZEB1 and SIP1. *Nature Cell Biol* 10:593–601.

Gregory P. A., Bracken C. P., Bert A. G., and Goodall G. J. 2008b. MicroRNAs as regulators of epithelial-mesenchymal transition. *Cell Cycle* 7:3112–3118.

Gressner A. M., Weiskirchen R., Breitkopf K., and Dooley S. 2002. Roles of TGF-beta in hepatic fibrosis. *Front Biosci* 7:d793–d807.

Grotegut S., von Schweinitz D., Christofori G., and Lehembre F. 2006. Hepatocyte growth factor induces cell scattering through MAPK/Egr-1-mediated upregulation of Snail. *EMBO. J.* 25:3534–3545.

Grunert S., Jechlinger M., and Beug H. 2003. Diverse cellular and molecular mechanisms contribute to epithelial plasticity and metastasis. *Nature Rev Mol Cell Biol* 4:657–665.

Guaita S., Puig I., Franci C. et al. 2002. Snail induction of epithelial to mesenchymal transition in tumor cells is accompanied by MUC1 repression and ZEB1 expression. *J. Biol. Chem.* 277:39209–39216.

Gustafsson M. V., Zheng X., Pereira T. et al. 2005. Hypoxia requires notch signaling to maintain the undifferentiated cell state. *Dev Cell* 9:617–628.

Gydee H., O'Neill J. T., Patel A. et al. 2004. Differentiated thyroid carcinomas from children and adolescents express IGF-I and the IGF-I receptor (IGF-I-R): cancers with the most intense IGF-I-R expression may be more aggressive. *Pediatr Res* 55:709–715.

Harris A. L. 2002. Hypoxia: a key regulatory factor in tumour growth. *Nature Rev Cancer* 2:38–47.

Hay E. D. 2005. The mesenchymal cell, its role in the embryo, and the remarkable signaling mechanisms that create it. *Dev Dyn* 233:706–720.

Hay E. D. and Zuk A. 1995. Transformations between epithelium and mesenchyme: normal, pathological, and experimentally induced. *Am J. Kidney. Dis.* 26:678–690.

Heldin C. H., Eriksson U., and Ostman A. 2002. New members of the platelet-derived growth factor family of mitogens. *Arch Biochem Biophys* 398:284–290.

Heldin C. H., Ostman A., and Ronnstrand L. 1998. Signal transduction via platelet-derived growth factor receptors. *Biochim Biophys Acta* 1378:F79–F113.

Higgins D. F., Kimura K., Bernhardt W. M. et al. 2007. Hypoxia promotes fibrogenesis *in vivo* via HIF-1 stimulation of epithelial-to-mesenchymal transition. *J. Clin. Invest.* 117:3810–3820.

Hlubek F., Spaderna S., Schmalhofer O. et al. 2007. Wnt/FZD signaling and colorectal cancer morphogenesis. *Front Biosci* 12:458–470.

Huang Q., Gumireddy K., Schrier M. et al. 2008. The microRNAs miR-373 and miR-520c promote tumour invasion and metastasis. *Nature Cell Biol* 10:202–210.

Huelsken J. and Behrens J. 2002. The Wnt signalling pathway. *J. Cell. Sci.* 115:3977–3978.

Hugo H., Ackland M. L., Blick T. et al. 2007. Epithelial–mesenchymal and mesenchymal–epithelial transitions in carcinoma progression. *J. Cell. Physiol.* 213:374–383.

Husemann Y., Geigl J. B., Schubert F. et al. 2008. Systemic spread is an early step in breast cancer. *Cancer Cell* 13:58–68.

Hutvagner G., McLachlan J., Pasquinelli A. E. et al. 2001. A cellular function for the RNA-interference enzyme Dicer in the maturation of the let-7 small temporal RNA. *Science* 293:834–838.

Illman S. A., Lehti K., Keski-Oja J., and Lohi J. 2006. Epilysin (MMP-28) induces TGF-beta mediated epithelial to mesenchymal transition in lung carcinoma cells. *J. Cell. Sci.* 119:3856–3865.

Irie H. Y., Pearline R. V., Grueneberg D. et al. 2005. Distinct roles of Akt1 and Akt2 in regulating cell migration and epithelial-mesenchymal transition. *J. Cell. Biol.* 171:1023–1034.

Jain R., Shaul P. W., Borok Z., and Willis B. C. 2007. Endothelin-1 induces alveolar epithelial-mesenchymal transition through endothelin type A receptor-mediated production of TGF-beta1. *Am. J. Respir. Cell. Mol. Biol.* 37:38–47.

Jamora C. and Fuchs E. 2002. Intercellular adhesion, signalling and the cytoskeleton. *Nature Cell Biol* 4:E101–E108.

Janda E., Lehmann K., Killisch I. et al. 2002a. Ras and TGF[beta] cooperatively regulate epithelial cell plasticity and metastasis: dissection of Ras signaling pathways. *J. Cell. Biol.* 156:299–313.

Janda E., Litos G., Grunert S., Downward J., and Beug H. 2002b. Oncogenic Ras/Her-2 mediate hyperproliferation of polarized epithelial cells in 3D cultures and rapid tumor growth via the PI3K pathway. *Oncogene* 21:5148–5159.

Jeanes A., Gottardi C. J., and Yap A. S. 2008. Cadherins and cancer: how does cadherin dysfunction promote tumor progression? *Oncogene* 27:6920–6929.

Jechlinger M., Grunert S., Tamir I. H. et al. 2003. Expression profiling of epithelial plasticity in tumor progression. *Oncogene* 22:7155–7169.

Jechlinger M., Sommer A., Moriggl R. et al. 2006. Autocrine PDGFR signaling promotes mammary cancer metastasis. *J. Clin. Invest.* 116:1561–1570.

Jiang Y., Wang L., Gong W. et al. 2004. A high expression level of insulin-like growth factor I receptor is associated with increased expression of transcription factor Sp1 and regional lymph node metastasis of human gastric cancer. *Clin Exp Metastasis* 21:755–764.

Jin G. and Howe P. H. 1997. Regulation of clusterin gene expression by transforming growth factor beta. *J. Biol. Chem.* 272:26620–26626.

Jin G. and Howe P. H. 1999. Transforming growth factor beta regulates clusterin gene expression via modulation of transcription factor c-Fos. *Eur. J. Biochem.* 263:534–542.

Jorda M., Olmeda D., Vinyals A. et al. 2005. Upregulation of MMP-9 in MDCK epithelial cell line in response to expression of the Snail transcription factor. *J. Cell. Sci.* 118:3371–3385.

Kang P. and Svoboda K. K. 2005. Epithelial-mesenchymal transformation during craniofacial development. *J. Dent. Res.* 84:678–690.

Kang Y. K., Hong S. W., Lee H. and Kim W. H. 2004. Overexpression of clusterin in human hepatocellular carcinoma. *Hum Pathol* 35:1340–1346.

Katoh M. and Katoh M. 2006. Cross-talk of WNT and FGF signaling pathways at GSK3beta to regulate beta-catenin and SNAIL signaling cascades. *Cancer Biol Ther* 5:1059–1064.

Kawada M., Inoue H., Masuda T., and Ikeda D. 2006. Insulin-like growth factor I secreted from prostate stromal cells mediates tumor-stromal cell interactions of prostate cancer. *Cancer Res* 66:4419–4425.

Kemler R., Hierholzer A., Kanzler B. et al. 2004. Stabilization of beta-catenin in the mouse zygote leads to premature epithelial-mesenchymal transition in the epiblast. *Development* 131:5817–5824.

Khwaja A., Lehmann K., Marte B. M., and Downward J. 1998. Phosphoinositide 3-kinase induces scattering and tubulogenesis in epithelial cells through a novel pathway. *J. Biol. Chem.* 273:18793–18801.

Kim K.K., Wei Y., Szekeres C. et al. 2009. Epithelial cell alpha3beta1 integrin links beta-catenin and Smad signaling to promote myofibroblast formation and pulmonary fibrosis. *J. Clin. Invest.* 119:213–224.

Kimelman D. 2006. Mesoderm induction: from caps to chips. *Nature Rev Genet* 7:360–372.

Kishigami S. and Mishina Y. 2005. BMP signaling and early embryonic patterning. *Cytokine Growth Factor Rev* 16:265–278.

Klattig J. and Englert C. 2007. The Mullerian duct: recent insights into its development and regression. *Sex Dev* 1:271–278.

Kong D., Wang Z., Sarkar S. H. et al. 2008. Platelet-derived growth factor-D overexpression contributes to epithelial-mesenchymal transition of PC3 prostate cancer cells. *Stem Cells* 26:1425–1435.

Kopan R. and Ilagan M. X. 2009. The canonical Notch signaling pathway: unfolding the activation mechanism. *Cell* 137:216–233.

Kopfstein L. and Christofori G. 2006. Metastasis: cell-autonomous mechanisms versus contributions by the tumor microenvironment. *Cell Mol Life Sci* 63:449–468.

Kouhara H., Hadari Y. R., Spivak-Kroizman T. et al. 1997. A lipid-anchored Grb2-binding protein that links FGF-receptor activation to the Ras/MAPK signaling pathway. *Cell* 89:693–702.

Krebs L. T., Xue Y., Norton C. R. et al. 2000. Notch signaling is essential for vascular morphogenesis in mice. *Genes Dev* 14:1343–1352.

Krug E. L., Mjaatvedt C. H., and Markwald R. R. 1987. Extracellular matrix from embryonic myocardium elicits an early morphogenetic event in cardiac endothelial differentiation. *Dev Biol* 120:348–355.

Kurisu S., Suetsugu S., Yamazaki D., Yamaguchi H. and Takenawa T. 2005. Rac-WAVE2 signaling is involved in the invasive and metastatic phenotypes of murine melanoma cells. *Oncogene* 24:1309–1319.

Kuriyama S. and Mayor R. 2008. Molecular analysis of neural crest migration. *Philos Trans R Soc Lond B Biol Sci* 363:1349–1362.

Lahsnig C., Mikula M., Petz M. et al. 2009. ILEI requires oncogenic Ras for the epithelial to mesenchymal transition of hepatocytes and liver carcinoma progression. *Oncogene* 28:638–650.

Larue L. and Bellacosa A. 2005. Epithelial-mesenchymal transition in development and cancer: role of phosphatidylinositol 3′ kinase/AKT pathways. *Oncogene* 24:7443–7454.

Lau S. H., Sham J. S., Xie D. et al. 2006. Clusterin plays an important role in hepatocellular carcinoma metastasis. *Oncogene* 25:1242–1250.

Lee Y., Ahn C., Han J. et al. 2003. The nuclear RNase III Drosha initiates microRNA processing. *Nature* 425:415–419.

Lehembre F., Yilmaz M., Wicki A. et al. 2008. NCAM-induced focal adhesion assembly: a functional switch upon loss of E-cadherin. *EMBO. J.* 27:2603–2615.

Lemmon M. A. 2009. Ligand-induced ErbB receptor dimerization. *Exp Cell Res* 315:638–648.

Lenferinh, A.E., Cantin, C., Nantel, A., Wang, E., Durocher, Y., Banville, M., Paul-Roc, B., Marcil, A., Wilson, M.R., and O'Connor-McCourt, M.D., Transcriptome profiling of a TGF-β induced epithelial-to-mesenchymal transition reveals extracellular clusterin as a target for therapeutic antibodies. *Oncogene* 2009 Nov 23 (E-pub).

Li M. X. and Liu B. C. 2007. Epithelial to mesenchymal transition in the progression of tubulointerstitial fibrosis. *Chinese. Med. J. (English)* 120:1925–1930.

Lin F. T., Lai Y. J., Makarova N., Tigyi G., and Lin W. C. 2007. The lysophosphatidic acid 2 receptor mediates down-regulation of Siva-1 to promote cell survival. *J. Biol. Chem.* 282:37759–37769.

Lo H. W., Hsu S. C., Xia W. et al. 2007. Epidermal growth factor receptor cooperates with signal transducer and activator of transcription 3 to induce epithelial-mesenchymal transition in cancer cells via up-regulation of TWIST gene expression. *Cancer Res* 67:9066–9076.

Lochter A., Sternlicht M. D., Werb Z., and Bissell M. J. 1998. The significance of matrix metalloproteinases during early stages of tumor progression. *Ann NY Acad Sci* 857:180–193.

Lopez T. and Hanahan D. 2002. Elevated levels of IGF-1 receptor convey invasive and metastatic capability in a mouse model of pancreatic islet tumorigenesis. *Cancer Cell* 1:339–353.

Ma L. and Weinberg R. A. 2008. MicroRNAs in malignant progression. *Cell Cycle* 7:570–572.

Mani S. A., Guo W., Liao M. J. et al. 2008. The epithelial-mesenchymal transition generates cells with properties of stem cells. *Cell* 133:704–715.

Markwald R. R., Fitzharris T. P., and Manasek F. J. 1977. Structural development of endocardial cushions. *Am. J. Anat.* 148:85–119.

Mathias R. A., Wang B., Ji H. et al. 2009. Secretome-based proteomic profiling of Ras-transformed MDCK cells reveals extracellular modulators of epithelial-mesenchymal transition. *J. Proteome. Res.* 8:2827–2837.

Mauro L., Salerno M., Morelli C. et al. 2003. Role of the IGF-I receptor in the regulation of cell-cell adhesion: implications in cancer development and progression. *J. Cell. Physiol.* 194:108–116.

McGuire J. K., Li Q., and Parks W. C. 2003. Matrilysin (matrix metalloproteinase-7) mediates E-cadherin ectodomain shedding in injured lung epithelium. *Am. J. Pathol.* 162:1831–1843.

Mezzano S. A., Ruiz-Ortega M., and Egido J. 2001. Angiotensin II and renal fibrosis. *Hypertension* 38:635–638.

Miettinen P. J., Ebner R., Lopez A. R., and Derynck R. 1994. TGF-beta induced transdifferentiation of mammary epithelial cells to mesenchymal cells: involvement of type I receptors. *J. Cell. Biol.* 127:2021–2036.

Miyake H., Nelson C., Rennie P. S., and Gleave M. E. 2000a. Acquisition of chemoresistant phenotype by overexpression of the antiapoptotic gene testosterone-repressed prostate message-2 in prostate cancer xenograft models. *Cancer Res* 60:2547–2554.

Miyake H., Nelson C., Rennie P. S., and Gleave M. E. 2000b. Testosterone-repressed prostate message-2 is an antiapoptotic gene involved in progression to androgen independence in prostate cancer. *Cancer Res* 60:170–176.

Monypenny J., Zicha D., Higashida C. et al. 2009. Cdc42 and Rac family GTPases regulate mode and speed but not direction of primary fibroblast migration during platelet-derived growth factor-dependent chemotaxis. *Mol Cell Biol* 29:2730–2747.

Morel A. P., Lievre M., Thomas C. et al. 2008. Generation of breast cancer stem cells through epithelial-mesenchymal transition. *PLoS ONE* 3:e2888.

Naber H. P., ten Dijke P., and Pardali E. 2008. Role of TGF-beta in the tumor stroma. *Curr Cancer Drug Targets* 8:466–472.

Nakajima Y., Yamagishi T., Hokari S., and Nakamura H. 2000. Mechanisms involved in valvuloseptal endocardial cushion formation in early cardiogenesis: roles of transforming growth factor (TGF)-beta and bone morphogenetic protein (BMP). *Anat Rec* 258:119–127.

Nakaya Y. and Sheng G. 2008. Epithelial to mesenchymal transition during gastrulation: an embryological view. *Dev Growth Differ* 50:755–766.

Nathke I.S. 2004. The adenomatous polyposis coli protein: the Achilles heel of the gut epithelium. *Annu Rev Cell Dev Biol* 20:337–366.

Negrini M. and Calin G. A. 2008. Breast cancer metastasis: a microRNA story. *Breast Cancer Res* 10:203.

Niimi H., Pardali K., Vanlandewijck M., Heldin C. H., and Moustakas A. 2007. Notch signaling is necessary for epithelial growth arrest by TGF-beta. *J. Cell. Biol.* 176:695–707.

Nishioka N., Yamamoto S., Kiyonari H. et al. 2008. Tead4 is required for specification of trophectoderm in pre-implantation mouse embryos. *Mech Dev* 125:270–283.

Norton L., Massague J. 2006. Is cancer a disease of self-seeding? *Nature Med* 12:875–878.

Novak A., Hsu S.C., Leung-Hagesteijn C. et al. 1998. Cell adhesion and the integrin-linked kinase regulate the LEF-1 and beta-catenin signaling pathways. *Proc Natl Acad Sci USA* 95:4374–4379.

O'Brien C. A., Kreso A., and Dick J. E. 2009. Cancer stem cells in solid tumors: an overview. *Semin Radiat Oncol* 19:71–77.

Oft M., Peli J., Rudaz C. et al. 1996. TGF-beta1 and Ha-Ras collaborate in modulating the phenotypic plasticity and invasiveness of epithelial tumor cells. *Genes Dev* 10:2462–2477.

Okagawa H., Markwald R. R., and Sugi Y. 2007. Functional BMP receptor in endocardial cells is required in atrioventricular cushion mesenchymal cell formation in chick. *Dev Biol* 306:179–192.

Orlichenko L. S. and Radisky D.C. 2008. Matrix metalloproteinases stimulate epithelial-mesenchymal transition during tumor development. *Clin Exp Metastasis* 25:593–600.

Ostman A. 2004. PDGF receptors-mediators of autocrine tumor growth and regulators of tumor vasculature and stroma. *Cytokine Growth Factor Rev* 15:275–286.

Ozen M., Giri D., Ropiquet F., Mansukhani A., and Ittmann M. 2001. Role of fibroblast growth factor receptor signaling in prostate cancer cell survival. *J. Natl. Cancer. Inst.* 93:1783–1790.

Paget S. 1989. The distribution of secondary growths in cancer of the breast. *Cancer Metastasis Rev* 8:98–101.

Park S. M. and Peter M. E. 2008. microRNAs and death receptors. *Cytokine Growth Factor Rev* 19:303–311.

Peinado H., Olmeda D., and Cano A. 2007. Snail, Zeb and bHLH factors in tumour progression: an alliance against the epithelial phenotype? *Nature Rev Cancer* 7:415–428.

Person A. D., Klewer S. E., and Runyan R. B. 2005. Cell biology of cardiac cushion development. *Int Rev Cytol* 243:287–335.

Pesce J., Kaviratne M., Ramalingam T. R. et al. 2006. The IL-21 receptor augments Th2 effector function and alternative macrophage activation. *J. Clin. Invest.* 116:2044–2055.

Peter M. E. 2009. Let-7 and miR-200 microRNAs: guardians against pluripotency and cancer progression. *Cell Cycle* 8:843–852.

Piek E., Moustakas A., Kurisaki A., Heldin C. H., and ten Dijke P. 1999. TGF-(beta) type I receptor/ALK-5 and Smad proteins mediate epithelial to mesenchymal transdifferentiation in NMuMG breast epithelial cells. *J. Cell. Sci.* 112 (Pt 24):4557–4568.

Polyak K. and Weinberg R. A. 2009. Transitions between epithelial and mesenchymal states: acquisition of malignant and stem cell traits. *Nature Rev Cancer* 9:265–273.

Postovit L. M., Margaryan N. V., Seftor E. A., and Hendrix M. J. 2008. Role of nodal signaling and the microenvironment underlying melanoma plasticity. *Pigment Cell Melanoma Res* 21:348–357.

Pucci S., Bonanno E., Pichiorri F., Angeloni C., and Spagnoli L. G. 2004. Modulation of different clusterin isoforms in human colon tumorigenesis. *Oncogene* 23:2298–2304.

Radisky D. C., Levy D. D., Littlepage L. E. et al. 2005. Rac1b and reactive oxygen species mediate MMP-3-induced EMT and genomic instability. *Nature* 436:123–127.

Redondo M., Esteban F., Gonzalez-Moles M. A. et al. 2006. Expression of the antiapoptotic proteins clusterin and bcl-2 in laryngeal squamous cell carcinomas. *Tumour Biol* 27:195–200.

Redondo M., Villar E., Torres-Munoz J. et al. 2000. Overexpression of clusterin in human breast carcinoma. *Am. J. Pathol.* 157:393–399.

Reiman R. A., Mauldin G. E., and Neal Mauldin G. 2008. A comparison of toxicity of two dosing schemes for doxorubicin in the cat. *J. Feline. Med. Surg.* 10:324–331.

Reya T. and Clevers H. 2005. Wnt signalling in stem cells and cancer. *Nature* 434:843–850.

Reya T., Morrison S.J., Clarke M. F., and Weissman I. L. 2001. Stem cells, cancer, and cancer stem cells. *Nature* 414:105–111.

Ribatti D., Mangialardi G., and Vacca A. 2006. Stephen Paget and the "seed and soil" theory of metastatic dissemination. *Clin Exp Med* 6:145–149.

Rosano L., Spinella F., Di Castro V. et al. 2006. Endothelin-1 is required during epithelial to mesenchymal transition in ovarian cancer progression. *Exp Biol Med (Maywood)* 231:1128–1131.

Rosano L., Spinella F., Di Castro V. et al. 2005. Endothelin-1 promotes epithelial-to-mesenchymal transition in human ovarian cancer cells. *Cancer Res* 65:11649–11657.

Sabbah M., Emami S., Redeuilh G. et al. 2008. Molecular signature and therapeutic perspective of the epithelial-to-mesenchymal transitions in epithelial cancers. *Drug Resist Update* 11:123–151.

Sahlgren C., Gustafsson M. V., Jin S., Poellinger L., and Lendahl U. 2008. Notch signaling mediates hypoxia-induced tumor cell migration and invasion. *Proc Natl Acad Sci USA* 105:6392–6397.

Sanchez-Elsner T., Botella L. M., Velasco B. et al. 2001. Synergistic cooperation between hypoxia and transforming growth factor-beta pathways on human vascular endothelial growth factor gene expression. *J. Biol. Chem.* 276:38527–38535.

Savagner P. 2001. Leaving the neighborhood: molecular mechanisms involved during epithelial-mesenchymal transition. *Bioessays* 23:912–923.

Schneller M. 2001. Identification of a candidate integrin-fraction associated with the activated form of the PDGF-receptor. *Biochem Biophys Res Commun* 281:595–602.

Seymour L., Dajee D., and Bezwoda W. R. 1993. Tissue platelet derived-growth factor (PDGF) predicts for shortened survival and treatment failure in advanced breast cancer. *Breast Cancer Res Treat* 26:247–252.

Shannan B., Seifert M., Boothman D. A., Tilgen W., and Reichrath J. 2006a. Clusterin and DNA repair: a new function in cancer for a key player in apoptosis and cell cycle control. *J. Mol. Histol.* 37:183–188.

Shannan B., Seifert M., Boothman D. A., Tilgen W., and Reichrath J. 2007. Clusterin over-expression modulates proapoptotic and antiproliferative effects of 1,25(OH)2D3 in prostate cancer cells *in vitro. J Steroid Biochem Mol Biol* 103:721–725.

Shannan B., Seifert M., Leskov K. et al. 2006b. Clusterin (CLU) and melanoma growth: CLU is expressed in malignant melanoma and 1,25-dihydroxyvitamin D3 modulates expression of CLU in melanoma cell lines *in vitro. Anticancer Res* 26:2707–2716.

Shannan B., Seifert M., Leskov K. et al. 2006c. Challenge and promise: roles for clusterin in pathogenesis, progression and therapy of cancer. *Cell Death Differ* 13:12–19.

Shook D., and Keller R. 2003. Mechanisms, mechanics and function of epithelial-mesenchymal transitions in early development. *Mech Dev* 120:1351–1383.

Shrader M., Pino M. S., Brown G. et al. 2007. Molecular correlates of gefitinib responsiveness in human bladder cancer cells. *Mol Cancer Ther* 6:277–285.

Sleeman K. E., Kendrick H., Ashworth A., Isacke C. M., and Smalley M. J. 2006. CD24 staining of mouse mammary gland cells defines luminal epithelial, myoepithelial/basal and non-epithelial cells. *Breast Cancer Res* 8:R7.

Snarr B. S., Kern C. B., and Wessels A. 2008. Origin and fate of cardiac mesenchyme. *Dev Dyn* 237:2804–2819.

Sok J. C., Coppelli F. M., Thomas S. M. et al. 2006. Mutant epidermal growth factor receptor (EGFRvIII) contributes to head and neck cancer growth and resistance to EGFR targeting. *Clin Cancer Res* 12:5064–5073.

Somasiri A., Wu C., Ellchuk T., Turley S., and Roskelley C. D. 2000. Phosphatidylinositol 3-kinase is required for adherens junction-dependent mammary epithelial cell spheroid formation. *Differentiation* 66:116–125.

Sorkin A. and Goh L. K. 2009. Endocytosis and intracellular trafficking of ErbBs. *Exp Cell Res* 315:683–696.

Stearns M., Tran J., Francis M. K., Zhang H., and Sell C. 2005. Activated Ras enhances insulin-like growth factor I induction of vascular endothelial growth factor in prostate epithelial cells. *Cancer Res* 65:2085–2088.

Stevens M. V., Broka D. M., Parker P. et al. 2008. MEKK3 initiates transforming growth factor beta 2-dependent epithelial-to-mesenchymal transition during endocardial cushion morphogenesis. *Circ Res* 103:1430–1440.

Streuli C. H. and Akhtar N. 2009. Signal co-operation between integrins and other receptor systems. *Biochem. J.* 418:491–506.

Sun S., Ning X., Zhang Y. et al. 2009. Hypoxia-inducible factor-1alpha induces Twist expression in tubular epithelial cells subjected to hypoxia, leading to epithelial-to-mesenchymal transition. *Kidney Int* 75:1278–1287.

Tam P. P. and Beddington R. S. 1987. The formation of mesodermal tissues in the mouse embryo during gastrulation and early organogenesis. *Development* 99:109–126.

Tam P. P., Williams E. A., and Chan W. Y. 1993. Gastrulation in the mouse embryo: ultrastructural and molecular aspects of germ layer morphogenesis. *Microsc Res Tech* 26:301–328.

Tarin D., Thompson E. W., and Newgreen D. F. 2005. The fallacy of epithelial mesenchymal transition in neoplasia. *Cancer Res* 65:5996-6000; discussion 00-1.

Thiery J. P. 2002. Epithelial-mesenchymal transitions in tumour progression. *Nature Rev Cancer* 2:442–454.

Thiery J. P. and Chopin D. 1999. Epithelial cell plasticity in development and tumor progression. *Cancer Metastasis Rev* 18:31–42.

Thompson E. W., Newgreen D. F., and Tarin D. 2005. Carcinoma invasion and metastasis: a role for epithelial-mesenchymal transition? *Cancer Res* 65:5991-5; discussion 95.

Thomson J. A., Itskovitz-Eldor J., Shapiro S. S. et al. 1998. Embryonic stem cell lines derived from human blastocysts. *Science* 282:1145–1147.

Townsend T. A., Wrana J. L., Davis G. E., and Barnett J. V. 2008. Transforming growth factor-beta-stimulated endocardial cell transformation is dependent on Par6c regulation of RhoA. *J. Biol. Chem.* 283:13834–13841.

Trelstad R. L., Hay E. D., and Revel J. D. 1967. Cell contact during early morphogenesis in the chick embryo. *Dev Biol* 16:78–106.

Trelstad R. L., Revel J. P., and Hay E. D. 1966. Tight junctions between cells in the early chick embryo as visualized with electron microscopy. *J. Cell. Biol.* 31:C6–C10.

Trougakos I. P. and Gonos E. S. 2002. Clusterin/apolipoprotein J in human aging and cancer. *Int. J. Biochem. Cell. Biol.* 34:1430–1448.

Trougakos I. P., So A., Jansen B., Gleave M. E., and Gonos E. S. 2004. Silencing expression of the clusterin/apolipoprotein j gene in human cancer cells using small interfering RNA induces spontaneous apoptosis, reduced growth ability, and cell sensitization to genotoxic and oxidative stress. *Cancer Res* 64:1834–1842.

Ullmann U., In't Veld P., Gilles C. et al. 2007. Epithelial-mesenchymal transition process in human embryonic stem cells cultured in feeder-free conditions. *Mol Hum Reprod* 13:21–32.

Valcourt U., Kowanetz M., Niimi H., Heldin C. H., and Moustakas A. 2005. TGF-beta and the Smad signaling pathway support transcriptomic reprogramming during epithelial-mesenchymal cell transition. *Mol Biol Cell* 16:1987–2002.

Valles A. M., Beuvin M., and Boyer B. 2004. Activation of Rac1 by paxillin-Crk-DOCK180 signaling complex is antagonized by Rap1 in migrating NBT-II cells. *J. Biol. Chem.* 279:44490–44496.

van Wijk B., Moorman A. F., and van den Hoff M. J. 2007. Role of bone morphogenetic proteins in cardiac differentiation. *Cardiovasc Res* 74:244–255.

Verghese E. T., Hanby A. M., Speirs V., and Hughes T. A. 2008. Small is beautiful: microRNAs and breast cancer—where are we now? *J. Pathol.* 215:214–221.

Vestweber D., Gossler A., Boller K., and Kemler R. 1987. Expression and distribution of cell adhesion molecule uvomorulin in mouse preimplantation embryos. *Dev Biol* 124:451–456.

Viebahn C. 1995. Epithelio-mesenchymal transformation during formation of the mesoderm in the mammalian embryo. *Acta Anat (Basel)* 154:79–97.

Vincan E. and Barker N. 2008. The upstream components of the Wnt signalling pathway in the dynamic EMT and MET associated with colorectal cancer progression. *Clin Exp Metastasis* 25:657–663.

Vincan E., Brabletz T., Faux M. C., and Ramsay R. G. 2007. A human three-dimensional cell line model allows the study of dynamic and reversible epithelial-mesenchymal and mesenchymal-epithelial transition that underpins colorectal carcinogenesis. *Cells Tissues Organs* 185:20–28.

Vincan E., Whitehead R. H., and Faux M. C. 2008. Analysis of Wnt/FZD-mediated signalling in a cell line model of colorectal cancer morphogenesis. *Methods Mol Biol* 468:263–273.

Waerner T., Alacakaptan M., Tamir I. et al. 2006. ILEI: a cytokine essential for EMT, tumor formation, and late events in metastasis in epithelial cells. *Cancer Cell* 10:227–339.

Waldmann J., Feldmann G., Slater E. P. et al. 2008. Expression of the zinc-finger transcription factor Snail in adrenocortical carcinoma is associated with decreased survival. *Br. J. Cancer.* 99:1900–1907.

Wanami L. S., Chen H. Y., Peiro S., Garcia de Herreros A., and Bachelder R. E. 2008. Vascular endothelial growth factor-A stimulates Snail expression in breast tumor cells: implications for tumor progression. *Exp Cell Res* 314:2448–2453.

Wang Z., Li Y., Banerjee S., and Sarkar F. H. 2008. Exploitation of the Notch signaling pathway as a novel target for cancer therapy. *Anticancer Res* 28:3621–3630.

Wells A., Yates C., and Shepard C. R. 2008. E-cadherin as an indicator of mesenchymal to epithelial reverting transitions during the metastatic seeding of disseminated carcinomas. *Clin Exp Metastasis* 25:621–628.

Wessels A. and Perez-Pomares J. M. 2004. The epicardium and epicardially derived cells (EPDCs) as cardiac stem cells. *Anat Rec A Discov Mol Cell Evol Biol* 276:43–57.

Wey J. S., Fan F., Gray M. J. et al. 2005. Vascular endothelial growth factor receptor-1 promotes migration and invasion in pancreatic carcinoma cell lines. *Cancer* 104:427–438.

Wienholds E., and Kloosterman W. P., Miska E. et al. 2005. MicroRNA expression in zebrafish embryonic development. *Science* 309:310–311.

Wienholds E. and Plasterk R. H. 2005. MicroRNA function in animal development. *FEBS Lett* 579:5911–5922.

Willis B. C. and Borok Z. 2007. TGF-beta-induced EMT: mechanisms and implications for fibrotic lung disease. *Am J Physiol Lung Cell Mol Physiol* 293:L525–L534.

Wrighton K. H., Lin X., and Feng X. H. 2009. Phospho-control of TGF-beta superfamily signaling. *Cell Res* 19:8–20.

Wynn T. A. 2008. Cellular and molecular mechanisms of fibrosis. *J. Pathol.* 214:199–210.

Wynn T. A., Jankovic D., Hieny S., Cheever A. W., and Sher A. 1995. IL-12 enhances vaccine-induced immunity to *Schistosoma mansoni* in mice and decreases T helper 2 cytokine expression, IgE production, and tissue eosinophilia. *J. Immunol.* 154:4701–4709.

Xie D., Lau S. H., Sham J. S. et al. 2005a. Up-regulated expression of cytoplasmic clusterin in human ovarian carcinoma. *Cancer* 103:277–283.

Xie D., Sham J. S., Zeng W. F. et al. 2005b. Oncogenic role of clusterin overexpression in multistage colorectal tumorigenesis and progression. *World. J. Gastroenterol.* 11:3285–3289.

Xu J., Lamouille S., and Derynck R. 2009. TGF-beta-induced epithelial to mesenchymal transition. *Cell Res* 19:156–172.

Yamagishi T., Ando K., and Nakamura H. 2009. Roles of TGFbeta and BMP during valvulo-septal endocardial cushion formation. *Anat Sci Int.* 84(3):77–87.

Yamazaki D., Kurisu S., and Takenawa T. 2009. Involvement of Rac and Rho signaling in cancer cell motility in 3D substrates. *Oncogene* 28:1570–1583.

Yan W., Fu Y., Tian D. et al. 2009. PI3 kinase/Akt signaling mediates epithelial-mesenchymal transition in hypoxic hepatocellular carcinoma cells. *Biochem Biophys Res Commun* 382:631–636.

Yang A. D., Camp E. R., Fan F. et al. 2006. Vascular endothelial growth factor receptor-1 activation mediates epithelial to mesenchymal transition in human pancreatic carcinoma cells. *Cancer Res* 66:46–51.

Yates C. C., Shepard C. R., Stolz D. B., and Wells A. 2007. Co-culturing human prostate carcinoma cells with hepatocytes leads to increased expression of E-cadherin. *Br. J. Cancer.* 96:1246–1252.

Zavadil J., Cermak L., Soto-Nieves N., and Bottinger E. P. 2004. Integration of TGF-beta/Smad and Jagged1/Notch signalling in epithelial-to-mesenchymal transition. *EMBO. J.* 23:1155–1165.

Zhang H., Akman H. O., Smith E. L. et al. 2003. Cellular response to hypoxia involves signaling via Smad proteins. *Blood* 101:2253–2260.

Zipori D. 2004. Mesenchymal stem cells: harnessing cell plasticity to tissue and organ repair. *Blood Cells Mol Dis* 33:211–215.

Zipori D. 2006. The stem state: mesenchymal plasticity as a paradigm. *Curr Stem Cell Res Ther* 1:95–102.

Ziyadeh F.N. 2008. Different roles for TGF-beta and VEGF in the pathogenesis of the cardinal features of diabetic nephropathy. *Diabetes Res Clin Pract* 82 (Suppl 1):S38–S41.

CHAPTER 14

Tumors and Their Microenvironments

Nicholas R. Bertos and Morag Park

CONTENTS

14.1 INTRODUCTION

14.1.1 The Tumor Microenvironment

It is becoming increasingly evident that the initiation and progression of tumors are dependent not only upon factors intrinsic to the tumor itself, but are also significantly influenced by the nature of the environment surrounding the lesion, here termed the tumor microenvironment or stroma and defined generally as those elements originally located on the distal side of the basement membrane in normal tissue. Multiple elements therein, both cellular and structural, act on and are acted on by the tumor in a dynamic manner. These interactions can play important roles in cancer initiation, progression, and invasion, and represent important areas for therapeutic intervention in conjunction with strategies targeting the tumor mass per se. In this chapter, we will focus on the role of the tumor microenvironment as it pertains to breast cancer as a model system.

14.1.2 Breast Cancer

Among solid tumors, breast cancer has been one of the most intensively studied malignancies to date. This is due to several features: it has a high incidence, and there is a significant but not inescapable mortality associated with this disease, which both provides additional incentive for research and permits the identification of prognostic factors. Surgical resection is a common therapeutic modality, permitting retrieval of samples for study. Multiple subtypes have been recognized at both the histological and molecular levels, rendering the identification of markers to direct targeted treatments, as well as the development of novel therapeutics, a pressing clinical issue.

The overwhelming majority of breast cancers arise from the epithelial cells lining the lobules and ducts of the breast. Three main subtypes of such tumors have initially been identified by immunohistochemistry, and the general validity of this classification scheme has been confirmed at the gene expression level by large-scale microarray-based gene profiling studies (Perou et al. 2000; Sorlie et al. 2001; van 't Veer et al. 2002). Briefly, the estrogen receptor (ER)-positive subtype as defined by immunohistochemical techniques essentially corresponds to the luminal subtype at the gene expression level, while the human epidermal growth factor receptor 2 (HER2)-positive subtype can also be identified by both approaches. The triple-negative subtype, which does not overexpress ER, HER2, or the progesterone receptor (PR) as assessed by immunohistochemistry, broadly corresponds to the basal subtype as defined by gene expression, although some tumors assigned to the basal class exhibit elevated expression of ER, PR, or HER2.

Each subtype is associated with differences at the biological level, as well as in prognosis and treatment targeting. ER-positive disease, considered to be associated with the most favorable overall prognosis, is driven by estrogen-dependent signaling, which can be targeted using estrogen (ant)agonists or through manipulation of estrogen biosynthesis using aromatase inhibitors (Eneman, Wood, and Muss 2004; Jordan and Brodie 2007; Patel, Sharma, and Jordan 2007). HER2-positive disease is treated using the monoclonal antibody trastuzumab, which binds to the orphan receptor HER2 (also known as ErbB2 or neu; Lewis et al. 1993). The mechanism of action of trastuzumab is not completely clear at this time (Hudis 2007; Valabrega, Montemurro, and Aglietta 2007); its binding to HER2 is thought to abrogate downstream signaling pathways (Delord et al. 2005) while also potentially targeting HER2-expressing cells for destruction via the immune system (Clynes et al. 2000; Cooley et al. 1999; Gennari et al. 2004), while its role in HER2 downregulation remains controversial (Austin et al. 2004; Sarup et al. 1991; Valabrega et al. 2005). Currently, there is no specific targeted therapy for the basal or triple-negative subtype (Rakha and Ellis 2009), which is associated with the worst prognosis among the three variants (Nishimura and Arima 2008).

While the majority of previous studies have addressed breast cancer either at the level of the tumor cells or at that of bulk tumors, where epithelial cells represent the predominant element, investigations of the cellular context within which breast cancers originate and progress are now beginning to uncover the role played by the microenvironment. Such studies have established not only that the tumor and its microenvironment exist and progress in a linked manner, but that features of the microenvironment are critical in determining disease course and ultimate outcome.

14.2 COMPONENTS OF THE MICROENVIRONMENT

14.2.1 Cellular Elements

In breast cancer, the microenvironment consists of both the preexisting stromal elements, the phenotype of which may be modulated by the epithelial tumor cells, and those recruited by the presence of the tumor. Cell types present in the tumor-adjacent space include components of the vasculature (endothelial cells and pericytes); immune cells of the hematopoietic lineage, including macrophages, T cells, and mast cells; fibroblasts; and, especially prevalent in the case of mammary tumors, adipocytes (see Figure 14.1).

Additionally, it is suggested that breast tumors actively recruit bone marrow-derived mesenchymal stem cells, which can differentiate into a variety of cell types, including cancer-associated fibroblasts (Mishra et al. 2008), leading to the promotion of a metastatic phenotype (Karnoub et al. 2007).

14.2.2 Structural Elements

The extracellular matrix (ECM) is primarily deposited by fibroblasts. It comprises both macromolecules, such as collagen, and polysaccharides, for example, hyaluronan, which are often organized into distinct structures. The most recognizable structure is the basement membrane, which under normal conditions separates epithelial cells from the underlying stromal compartment. The ECM both acts as a physical supporting matrix for the

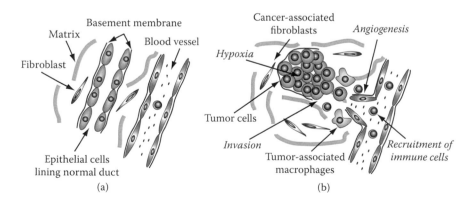

FIGURE 14.1 Schematic depiction of key elements of the microenvironment. (a) Components of the microenvironment in the normal breast; (b) cell types and processes that comprise the tumor microenvironment in breast cancer.

cells embedded within it, and also dynamically communicates with the cells with which it is in contact through multiple mechanisms, for example, integrin-mediated signaling. Elements of these signaling interactions have been shown to function in both the initiation and the maintenance of tumors (Ghajar and Bissell 2008).

14.3 IMMUNE CELLS

14.3.1 Macrophages

Macrophages generally comprise the predominant element among the tumor-associated immune cell population (Balkwill, Charles, and Mantovani 2005; Balkwill and Mantovani 2001; Coussens and Werb 2002). Under normal conditions, macrophages are generally considered to act as positive immune mediators, acting both directly in cell and pathogen killing and indirectly as antigen-presenting cells (APCs) for exogenous antigens. However, it has been demonstrated in multiple studies that the presence of an elevated number of macrophages in close proximity to a tumor is commonly associated with poor disease outcome, both in solid cancers in general and breast cancer in particular (Bingle, Brown, and Lewis 2002; Leek and Harris 2002). This apparent paradox may be explained by the multiple context-dependent functional roles that can be played by macrophages; in the case of breast cancer, the balance generally inclines toward tumor promotion, rather than antitumor activity.

Genetic studies conducted using mouse models in which macrophages were absent (generated by ablation of the macrophage growth factor CSF-1) demonstrated that there were at least two stages at which macrophages exert a tumor-promoting activity (Lin et al. 2001, 2002). Essentially, these entail supporting the transition from a benign to a malignant phenotype, as well as promoting the metastatic capacity of late-stage disease, and can be explained via a model in which macrophages act to increase tumor cell invasion, initially through the basement membrane and ultimately into the circulatory system, leading to distant metastasis (Pollard 2008).

Macrophages found in association with tumors are derived from monocytes recruited from the circulation via the production of chemoattractants, which are secreted either by the tumor or by adjacent stromal cells. These include the chemokines CCL2 (Bottazzi et al. 1983; Matsushima et al. 1989), as well as CCL5 and CXCL1, among others (Arenberg et al. 2000; Balkwill 2004; Mantovani et al. 2004a,b). Beyond chemokines, other factors released at tumor sites, including vascular endothelial growth factor (VEGF), platelet-derived growth factor (PDGF), transforming growth factor-β (TGF-ß), and macrophage colony-stimulating factor (M-CSF), as well as bioactive peptides released through the actions of secreted proteases on the extracellular matrix, also act as monocyte/macrophage chemoattractants (Coussens and Werb 2002).

Underlying the protumor role of these cells is the polarization of tumor-associated macrophages (TAMs) toward an M2 rather than an M1 phenotype (Mantovani et al. 2002; Sica et al. 2008). Classically activated macrophages, corresponding to the M1 type, promote Th1-type T-cell activation and killing of intracellular pathogens, while M2-activated (also known as alternately activated) macrophages are involved in Th1-type T cell suppression,

Th2-type T cell activation, immunosuppression, wound healing, and tissue remodeling. Investigation of TAMs has demonstrated that, in general, they correspond more closely to the M2 phenotype (Mantovani et al. 2002); among the chief effectors of this polarization are IL-10 and hypoxia, both present in the tumor microenvironment (Mantovani et al. 2004a,b; Sica et al. 2008). This selective pressure toward a tumor-promoting phenotype has been termed "immunoediting" (Lewis and Pollard 2006). Interestingly, M2-activated macrophages secrete IL-10 (Anderson and Mosser 2002), thus creating a positive feedback loop which promotes the polarization of additional macrophages along this axis.

At the tumor site, TAMs play multiple roles that contribute to tumor survival and progression. These include stimulation of angiogenesis, matrix remodeling, suppression of the immune response, secretion of growth factors that support tumor cell proliferation, and promotion of metastasis via cellular cooperation (Condeelis and Pollard 2006). This latter activity has recently come under intense investigation, leading to the identification of a paracrine loop model of interactions between tumor cells and macrophages that mediate tumor cell migration toward and intravasation into the vasculature. Using mouse models and a microneedle approach, it was demonstrated that tumor cells, which bear epidermal growth factor (EGF) receptors, migrate along a gradient of increasing EGF, while CSF-1 receptor-expressing macrophages migrate toward higher CSF-1 concentrations (Wyckoff et al. 2004). Tumor cells secrete CSF-1, which stimulates macrophage secretion of EGF, thus directing macrophages to migrate from the vasculature toward the tumor, while tumor cells follow the reverse route (Goswami et al. 2005). Intravital imaging studies reveal that motility and intravasation of tumor cells occur predominantly at sites where perivascular TAMs are present (Wyckoff et al. 2007). Genetic ablation of CSF-1 or pharmacological blockage of EGF receptor activity directly reduced the number of tumor cells in the circulation downstream of the tumor (Wyckoff et al. 2007), demonstrating a requirement for both elements of this loop in the migration and intravasation process.

14.3.2 T cells

As well as macrophages, the population of T cells neighboring and/or recruited to tumors undergoes modifications, the nature of which has implications for disease progression. The presence of Th1-type T cells, involved in activation of a cytotoxic T-cell response and upregulation of antigen-presenting cells (Knutson and Disis 2005), is generally associated with good outcome in breast, lung, and colon cancers (Finak et al. 2008; Hiraoka et al. 2006; Pages et al. 2005). On the other hand, decreased prosurvival signaling for T lymphocytes is observed in stroma from poor-outcome breast cancer patients (Finak et al. 2008). Additionally, it appears that successfully proliferating tumors recruit a mechanism normally used to mediate immune self-tolerance to evade immune surveillance. Regulatory T cells, known as Tregs, block the activity of multiple immune effectors (Lan et al. 2005), and it has been shown that levels of tumor-infiltrating Tregs are correlated with factors prognostic for poor outcome (Bohling and Allison 2008) and with shorter relapse-free and overall survival times (Bates et al. 2006). It is also interesting to note that the extent of pathological response to neoadjuvant (presurgery) chemotherapy also corresponds to the specific effect of such therapy on Treg cell numbers (Ladoire et al. 2008), suggesting that an antitumor immune response can be induced by pharmacological intervention only in the absence of negative regulatory signals originating from the Treg cell population.

Interestingly, it has been demonstrated that the stroma of high-grade breast tumors exhibit a strong immune response signature, including markers of interferon signaling and immune cell activation (Ma et al. 2009); on the other hand, another immune response signature has been identified as characteristic of a subset of ER-negative breast tumors with significantly improved outcome (Teschendorff et al. 2006, 2007). The opposite roles played by different immune system-related processes and components as described here may help to explain these apparently contradictory results.

14.4 FIBROBLASTS

Under nontumor conditions, the principal function of fibroblasts is in extracellular matrix (ECM) deposition and remodeling, primarily in the contexts of development and wound healing. In the tumor microenvironment, cancer-associated fibroblasts (CAFs), which display a myofibroblast phenotype, promote tumor growth and invasion via multiple mechanisms, several of which are similar to those previously described for TAMs (Radisky and Radisky 2007). As tumors progress toward an invasive phenotype, fibroblasts increase in both abundance and activity (Sappino et al. 1988), and the resulting deposition of fibrotic ECM both leads to increased disruption of the organization of the original epithelial structures and promotes proliferation (Kalluri and Zeisberg 2006).

Additionally, CAFs secrete factors that promote tumor cell survival and metastasis, including stromal-derived factor-1 (SDF-1, alternately termed CXCL12), which binds to and activates the cytokine receptor CXCR4 expressed on tumor cells (Orimo et al. 2005), promoting metastasis of breast cancer to bone and lung (Muller et al. 2001). Furthermore, secretion of SDF-1 acts to recruit endothelial progenitor cells to the tumor mass, leading to accelerated angiogenesis (Orimo and Weinberg 2006).

Secretion of matrix metalloproteases (MMPs) by CAFs also plays significant roles in promoting tumor progression. Urokinase plasminogen activator (uPA, also known as PLAU) is activated after binding to its membrane-bound receptor (uPAR, or PLAUR), which is thought to be expressed on both stromal and tumor cells (Giannopoulou et al. 2007), although conflicting reports exist regarding the tumoral vs. stromal localization of its expression (Hurd et al. 2007; Meng et al. 2006; Nielsen et al. 2007). Subsequently, it can exert its effects on the ECM via activation of the broad-spectrum protease plasmin, which acts to degrade ECM components, activate MMPs, and release latent bioactive peptides such as TGF-β.

The production of uPA in stromal cells can be stimulated by the tumor cells via the cell surface glycoprotein EMMPRIN (extracellular MMP inducer) (Quemener et al. 2007), while uPAR-dependent signaling in tumor cells may prime these to exhibit a proliferative response to EGF. Interestingly, signaling via the hepatocyte growth factor (HGF) receptor, Met, which is overexpressed in a subset of breast cancers with poor outcomes (Garcia et al. 2007a, 2007b), and which has been linked with increased tumorigenicity and invasiveness (Jeffers et al. 1996b, Rong et al. 1993, 1994), induces elevated expression of both uPA and uPAR, thus enhancing the generation of plasmin (Jeffers et al. 1996; Pepper et al. 1992).

It has been demonstrated that the vast majority of gene expression changes occur in the transition from a normal to a preinvasive (ductal carcinoma *in situ*, or DCIS) breast tumor stroma; these consist primarily of genes associated with extracellular matrix components or metallopeptidase activity, while progression from preinvasive to invasive tumor growth is accompanied by increased expression of a number of MMPs in tumor-associated stroma (Ma et al. 2009). These observations serve to further illustrate the complex and intimate nature of the interactions between the tumor and its microenvironment involved in disease progression.

Stromal fibroblasts have been identified as sites for estrogen production as assessed via inducible expression of aromatase, a key enzyme in the estrogen biosynthetic pathway (Santen et al. 1997, 1998; Santner et al. 1997). This suggests that estrogen-dependent tumor cells may respond to this local production in a paracrine manner.

There have been divergent reports in the literature regarding the question of whether CAFs and stromal cells in general exhibit genetic alterations with respect to normal fibroblasts (Allinen et al. 2004; Fukino et al. 2004, 2007; Hill et al. 2005; Hu et al. 2005; Kurose et al. 2002; Lafkas et al. 2008; Moinfar et al. 2000; Orimo and Weinberg 2006). Currently, it appears that the features of CAFs are more likely to result from alterations at the epigenetic, rather than the genetic, level. However, it is clear that the protein and gene expression profiles of CAFs are distinct from those of fibroblasts in nontumor settings (Hawsawi et al. 2008; Sadlonova et al. 2009; Singer et al. 2008). CAFs exhibit increased expression of genes encoding tumor-promoting cytokines and matrix-associated proteins (Singer et al. 2008), as well as increased expression of the proliferation marker Ki-67 (Hawsawi et al. 2008).

A recent gene expression profiling study from our group suggests that gene expression profiles collected from morphologically normal regions of epithelium and stroma adjacent to breast tumors are indistinguishable from those isolated from normal mammary tissue (Finak et al. 2006), suggesting that stromal changes are likely to be localized to those regions most intimately associated with the tumor site.

It is also interesting to note that a comparison of gene expression profiles between fibroblasts isolated from interlobular vs. intralobular normal mammary stroma yielded no significant differences, while validation at the protein level by immunohistochemical methods revealed that several of the proteins investigated were differentially expressed between the two groups (Fleming et al. 2008). This accentuates the importance of deriving information at multiple levels, and suggests that gene expression profiling, while useful in identifying overall pathways and functional assemblies of genes involved in specific processes, may not always reflect the level and functional activity of individual genes of interest.

The activities of CAFs are similar to those seen in the fibroblast response to inflammation. A gene expression signature reflecting the response of normal fibroblasts to serum (Chang et al. 2004), termed the wound-response signature, has independent prognostic value when applied to whole-tumor datasets (Chang et al. 2005), supporting the hypothesis that increased CAF activation and the inflammatory phenotype are associated with promotion of tumor progression.

14.5 ADIPOCYTES

Adipocytes represent one of the most abundant cell types in normal mammary tissue. Weight gain and postmenopausal obesity constitute a major breast cancer risk factor in postmenopausal women (Calle et al. 2003), and obesity has been linked to poor disease outcome across all age groups (Stephenson and Rose 2003). It has previously been demonstrated that the normal formation of mammary ducts requires the presence of adipose tissue (Howlett and Bissell 1993; Huss and Kratz 2001; Zangani et al. 1999). The proliferation of estrogen-dependent tumor cell lines is dependent on the local presence of adipocytes both in three-dimensional culture (Manabe et al. 2003) and in mouse models (Elliott et al. 1992), suggesting that adipocyte-secreted factors can affect the growth and development of specific subtypes of breast tumor cells.

Adipose tissue represents an important local source of estrogen, although it is currently unclear as to whether stromal or epithelial tumor cells are the primary producers of estrogen in the breast tumor environment (Miki, Suzuki, and Sasano 2007; Suzuki et al. 2008). Beyond steroid hormones, other factors secreted by adipocytes have also been shown to influence tumor cell behavior. These include collagen VI, a component of the ECM that has been shown to stabilize β-catenin and cyclin D1, as well as activate GSK3β- and Akt-dependent signaling pathways (Iyengar et al. 2003, 2005); the peptide lectin, which can activate the Jak/STAT3, ERK1/2, and/or PI3K pathways, enhance cyclin D1 and E-cadherin expression, and induce trans-activation of both HER2 and the EGF receptor (EGFR), as well as stimulating the expression of both aromatase and VEGF (Catalano et al. 2003, 2004, 2009; Cirillo et al. 2008; Mauro et al. 2007); and adiponectin, levels of which are inversely related to obesity (Trujillo and Scherer 2005), and which exerts antiproliferative and proapoptotic effects on breast tumor cell lines, as well as affecting angiogenesis (Dieudonne et al. 2006; Dos Santos et al. 2008; Landskroner-Eiger et al. 2009).

Tumor cells and the tumor microenvironment may also act to modulate the behavior of adipocytes; for example, adipocytes proximal to the invasive front of the tumor express the metalloprotease MMP11, which is a potent negative regulator of adipogenesis (Andarawewa et al. 2005). Thus, the role of adipocytes as participants in tumor initiation and progression is currently an area of active investigation.

14.6 ANGIOGENESIS

Since this topic is covered in detail in another chapter of this work in the context of tumor progression (Chapter 15), it will not be addressed in detail here. However, it should be noted that the complex web of interrelationships between stromal and epithelial elements contains many factors that impinge upon angiogenesis, from secretion of VEGF by macrophages to the recruitment of blood vessels as a consequence of tumor hypoxia.

One interesting interaction is that between the physical structure of the extracellular matrix and the response of the tumor vasculature to cessation of VEGF inhibition. While pharmacological blockade of VEGF signaling leads to the loss of a majority of tumor blood vessels and partial restoration of a normal vascular pattern (Bergers et al. 2003; Inai et al. 2004; Jain 2001, 2005), the pericytes and basement membranes originally associated with the regressed vessels remain present in the stroma (Inai et al. 2004). Examination of the consequences of inhibitor cessation in mouse models reveals that subsequent revascularization is rapid and utilizes these preexisting structures, such that within 7 days the tumor vasculature has returned to the pretreatment state (Mancuso et al. 2006). This accentuates the importance of considering all aspects of stromal biology and structure when designing effective therapeutic combinations.

14.7 THE EXTRACELLULAR MATRIX

As a tumor progresses, sequential changes in the architecture of the extracellular matrix become evident, the most obvious of which is the breach of the basement membrane that defines the transition from a benign to a malignant phenotype. The basement membrane, composed of laminins, is primarily elaborated by the myoepithelial cells located between the polarized luminal epithelium, from which most tumors arise, and the basement membrane. Interestingly, myoepithelial cells derived from normal breast tissue, but not those derived from breast tumor tissue, can recreate normal-appearing duct-like structures when co-cultured with primary breast luminal epithelial cells (Gudjonsson et al. 2002), while myoepithelial cells isolated from normal breast tissue could reverse the invasive potential of cells bearing a DCIS-like phenotype (Hu et al. 2008). The primary difference between these two populations was identified as a failure to synthesize laminin-1 (Gudjonsson et al. 2002), pointing to the importance of ECM elements in the maintenance of cell polarity and tissue organization.

Beyond the role of the basement membrane in the early steps of tumor progression, other ECM elements play important roles. Breast tumor tissue is physically more rigid or stiff than normal mammary tissue (Huang and Ingber 2005; Paszek et al. 2005). This increased stiffness can induce clustering of β1-integrins and activation of the Rho GTPase and ERK pathways to induce a DCIS-like phenotype in mammary epithelial cells (Kass et al. 2007; Paszek et al. 2005), while β1-integrin deletion significantly reduces tumor formation in a mouse model (White et al. 2004).

The temporal sequence of these events and their interdependence, however, is not completely clear at this time. In a mouse model where stromal collagen was increased in mammary tissue, tumor formation and invasion were significantly increased (Provenzano et al. 2008), while dense breast tissue has long been known to be a significant risk factor for development of breast cancer in humans (Boyd et al. 2002). Additionally, fibers in the ECM can act as tracks along which tumor cells can migrate in a linear manner, often following a chemotactic gradient (Condeelis and Segall 2003; Provenzano et al. 2006; Wang et al. 2002). Tumor explants tend to reorganize a collagen matrix into a radial pattern, facilitating their spread (Provenzano et al. 2006), once again illustrating the intimacy of the mutual interactions through which tumor cells modify their microenvironment to drive progression.

In one study, matrices deposited by tumor-associated fibroblasts were compared to those generated by a tissue culture cell line to determine the nature of their effects on a set of breast epithelial cells with varying degrees of tumorigenicity and invasiveness (Castello-Cros et al. 2009). It was observed that the two matrices induced different responses in the cell lines investigated, further supporting the hypothesis that tumor ECM is significantly different from the normal stromal form, and that these differences can affect tumor cell behavior.

At the level of gene expression profiling, examination of the expression patterns of ECM-related genes in whole tumor-derived datasets (containing information from both epithelial and stromal components) identified four ECM subclasses that correlated with differences in disease outcome (Bergamaschi et al. 2008). While good-outcome tumors exhibited increased expression of serpin-family protease inhibitors, poor-outcome tumors possessed high levels of expression of integrins and metallopeptidases, along with low levels of laminin chain expression (Bergamaschi et al. 2008). Other studies have also found that individual elements of the ECM are correlated with clinical differences: the expression level of collagen α(XI) is decreased in tumor vs. normal stroma, and was further decreased in the stroma of metastasized vs. nonmetastasized breast tumors (Halsted et al. 2008). Examination of the levels of various ECM components before and after neoadjuvant treatment revealed that the basement membrane protein collagen IV was increased while that of syndecan-1 was decreased after treatment, while tenascin-C levels tended to be increased in the nonresponder patient subset (Tokes et al. 2009). It is also interesting to note that investigations of polymorphisms in the ECM proteoglycans decorin and lumican identified a weak link between one variant of lumican and an increased risk of breast cancer (Kelemen et al. 2008). This area warrants further investigation to establish the links between individual variations in ECM components and changes in both cancer risk and response to therapy.

14.8 MICROENVIRONMENT OF THE METASTATIC NICHE

In breast cancer, the vast majority of disease-associated deaths are due to the effects of metastases to distant organs, for example, bone, brain, lung, or liver. This pattern is characteristic of breast cancer; other tumor types exhibit different preferences for sites of metastasis, suggesting that the compatibility of local microenvironments at distant sites may play a role in determining the location of successful metastatic spread. Referred to as the seed-and-soil model, this was first proposed by Stephen Paget in 1889 (Paget 1889). In the case of breast cancer, it has become apparent that distinct cell populations in the primary tumor possess features driving preferential organ targeting, and that the microenvironment plays an important role in this process (Joyce and Pollard 2009). For instance, metastasis to bone involves the chemokine receptor CXR4, involved in homing, as well as elements involved in bone collagen matrix degradation, osteoclast activation, and angiogenesis (Lu and Kang 2007), while lung homing may involve a transmembrane domain of metadherin (Brown and Ruoslahti 2004). Further studies of the molecular features governing site-specific metastatic targeting have demonstrated, among others, that TGF-ß-dependent signaling plays a key role in mediating bone-directed metastasis (Kang et al. 2003; Mourskaia et al. 2009).

The trafficking of cells and information between the initial tumor and distant sites appear to also be bidirectional. Soluble factors secreted by primary tumors can act to instruct cells at distant sites to prepare patches rich in fibronectin prior to the implantation and proliferation of tumor cells at these sites (Kaplan, Psaila, and Lyden 2006; Kaplan et al. 2005; Psaila et al. 2006). Following implantation, tumor cells again act to modulate their microenvironment to promote further growth; the time required for successful completion of this task may form one of the elements underlying tumor dormancy. Secretion of osteopontin by instigating tumors leads to the recruitment of bone marrow cells into the stroma of distant, previously indolent tumors in a mouse model of successive tumor implantation, leading to tumor outgrowth (McAllister et al. 2008). Modulation of the microenvironment at the potential metastatic site can affect the metastatic process; for example, the administration of bisphosphonates, which inhibit bone resorption among a host of other effects, has been shown to decrease the incidence of bone metastases in breast cancer patients (Coleman 2009; Lipton 2008).

14.9 GLOBAL CHARACTERIZATION OF THE STROMAL MICROENVIRONMENT

Recently, several studies have been performed that focused on the stromal elements present in breast cancers (Allinen et al. 2004; Bacac et al. 2006; Boersma et al. 2008; Casey et al. 2009; Farmer et al. 2009; Finak et al. 2008; Ma et al. 2009). These studies have revealed that tumor stroma contains information that is related to disease course and outcome (Bacac et al. 2006, Finak et al. 2008) and resistance to therapy (Farmer et al. 2009), as well as establishing that some cancer subtypes, such as inflammatory breast cancer, are associated with distinct stromal features (Boersma et al. 2008).

Laser capture microdissection and gene expression profiling of stromal tissue from a multistage mouse model of prostate carcinogenesis revealed that genes associated with endopeptidase activity, as well as others encoding structural matrix components, uPAR and growth factor receptors, were differentially expressed between early and late stages of tumor progression (Bacac et al. 2006). Interestingly, clustering of samples using this list of genes led to the generation of patient cohorts with significantly different overall outcomes in human prostate or breast cancer patient datasets (Bacac et al. 2006).

In another attempt to unravel some of the interactions occurring between tumor and stromal cells, a variety of breast cancer-derived cell lines were co-cultivated with stromal fibroblasts and the gene expression profiles of the co-cultures compared with those of either cell type alone (Buess et al. 2007). Interestingly, co-cultivation of ER-negative, but not ER-positive, breast cancer cell lines with fibroblasts resulted in the induction of a set of interferon-associated genes, which occurred predominantly in the epithelial cells, apparently in response to the secretion of type I interferons by the co-cultured fibroblasts. Clustering of external breast cancer gene expression datasets using this gene list led to the separation of the samples into two clusters with significant differences in outcome (Buess et al. 2007); however, the interferon-response signature is closely correlated with tumor ER status, which is in turn intimately related to outcome. These results support a model in which tumor and stroma mutually influence each other, and additionally reveal that tumor subtype can play a role in determining stroma responses.

The observation that there are multiple similarities between the tumor environment and physiological wound healing prompted the hypothesis that the tumor stroma represents a disturbed version of the wound healing response (Dvorak 1986). Following this lead, a 512-member gene expression signature of the fibroblast response to serum, which integrates many processes involved in wound healing, including plasminogen activation (Iyer et al. 1999), was derived and compared to human tumor gene expression datasets (Chang et al. 2004, 2005). Interestingly, the molecular features of the wound response were present in a subset of breast, lung, and gastric cancer samples both before and after chemotherapy; patients whose samples fell within this subset exhibited significantly worse clinical outcomes than those whose samples did not exhibit the wound-response signature (Chang et al. 2004, 2005). There was a significant association between the basal breast tumor subtype and the wound-response signature (Chang et al. 2004, 2005); however, combination of the wound-response signature with predictors derived primarily from tumor epithelial tissues led to improved risk stratification versus the outputs of either predictor alone (Chang et al. 2005), suggesting that at least some elements of the wound-response signature represent information emanating primarily from the stromal compartment.

In a recent study from our group, a predictor of disease outcome (stroma-derived prognostic predictor, or SDPP) was generated from isolated tumor-adjacent stroma from 53 primary breast tumors (Finak et al. 2008). A key finding of this investigation was that signals reflecting distinct biological responses could be identified as differentially represented between patient groups with distinct outcome profiles. These include markers of a Th1-type T cell response (e.g., CD8A, CD247, CD3D, and GZMA), which are highly expressed in samples of the good-outcome group, consistent with previous findings that an increased Th1-type response is correlated with good outcome in lung and colon cancer (Hiraoka et al. 2006; Pages et al. 2005). Conversely, members of the poor-outcome group exhibited reduced Th1-type signals, as well as decreased levels of chemokines promoting the recruitment and survival of NK and T cells (i.e., CXCL14 and GIMAP5). Additionally, the poor-outcome group contained signals indicative of angiogenic, hypoxic, and TAM-associated responses. These include elevated expression of adrenomedullin (ADM), MMP1, and osteopontin (OPN, also known as SPP1), also seen in the transcriptome of hypoxic monocytes and macrophages (Bosco et al. 2006), as well as interleukin-8 (IL8), which can enhance proliferation of endothelial cells (Li et al. 2003) and metalloproteases such as MMP1 and MMP12, known to be involved in macrophage-mediated tissue remodeling.

While individual stroma-derived single genes or specific response signatures have poor (Gruber et al. 2004; Uzzan et al. 2004) or subtype-restricted (Teschendorff et al. 2007) prognostic value, the integration of separate readouts of multiple distinct biological responses, including hypoxia, angiogenesis, and differential immune responses in the tumor microenvironment permits the effective and accurate prediction of outcome, for example, within the SDPP. This further reinforces the hypothesis that the evolutionary path of individual breast tumors is not completely determined by features of the tumor cells themselves, but rather can be differentially modified by inputs from multiple processes and cell types present in the microenvironment.

While the majority of studies have focused on the generation of predictors of overall prognosis, stromal elements have also been implicated as pure determinants of response to treatment. Using a metagene-based approach, it has been reported that a gene set representative of reactive stroma, primarily reflecting a fibroblast response, was overexpressed in samples from ER-negative breast cancer patients whose tumors demonstrated resistance to neoadjuvant chemotherapy (Farmer et al. 2009), suggesting that a particular configuration of tumor-stroma communication may generate an intrinsic resistance to specific therapeutic modalities.

Interestingly, in most cases stromal classes appear to be relatively independent of other clinical variables, including the subtype of the associated epithelial tumor, suggesting that the spectrum of breast cancers includes at least two mutually orthogonal axes. The location of a specific tumor/stroma combination in this landscape encodes important information regarding the processes active in each of the two tissue types, as well as the set of mutual interactions occurring between them. A more complete understanding of these factors will be critical both to selection of the optimal combinations of treatment regimens and to the development of novel therapeutic modalities to address these interactions.

14.10 SUMMARY

In contrast to the tumor mass, cells present in the stromal environment are generally thought to not have undergone significant genetic changes. Thus, it has been proposed that these may comprise a stable population critical for tumor progression, and that therefore the stroma represents a viable target for therapeutic intervention (Jain 2005; Joyce 2005; Micke and Ostman 2004; Orimo and Weinberg 2006). However, from the information presented above, it is evident that the composition, abundance, and activities of multiple stromal elements will vary on a patient to patient basis. These variations are not necessarily linked to the subtype of the primary tumor, which is currently the primary guide used for selection of tailored treatment protocols. Thus, we suggest that specific combinations of epithelial and stromal elements constitute more appropriate definitions of individualized tumor types, and that therapies targeting both the tumor and the microenvironment may prove more efficacious than those specific to one aspect alone.

It is also interesting to note that beyond patient-specific differences in stromal phenotype, broad differences have been identified between members of different ethnic groups. A recent analysis of differential gene expression in isolated stroma from African American vs. European American breast cancer patients has demonstrated increased expression of genes involved in angiogenesis and response to DNA damage stimuli in the African American cohort. It is likely that similar differences would also be observed when comparing other ethnic groups, or among individuals exposed to different macroenvironmental conditions.

Since stromal elements appear to play important roles in determining the success of a particular tumor in progressing along the path to frank malignancy, it is possible that manipulation at the level of the microenvironment will prove to be more successful in the early stages of tumor development. This hypothesis is supported by the results of studies comparing the expression profiles of preinvasive to invasive breast tumors, which demonstrated that the majority of changes in stromal gene expression occurred prior to the acquisition of an invasive phenotype (Ma et al. 2009; Schedin and Borges 2009).

ACKNOWLEDGMENTS

Space constraints render it impossible to adequately acknowledge all of the contributions made to this field by the many key individuals and groups who have studied different aspects of this research area in detail, and have made it necessary to refer the reader to reviews in many cases where the primary literature is very large. The authors apologize in advance for any omissions.

This work was supported by a postdoctoral fellowship from the Research Institute of the McGill University Health Centre (to N.B.), as well as by a Team Grant from the Canadian Institutes of Health Research (Grant No. CTP-79857 to M.P.). M.P. holds the Diane and Sal Guerrera Chair in Cancer Genetics at McGill University.

REFERENCES

Allinen M., Beroukhim R., Cai L. et al. 2004. Molecular characterization of the tumor microenvironment in breast cancer. *Cancer Cell* 6:17–32.

Andarawewa K. L., Motrescu E. R., Chenard M. P. et al. 2005. Stromelysin-3 is a potent negative regulator of adipogenesis participating to cancer cell-adipocyte interaction/crosstalk at the tumor invasive front. *Cancer Res* 65:10862–10871.

Anderson C. F. and Mosser D. M. 2002. A novel phenotype for an activated macrophage: the type 2 activated macrophage. *J Leukoc Biol* 72:101–106.

Arenberg D. A., Keane M. P., DiGiovine B. et al. 2000. Macrophage infiltration in human non-small-cell lung cancer: the role of CC chemokines. *Cancer Immunol Immunother* 49:63–70.

Austin C. D., De Maziere A. M., Pisacane P.I. et al. 2004. Endocytosis and sorting of ErbB2 and the site of action of cancer therapeutics trastuzumab and geldanamycin. *Mol Biol Cell* 15:5268–5282.

Bacac M., Provero P., Mayran N. et al. 2006. A mouse stromal response to tumor invasion predicts prostate and breast cancer patient survival. *PLoS ONE* 1:e32.

Balkwill F. 2004. Cancer and the chemokine network. *Nature Rev Cancer* 4:540–550.

Balkwill F., Charles K. A., and Mantovani A. 2005. Smoldering and polarized inflammation in the initiation and promotion of malignant disease. *Cancer Cell* 7:211–217.

Balkwill F. and Mantovani A. 2001. Inflammation and cancer: back to Virchow? *Lancet* 357:539–545.

Bates G.J., Fox S. B., Han C. et al. 2006. Quantification of regulatory T cells enables the identification of high-risk breast cancer patients and those at risk of late relapse. *J Clin Oncol* 24:5373–5380.

Bergamaschi A., Tagliabue E., Sorlie T. et al. 2008. Extracellular matrix signature identifies breast cancer subgroups with different clinical outcome. *J Pathol* 214:357–367.

Bergers G., Song S., Meyer-Morse N., Bergsland E., and Hanahan D. 2003. Benefits of targeting both pericytes and endothelial cells in the tumor vasculature with kinase inhibitors. *J Clin Invest* 111:1287–1295.

Bingle L., Brown N. J., and Lewis C. E. 2002. The role of tumour-associated macrophages in tumour progression: implications for new anticancer therapies. *J Pathol* 196:254–265.

Boersma B. J., Reimers M., Yi M. et al. 2008. A stromal gene signature associated with inflammatory breast cancer. *Int J Cancer* 122:1324–1332.

Bohling S. D. and Allison K. H. 2008. Immunosuppressive regulatory T cells are associated with aggressive breast cancer phenotypes: a potential therapeutic target. *Mod Pathol* 21:1527–1532.

Bosco M. C., Puppo M., Santangelo C. et al. 2006. Hypoxia modifies the transcriptome of primary human monocytes: modulation of novel immune-related genes and identification of CC-chemokine ligand 20 as a new hypoxia-inducible gene. *J Immunol* 177:1941–1955.

Bottazzi B., Polentarutti N., Acero R. et al. 1983. Regulation of the macrophage content of neoplasms by chemoattractants. *Science* 220:210–212.

Boyd N. F., Dite G.S., Stone J. et al. 2002. Heritability of mammographic density, a risk factor for breast cancer. *N Engl J Med* 347:886–894.

Brown D. M. and Ruoslahti E. 2004. Metadherin, a cell surface protein in breast tumors that mediates lung metastasis. *Cancer Cell* 5:365–374.

Buess M., Nuyten D. S., Hastie T. et al. 2007. Characterization of heterotypic interaction effects *in vitro* to deconvolute global gene expression profiles in cancer. *Genome Biol* 8:R191.

Calle E. E., Rodriguez C., Walker-Thurmond K., and Thun M. J. 2003. Overweight, obesity, and mortality from cancer in a prospectively studied cohort of U.S. adults. *N Engl J Med* 348:1625–1638.

Casey T., Bond J., Tighe S. et al. 2009. Molecular signatures suggest a major role for stromal cells in development of invasive breast cancer. *Breast Cancer Res Treat* 114:47–62.

Castello-Cros R., Khan D. R., Simons J., Valianou M., and Cukierman E. 2009. Staged stromal extracellular 3D matrices differentially regulate breast cancer cell responses through PI3K and beta1-integrins. *BMC Cancer* 9:94.

Catalano S., Giordano C., Rizza P. et al. 2009. Evidence that leptin through STAT and CREB signaling enhances cyclin D1 expression and promotes human endometrial cancer proliferation. *J Cell Physiol* 218:490–500.

Catalano S., Marsico S., Giordano C. et al. 2003. Leptin enhances, via AP-1, expression of aromatase in the MCF-7 cell line. *J Biol Chem* 278:28668–28676.

Catalano S., Mauro L., Marsico S. et al. 2004. Leptin induces, via ERK1/ERK2 signal, functional activation of estrogen receptor alpha in MCF-7 cells. *J Biol Chem* 279:19908–19915.

Chang H. Y., Nuyten D. S. A., Sneddon J. B. et al. 2005. Robustness, scalability, and integration of a wound-response gene expression signature in predicting breast cancer survival. *Proc Natl Acad Sci USA* 102:3738–3743.

Chang H. Y., Sneddon J. B., Alizadeh A. A. et al. 2004. Gene expression signature of fibroblast serum response predicts human cancer progression: similarities between tumors and wounds. *PLoS Biol* 2:E7.

Cirillo D., Rachiglio A. M., la Montagna R., Giordano A., and Normanno N. 2008. Leptin signaling in breast cancer: an overview. *J Cell Biochem* 105:956–964.

Clynes R. A., Towers T. L., Presta L. G., and Ravetch J. V. 2000. Inhibitory Fc receptors modulate *in vivo* cytoxicity against tumor targets. *Nature Med* 6:443–446.

Coleman R. E. 2009. Adjuvant bisphosphonates in breast cancer: are we witnessing the emergence of a new therapeutic strategy? *Eur J Cancer* 45:1909–1915.

Condeelis J. and Pollard J. W. 2006. Macrophages: obligate partners for tumor cell migration, invasion, and metastasis. *Cell* 124:263–266.

Condeelis J. and Segall J. E. 2003. Intravital imaging of cell movement in tumours. *Nature Rev Cancer* 3:921–930.

Cooley S., Burns L. J., Repka T., and Miller J. S. 1999. Natural killer cell cytotoxicity of breast cancer targets is enhanced by two distinct mechanisms of antibody-dependent cellular cytotoxicity against LFA-3 and HER2/neu. *Exp Hematol* 27:1533–1541.

Coussens L. M. and Werb Z. 2002. Inflammation and cancer. *Nature* 420:860–867.

Delord J. P., Allal C., Canal M. et al. 2005. Selective inhibition of HER2 inhibits AKT signal transduction and prolongs disease-free survival in a micrometastasis model of ovarian carcinoma. *Ann Oncol* 16:1889–1897.

Dieudonne M. N., Bussiere M., Dos Santos E. et al. 2006. Adiponectin mediates antiproliferative and apoptotic responses in human MCF7 breast cancer cells. *Biochem Biophys Res Commun* 345:271–279.

Dos Santos E., Benaitreau D., Dieudonne M. N. et al. 2008. Adiponectin mediates an antiproliferative response in human MDA-MB 231 breast cancer cells. *Oncol Rep* 20:971–977.

Dvorak H. F. 1986. Tumors: wounds that do not heal. Similarities between tumor stroma generation and wound healing. *N Engl J Med* 315:1650–1659.

Elliott B. E., Tam S. P., Dexter D., and Chen Z. Q. 1992. Capacity of adipose tissue to promote growth and metastasis of a murine mammary carcinoma: effect of estrogen and progesterone. *Int J Cancer* 51:416–424.

Eneman J. D., Wood M. E., and Muss H. B. 2004. Selecting adjuvant endocrine therapy for breast cancer. *Oncology (Williston Park)* 18:1733–1744, discussion 44–45, 48, 51–54.

Farmer P., Bonnefoi H., Anderle P. et al. 2009. A stroma-related gene signature predicts resistance to neoadjuvant chemotherapy in breast cancer. *Nature Med* 15:68–74.

Finak G., Bertos N., Pepin F. et al. 2008. Stromal gene expression predicts clinical outcome in breast cancer. *Nature Med* 14:518–527.

Finak G., Sadekova S., Pepin F. et al. 2006. Gene expression signatures of morphologically normal breast tissue identify basal-like tumors. *Breast Cancer Res* 8:R58.

Fleming J. M., Long E. L., Ginsburg E. et al. 2008. Interlobular and intralobular mammary stroma: genotype may not reflect phenotype. *BMC Cell Biol* 9:46.

Fukino K., Shen L., Matsumoto S. et al. 2004. Combined total genome loss of heterozygosity scan of breast cancer stroma and epithelium reveals multiplicity of stromal targets. *Cancer Res* 64:7231–7236.

Fukino K., Shen L., Patocs A., Mutter G. L., and Eng C. 2007. Genomic instability within tumor stroma and clinicopathological characteristics of sporadic primary invasive breast carcinoma. *JAMA* 297:2103–2111.

Garcia S., Dales J. P., Charafe-Jauffret E. et al. 2007a. Overexpression of c-Met and of the transducers PI3K, FAK and JAK in breast carcinomas correlates with shorter survival and neoangiogenesis. *Int J Oncol* 31:49–58.

Garcia S., Dales J. P., Charafe-Jauffret E. et al. 2007b. Poor prognosis in breast carcinomas correlates with increased expression of targetable CD146 and c-Met and with proteomic basal-like phenotype. *Hum Pathol* 38:830–841.

Gennari R., Menard S., Fagnoni F. et al. 2004. Pilot study of the mechanism of action of preoperative trastuzumab in patients with primary operable breast tumors overexpressing HER2. *Clin Cancer Res* 10:5650–5655.

Ghajar C. M. and Bissell M. J. 2008. Extracellular matrix control of mammary gland morphogenesis and tumorigenesis: insights from imaging. *Histochem Cell Biol* 130:1105–1118.

Giannopoulou I., Mylona E., Kapranou A. et al. 2007. The prognostic value of the topographic distribution of uPAR expression in invasive breast carcinomas. *Cancer Lett* 246:262–267.

Goswami S., Sahai E., Wyckoff J. B. et al. 2005. Macrophages promote the invasion of breast carcinoma cells via a colony-stimulating factor-1/epidermal growth factor paracrine loop. *Cancer Res* 65:5278–5283.

Gruber G., Greiner R. H., Hlushchuk R. et al. 2004. Hypoxia-inducible factor 1 alpha in high-risk breast cancer: an independent prognostic parameter? *Breast Cancer Res* 6:R191–R198.

Gudjonsson T., Ronnov-Jessen L., Villadsen R. et al. 2002. Normal and tumor-derived myoepithelial cells differ in their ability to interact with luminal breast epithelial cells for polarity and basement membrane deposition. *J Cell Sci* 115:39–50.

Halsted K. C., Bowen K. B., Bond L. et al. 2008. Collagen alpha1(XI) in normal and malignant breast tissue. *Mod Pathol* 21:1246–1254.

Hawsawi N. M., Ghebeh H., Hendrayani S. F. et al. 2008. Breast carcinoma-associated fibroblasts and their counterparts display neoplastic-specific changes. *Cancer Res* 68:2717–2725.

Hill R., Song Y., Cardiff R. D., and Van Dyke T. 2005. Selective evolution of stromal mesenchyme with p53 loss in response to epithelial tumorigenesis. *Cell* 123:1001–1011.

Hiraoka K., Miyamoto M., Cho Y. et al. 2006. Concurrent infiltration by CD8+ T cells and CD4+ T cells is a favourable prognostic factor in non-small-cell lung carcinoma. *Br J Cancer* 94:275–280.

Howlett A. R. and Bissell M.J. 1993. The influence of tissue microenvironment (stroma and extracellular matrix) on the development and function of mammary epithelium. *Epithelial Cell Biol* 2:79–89.

Hu M., Yao J., Cai L. et al. 2005. Distinct epigenetic changes in the stromal cells of breast cancers. *Nature Genet* 37:899–905.

Hu M., Yao J., Carroll D.K. et al. 2008. Regulation of *in situ* to invasive breast carcinoma transition. *Cancer Cell* 13:394–406.

Huang S. and Ingber D. E. 2005. Cell tension, matrix mechanics, and cancer development. *Cancer Cell* 8:175–176.

Hudis C. A. 2007. Trastuzumab: mechanism of action and use in clinical practice. *N Engl J Med* 357:39–51.

Hurd T. C., Sait S., Kohga S. et al. 2007. Plasminogen activator system localization in 60 cases of ductal carcinoma *in situ*. *Ann Surg Oncol* 14:3117–3124.

Huss F.R. and Kratz G. 2001. Mammary epithelial cell and adipocyte co-culture in a 3-D matrix: the first step towards tissue-engineered human breast tissue. *Cells Tissues Organs* 169:361–367.

Inai T., Mancuso M., Hashizume H. et al. 2004. Inhibition of vascular endothelial growth factor (VEGF) signaling in cancer causes loss of endothelial fenestrations, regression of tumor vessels, and appearance of basement membrane ghosts. *Am J Pathol* 165:35–52.

Iyengar P., Combs T. P., Shah S. J. et al. 2003. Adipocyte-secreted factors synergistically promote mammary tumorigenesis through induction of anti-apoptotic transcriptional programs and proto-oncogene stabilization. *Oncogene* 22:6408–6423.

Iyengar P., Espina V., Williams T. W. et al. 2005. Adipocyte-derived collagen VI affects early mammary tumor progression *in vivo*, demonstrating a critical interaction in the tumor/stroma microenvironment. *J Clin Invest* 115:1163–1176.

Iyer V. R., Eisen M. B., Ross D. T. et al. 1999. The transcriptional program in the response of human fibroblasts to serum. *Science* 283:83–87.

Jain R. K. 2001. Normalizing tumor vasculature with anti-angiogenic therapy: a new paradigm for combination therapy. *Nature Med* 7:987–989.

Jain R. K. 2005. Normalization of tumor vasculature: an emerging concept in anti-angiogenic therapy. *Science* 307:58–62.

Jeffers M., Rong S., Anver M., and Vande Woude G.F. 1996. Autocrine hepatocyte growth factor/ scatter factor-Met signaling induces transformation and the invasive/metastastic phenotype in C127 cells. *Oncogene* 13:853–856.

Jeffers M., Rong S., and Vande Woude G.F. 1996. Enhanced tumorigenicity and invasion-metastasis by hepatocyte growth factor/scatter factor-met signalling in human cells concomitant with induction of the urokinase proteolysis network. *Mol Cell Biol* 16:1115–1125.

Jeffers M., Rong S., and Woude G.F. 1996. Hepatocyte growth factor/scatter factor-Met signaling in tumorigenicity and invasion/metastasis. *J Mol Med* 74:505–513.

Jordan V. C. and Brodie A. M. 2007. Development and evolution of therapies targeted to the estrogen receptor for the treatment and prevention of breast cancer. *Steroids* 72:7–25.

Joyce J. A. 2005. Therapeutic targeting of the tumor microenvironment. *Cancer Cell* 7:513–520.

Joyce J. A. and Pollard J. W. 2009. Microenvironmental regulation of metastasis. *Nature Rev Cancer* 9:239–252.

Kalluri R. and Zeisberg M. 2006. Fibroblasts in cancer. *Nature Rev Cancer* 6:392–401.

Kang Y., Siegel P. M., Shu W. et al. 2003. A multigenic program mediating breast cancer metastasis to bone. *Cancer Cell* 3:537–549.

Kaplan R. N., Psaila B., and Lyden D. 2006. Bone marrow cells in the "pre-metastatic niche": within bone and beyond. *Cancer Metastasis Rev* 25:521–529.

Kaplan R. N., Riba R. D., Zacharoulis S. et al. 2005. VEGFR1-positive haematopoietic bone marrow progenitors initiate the pre-metastatic niche. *Nature* 438:820–827.

Karnoub A. E., Dash A. B., Vo A. P. et al. 2007. Mesenchymal stem cells within tumour stroma promote breast cancer metastasis. *Nature* 449:557–563.

Kass L., Erler J. T., Dembo M., and Weaver V. M. 2007. Mammary epithelial cell: influence of extracellular matrix composition and organization during development and tumorigenesis. *Int J Biochem Cell Biol* 39:1987–1994.

Kelemen L. E., Couch F. J., Ahmed S. et al. 2008. Genetic variation in stromal proteins decorin and lumican with breast cancer: investigations in two case-control studies. *Breast Cancer Res* 10:R98.

Knutson K. L. and Disis M .L. 2005. Tumor antigen-specific T helper cells in cancer immunity and immunotherapy. *Cancer Immunol Immunother* 54:721–728.

Kurose K., Gilley K., Matsumoto S. et al. 2002. Frequent somatic mutations in PTEN and TP53 are mutually exclusive in the stroma of breast carcinomas. *Nature Genet* 32:355–357.

Ladoire S., Arnould L., Apetoh L. et al. 2008. Pathologic complete response to neoadjuvant chemotherapy of breast carcinoma is associated with the disappearance of tumor-infiltrating foxp3+ regulatory T cells. *Clin Cancer Res* 14:2413–2420.

Lafkas D., Trimis G., Papavassiliou A. G., and Kiaris H. 2008. P53 mutations in stromal fibroblasts sensitize tumors against chemotherapy. *Int J Cancer* 123:967–971.

Lan R. Y., Ansari A. A., Lian Z. X., and Gershwin M. E. 2005. Regulatory T cells: development, function and role in autoimmunity. *Autoimmun Rev* 4:351–363.

Landskroner-Eiger S., Qian B., Muise E. S. et al. 2009. Proangiogenic contribution of adiponectin toward mammary tumor growth *in vivo*. *Clin Cancer Res* 15:3265–3276.

Leek R. D. and Harris A. L. 2002. Tumor-associated macrophages in breast cancer. *J Mammary Gland Biol Neoplasia* 7:177–189.

Lewis C. E. and Pollard J.W. 2006. Distinct role of macrophages in different tumor microenvironments. *Cancer Res* 66:605–612.

Lewis G. D., Figari I., Fendly B. et al. 1993. Differential responses of human tumor cell lines to anti-p185HER2 monoclonal antibodies. *Cancer Immunol Immunother* 37:255–263.

Li A., Dubey S., Varney M. L., Dave B. J., and Singh R. K. 2003. IL-8 directly enhanced endothelial cell survival, proliferation, and matrix metalloproteinases production and regulated angiogenesis. *J Immunol* 170:3369–3376.

Lin E. Y., Gouon-Evans V., Nguyen A. V., and Pollard J.W. 2002. The macrophage growth factor CSF-1 in mammary gland development and tumor progression. *J Mammary Gland Biol Neoplasia* 7:147–162.

Lin E. Y., Nguyen A. V., and Russell R. G., and Pollard J. W. 2001. Colony-stimulating factor 1 promotes progression of mammary tumors to malignancy. *J Exp Med* 193:727–740.

Lipton A. 2008. Emerging role of bisphosphonates in the clinic: antitumor activity and prevention of metastasis to bone. *Cancer Treat Rev* 34 (Suppl 1):S25–S30.

Lu X. and Kang Y. 2007. Organotropism of breast cancer metastasis. *J Mammary Gland Biol Neoplasia* 12:153–162.

Ma X.J., Dahiya S., Richardson E., Erlander M., and Sgroi D.C. 2009. Gene expression profiling of the tumor microenvironment during breast cancer progression. *Breast Cancer Res* 11:R7.

Manabe Y., Toda S., Miyazaki K., and Sugihara H. 2003. Mature adipocytes, but not preadipocytes, promote the growth of breast carcinoma cells in collagen gel matrix culture through cancer-stromal cell interactions. *J Pathol* 201:221–228.

Mancuso M. R., Davis R., Norberg S. M. et al. 2006. Rapid vascular regrowth in tumors after reversal of VEGF inhibition. *J Clin Invest* 116:2610–2621.

Mantovani A., Allavena P., Sozzani S. et al. 2004a. Chemokines in the recruitment and shaping of the leukocyte infiltrate of tumors. *Semin Cancer Biol* 14:155–160.

Mantovani A., Sica A., Sozzani S. et al. 2004b. The chemokine system in diverse forms of macrophage activation and polarization. *Trends Immunol* 25:677–686.

Mantovani A., Sozzani S., Locati M., Allavena P., and Sica A. 2002. Macrophage polarization: tumor-associated macrophages as a paradigm for polarized M2 mononuclear phagocytes. *Trends Immunol* 23:549–555.

Matsushima K., Larsen C. G., DuBois G. C., and Oppenheim J.J. 1989. Purification and characterization of a novel monocyte chemotactic and activating factor produced by a human myelomonocytic cell line. *J Exp Med* 169:1485–1490.

Mauro L., Catalano S., Bossi G. et al. 2007. Evidences that leptin up-regulates E-cadherin expression in breast cancer: effects on tumor growth and progression. *Cancer Res* 67:3412–3421.

McAllister S. S., Gifford A. M., Greiner A. L. et al. 2008. Systemic endocrine instigation of indolent tumor growth requires osteopontin. *Cell* 133:994–1005.

Meng S., Tripathy D., Shete S. et al. 2006. uPAR and HER-2 gene status in individual breast cancer cells from blood and tissues. *Proc Natl Acad Sci USA* 103:17361–17365.

Micke P., and Ostman A. 2004. Tumour-stroma interaction: cancer-associated fibroblasts as novel targets in anti-cancer therapy? *Lung Cancer* 45 (Suppl 2):S163–S175.

Miki Y., Suzuki T., and Sasano H. 2007. Controversies of aromatase localization in human breast cancer: stromal versus parenchymal cells. *J Steroid Biochem Mol Biol* 106:97–101.

Mishra P. J., Humeniuk R., Medina D. J. et al. 2008. Carcinoma-associated fibroblast-like differentiation of human mesenchymal stem cells. *Cancer Res* 68:4331–4339.

Moinfar F., Man Y. G., Arnould L. et al. 2000. Concurrent and independent genetic alterations in the stromal and epithelial cells of mammary carcinoma: implications for tumorigenesis. *Cancer Res* 60:2562–2566.

Mourskaia A. A., Dong Z., Ng S. et al. 2009. Transforming growth factor-beta1 is the predominant isoform required for breast cancer cell outgrowth in bone. *Oncogene* 28:1005–1015.

Muller A., Homey B., Soto H. et al. 2001. Involvement of chemokine receptors in breast cancer metastasis. *Nature* 410:50–56.

Nielsen B. S., Rank F., Illemann M., Lund L. R., and Dano K. 2007. Stromal cells associated with early invasive foci in human mammary ductal carcinoma *in situ* coexpress urokinase and urokinase receptor. *Int J Cancer* 120:2086–2095.

Nishimura R. and Arima N. 2008. Is triple negative a prognostic factor in breast cancer? *Breast Cancer* 15:303–308.

Orimo A., Gupta P. B., Sgroi D. C. et al. 2005. Stromal fibroblasts present in invasive human breast carcinomas promote tumor growth and angiogenesis through elevated SDF-1/CXCL12 secretion. *Cell* 121:335–348.

Orimo A. and Weinberg R. A. 2006. Stromal fibroblasts in cancer: a novel tumor-promoting cell type. *Cell Cycle* 5:1597–1601.

Pages F., Berger A., Camus M. et al. 2005. Effector memory T cells, early metastasis, and survival in colorectal cancer. *N Engl J Med* 353:2654–2666.

Paget S. 1889. The distribution of secondary growths in cancer of the breast. *Cancer Metastasis Rev* 8:98–101.

Paszek M. J., Zahir N., Johnson K. R. et al. 2005. Tensional homeostasis and the malignant phenotype. *Cancer Cell* 8:241–254.

Patel R. R., Sharma C. G., and Jordan V. C. 2007. Optimizing the antihormonal treatment and prevention of breast cancer. *Breast Cancer* 14:113–122.

Pepper M. S., Matsumoto K., Nakamura T., Orci L., and Montesano R. 1992. Hepatocyte growth factor increases urokinase-type plasminogen activator (u-PA) and u-PA receptor expression in Madin-Darby canine kidney epithelial cells. *J Biol Chem* 267:20493–20496.

Perou C. M., Sorlie T., Eisen M. B. et al. 2000. Molecular portraits of human breast tumours. *Nature* 406:747–752.

Pollard J. W. 2008. Macrophages define the invasive microenvironment in breast cancer. *J Leukoc Biol* 84:623–630.

Provenzano P. P., Eliceiri K. W., Campbell J. M. et al. 2006. Collagen reorganization at the tumor-stromal interface facilitates local invasion. *BMC Med* 4:38.

Provenzano P. P., Inman D. R., Eliceiri K. W. et al. 2008. Collagen density promotes mammary tumor initiation and progression. *BMC Med* 6:11.

Psaila B., Kaplan R. N., Port E. R., and Lyden D. 2006. Priming the "soil" for breast cancer metastasis: the pre-metastatic niche. *Breast Dis* 26:65–74.

Quemener C., Gabison E. E., Naimi B. et al. 2007. Extracellular matrix metalloproteinase inducer up-regulates the urokinase-type plasminogen activator system promoting tumor cell invasion. *Cancer Res* 67:9–15.

Radisky E. S. and Radisky D. C. 2007. Stromal induction of breast cancer: inflammation and invasion. *Rev Endocr Metab Disord* 8:279–287.

Rakha E. A. and Ellis I. O. 2009. Triple-negative/basal-like breast cancer: review. *Pathology* 41:40–47.

Rong S., Oskarsson M., Faletto D. et al. 1993. Tumorigenesis induced by coexpression of human hepatocyte growth factor and the human met protooncogene leads to high levels of expression of the ligand and receptor. *Cell Growth Differ* 4:563–569.

Rong S., Segal S., Anver M., Resau J. H., and Vande Woude G. F. 1994. Invasiveness and metastasis of NIH 3T3 cells induced by Met-hepatocyte growth factor/scatter factor autocrine stimulation. *Proc Natl Acad Sci USA* 91:4731–4735.

Sadlonova A., Bowe D. B., Novak Z. et al. 2009. Identification of molecular distinctions between normal breast-associated fibroblasts and breast cancer-associated fibroblasts. *Cancer Microenviron.*

Santen R. J., Martel J., Hoagland M. et al. 1998. Demonstration of aromatase activity and its regulation in breast tumor and benign breast fibroblasts. *Breast Cancer Res Treat* 49 (Suppl 1):S93–S9; discussion S109–S119.

Santen R. J., Santner S. J., Pauley R. J. et al. 1997. Estrogen production via the aromatase enzyme in breast carcinoma: which cell type is responsible? *J Steroid Biochem Mol Biol* 61:267–271.

Santner S. J., Pauley R. J., Tait L., Kaseta J., and Santen R.J. 1997. Aromatase activity and expression in breast cancer and benign breast tissue stromal cells. *J Clin Endocrinol Metab* 82:200–208.

Sappino A. P., Skalli O., Jackson B., Schurch W., and Gabbiani G. 1988. Smooth-muscle differentiation in stromal cells of malignant and non-malignant breast tissues. *Int J Cancer* 41:707–712.

Sarup J. C., Johnson R. M., King K. L. et al. 1991. Characterization of an anti-p185HER2 monoclonal antibody that stimulates receptor function and inhibits tumor cell growth. *Growth Regul* 1:72–82.

Schedin P. and Borges V. 2009. Breaking down barriers: the importance of the stromal microenvironment in acquiring invasiveness in young women's breast cancer. *Breast Cancer Res* 11:102.

Sica A., Larghi P., Mancino A. et al. 2008. Macrophage polarization in tumour progression. *Semin Cancer Biol* 18:349–355.

Singer C. F., Gschwantler-Kaulich D., Fink-Retter A. et al. 2008. Differential gene expression profile in breast cancer-derived stromal fibroblasts. *Breast Cancer Res Treat* 110:273–281.

Sorlie T., Perou C. M., Tibshirani R. et al. 2001. Gene expression patterns of breast carcinomas distinguish tumor subclasses with clinical implications. *Proc Natl Acad Sci USA* 98:10869–10874.

Stephenson G. D. and Rose D. P. 2003. Breast cancer and obesity: an update. *Nutr Cancer* 45:1–16.

Suzuki T., Miki Y., Akahira J. et al. 2008. Aromatase in human breast carcinoma as a key regulator of intratumoral sex steroid concentrations. *Endocr J* 55:455–463.

Teschendorff A. E., Miremadi A., Pinder S.E., Ellis I. O., and Caldas C. 2007. An immune response gene expression module identifies a good prognosis subtype in estrogen receptor negative breast cancer. *Genome Biol* 8:R157.

Teschendorff A. E., Naderi A., Barbosa-Morais N. L., and Caldas C. 2006. PACK: Profile Analysis using Clustering and Kurtosis to find molecular classifiers in cancer. *Bioinformatics* 22:2269–2275.

Tokes A. M., Szasz A. M., Farkas A. et al. 2009. Stromal matrix protein expression following preoperative systemic therapy in breast cancer. *Clin Cancer Res* 15:731–739.

Trujillo M. E. and Scherer P. E. 2005. Adiponectin: journey from an adipocyte secretory protein to biomarker of the metabolic syndrome. *J Intern Med* 257:167–175.

Uzzan B., Nicolas P., Cucherat M., and Perret G. Y. 2004. Microvessel density as a prognostic factor in women with breast cancer: a systematic review of the literature and meta-analysis. *Cancer Res* 64:2941–2955.

Valabrega G., Montemurro F., and Aglietta M. 2007. Trastuzumab: mechanism of action, resistance and future perspectives in HER2-overexpressing breast cancer. *Ann Oncol* 18:977–984.

Valabrega G., Montemurro F., Sarotto I. et al. 2005. TGFalpha expression impairs trastuzumab-induced HER2 downregulation. *Oncogene* 24:3002–3010.

van 't Veer L. J., Dai H., van de Vijver M. J. et al. 2002. Gene expression profiling predicts clinical outcome of breast cancer. *Nature* 415:530–536.

Wang W., Wyckoff J. B., Frohlich V. C. et al. 2002. Single cell behavior in metastatic primary mammary tumors correlated with gene expression patterns revealed by molecular profiling. *Cancer Res* 62:6278–6288.

White D. E., Kurpios N.A., Zuo D. et al. 2004. Targeted disruption of beta1-integrin in a transgenic mouse model of human breast cancer reveals an essential role in mammary tumor induction. *Cancer Cell* 6:159–170.

Wyckoff J., Wang W., Lin E. Y. et al. 2004. A paracrine loop between tumor cells and macrophages is required for tumor cell migration in mammary tumors. *Cancer Res* 64:7022–7029.

Wyckoff J. B., Wang Y., Lin E. Y. et al. 2007. Direct visualization of macrophage-assisted tumor cell intravasation in mammary tumors. *Cancer Res* 67:2649–2656.

Zangani D., Darcy K. M., Shoemaker S., and Ip M. M. 1999. Adipocyte-epithelial interactions regulate the *in vitro* development of normal mammary epithelial cells. *Exp Cell Res* 247:399–409.

CHAPTER 15

Tumor Angiogenesis

Cell-Microenvironment Interactions

Ally Pen, Danica B. Stanimirovic, and Maria J. Moreno

CONTENTS

15.1 INTRODUCTION

In the early 1970s, a seminal paper in the *New England Journal of Medicine* proposed the hypothesis that tumor growth is angiogenesis dependent (Folkman 1971). According to this hypothesis, during the initial avascular phase, solid tumors obtain nutrients and oxygen from the host vasculature through passive diffusion. During this phase, cell proliferation and cell death are balanced and tumors do not exceed a few cubic millimeters. To support continuing growth, tumors acquire an angiogenic phenotype and recruit new vasculature (Hanahan and Folkman 1996). The dependence of tumor growth and metastasis on angiogenesis has now been widely accepted and has provided a powerful rationale for the development of anti-angiogenic strategies (Folkman 1971).

Angiogenesis is a complex process that is highly dependent on the tumor microenvironment (Nikitenko 2009). Tumors are composed of different cell types, including tumor cells and stromal cells (i.e., fibroblasts, inflammatory cells, pericytes, hematopoietic cells, and endothelial cells), and extracellular matrix (ECM) components (Jung et al. 2002). It is now apparent that these different cellular and acellular constituents interact with each other in a complex manner to induce and control the expression of mediators that determine the angiogenic phenotype of a particular tumor (Carmeliet and Jain 2000; Jung et al. 2002). The discovery of different pathways involved in tumor angiogenesis has led to the identification of specific therapeutic targets with potential anti-tumorigenic capacity. This chapter will focus on the complex molecular network involved in the cross-talk between different tumor cell types and the tumor microenvironment during the process of tumor neovascularization (Figure 15.1).

15.2 TUMOR VASCULATURE

The normal blood vasculature is organized spatially and branches in a hierarchical fashion (arteries/veins, arterioles/venules, and capillaries) to provide adequate oxygen and nutrients to tissues in the body (Carmeliet 2000). Capillaries consist of endothelial cells (ECs) and pericytes, both embedded in the same basement membrane (BM). Arterioles and venules have an additional coverage of smooth muscle cells (SMCs). In larger vessels, veins are irregularly covered by SMCs while arteries are densely coated with multiple layers of SMCs called media, and with a large population of fibroblasts, elastic and collagenous fibers, called adventitia; these specialized layers regulate vessel diameter and blood flow (Jain 2003). In contrast to normal vessels, tumor vessels are heterogeneous, tortuous, and disorganized, and do not follow a hierarchical branching pattern (Jain 2003). Tumor vasculature exhibits significant abnormalities in cellular (ECs and pericytes) and acellular (BM) constituents. Tumor vessel walls are characterized by an increased number of endothelial fenestrae, vesicles, and transcellular holes, and wide intercellular junctions resulting in increased vessel permeability (Dvorak 2006; Hashizume et al. 2000). Pericytes in tumors present structural abnormalities that distinguish them from those in quiescent vessels and also display a loose attachment to ECs with some cytoplasmic processes infiltrating the tumor parenchyma (Morikawa et al. 2002). The pericyte coverage of tumor vessels appears to be dependent not only on the tumor types but also on the specificity markers used to identify pericytes/SMCs (Morikawa et al. 2002). Some studies described abundant pericyte coverage, whereas others showed reduced pericyte density in tumor vessels compared to normal vessels (Bergers and Song 2005). In addition to abnormal pericytes, tumor vessels display an irregular BM, sometimes discontinuous or absent, with variable thickness, and, in some regions, projecting into the tumor parenchyma (Baluk et al. 2003; Carmeliet and Jain 2000; Jain et al. 2007). The imbalance of pro- and anti-angiogenic factors and mechanical stress generated by tumor cells compressing vessels are considered key contributors of the abnormal tumor vessel phenotype (Jain 2005). The ultrastructural alterations in tumor vessels result in a chaotic and variable blood flow that hinders the efficacy of drug delivery (Fukumura and Jain 2007).

Phenotypically abnormal tumor vasculature is also characterized by an abnormal gene/protein expression (St. Croix et al. 2000). Analyses of transcriptome/proteome of tumor-associated vessels in recent years have yielded several novel selective molecular biomarkers. For example, *in vivo* selection of phage display libraries recovered novel peptide sequences (i.e., containing RGD and NGR motifs) that specifically recognize tumor endothelium in a human breast cancer xenograft mouse model (Arap, Pasqualini, and Ruoslahti 1998). Substractive proteomic studies coupled with bioinformatic approaches led to the identification of EC surface or secreted proteins upregulated in solid tumors (Oh et al. 2004). Using immunopurification and serial analysis of gene expression St. Croix and colleagues (2000) identified 46 transcripts, termed tumor endothelial markers (TEMs), that are elevated in malignant colorectal vessels. A transcriptional profiling study using a combination of laser capture microdissection coupled with microarray analyses resulted in the identification of insulin-like growth factor binding protein 7 (IGFBP7) as a selective biomarker of glioblastoma vessels, induced in and secreted by tumor ECs and involved in the late phase of angiogenesis (Pen et al. 2007, 2008). These and other (Bhati et al. 2008; Seaman et al. 2007; St. Croix et al. 2000) molecular profiling studies discovered many distinctive molecular features of tumor-derived endothelium that could be exploited for molecular diagnosis, imaging, and selective therapeutic targeting of angiogenic tumor vessels.

15.3 MECHANISMS OF TUMOR NEOVASCULARIZATION

Solid tumors develop new vessels through several mechanisms: sprouting angiogenesis, intussusception, cooption, vasculogenesis, mosaic vessels, and vasculogenic mimicry (Auguste et al. 2005; Carmeliet and Jain 2000). Sprouting angiogenesis, the growth of new blood vessels from pre-existing ones, is the most extensively studied process and involves BM degradation, EC proliferation and migration, lumen formation, and vessel stabilization (Carmeliet 2000). The sequential steps of sprouting angiogenesis will be reviewed in detail in sections below. Until recently, sprouting angiogenesis was considered the sole mechanism of tumor vascularization. However, it has become apparent that other mechanisms also participate in enhancing a network of vessels carrying blood and nutrient supply within the tumors.

A subset of tumors, including lung and brain cancers, can coopt existing host vessels at early stages of tumor development to gain access to oxygen and nutrients (Holash et al. 1999; Leenders, Kusters, and de Waal 2002). Coopted ECs interact with tumor cells through the angiopoietin-2/Tie2 pathway, resulting in vessel regression triggering hypoxia, which, in turn, stimulates vascular endothelial growth factor (VEGF) expression and initiates sprouting angiogenesis (Holash et al. 1999). Although vessel cooption generally occurs at early stages of tumorigenesis, further evidence suggests that cooption could persist during the whole process of tumor growth (Dome et al. 2002).

A variant of sprouting angiogenesis is intussusception—growth within itself—which involves the insertion of tissue pillars into the lumen of blood vessels to form ancillary vascular branches (Djonov, Baum, and Burri 2003). Intussusceptive microvascular growth was first described in the developing microvasculature of neonatal rat lung (Caduff, Fischer, and Burri 1986), and subsequently validated *in vivo* in the chick chorioallantoic membrane using video microscopy (Patan, Haenni, and Burri 1993). Since then, intussusception has been demonstrated in pathological vascular growth in colon and breast cancers (Djonov, Andres, and Ziemiecki 2001; Patan, Munn, and Jain 1996). Intussusception enables fast formation of new blood vessels and is metabolically less demanding than sprouting since it does not rely on EC proliferation or BM degradation for invasion (Djonov, Baum, and Burri 2003). Phases implicated in the intussusceptive angiogenesis are described in detail by Burri, Hlushchuk, and Djonov (2004). Shear stress and increased blood flow rate have been suggested to induce intussusception through a mechanotransduction system involving PECAM/CD31, which leads to the activation of adhesion molecules, angiogenic factors, and eNOS (Fisher et al. 2001). Intussusception is mediated by EC-EC and EC-pericyte interactions (Djonov et al. 2000). Therefore, molecules that participate in cell-cell interactions, including PDGF-BB, angiopoietins, Tie-2, TGF-β, ephrins and eprinB receptor are proposed to play important roles in the process of intussusception (Burri, Hlushchuk, and Djonov 2004). For example, over-expression of Ang-1 in a transgenic mouse model results in the formation of blood vessels resembling those formed by intussusception (Thurston et al. 1999).

Another mechanism by which tumors can acquire new vasculature is through vasculogenesis, defined as the *in situ* differentiation of ECs from angioblasts or endothelial progenitor cells (EPCs). EPCs, first isolated by Asahara and colleagues, were shown to express several endothelial markers, including CD34, CD31, vascular endothelial growth factor receptor-2 (VEGFR-2), and Tie-2 (Asahara et al. 1997). Studies using transgenic mice expressing green fluorescent protein (GFP) or β-galactosidase (Z-lac) subsequently demonstrated that circulating EPCs are incorporated into the angiogenic vasculature of growing tumors (Lyden et al. 2001; Nolan et al. 2007). EPC mobilization, recruitment, homing, and incorporation into tumors are complex processes regulated by several growth factors secreted by the tumor microenvironment. VEGF and placenta growth factor (PlGF) stimulate EPCs mobilization from the bone marrow by binding to VEGFR-2 expressed on the surface of EPCs (Asahara et al. 1999; Hattori et al. 2002). Metalloproteases, particularly MMP-9 produced by bone marrow cells, releases soluble kit ligand, which, in turn, promotes EPC proliferation, migration, and mobilization (Heissig et al. 2002). In addition, Ang-1, stromal cell-derived factor-1 (SDF-1), granulocyte colony-stimulating factor (G-CSF), and granulocyte-macrophage colony-stimulating factor (GM-CSF) have all been identified as bone marrow stem cell mobilizing factors (Moore et al. 2001; Orimo et al. 2005; Takahashi et al. 1999). Recently, the IGF2/IGFR2 system has been shown to participate in the activation and homing of EPCs to ischemic sites (Maeng et al. 2009). Incorporation of EPCs into the endothelial lining of tumor vessels mobilizes interactions among several adhesion molecules, including P-selectin, E-selectin, and integrins (Jin et al. 2006; Vajkoczy et al. 2003). However, the contribution of EPCs to tumor neovascularization remains controversial. Some studies showed that up to 50% of circulating EPCs were integrated into newly formed vessels (Garcia-Barros et al. 2003), whereas others reported a relatively small contribution of bone marrow stem cells to tumor vascularization (Machein et al. 2003; Peters et al. 2005).

Vasculogenic mimicry was first proposed as a mechanism of tumor neovascularization by Maniotis and colleagues (1999) to describe the ability of aggressive melanoma cells to dedifferentiate into an endothelial phenotype and to organize into vascular channel-like structures, thereby providing a secondary circulation system independent of angiogenesis. Evidence suggests that the tumor-lined structures form functional channels that contribute to tumor circulation since they are capable of conducting dyes and are perfused with blood (Maniotis et al. 1999). Blood flow has also been detected in these channels using magnetic resonance imaging techniques (Shirakawa et al. 2002). The mechanisms underlying the process of vasculogenic mimicry are still poorly understood. Microarray analyses and functional studies identified VE-cadherin, focal adhesion molecule (FAK), phosphatidylinositol-3-kinase (PI3K), MMPs, and ephrin receptors as mediators promoting vasculogenic mimicry (Hess et al. 2003, 2006, 2007). Recently, Petty and colleagues (2007) discovered migration-inducing protein 7 (Mig-7) as an important contributor and a specific marker of vasculogenic mimicry. Vasculogenic mimicry has been reported in several aggressive tumors associated with poor prognosis, including melanoma, breast, prostate, lung, and ovarian carcinomas (Hendrix et al. 2003).

In contrast to vascular mimicry, where vascular tubes are completely formed by tumor cells, in mosaic vessels, both ECs and tumor cells are co-localized in the walls of tumor vessels (Chang et al. 2000). Using the endothelial markers CD31 and CD105 to identify ECs and green fluorescent protein-labeled to identify tumor cells, Chang and colleagues (2000) demonstrated that 15% of the vessels in a colon carcinoma xenograft are mosaic vessels. Mosaic vessels can be formed either by the detachment of vessel-lining ECs or by the excessive growth of vessels with insufficient EC proliferation and coverage, in both cases resulting in exposure of the underlying tumor cells to the vessel lumen (Chang et al. 2000). Another proposed mechanism of mosaic vessels is the loss of EC markers. Using confocal microscopy, di Tomaso and colleagues (2005) demonstrated that the majority of "mosaic vessels" in a mouse model of colon cancer have a continuous, thin endothelial lining, but lacked CD31 and CD105 immunoreactivity, suggesting EC dedifferentiation and perhaps their acquisition of tumor-specific markers.

Accumulating evidence also suggests a role for lymphatic vessels in tumor progression and metastasis (Pepper et al. 2003). As this topic will not be discussed here, readers are referred to these excellent reviews dealing with the subject of lymphangiogenesis in tumor growth (Cueni and Detmar 2008; Stacker et al. 2002).

Tumor vascularization does not occur exclusively through one process, but rather through several described mechanisms dependent on the tumor context (Carmeliet and Jain 2000; Dome et al. 2007). For example, the microcirculation of uveal melanoma contains preexisting normal vessels, mosaic vessels, and angiogenic vessels (Chen et al. 2002). Highly vascularized glioblastoma tumors exploit sprouting angiogenesis, cooption, intussusception, and vasculogenesis to recruit blood vessels (Jain et al. 2007). Recently, Hlushchuk and colleagues (2008) showed that, after treatment with anti-angiogenic agent or irradiation, sprouting angiogenesis was replaced by intussusceptive angiogenesis in a mammary allograft tumor model, suggesting that intussusception can account for the development of anti-angiogenic resistance. The redundant nature of these processes presents an enormous challenge for developing efficacious anti-angiogenic therapy—targeting only one process of vascularization (e.g., sprouting angiogenesis) would still leave intact other perfusion mechanism(s), resulting in an ineffective treatment.

Among the different mechanisms of tumor neovascularization described, sprouting angiogenesis has been the most extensively studied (Carmeliet 2000; Carmeliet and Jain

2000). Sprouting angiogenesis is primarily triggered by hypoxia, a reduction in the normal level of tissue oxygen tension (Harris 2002). Intratumoral hypoxia is a hallmark of the metabolic environment in solid cancers, correlates with poor prognosis, and is a frequent cause of failure of radiotherapy (Harris 2002). In solid cancers, tissue hypoxia results from combined effects of the increased metabolic activity and oxygen consumption by the proliferating cells, structural and functional abnormalities of oxygen-supplying tumor vasculature, and the increased tumor cell distance from the existing capillaries, leading to the formation of hypovascular regions that become chronically hypoxic (Fukumura and Jain 2007; Vaupel and Mayer 2007). Blood flow in tumor vessels is often irregular, and some tumor regions are starved for oxygen periodically, resulting in "acute hypoxia" or "perfusion-limited hypoxia" (Brown and Giaccia 1998; Dewhirst 1998). Cancer cells adapt to a hypoxic environment by shifting to anaerobic metabolism (Dang and Semenza 1999), inducing erythropoietin production and by promoting angiogenesis and cell survival programs (Harris 2002). These adaptive responses to hypoxia are principally regulated by the hypoxia-inducible factor-1 (HIF-1). However, in tumors, HIF-1α expression is also upregulated by a wide range of growth factors and cytokines secreted by tumoral cells, including epidermal growth factor, insulin, insulin-like growth factors, interleukin 1β, and tumor necrosis factor (TNF)-α (Feldser et al. 1999; Sandau et al. 2001; Thornton et al. 2000; Zhong et al. 2000), by the expression of oncogenic pathways, such as mutant Ras and Src kinase pathways, and by tumor suppressor gene mutations, including p53 and PTEN (Ravi et al. 2000; Zundel et al. 2000).

In hypoxic environments, activation of the HIF-mediated transcription initiates the production of a number of angiogenic factors, including VEGF, VEGF receptors-1 (VEGFR-1/flt-1) and -2 (VEGFR-2/flk-1/KDR), angiopoietins (ANG-1 and -2), TIE-2 receptor, platelet-derived growth factor B (PDGF-B), and various matrix metalloproteinases (Forsythe et al. 1996; Harris 2002; Lal et al. 2001). Among a multitude of HIF-inducible mediators, VEGF is particularly important for tumor angiogenesis as it exhibits a very potent angiogenic activity, is more specific for ECs than other growth factors, and is markedly upregulated in the vast majority of human tumors (Ferrara 2005). The VEGF family of growth factors consists of six members (VEGF-A to -E and PlGF) that bind with different affinity to three cell surface tyrosine kinase receptors (VEGFR1 to 3) and/or to nonsignaling coreceptors, neuropillins (Ferrara 2004). VEGF not only induces mitogenic, chemotactic, and prosurvival effects in cultured endothelial cells (Ferrara, Gerber, and LeCouter 2003) but is also a potent vasodilator of existing vessels (Ferrara and Davis-Smyth 1997). A novel C-terminal splice variant of VEGF, VEGF165b, has been identified (Bates et al. 2002). This isoform is more frequently expressed in normal than in malignant tissues (Bates et al. 2002) and, contrary to VEGF, is a potent inhibitor of blood vessel and tumor growth (Konopatskaya et al. 2006; Rennel et al. 2008).

The critical roles of HIF (Brahimi-Horn, Chiche, and Pouyssegur 2007; Harris 2002; Pugh and Ratcliffe 2003; Semenza 2003) and the VEGF/VEGFR system (Ferrara 2005; Ferrara, Gerber, and LeCouter 2003) in sprouting angiogenesis have been extensively reviewed elsewhere. This chapter will instead focus on summarizing more recent understanding of temporal regulation and functional consequences of various processes involved in sprouting angiogenesis. Sprouting angiogenesis is a multistep process characterized by two phases: the activation (early) and the resolution (late) phase (Kalluri 2003; Pepper 1997). Each phase is modulated by specific interactions among different tumor and vascular cell types and by mediators released into the tumor microenvironment.

15.4 MODULATORS OF SPROUTING ANGIOGENESIS

15.4.1 Smooth Muscle Cell and Pericyte Detachment

One of the first events that occurs during initiation of sprouting is the detachment of smooth muscle cells and pericytes from vessels, enabling the access of angiogenic inducers to ECs. The angiopoietin (Ang)-Tie-receptor system has been shown to play an important role in this process (Lauren, Gunji, and Alitalo 1998).

Ang-1 and Ang-2 are secreted glycoproteins with approximately 60% identity in amino acid sequence. Both bind to the tyrosine kinase Tie-2 receptor with similar affinity, but trigger antagonistic effects on the receptor (Maisonpierre et al. 1993). Tie-2 receptors are expressed in ECs, in proangiogenic bone marrow-derived monocyte/macrophages, and in pericytes (De Palma et al. 2005). The ligand(s) for the Tie-1 receptor are currently unknown. Whereas Ang-1 is mainly expressed in perivascular and mural cells and contributes to vessel stabilization by inducing pericyte attachment (Hawighorst et al. 2002), Ang-2 is expressed by ECs only in localized regions of vascular remodeling (Maisonpierre et al., 1997) and antagonizes Ang-1 activity resulting in vascular smooth muscle cells and pericyte dissociation from vessels (Maisonpierre et al. 1997). In the presence of VEGF, Ang-2 facilitates vessel sprouting (Asahara et al. 1998) and intussusception (Patan et al. 1992), while in the absence of VEGF, Ang-2 induces regression of blood vessels (Maisonpierre et al. 1997). Ang-2 is the earliest marker of tumor vacularization, prior to VEGF induction (Holash et al. 1999) and is specifically expressed in coopted glioma vessels (Holash et al. 1999) and at the tips of growing capillaries (Acker, Beck, and Plate 2001; Maisonpierre et al. 1997), possibly in response to hypoxia and/or angiogenic growth factors (Oh et al. 1999; Yuan, Yang, and Woolf 2000).

Interestingly, Greenberg and colleagues (2008) have recently demonstrated that VEGF inhibits the PDGF-induced pericyte coverage of nascent vascular sprouts, leading to vessel destabilization; this effect is the consequence of a novel VEGFR2–PDGFR complex being formed in VSMC that inhibited the phosphorylation of PDGFR (Greenberg et al. 2008). These studies suggest that both Ang-2 and VEGF have a complex spatiotemporal influence on pericyte denudation from the vessels during early phases of angiogenic sprouting.

15.4.2 Basement Membrane Degradation

The vascular basal lamina consists of a meshwork of proteins, including various members of the collagen and the laminin families, thrombospondins, fibronectins, and a variety of proteoglycans and other glycoproteins that self-assemble into organized BM. The BM provides a scaffold that maintains the organization of the ECs into blood vessels (Kalluri 2003). Extracellular matrix (ECM) components of the BM provide stimuli for maintaining EC quiescence and triggering EC proliferation and survival as well as guidance cues for EC migration (Davis and Senger 2005). The BM also presents a physical barrier for new vessel formation—to migrate away from vessels and to form neovascular networks, ECs have to degrade the surrounding ECM. During the process of early angiogenesis, ECs acquire an invasive phenotype characterized by increased expression and secretion of proteases, among which metalloproteinases (MMPs), also called matrixins, are particularly important (Lynch and Matrisian 2002). MMPs are a family of over 20 zinc-containing endopeptidases subdivided into at least five groups based on their structure and/or substrate specificities: matrilysins, collagenases, stromelysins, gelatinases, and the membrane-type MMPs (MT-MMPs). The MT-MMPs are bound to the cell surface and degrade gelatin, fibronectin, aggrecan, and other ECM substrates (Vihinen and Kahari 2002) and play a critical role in EC invasion and lumen formation. Various other MMPs, including MMP-3, MMP-9, and MMP-10, have also been shown to degrade BM matrix components (Heissig et al. 2003). MMPs are generally abundantly expressed in all human cancers, where their expression and activity levels correlate with tumor invasiveness and poor prognosis (Vihinen and Kahari 2002). During the process of BM proteolysis, MMPs release active VEGF from the ECM, expose cryptic pro-angiogenic integrin binding sites in the ECM, and inactivate constitutive inhibitors of EC sprouting such as TIMP-3 (Rundhaug 2005). Pro-angiogenic activity of MMPs is also facilitated through cleavage of both BM perlecan which results in the release of bFGF, and VE-cadherin which loosens endothelial cell-cell adhesion (Rundhaug 2005). However, proteolytic actions of some MMPs could also expose cryptic domains of ECM molecules that exhibit anti-angiogenic properties (e.g., endostatin and angiostatin; Sottile 2004), or cleave the ligand-binding domains of FGFR-1 and uPAR resulting in inhibition of angiogenesis (Koolwijk et al. 2001; Levi et al. 1996; O'Reilly et al. 1999).

15.4.3 Endothelial Cell Migration and Proliferation

After mural cell detachment and BM degradation, ECs of vessels undergoing the early phase of angiogenesis receive and process a variety of stimuli originating from the cellular and acellular microenvironment and respond to these stimuli by proliferation and migration.

After recent discovery that ECs in the sprouting vasculature are phenotypically subdivided into various specialized types, each exhibiting distinct cellular fate specification and defined role in vascular sprouting (De Smet et al. 2009; Horowitz and Simons 2008), a new model of vessel branching has emerged. Such cellular and temporal specification is necessary to maintain the integrity of the vascular network during angiogenic branching. Only specific ECs within the capillary are "tasked" with and capable of initiating angiogenic sprouting. These cells, called tip cells, occupy the leading position during vessel growth, they are highly polarized and project numerous filopodia that react to the VEGF-A gradients and direct migration toward; the angiogenic stimuli (Gerhardt et al. 2003; Ruhrberg et al. 2002). Tip cells have a specific molecular signature characterized, among other features, by the expression of VEGFR-2, VEGFR-3, PDGF-BB, delta-like ligand 4 (Dll4), neuropilin-1, and Unc5b (De Smet et al. 2009; Gerhardt et al. 2003). VEGF initiates the selection and induction of the tip ECs upon binding to VEGFR-2, while at the same time prevents the activation of the neighboring ECs called stalk cells. This lateral inhibition is mediated by Dll4/Notch signaling. Activation of VEGFR-2 by VEGF stimulates the expression of Dll4 in the tip cells (Roca and Adams 2007), which binds to Notch on the neighboring stalk cells, resulting in downregulation of VEGFR-2 and its co-receptor neuropilin-1 (Williams et al. 2006) in stalk cells. This prevents stalk cell transition to the active state, restricting the number of emerging tip cells during the branching process. While tip cells become polarized and project filopodia in the leading front, their rear part maintains contact with trailing stalk cells, thus avoiding the disintegration of the forming vascular structure. VEGFR3 expression, which in the adult is restricted to the lymphatic endothelium, reappears in the filopodia of the tip cells during sprouting (Tammela et al. 2008) and increases vascular branching. In contrast, VEGFR-3 expression is downregulated by Notch in the stalk cells (Tammela et al. 2008).

The migratory tip cells acquire invasive properties and upregulate membrane type-1 matrix metalloproteinase (MT1-MMP/MMP14). EC-mural cell interactions direct MT1-MMP expression to the neovessel tip, which results in partial degradation of the BM at the leading edge of the developing vasculature (Yana et al. 2007). MT1-MMP expression is downregulated in the stalk cells.

Stalk cells, contrary to tip cells, do not extend filopodia (Gerhardt et al. 2003). Whereas tip cells depend on ECM VEGF-A gradient for cell migration, stalk cells proliferate, elongate, form a lumen, and connect to circulation in a VEGF-A concentration-dependent manner (Gerhardt et al. 2003). These cells express specific molecular markers, including Jagged1, Dll1, and Robo4 (Roca and Adams 2007). The role of these factors and the molecular mechanisms involved in cell-cell contacts that stabilize stalk cells still remain unclear.

Recently, a third EC type, phalanx cells, has been described (De Smet et al. 2009). These are the most quiescent type of ECs which form a cobblestone monolayer and align "as in a phalanx formation of the ancient Greek soldiers." They are embedded in a thick BM, covered by pericytes, form a tight barrier, and migrate and proliferate poorly in response to VEGF. Various signaling pathways have been proposed to modulate EC quiescence and low turnover—it is, however, still not clear whether all of them are implicated in the quiescent phenotype of phalanx ECs.

15.4.4 Lumen Formation

ECs initially develop tube-like structures lacking a lumen; as they migrate into the surrounding ECM, they organize into solid cords and form a lumen (Carmeliet 2000). The lumen can be created either by intercellular canalization through membrane apposition of two different ECs or by fusion of intracellular vacuoles, which are formed through pinocytic uptake of plasma membrane (Davis and Camarillo 1996; Egginton and Gerritsen 2003). The latter is currently thought to be the major mechanism of EC lumen formation (Kamei et al. 2006). Lumen formation is regulated by several factors produced in the angiogenic microenvironment. While VEGF121 and VEGF165 increase lumen formation, VEGF189, Ang-1 (Suri et al. 1998) and thrombospondin-1 decrease lumen diameter (Carmeliet 2000). Multiple integrins also participate in EC lumenogenesis; however, their contribution is dependent on the ECM context in which the lumen formation is occurring (Davis, Koh, and Stratman 2007). For example, Bayless and colleagues (2000) have reported the involvement of $\alpha5\beta1$ and $\alpha v\beta3$ integrins in EC lumen formation in a 3D fibrin matrix, whereas integrin $\alpha2\beta1$, Cdc42, Rac1, and MT1-MMP have been shown to coordinate lumenogenesis in 3D collagen matrices (Davis, Koh, and Stratman 2007).

During tumor angiogenesis, an abundant provisional extracellular matrix (i.e., fibrin, fibronectin, and vitronectin) is deposited (Senger 1996), and participates in regulation of neovascular lumenogenesis.

15.4.5 Pericyte Recruitment and Basement Membrane Reconstitution

Once the lumen of new vessels is formed, the process of angiogenesis enters into a late phase. During late phase angiogenesis, mural cells (pericytes and smooth muscle cells) are recruited and integrated into the vascular wall, leading to quiescence of both mural cells and ECs and subsequent BM reconstitution (Ergun et al. 2006). Pericyte recruitment around ECs is regulated by four pathways: platelet-derived growth factor B (PDGFB)/PDGF receptor-β (PDGFR-β), sphingosine-1-phosphate (S1P)/endothelial differentiation sphingolipid G-protein-coupled receptor-1 (EDG1), angiopoietin 1 (Ang-1)/Tie2, and transforming growth factor β type I (TGF-β1) (Chantrain et al. 2006). These receptor/ligand systems promote pericyte proliferation and migration toward ECs (PDGF-B/PDGF-β, Ang-1/Tie-2) (von Tell, Armulik, and Betsholtz 2006), stimulate differentiation of mesenchymal cells into a pericyte-like phenotype (TGF-β/TGF-β receptors) (Hirschi, Rohovsky, and D'Amore 1998), and/or mediate endothelial-pericyte adhesion (S1P/EDG1) (Paik et al. 2004). The EC-pericyte interface contains peg-and-socket contacts rich in N-cadherin, β-catenin-based adherent junctions, which contribute to the transmission of mechanical contractile forces from the pericytes to the endothelium (Gerhardt et al. 2003). Once the pericytes are recruited into the vessel wall, both pericytes and ECs contribute to vessel stabilization, in part by producing ECM components which assemble into a vascular BM (Jain et al. 2007). In parallel, microenvironmental signals are modified to attenuate proteolytic processes. For example, TGF-β1/ALK5 signaling has been shown to prevent the degradation of the provisional matrix around nascent vessels by inducing plasminogen-activator inhibitor-1 (PAI-1) (Chantrain et al. 2006). ECM provides structural and organizational stability for vascular endothelium predominantly through adhesive interaction with integrins on the surface of ECs and pericytes (Davis and Senger 2005). Laminin (likely 10 and 8) interaction with integrins (α6β1 and α3β1) facilitates vessel stabilization (Davis and Senger 2005). While BM degradation during the early phase of angiogenesis provides proliferative cues to ECs, the reconstituted BM matrix provides growth-arrest signals (Kalluri 2003). Intact ECM molecules and/or ECM proteolytic fragments are revealed to resolve angiogenesis; thrombospondins, endostatin, arresten, canstatin, and tumstatin all exhibit anti-angiogenic properties (Sottile 2004).

15.4.6 Vessel Maturation

During late phase angiogenesis, vessel walls and network structures undergo further remodeling processes to establish a functional vascular system. These processes include formation of cell-cell junctions, arterio-venous determination, development of a hierarchical vessel network, and tissue- and organ-specific specialization (Jain 2003). This process is called vessel maturation. Homotypic (EC-EC) and heterotypic (EC-pericyte) junctions and gap junctions are formed to control vessel permeability (Jain 2003). Brain and retinal vessels differ from the peripheral vessels in that they express tight junctions that are organized in a complex way to form strictly controlled blood-brain and blood-retinal barriers (Wolburg and Lippoldt 2002). The arterial-venous fate of ECs is primarily determined through shear-stress influences of blood flow; the Notch pathway and ephrins are implicated in this process (Rossant and Howard 2002). In contrast to physiological angiogenesis, tumor vessels remain structurally abnormal and do not differentiate into macro- and microvasculature; therefore, tumor vessels can achieve stabilization, but do not attain complete maturation (Ergun et al. 2006).

Tumor angiogenesis research has been predominantly focused on early phases of angiogenesis involving EC proliferation and migration. However, better understanding of the late phases of angiogenesis has opened new avenues in research related to tumor vessel stabilization.

Compelling evidence suggests that pericytes are critical for the establishment and maintenance of tumor vessel integrity (Bergers and Song 2005). PDGFR-β and PDGF retention motifs are critical for pericyte recruitment to tumor vessels and their integration into tumor vascular walls, respectively (Abramsson, Lindblom, and Betsholtz 2003). However, the effect of vessel stabilization by pericyte recruitment on tumor growth remains controversial and appears to be dependent on the tissue/tumor context. Some published literature suggests that pericyte recruitment and vessel stabilization support tumor growth and that inhibiting this process may be a viable anti-angiogenic strategy. For example, inhibition of EDG1 by RNA interference in lung cancer results in a strong reduction of pericyte coverage and a dramatic reduction of tumor growth (Chae et al. 2004). Attenuated angiogenesis and reduced tumor growth rate were also achieved by reducing pericyte attachment to tumor vessels with the tyrosine receptor kinase inhibitor SU6668 (Bergers et al. 2003). In contrast, one body of evidence suggests that pericyte recruitment and tumor vessel stabilization reduce tumor growth. For example, over-expression of Ang-1 in several cancer cells results in increased pericyte coverage around microvessels and inhibition of xenograft tumor growth (Hawighorst et al. 2002; Tian et al. 2002). It has been proposed that the judicious application of anti-angiogenic agents can normalize the tumor vessels, providing an opportunity to improve the efficiency of radiation and chemotherapy (Jain 2005). Accumulating experimental and clinical data demonstrated that tumor vessels can be transiently normalized under anti-angiogenic therapy (Ergun et al. 2006; Tong et al. 2004; Winkler et al. 2004); these "normalized" vessels are less permeable, less tortuous, less dilated, and exhibit more normal BM thickness and higher perivascular cell coverage (Tong et al. 2004; Winkler et al. 2004). This normalization of tumor vessels is accompanied by the normalization of the tumor microenvironment, including decreased interstitial fluid pressure, increased tumor oxygenation, and neutralized pH (Tong et al. 2004; Winkler et al. 2004), creating a window of opportunity to improve drug delivery and the efficacy of radiation therapy. Recent clinical trials also support the concept of tumor vessel normalization (Fukumura and Jain 2007). For example, using MRI techniques, Batchelor and colleagues (2007) demonstrated that in recurrent glioblastoma patient treatment with the tyrosine kinase inhibitor for VEGFRs and PDGFRs, AZD2171, causes a transient normalization of tumor vessels. Although tumor vessel size and tumor volume were reduced up to 28 days after the start of AZD2171 treatment, the recurrence of abnormal large vessels was observed at 56 days. Nevertheless, a clinical benefit was observed due to a long-lasting reduction of vasogenic edema and lower corticosteroid use (Batchelor et al. 2007).

The overall effect of vessel stabilization on tumor growth and metastasis remains controversial. Hence, better understanding of mechanisms underlying these processes is required before translating vessel stabilization/normalization or vessel destabilization therapies into the clinic.

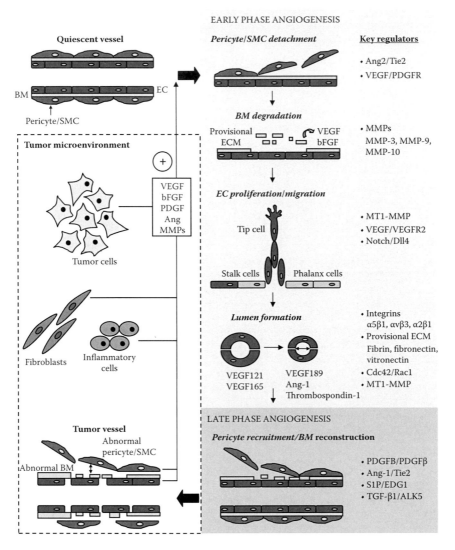

FIGURE 15.1 (See color insert following page 332.) Schematic illustration of the different steps and key modulators involved in the process of tumor angiogenesis. The transition from a prevascular to a vascularized tumor phenotype is tightly regulated by the balance between pro-and anti-angiogenic factors secreted by the tumor cells and/or stromal cells—including endothelial cells, pericytes, smooth muscle cells and fibroblasts, and by infiltrating cells of the immune system. Each phase during this process is specifically modulated by temporal and spatial interactions between the tumor and vascular cell types and the mediators released into the tumor microenvironment. **BM**: basement membrane; **EC**: endothelial cells; **SMC**: smooth muscle cells; **ECM**: extracellular matrix; MMPs: matrix metalloproteinases.

15.5 ANTI-ANGIOGENIC STRATEGIES TARGETING TUMOR MICROENVIRONMENT

Over the past two decades, research on tumor angiogenesis has resulted in the development of anti-angiogenic therapies for cancer treatment. The majority of anti-angiogenic therapies block the function of a specific growth factor or a receptor, the VEGF ligand-receptor system being the target of many anti-angiogenic agents. Bevacizumab, an anti-VEGF-A monoclonal antibody, was the first anti-angiogenic biologic drug approved by the U.S. Food and Drug Administration for the treatment of metastatic colorectal, non-small cell lung, and breast cancers (Ferrara, Hillan, and Novotny 2005). Small molecule tyrosine kinase inhibitors, sorefenib (Nexavar, Bayer) and sunitinib (Sutent, Pfizer), targeting preferentially VEGFR2, have been approved for the treatment of several cancers, especially renal carcinoma (Folkman 2007). However, both preclinical and clinical studies demonstrated that these VEGF pathway inhibitors failed to offer a substantial survival advantage in most patients (Shojaei and Ferrara 2007). Recently, Bergers and Hanahan (2008) postulated two modes of tumor resistance that can be encountered in response to anti-angiogenic therapies: evasive and intrinsic resistance. Evasive resistance was defined as the ability of a tumor to adapt and evade therapeutic effect after an initial response phase through the activation of several mechanisms, including the release of redundant angiogenic mediators by the tumor microenvironment, increase of protective pericyte coverage, and/or activation of alternative types of vascularization (e.g., cooption, lymphangiogenesis, etc.) (Bergers and Hanahan 2008). In contrast, intrinsic resistance is referred to as the absence of any beneficial effect of an anti-angiogenic treatment, probably due to the activation of some resistance mechanisms during premalignant stages (Bergers and Hanahan 2008). Since the resistance to anti-angiogenic therapy is imparted by the tumor microenvironment, current rethinking of anti-angiogenic approaches emphasizes the need for pleiotropic, multifunctional effectors to achieve either a total vessel regression or to promote a complete vessel stabilization. The combination of anti-angiogenic therapy with conventional chemotherapy and radiotherapy is increasingly used since the anti-angiogenic therapy can normalize tumor vessels, thereby enhancing delivery and efficacy of cytotoxic drugs (Jain 2005). combination of anti-invasive and anti-metastatic therapies, that is, drugs targeting hepatocyte growth factor (HGF-MET) (Wang et al. 2003) and insulin-like growth factor (IGF) (Feng and Dimitrov 2008) with anti-angiogenic therapy (Bergers and Hanahan 2008), as well as approaches designed to simultaneously target complex interactions among tumor cells, tumor vasculature, and the surrounding tumor microenvironment will become leading future strategies to develop novel and more efficacious anti-cancer therapies.

REFERENCES

Abramsson A., Lindblom P., and Betsholtz C. 2003. Endothelial and nonendothelial sources of PDGF-B regulate pericyte recruitment and influence vascular pattern formation in tumors. *J. Clin. Invest.* 112:1142–1151.

Acker T., Beck H., and Plate K. H. 2001. Cell type specific expression of vascular endothelial growth factor and angiopoietin-1 and -2 suggests an important role of astrocytes in cerebellar vascularization. *Mechanisms Dev* 108:45–57.

Arap W., Pasqualini R., and Ruoslahti E. 1998. Cancer treatment by targeted drug delivery to tumor vasculature in a mouse model. *Science* 279:377–380.

Asahara T., Chen D., Takahashi T. et al. 1998. Tie2 receptor ligands, angiopoietin-1 and angiopoietin-2, modulate VEGF-induced postnatal neovascularization. *Circul Res* 83:233–240.

Asahara T., Murohara T., Sullivan A. et al. 1997. Isolation of putative progenitor endothelial cells for angiogenesis. *Science* 275:964–967.

Asahara T., Takahashi T., Masuda H. et al. 1999. VEGF contributes to postnatal neovascularization by mobilizing bone marrow-derived endothelial progenitor cells. *EMBO. J.* 18:3964–3972.

Auguste P., Lemiere S., Larrieu-Lahargue F., and Bikfalvi A. 2005. Molecular mechanisms of tumor vascularization. *Crit Rev Oncol Hematol* 54:53–61.

Baluk P., Morikawa S., Haskell A., Mancuso M., and McDonald D.M. 2003. Abnormalities of basement membrane on blood vessels and endothelial sprouts in tumors. *Am. J. Pathol.* 163:1801–1815.

Batchelor T. T., Sorensen A. G., di Tomaso E. et al. 2007. AZD2171, a pan-VEGF receptor tyrosine kinase inhibitor, normalizes tumor vasculature and alleviates edema in glioblastoma patients. *Cancer Cell* 11:83–95.

Bates D. O., Cui T. G., Doughty J. M. et al. 2002. VEGF165b, an inhibitory splice variant of vascular endothelial growth factor, is down-regulated in renal cell carcinoma. *Cancer Res* 62:4123–4131.

Bayless K. J., Salazar R., and Davis G. E. 2000. RGD-dependent vacuolation and lumen formation observed during endothelial cell morphogenesis in three-dimensional fibrin matrices involves the alpha(v)beta(3) and alpha(5)beta(1) integrins. *Am. J. Pathol.* 156:1673–1683.

Bergers G. and Hanahan D. 2008. Modes of resistance to anti-angiogenic therapy. *Nature Rev Cancer* 8:592–603.

Bergers G. and Song S. 2005. The role of pericytes in blood-vessel formation and maintenance. *Neuro-oncol.* 7:452–464.

Bergers G., Song S., Meyer-Morse N., Bergsland E., and Hanahan D. 2003. Benefits of targeting both pericytes and endothelial cells in the tumor vasculature with kinase inhibitors. *J. Clin. Invest.* 111:1287–1295.

Bhati R., Patterson C., Livasy C. A. et al. 2008. Molecular characterization of human breast tumor vascular cells. *Am. J. Pathol.* 172:1381–1390.

Brahimi-Horn M. C., Chiche J., and Pouyssegur J. 2007. Hypoxia and cancer. *J. Mol. Med.* 85:1301–1307.

Brown J. M. and Giaccia A. J. 1998. The unique physiology of solid tumors: opportunities (and problems) for cancer therapy. *Cancer Res* 58:1408–1416.

Burri P. H., Hlushchuk R., and Djonov V. 2004. Intussusceptive angiogenesis: its emergence, its characteristics, and its significance. *Dev Dyn* 231:474–488.

Caduff J. H., Fischer L. C., and Burri P. H. 1986. Scanning electron microscope study of the developing microvasculature in the postnatal rat lung. *Anat Rec* 216:154–164.

Carmeliet P. 2000. Mechanisms of angiogenesis and arteriogenesis. *Nature Med* 6:389–395.

Carmeliet P. and Jain R. K. 2000. Angiogenesis in cancer and other diseases. *Nature* 407:249–257.

Chae S. S., Paik J. H., Furneaux H., and Hla T. 2004. Requirement for sphingosine 1-phosphate receptor-1 in tumor angiogenesis demonstrated by *in vivo* RNA interference. *J. Clin. Invest.* 114:1082–1089.

Chang Y. S., di Tomaso E., and McDonald D. M. et al. 2000. Mosaic blood vessels in tumors: frequency of cancer cells in contact with flowing blood. *Proc Natl Acad Sci USA* 97:14608–14613.

Chantrain C. F., Henriet P., Jodele S. et al. 2006. Mechanisms of pericyte recruitment in tumour angiogenesis: a new role for metalloproteinases. *Eur. J. Cancer.* 42:310–318.

Chen X., Maniotis A. J., Majumdar D., Pe'er J., and Folberg R. 2002. Uveal melanoma cell staining for CD34 and assessment of tumor vascularity. *Invest Ophthalmol Vis Sci* 43:2533–2539.

Cueni L. N. and Detmar M. 2008. The lymphatic system in health and disease. *Lymphat Res Biol* 6:109–122.

Dang C. V. and Semenza G. L. 1999. Oncogenic alterations of metabolism. *Trends Biochem Sci* 24:68–72.

Davis G. E. and Camarillo C. W. 1996. An alpha 2 beta 1 integrin-dependent pinocytic mechanism involving intracellular vacuole formation and coalescence regulates capillary lumen and tube formation in three-dimensional collagen matrix. *Exp Cell Res* 224:39–51.

Davis G. E., Koh W., and Stratman A. N. 2007. Mechanisms controlling human endothelial lumen formation and tube assembly in three-dimensional extracellular matrices. *Birth Defects Res C Embryol Today* 81:270–285.

Davis G. E. and Senger D. R. 2005. Endothelial extracellular matrix: biosynthesis, remodeling, and functions during vascular morphogenesis and neovessel stabilization. *Circul Res* 97:1093–1107.

De Palma M., Venneri M. A., Galli R. et al. 2005. Tie2 identifies a hematopoietic lineage of proangiogenic monocytes required for tumor vessel formation and a mesenchymal population of pericyte progenitors. *Cancer Cell* 8:211–226.

De Smet F., Segura I., De Bock K., Hohensinner P. J., and Carmeliet P. 2009. Mechanisms of vessel branching: filopodia on endothelial tip cells lead the way. *Arteriosclerosis, Thrombosis Vascular Biol* 29: 639–649.

Dewhirst M. W. 1998. Concepts of oxygen transport at the microcirculatory level. *Semin Radiat Oncol* 8:143–150.

di Tomaso E., Capen D., Haskell A. et al. 2005. Mosaic tumor vessels: cellular basis and ultrastructure of focal regions lacking endothelial cell markers. *Cancer Res* 65:5740–5749.

Djonov V., Andres A. C., and Ziemiecki A. 2001. Vascular remodelling during the normal and malignant life cycle of the mammary gland. *Microsc Res Tech* 52:182–189.

Djonov V., Baum O., and Burri P. H. 2003. Vascular remodeling by intussusceptive angiogenesis. *Cell Tissue Res* 314:107–117.

Djonov V., Schmid M., Tschanz S. A., and Burri P. H. 2000. Intussusceptive angiogenesis: its role in embryonic vascular network formation. *Circul Res* 86:286–292.

Dome B., Hendrix M. J., Paku S., Tovari J., and Timar J. 2007. Alternative vascularization mechanisms in cancer: pathology and therapeutic implications. *Am. J. Pathol.* 170:1–15.

Dome B., Paku S., Somlai B., and Timar J. 2002. Vascularization of cutaneous melanoma involves vessel co-option and has clinical significance. *J. Pathol.* 197:355–362.

Dvorak H. F. 2006. Discovery of vascular permeability factor (VPF). *Exp Cell Res* 312:522–526.

Egginton S. and Gerritsen M. 2003. Lumen formation: *in vivo* versus *in vitro* observations. *Microcirculation* 10:45–61.

Ergun S., Tilki D., Oliveira-Ferrer L., Schuch G., and Kilic N. 2006. Significance of vascular stabilization for tumor growth and metastasis. *Cancer Lett* 238:180–187.

Feldser D., Agani F., Iyer N. V. et al. 1999. Reciprocal positive regulation of hypoxia-inducible factor 1alpha and insulin-like growth factor 2. *Cancer Res* 59:3915–3918.

Feng Y. and Dimitrov D. S. 2008. Monoclonal antibodies against components of the IGF system for cancer treatment. *Curr Opin Drug Discov Dev* 11:178–185.

Ferrara N. 2004. Vascular endothelial growth factor: basic science and clinical progress. *Endocr Rev* 25:581–611.

Ferrara N. 2005. The role of VEGF in the regulation of physiological and pathological angiogenesis. *EXS*:209–231.

Ferrara N. and Davis-Smyth T. 1997. The biology of vascular endothelial growth factor. *Endocr Rev* 18:4–25.

Ferrara N., Gerber H. P., and LeCouter J. 2003. The biology of VEGF and its receptors. *Nature Med* 9:669–676.

Ferrara N., Hillan K. J., and Novotny W. 2005. Bevacizumab (Avastin), a humanized anti-VEGF monoclonal antibody for cancer therapy. *Biochem Biophys Res Commun* 333:328–335.

Fisher A. B., Chien S., Barakat A.I., and Nerem R. M. 2001. Endothelial cellular response to altered shear stress. *Am. J. Physiol. Lung. Cell. Mol. Physiol.* 281:L529–L533.

Folkman J. 1971. Tumor angiogenesis: therapeutic implications. *N. Engl. J. Med.* 285:1182–1186.

Folkman J. 2007. Angiogenesis: an organizing principle for drug discovery? *Nature Rev Drug Discov* 6:273–286.

Forsythe J. A., Jiang B. H., Iyer N. V. et al. 1996. Activation of vascular endothelial growth factor gene transcription by hypoxia-inducible factor 1. *Mol Cell Biol* 16:4604–4613.

Fukumura D. and Jain R. K. 2007b. Tumor microvasculature and microenvironment: targets for anti-angiogenesis and normalization. *Microvasc Res* 74:72–84.

Garcia-Barros M., Paris F., Cordon-Cardo C. et al. 2003. Tumor response to radiotherapy regulated by endothelial cell apoptosis. *Science* 300:1155–1159.

Gerhardt H. and Betsholtz C. 2003. Endothelial-pericyte interactions in angiogenesis. *Cell Tissue Res* 314:15–23.

Gerhardt H. and Golding M., Fruttiger M. et al. 2003. VEGF guides angiogenic sprouting utilizing endothelial tip cell filopodia. *J. Cell. Biol.* 161:1163–1177.

Greenberg J. I., Shields D. J., Barillas S. G. et al. 2008. A role for VEGF as a negative regulator of pericyte function and vessel maturation. *Nature* 456:809–813.

Hanahan D. and Folkman J. 1996. Patterns and emerging mechanisms of the angiogenic switch during tumorigenesis. *Cell* 86:353–364.

Harris A. L. 2002. Hypoxia: a key regulatory factor in tumour growth. *Nature Rev Cancer* 2:38–47.

Hashizume H., Baluk P., Morikawa S. et al. 2000. Openings between defective endothelial cells explain tumor vessel leakiness. *Am. J. Pathol.* 156:1363–1380.

Hattori K., Heissig B., Wu Y. et al. 2002. Placental growth factor reconstitutes hematopoiesis by recruiting VEGFR1(+) stem cells from bone-marrow microenvironment. *Nature Med* 8:841–849.

Hawighorst T., Skobe M., Streit M. et al. 2002. Activation of the tie2 receptor by angiopoietin-1 enhances tumor vessel maturation and impairs squamous cell carcinoma growth. *Am. J. Pathol.* 160:1381–1392.

Heissig B., Hattori K., Dias S. et al. 2002. Recruitment of stem and progenitor cells from the bone marrow niche requires MMP-9 mediated release of kit-ligand. *Cell* 109:625–637.

Heissig B., Hattori K., Friedrich M., Rafii S., and Werb Z. 2003. Angiogenesis: vascular remodeling of the extracellular matrix involves metalloproteinases. *Current Opin Hematol* 10:136–141.

Hendrix M. J., Seftor E. A., Hess A. R., and Seftor R. E. 2003. Vasculogenic mimicry and tumour-cell plasticity: lessons from melanoma. *Nature Rev Cancer* 3:411–421.

Hess A. R., Margaryan N. V., Seftor E. A., and Hendrix M. J. 2007. Deciphering the signaling events that promote melanoma tumor cell vasculogenic mimicry and their link to embryonic vasculogenesis: role of the Eph receptors. *Dev Dyn* 236:3283–3296.

Hess A. R., Seftor E. A., Gruman L. M. et al. 2006. VE-cadherin regulates EphA2 in aggressive melanoma cells through a novel signaling pathway: implications for vasculogenic mimicry. *Cancer Biol Ther* 5:228–233.

Hess A. R., Seftor E. A., Seftor R. E., and Hendrix M. J. 2003. Phosphoinositide 3-kinase regulates membrane type 1-matrix metalloproteinase (MMP) and MMP-2 activity during melanoma cell vasculogenic mimicry. *Cancer Res* 63:4757–4762.

Hirschi K. K., Rohovsky S. A., and D'Amore P. A. 1998. PDGF, TGF-beta, and heterotypic cell-cell interactions mediate endothelial cell-induced recruitment of 10T1/2 cells and their differentiation to a smooth muscle fate. *J. Cell. Biol.* 141:805–814.

Hlushchuk R., Riesterer O., Baum O. et al. 2008. Tumor recovery by angiogenic switch from sprouting to intussusceptive angiogenesis after treatment with PTK787/ZK222584 or ionizing radiation. *Am. J. Pathol.* 173:1173–1185.

Holash J., Maisonpierre P.C., Compton D. et al. 1999. Vessel cooption, regression, and growth in tumors mediated by angiopoietins and VEGF. *Science* 284:1994–1998.

Horowitz A. and Simons M. 2008. Branching morphogenesis. *Circul Res* 103:784–795.

Jain R. K. 2003. Molecular regulation of vessel maturation. *Nature Med* 9:685–693.

Jain R. K. 2005. Normalization of tumor vasculature: an emerging concept in anti-angiogenic therapy. *Science* 307:58–62.

Jain R. K., di Tomaso E., Duda D.G. et al. 2007. Angiogenesis in brain tumours. *Nature Rev Neurosci* 8:610–622.

Jin H., Aiyer A., Su J. et al. 2006. A homing mechanism for bone marrow-derived progenitor cell recruitment to the neovasculature. *J. Clin. Invest.* 116:652–662.

Jung Y. D., Ahmad S. A., Liu W. et al. 2002. The role of the microenvironment and intercellular crosstalk in tumor angiogenesis. *Semin Cancer Biol* 12:105–112.

Kalluri R. 2003. Basement membranes: structure, assembly and role in tumour angiogenesis. *Nature Rev Cancer* 3:422–433.

Kamei M., Saunders W. B., Bayless K. J. et al. 2006. Endothelial tubes assemble from intracellular vacuoles *in vivo*. *Nature* 442:453–456.

Konopatskaya O., Churchill A. J., Harper S.J., Bates D. O., and Gardiner T. A. 2006. VEGF165b, an endogenous C-terminal splice variant of VEGF, inhibits retinal neovascularization in mice. *Mol Vis* 12:626–632.

Koolwijk P., Sidenius N., Peters E. et al. 2001. Proteolysis of the urokinase-type plasminogen activator receptor by metalloproteinase-12: implication for angiogenesis in fibrin matrices. *Blood* 97:3123–3131.

Lal A., Peters H., St. Croix B. et al. 2001. Transcriptional response to hypoxia in human tumors. *J. Natl. Cancer. Inst.* 93:1337–1343.

Lauren J., Gunji Y., and Alitalo K. 1998. Is angiopoietin-2 necessary for the initiation of tumor angiogenesis? *Am. J. Pathol.* 153:1333–1339.

Leenders W. P., Kusters B., and de Waal R. M. 2002. Vessel co-option: how tumors obtain blood supply in the absence of sprouting angiogenesis. *Endothelium* 9:83–87.

Levi E., Fridman R., Miao H. Q. et al. 1996. Matrix metalloproteinase 2 releases active soluble ectodomain of fibroblast growth factor receptor 1. *Proc Natl Acad Sci USA* 93:7069–7074.

Lyden D., Hattori K., Dias S. et al. 2001. Impaired recruitment of bone-marrow-derived endothelial and hematopoietic precursor cells blocks tumor angiogenesis and growth. *Nature Med* 7:1194–1201.

Lynch C. C. and Matrisian L. M. 2002. Matrix metalloproteinases in tumor-host cell communication. *Differentiation Res Biol Diversity* 70:561–573.

Machein M. R., Renninger S., de Lima-Hahn E., and Plate K.H. 2003. Minor contribution of bone marrow-derived endothelial progenitors to the vascularization of murine gliomas. *Brain Pathol* 13:582–597.

Maeng Y. S., Choi H. J., Kwon J. Y. et al. 2009. Endothelial progenitor cell homing: prominent role of the IGF2-IGF2R-PLCbeta2 axis. *Blood* 113:233–243.

Maisonpierre P. C., Goldfarb M., Yancopoulos G. D., and Gao G. 1993. Distinct rat genes with related profiles of expression define a TIE receptor tyrosine kinase family. *Oncogene* 8:1631–1637.

Maisonpierre P. C., Suri C., Jones P. F. et al. 1997. Angiopoietin-2, a natural antagonist for Tie2 that disrupts *in vivo* angiogenesis. *Science (New York)* 277:55–60.

Maniotis A. J., Folberg R., Hess A. et al. 1999. Vascular channel formation by human melanoma cells *in vivo* and *in vitro*: vasculogenic mimicry. *Am. J. Pathol.* 155:739–752.

Moore M. A., Hattori K., Heissig B. et al. 2001. Mobilization of endothelial and hematopoietic stem and progenitor cells by adenovector-mediated elevation of serum levels of SDF-1, VEGF, and angiopoietin-1. *Ann NY Acad Sci* 938:36–45; discussion 45–47.

Morikawa S., Baluk P., Kaidoh T. et al. 2002. Abnormalities in pericytes on blood vessels and endothelial sprouts in tumors. *Am. J. Pathol.* 160:985–1000.

Nikitenko L. L. 2009. Vascular endothelium in cancer. *Cell Tissue Res* 335:223–240.

Nolan D. J., Ciarrocchi A., Mellick A. S. et al. 2007. Bone marrow-derived endothelial progenitor cells are a major determinant of nascent tumor neovascularization. *Genes Dev* 21:1546–1558.

O'Reilly M. S., Wiederschain D., Stetler-Stevenson W. G., Folkman J., and Moses M. A. 1999. Regulation of angiostatin production by matrix metalloproteinase-2 in a model of concomitant resistance. *J. Biol. Chem.* 274:29568–29571.

Oh H., Takagi H., Suzuma K. et al. 1999. Hypoxia and vascular endothelial growth factor selectively up-regulate angiopoietin-2 in bovine microvascular endothelial cells. *J. Biol. Chem.* 274:15732–15739.

Oh P., Li Y., Yu J. et al. 2004. Subtractive proteomic mapping of the endothelial surface in lung and solid tumours for tissue-specific therapy. *Nature* 429:629–635.

Orimo A., Gupta P. B., Sgroi D. C. et al. 2005. Stromal fibroblasts present in invasive human breast carcinomas promote tumor growth and angiogenesis through elevated SDF-1/CXCL12 secretion. *Cell* 121:335–348.

Paik J. H., Skoura A., Chae S.S. et al. 2004. Sphingosine 1-phosphate receptor regulation of N-cadherin mediates vascular stabilization. *Genes Dev* 18:2392–2403.

Patan S., Alvarez M. J., Schittny J. C., and Burri P. H. 1992. Intussusceptive microvascular growth: a common alternative to capillary sprouting. *Arch Histol Cytol* 55 (Suppl:)65–75.

Patan S., Haenni B., and Burri P. H. 1993. Evidence for intussusceptive capillary growth in the chicken chorio-allantoic membrane (CAM). *Anat Embryol (Berlin)* 187:121–130.

Patan S., Munn L. L., and Jain R. K. 1996. Intussusceptive microvascular growth in a human colon adenocarcinoma xenograft: a novel mechanism of tumor angiogenesis. *Microvasc Res* 51:260–272.

Pen A., Moreno M. J., Durocher Y., Deb-Rinker P., and Stanimirovic D. B. 2008. Glioblastoma-secreted factors induce IGFBP7 and angiogenesis by modulating Smad-2-dependent TGF-beta signaling. *Oncogene* 27:6834–6844.

Pen A., Moreno M. J., Martin J., and Stanimirovic D. B. 2007. Molecular markers of extracellular matrix remodeling in glioblastoma vessels: microarray study of laser-captured glioblastoma vessels. *Glia* 55:559–572.

Pepper M. S. 1997. Transforming growth factor-beta: vasculogenesis, angiogenesis, and vessel wall integrity. *Cytokine Growth Factor Rev* 8:21–43.

Pepper M. S., Tille J. C., Nisato R., and Skobe M. 2003. Lymphangiogenesis and tumor metastasis. *Cell Tissue Res* 314:167–177.

Peters B. A., Diaz L. A., Polyak K. et al. 2005. Contribution of bone marrow-derived endothelial cells to human tumor vasculature. *Nature Med* 11:261–262.

Petty A. P., Garman K. L., Winn V. D., Spidel C. M., and Lindsey J.S. 2007. Overexpression of carcinoma and embryonic cytotrophoblast cell-specific Mig-7 induces invasion and vessel-like structure formation. *Am. J. Pathol.* 170:1763–1780.

Pugh C. W. and Ratcliffe P. J. 2003. Regulation of angiogenesis by hypoxia: role of the HIF system. *Nature Med* 9:677–684.

Ravi R., Mookerjee B., Bhujwalla Z. M. et al. 2000. Regulation of tumor angiogenesis by p53-induced degradation of hypoxia-inducible factor 1alpha. *Genes Dev* 14:34–44.

Rennel E. S., Hamdollah-Zadeh M. A., Wheatley E. R. et al. 2008. Recombinant human VEGF165b protein is an effective anti-cancer agent in mice. *Eur. J. Cancer.* 44:1883–1894.

Roca C. and Adams R. H. 2007. Regulation of vascular morphogenesis by Notch signaling. *Genes Dev* 21:2511–2524.

Rossant J. and Howard L. 2002. Signaling pathways in vascular development. *Annu Rev Cell Dev Biol* 18:541–573.

Ruhrberg C., Gerhardt H., Golding M. et al. 2002. Spatially restricted patterning cues provided by heparin-binding VEGF-A control blood vessel branching morphogenesis. *Genes Dev* 16:2684–2698.

Rundhaug J. E. 2005. Matrix metalloproteinases and angiogenesis. *J. Cell. Mol. Med* 9:267–285.

Sandau K. B., Zhou J., Kietzmann T., and Brune B. 2001. Regulation of the hypoxia-inducible factor 1alpha by the inflammatory mediators nitric oxide and tumor necrosis factor-alpha in contrast to desferroxamine and phenylarsine oxide. *J. Biol. Chem.* 276:39805–39811.

Seaman S., Stevens J., Yang M. Y. et al. 2007. Genes that distinguish physiological and pathological angiogenesis. *Cancer Cell* 11:539–554.

Semenza G. L. 2003. Targeting HIF-1 for cancer therapy. *Nature Rev Cancer* 3:721–732.

Senger D. R. 1996. Molecular framework for angiogenesis: a complex web of interactions between extravasated plasma proteins and endothelial cell proteins induced by angiogenic cytokines. *Am. J. Pathol.* 149:1–7.

Shirakawa K., Kobayashi H., Heike Y. et al. 2002. Hemodynamics in vasculogenic mimicry and angiogenesis of inflammatory breast cancer xenograft. *Cancer Res* 62:560–566.

Shojaei F. and Ferrara N. 2007. Anti-angiogenic therapy for cancer: an update. *Cancer. J.*13:345–348.

Sottile J. 2004. Regulation of angiogenesis by extracellular matrix. *Biochim Biophys Acta* 1654:13–22.

St. Croix B., Rago C., Velculescu V. et al. 2000. Genes expressed in human tumor endothelium. *Science* 289:1197–1202.

Stacker S. A., Achen M. G., Jussila L., Baldwin M.E., and Alitalo K. 2002. Lymphangiogenesis and cancer metastasis. *Nature Rev Cancer* 2:573–583.

Suri C., McClain J., Thurston G. et al. 1998. Increased vascularization in mice overexpressing angiopoietin-1. *Science* 282:468–471.

Takahashi T., Kalka C., Masuda H. et al. 1999. Ischemia- and cytokine-induced mobilization of bone marrow-derived endothelial progenitor cells for neovascularization. *Nature Med* 5:434–438.

Tammela T., Zarkada G., Wallgard E. et al. 2008. Blocking VEGFR-3 suppresses angiogenic sprouting and vascular network formation. *Nature* 454:656–660.

Thornton R. D., Lane P., Borghaei R. C. et al. 2000. Interleukin 1 induces hypoxia-inducible factor 1 in human gingival and synovial fibroblasts. *Biochem. J.* 350 (Pt 1):307–312.

Thurston G., Suri C., Smith K. et al. 1999. Leakage-resistant blood vessels in mice transgenically overexpressing angiopoietin-1. *Science* 286:2511–2514.

Tian S., Hayes A. J., Metheny-Barlow L. J., and Li L. Y. 2002. Stabilization of breast cancer xenograft tumour neovasculature by angiopoietin-1. *Br. J. Cancer.* 86:645–651.

Tong R. T., Boucher Y., Kozin S. V. et al. 2004. Vascular normalization by vascular endothelial growth factor receptor 2 blockade induces a pressure gradient across the vasculature and improves drug penetration in tumors. *Cancer Res* 64:3731–3736.

Vajkoczy P., Blum S., Lamparter M. et al. 2003. Multistep nature of microvascular recruitment of ex vivo-expanded embryonic endothelial progenitor cells during tumor angiogenesis. *J. Exp. Med.* 197:1755–1765.

Vaupel P. and Mayer A. 2007. Hypoxia in cancer: significance and impact on clinical outcome. *Cancer Metastasis Rev* 26:225–239.

Vihinen P. and Kahari V. M. 2002. Matrix metalloproteinases in cancer: prognostic markers and therapeutic targets. *Int. J. Cancer.* 99:157–166.

von Tell D., Armulik A., and Betsholtz C. 2006. Pericytes and vascular stability. *Exp Cell Res* 312:623–629.

Wang X., Le P., Liang C. et al. 2003. Potent and selective inhibitors of the Met [hepatocyte growth factor/scatter factor (HGF/SF) receptor] tyrosine kinase block HGF/SF-induced tumor cell growth and invasion. *Mol Cancer Ther* 2:1085–1092.

Williams C. K., Li J.L., Murga M., Harris A. L., and Tosato G. 2006. Up-regulation of the Notch ligand Delta-like 4 inhibits VEGF-induced endothelial cell function. *Blood* 107:931–939.

Winkler F., Kozin S. V., Tong R. T. et al. 2004. Kinetics of vascular normalization by VEGFR2 blockade governs brain tumor response to radiation: role of oxygenation, angiopoietin-1, and matrix metalloproteinases. *Cancer Cell* 6:553–563.

Wolburg H. and Lippoldt A. 2002. Tight junctions of the blood-brain barrier: development, composition and regulation. *Vascul Pharmacol* 38:323–337.

Yana I., Sagara H., Takaki S. et al. 2007. Crosstalk between neovessels and mural cells directs the site-specific expression of MT1-MMP to endothelial tip cells. *J. Cell. Sci.* 120:1607–1614.

Yuan H.T., Yang S.P., and Woolf A.S. 2000. Hypoxia up-regulates angiopoietin-2, a Tie-2 ligand, in mouse mesangial cells. *Kidney Int* 58:1912–1919.

Zhong H., Chiles K., Feldser D. et al. 2000. Modulation of hypoxia-inducible factor 1alpha expression by the epidermal growth factor/phosphatidylinositol 3-kinase/PTEN/AKT/FRAP pathway in human prostate cancer cells: implications for tumor angiogenesis and therapeutics. *Cancer Res* 60:1541–1545.

Zundel W., Schindler C., Haas-Kogan D. et al. 2000. Loss of PTEN facilitates HIF-1-mediated gene expression. *Genes Dev* 14:391–396.

III

Data Resources and Software Tools
for Cancer Systems Biology

CHAPTER 16

Modeling Tools for Cancer Systems Biology

Wayne Materi and David S. Wishart

CONTENTS

16.1 INTRODUCTION

The complexity of cancer as a disease state is a reflection of the inherent complexity of the molecular interactions that govern cell metabolism, growth, and division. This is further complicated by the underlying heterogeneity of tumor tissue and the dependence of tumor growth on surrounding support cells (Burkert, Wright, and Alison 2006). Therefore, understanding the initiation, growth, and spread of cancer requires the use of mathematical and computational tools that can help researchers visualize and model these complex interactions both within and between cells. Modeling cancer is not new. Efforts aimed at mathematically modeling certain features of tumor growth date back to the 1920s but it is Burton who is largely credited with developing one of the first accurate tumor models (Burton 1966). His work not only explained the observed distribution of oxygen in a solid tumor (i.e., the necrotic core), but also the characteristic Gompertzian growth curve seen in solid tumors. Subsequent efforts in the 1970s and 1980s focused on modeling tumor invasion and metastasis using simplified cellular diffusion models or the examination of mechanical stress on tumor shape (Araujo and McElwain 2004). With the increasing availability of computers and with our improved understanding of molecular biology, cancer modeling during the late 1990s evolved to have a greater focus on simulating (albeit crudely) the molecular etiology of different cancers. Over the past decade, with the explosion of quantitative data coming from genomic, proteomic, and metabolomic experiments along with equally impressive data coming from advanced cellular and tissue imaging techniques, we have now reached the point where far more realistic computational modeling of both the molecular and cellular events (i.e., the systems biology) in tumor development is becoming a possibility (Alberghina Chiaradonna, and Vanoni 2004; Bugrim, Nikolskaya, and Nikolsky 2004; Hollywood, Brison, and Goodacre 2006; Ideker et al. 2001; Jares 2006).

Realistic cancer models require substantial detailed input information including gene, protein, and metabolite names, concentrations, locations, reactions, rate constants, and pathway connectivity. They may also require information about cell types, cell phases, cell dimensions, tumor dimensions, tumor mechanics, and cellular interactions. These modeling parameters or descriptors can often number in the hundreds or even the thousands. Therefore, modern approaches to simulating biological systems not only require high-end computers, but they also require a machine-readable language to translate the model descriptors into something that can be computationally processed. Many of today's high-end biological models are based on a standardized machine-readable language called the Systems Biology Markup Language (SBML; http://sbml.org) first proposed by Hucka, Finney and others (Finney and Hucka 2003; Hucka et al. 2003). SBML is a simple language for describing biological networks of chemical reactions, compartments, molecular species, parameters, and rules based on the widely accepted XML (eXtensible Markup Language) standard (Webb and White 2005). CellML is an example of another XML-based language, developed through the International Union of Physiological Sciences (IUPS) Human Physiome Project (Hunter, Robbins, and Noble 2002; Lloyd, Halstead, and Nielsen 2004). CellML models networks of interconnected components whose behavior is described by mathematical equations written in Content MathML. These features make CellML particularly amenable to modeling electrophysiological systems, though it readily incorporates chemical reactions and gene networks (Garny et al. 2008). However, while the SBML and CellML formats have proven to be both useful and popular, biological processes can be represented independent of these formats or even of their theoretical basis in ordinary differential equations, as discussed below.

In order to be generally useful to researchers or clinicians, simulations (computer models) of cancer need to provide at least some of the following: (1) a sound biological basis for the model, (2) an improved ability to visualize or represent complex processes, (3) verification of the model with previous data, (4) the ability to extrapolate or predict future experimental results, (5) the ability to identify missing biological components or processes, and (6) the capacity to perform in silico experiments to save time and expense. In this chapter we will focus on describing some of the leading theoretical approaches and software tools that meet many of these criteria. While we hope that this chapter provides a salient overview of many of the principles of computational modeling, a number of other excellent texts should be consulted as a general introduction to the field and for more detailed mathematical derivations of specific formulas (Alon 2007; Bolouri 2008; Demin and Goryanin

2009; Wilkinson 2006). The purpose of this chapter is to review some of the most common approaches to computational modeling of cancer, to discuss important issues in modeling, and to help the reader select from the wide variety of software available for modeling biological processes.

16.1.1 Basics of Biological Modeling

Biological modeling can be done at many different levels (nano-scale, micro-scale or macro-scale) for many different purposes (quantitative, qualitative, predictive, or explanatory) and applications (physiological, clinical, epidemiological, or ecological). Regardless of the scale, purpose, or application, the design and construction of a model should include the following steps: (1) converting biological knowledge to a formal representation, (2) translating this formal description to a selected mathematical or algorithmic form, (3) parameterizing this model using data from literature searches, databases, or experiments, (4) verifying the model by comparing to previous data, (5) performing an analysis of model robustness and sensitivity, (6) simulating the desired conditions and comparing to experimental results, and (7) refining and iterating. Building models of complex biological systems like cancer is an iterative process and no single source of data or single modeling approach will suffice to capture all the subtlety of the system. Several approaches to constructing any particular model are possible, depending on the availability of supporting quantitative and qualitative data and on the specificity of the desired conclusions from the simulation. This chapter will focus on the first two steps, so the reader should consult the above-mentioned texts for more details on the latter steps.

The relevant molecular components of the model and the structure of the genetic or protein-protein interaction network may be discovered through manual or automated surveys of existing literature and electronic databases (Arakawa et al. 2006; Feist et al. 2009; Wang, Lenferink, and O'Connor-McCourt 2007). Most modeling researchers are simultaneously confronted with an overwhelming quantity of data of one type and a frustrating lack of data of another type. For example, while comparative expression data (e.g., from microarray studies) are frequently available, reliable and relevant measures of such parameters as promoter strength, enzyme rate constants, diffusion constants, and protein concentrations can be difficult to find. Because of the paucity of some key data, many parameters may require estimation followed by subsequent refinement through multiple iterations of the model (Ideker et al. 2001b; Kunkel et al. 2004).

16.1.2 Graphical Representations of Biological Processes

Models are intended to accurately reflect actual biological processes and so the first step in developing a model is to completely and unambiguously define the selected processes. At the subcellular (nanoscale) level, these processes may involve millions or even billions of molecules, while similar numbers of cells may be involved in creating a cell-level (microscale) or tissue-level (macroscale) model. In all cases, it is important to capture the biologically or chemically relevant information such as metabolic reactions, genetic networks, signal transduction cascades, and intercellular interactions but it is not absolutely necessary during the process definition stage to assign quantitative values to any of these processes. Also, while it might be tempting for some to directly write out a series of ordinary differential equations (ODEs) to describe the various chemical reactions in the system, we strongly recommend that at least some attempt should always be made to produce a graphical layout of the system as an initial step.

Depending on the processes being modeled, a variety of formal or informal representations can be used to clarify the essential mechanisms. Informal representations (e.g., hand-drawn sketches) can be productively utilized to represent simple systems with few interacting components. Surprisingly insightful conclusions can often result upon conversion of such simple representations into mathematical or algorithmic form. For example, a relatively simple model based on Hanahan and Weinberg's "hallmarks of cancer" perspective (Hanahan and Weinberg 2000) analyzed the relative contribution of genetic instability to tumor growth compared to other important factors. Although genetic instability had not been included as a hallmark in Hanahan and Weinberg's original article, it was thought to be an important potentiating factor by many researchers. A $100 \times 100 \times 100$ grid, representing a million possible cells, was initialized with a single cell, a nutrient supply stream, a limited growth factor supply, and some simple rules governing cell division and mutational acquisition (Spencer et al. 2006). Cells could acquire mutations in genetic stability as well as in most of the Hanahan and Weinberg hallmarks (sensitivity to growth inhibitory signals, evasion of apoptosis, replication control, self-sufficiency of growth, and sustained angiogenic signaling). The simulation results demonstrated that genetic instability dominated the growth of early onset tumors while later growth was driven by the acquisition of limitless growth mutations.

For systems with a larger number of interacting components, formal graphical notation systems can be helpful in clearly describing the relevant biology. Several competing, canonical systems have been developed over the past decade to represent biological processes at different levels of detail and software has been written to facilitate the use of many of these (Table 16.1). The molecular interaction map (MIM) notation developed by Kohn and colleagues, the process diagram of Kitano and colleagues, and the Edinburgh Pathway Notation (EPN) of Goryanin and colleagues are among the most biologically complete graphical notation systems, providing symbols for most types of biomolecules and biologically relevant reactions (Kitano et al. 2005; Kohn et al. 2006b; Moodie et al. 2006). In addition to developing these graphical annotation systems, Kitano's group has developed the CellDesigner software to assist in the production of process diagrams and signaling pathways (Oda et al. 2005), while Goryanin's group has developed the Edinburgh Pathway Editor software package (Raza et al. 2008). Although Kohn's group has not implemented a computer program to support the construction of MIMs, it has been used in depicting complex pathways (Kohn et al. 2006a) and the CADLIVE program permits the construction of a slightly modified MIM-based diagram (Kurata et al. 2007).

Formal graphical notations are typically composed of nodes and directed arcs. The nodes represent biological components (or chemical species) such as proteins, metabolites, or genes, while the directed arcs represent interactions between the nodes. Most popular formalisms may be broken down into three major categories: (1) those that use a single generic node symbol for all species, (2) those that provide different node symbols for different

TABLE 16.1   Graphical and Mathematical Representation Formalisms

Formalism	Program	URL and Reference	Simulation Approach[a]
Indistinguished nodes	Edinburgh Pathway Editor (EPE)	http://www.csbe.ed.ac.uk/epe.php (Moodie et al. 2006)	ODE
	JDesigner	http://www.sys-bio.org/software/jdesigner.htm (Sauro et al. 2003)	ODE
	VCell	http://www.vcell.org (Loew and Schaff 2001)	Hybrid (ODE/Algebraic)
Distinguished nodes	BioUML	http://www.biouml.org/ (Kolpakov et al. 2007)	ODE
	Cell Illustrator	http://www.cellillustrator.com (Peleg et al. 2005)	FHPN
	CellWare	http://www.cellware.org (Dhar et al. 2004)	ODE
	CPN Tools	http://wiki.daimi.au.dk/cpntools/cpntools.wiki (Lee et al. 2006)	CPN
	Dizzy	http://magnet.systemsbiology.net/software/Dizzy/ (Ramsey et al. 2005)	ODE
	SmartCell	http://smartcell.embl.de/ (Ander et al. 2004)	ODE
Complex symbology	CADLIVE (MIM)	http://www.cadlive.jp/ (Kurata et al. 2007)	Hybrid (ODE/Algebraic)
	Cell Designer (Process Diagram)	http://www.celldesigner.org/index.html (Kitano et al. 2005)	ODE
	EPE 2.0 (SBGN)	http://www.csbe.ed.ac.uk/epe.php (Moodie et al. 2006)	ODE
	SimCell	http://wishart.biology.ualberta.ca/SimCell/ (Wishart et al. 2005)	DCA
Mathematical/algorithmic	Cell+ +	http://theileria.ccb.sickkids.ca/CellSim (Sanford et al. 2006)	DCA
	CellML	http://www.cellml.org/ (Bhalla and Ravi Iyengar 1999)	Hybrid (ODE/Algebraic)
	Dynetica	http://www.duke.edu/~you/Dynetica_page.htm (You et al. 2003)	ODE

[a]   ABM, agent based model; CPN, colored Petri net; DCA, dynamic cellular automata; FHPN, functional hybrid Petri net; ODE, ordinary differential equations.

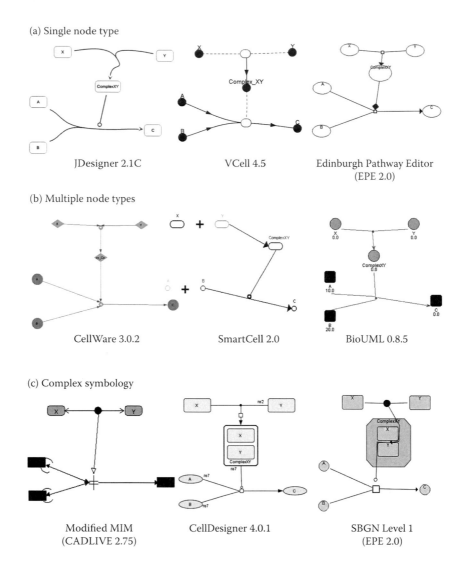

FIGURE 16.1 Various graphical representations by selected software of a basic biochemical reaction. Protein X and Protein Y interact noncovalently to form a protein complex that catalyzes the irreversible, covalent conversion of metabolites A and B into C. Three different classifications of diagrams are shown: single node types, multiple node types, and complex symbologies.

species, and (3) those that provide a complex symbology for different nodes, reactions, and interactions. Figure 16.1 illustrates the graphical notations for a simple biological reaction using several common formalisms. These notation systems may also be differentiated by their approach to specifying biochemical reactions between components. Some notation systems (e.g., JDesigner) employ different types of arcs to distinguish between different types of reactions, such as protein complex formation or catalysis. Other formalisms (e.g., CellDesigner, Edinburgh Pathway Editor) encapsulate reaction details within special purpose nodes and use different types of arcs only to show the way in which different species participate in the reaction (e.g., reactant, product, or catalyst).

Because of the lack of accepted international standards for detailed graphical notations, several groups have recently combined their efforts to produce the Systems Biology Graphical Notation (SBGN) standard (http://www.sbgn.org). The Edinburgh Pathway Editor (EPE 2.0) is capable of utilizing the SBGN formalism and work is under way to support SBGN within CellDesigner as well (Kitano et al. 2005).

Most of the graphical notation systems shown in Figure 16.1 may be used independently of any particular modeling approach but many of them either come with an integrated simulation engine or assume a particular simulator will be used. For example, JDesigner 2 is included in the Systems Biology Workbench (SBW) and is compatible with SBML while VCell, CellWare, BioUML, and SimCell use their own proprietary internal languages (though they can import and export SBML and/or CellML files). CellDesigner, EPE 2.0, and SmartCell are SBML compliant.

Different graphical notation systems have different representative capabilities and so the selection of any particular formalism will depend to some extent on the selected simulation method and the desired level of modeling detail. This will be based on the experience of the researcher and the purpose of the simulation as well as the availability of appropriate parametric data in the literature and databases. More detailed models may require detailed kinetic data on metabolic reaction rates, while less detailed approaches may only need the general topology of a select subset of genetic interactions. In general, the more detailed the molecular interactions described by the model, the greater the amount of quantitative data that will be required for an accurate simulation—and the less likely it is to be available. Figure 16.2 demonstrates the ways in which different notation systems represent a moderately complex signal transduction pathway with differing levels of detail.

16.2 APPROACHES TO COMPUTATIONAL MODELING

In the second stage of developing a computational model of cancer, the researcher must select a specific machine-interpretable approach to the simulation and then transform the formal description of the underlying biological processes into the appropriate format. The specific approach chosen for this step will depend entirely on decisions made earlier about the amount and quality of the quantitative data available to describe component interactions and on the detail required of the model's predictions. Three general approaches are available, including:

1. Discrete stochastic approaches, such as dynamic cellular automata (DCM), agent-based models (ABMs), or Petri nets

2. Continuous deterministic or stochastic approaches, such as systems of ordinary differential equations (ODEs), partial differential equations (PDEs), or colored Petri nets (CPNs)

3. Network approaches, such as gene regulatory networks (GRNs) or Boolean networks (BNs)

These three approaches require different types and different quantities of molecular interaction data to implement a model. For example, discrete stochastic approaches require

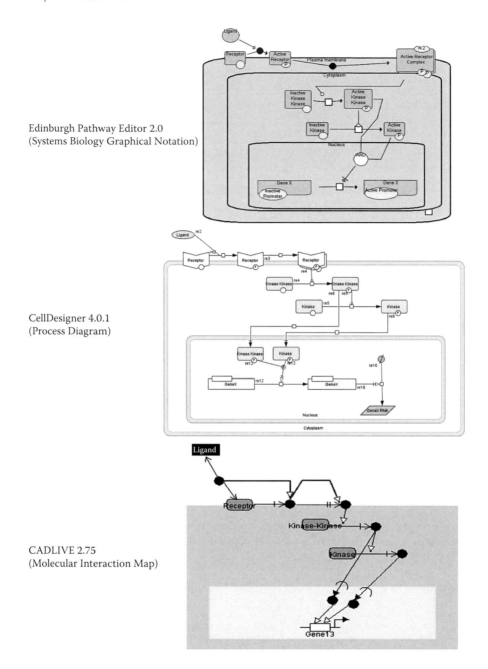

Edinburgh Pathway Editor 2.0
(Systems Biology Graphical Notation)

CellDesigner 4.0.1
(Process Diagram)

CADLIVE 2.75
(Molecular Interaction Map)

FIGURE 16.2 Some selected graphical representations of a signal transduction cascade, showing compartmentalization, phosphorylation, and gene transcription. The extra-cellular ligand binds to a plasma-membrane-bound receptor which autophosphorylates then homo-dimerizes. The activated complex then catalyzes the phosphorylation of a cytoplasmic kinase-kinase which subsequently activates a kinase. Both the kinase-kinase and the kinase translocate into the nucleus where they activate the transcription of Gene X. These diagrams demonstrate the expressive power of these formalisms. The EPE 2.0 diagram uses the SBGN notation while CADLIVE 2.75 uses a modification of the MIM notation. The CellDesigner notation is proprietary though closely related to SBGN.

precise data about localized concentration, diffusion rates, and probability of reaction upon molecular contact. By contrast, mass action kinetic reaction rates found in continuous approaches (e.g., ODEs and PDEs) can be shown to be statistically emergent properties of large collections of discrete molecules and have reduced precision compared to single-molecule parameters, especially for species with very low concentrations. Network approaches only require data specifying the presence of macromolecular interactions and the direction of the dependency relationships, though they may be supplemented with threshold activation and inhibition parameters as well. In most cases, less detailed data will be more easily obtained so, for example, interaction networks will be more available than kinetic reaction rates, which will be more available than localized macromolecular concentrations. Thus, modelers frequently must perform a "cost-benefit analysis" to optimize the amount of information attainable from their model with the minimum amount of detailed input data. Simulation engines incorporated in or supported by selected modeling programs are also shown in Table 16.1.

16.2.1 Issues in Computational Modeling

Complete and accurate representation of the detailed cellular processes at the molecular level needs to take into account a variety of issues. One of the most important of these is the effect of macromolecular crowding. Kinetic reaction rates are measured biochemically using dilute enzyme concentrations with excess reactants in well mixed solutions. In reality, the cellular space is crowded with macromolecular concentrations approaching 300 to 400 g/L. This translates to a macromolecular occupancy of 20% to 30% in the cytoplasmic space, which corresponds to a 15- to 400-fold increase in concentration compared to standard *in vitro* conditions (Ellis 2001). This crowding has the net effect of increasing actual equilibrium rate constants for macromolecular associations by two or three orders of magnitude. Macromolecular crowding also has a strong effect on diffusion within the cell and may cause transient fluctuations in local concentrations of both proteins and metabolites (Ellis 2001). Crowding also affects diffusion of macromolecules more than that of small molecules, so assumptions of a single diffusion rate for an entire compartment may be incorrect.

Another important issue in the accurate representation of biological processes is cellular topology and intracellular compartmentalization. Eukaryotic cells contain membrane-separated compartments such as the nucleus, mitochondria, endoplasmic reticulum, and endosomes. Many existing formalisms do not implicitly separate components into their appropriate compartments so they must be explicitly defined (where possible) along with intercompartmental transportation. Formalisms that support compartmentalization also permit the definition of components that span compartments. For some models, it might be important to define membranes as separate components because they can act as two-dimensional compartments, leading to increased localized concentrations of some macromolecules (Clegg 1984; Srere, Jones, and Mathews 1989).

As has been mentioned before, computer-based modeling of cancer can potentially span an incredible range of spatial and temporal scales, from nanometers (metabolites, proteins, DNA) to centimeters (tissues and organs) and from nanoseconds to years. A variety of theoretical and computational approaches have been derived to deal with this range of

TABLE 16.2   Approaches to Modeling Issues

Modeling Method	Macromolecular Crowding	Diffusion	Compartmentalization	Multiple Scales	Stochasticity
Discrete	Implicit	Implicit	Implicit/Explicit	Rapidly increased computational complexity	Implicit
Continuous	Fractal-like kinetics (Schnell and Turner 2004)	PDEs, Explicit	Explicit transport	Scales smoothly	SSA
Network	N/A	N/A	Explicit transport	Rapidly increased computational complexity	Implicit in BN thresholds and PLDEs

N/A, not applicable; BN, Boolean network; PDE, partial differential equations; PLDE, piecewise-linear differential equations; SSA, stochastic simulation algorithm.

"fine grain" to "coarse grain" phenomena, though few tools can accommodate more than a small portion of the complete range (Materi and Wishart 2007; Ridgway, Broderick, and Ellison 2006).

Finally, it is important to recognize that many important macromolecules, such as transcription factors, are present within the cell in very low concentrations (e.g., 1 to 10 nM or fewer than 100 molecules per cell). This can have a substantial effect on the selected approach to simulation as stochastic approaches are required to accurately reflect the low probability of component interactions at such concentrations. Table 16.2 describes how each of the major approaches handles these important issues.

16.2.2 Discrete Stochastic Approaches

Discrete stochastic models of biological processes attempt to represent every component of the model as a unique entity, just as every molecule in a cell is a unique object. One of the earliest (and still most common) discrete, stochastic methods is based on the cellular automata (CA) of von Neumann and Ulam (Rucker et al. 1997; Von Neumann 1966). CAs represent space as a finite lattice of two- or three-dimensional locations (cells) which can be occupied at most by a single component at a time, denoted by the state of the cell. Each cell interacts with its neighbors according to some well-defined and generally simple rules leading to new states with each discrete time step. So-called "lattice free" CAs shrink the size of each cell in the lattice and allow components to span individual lattice cells in an attempt to reflect the real physical size of various components.

Dynamic cellular automata (DCAs) extend the CA paradigm by incorporating Brownian movement of components through the lattice according to randomly generated positional moves (Wishart et al. 2005). More physically realistic models can be built by supplementing this Brownian motion with representations of real physical forces between components, such as the hydrophobic forces between individual phospholipid molecules in compartmental membranes (Broderick et al. 2005). DCA approaches have been used to model a wide variety of basic processes, including diffusion (Kier et al. 1997), micelle formation (Kier et al. 1996b), and enzyme kinetics (Kier et al. 1996a), among others. DCA models are particularly useful for simulating tissues because positions in the lattice can easily represent individual cells in the tissue. For example, a lattice-free biophysical DCA model has been used to simulate cell- and tissue-shape changes under adhesion and deformation pressures in a developing tumor (Galle, Loeffler, and Drasdo 2005; Thiery 2002). The model simulated the behavior of about 10^4 cells in a 3-D monolayer and showed that the strength of the adhesive interaction between cells and the underlying extracellular matrix was critical in inhibiting the formation of growth above the epithelium.

Petri nets are a second example of the discrete stochastic approach to biological model development (Reddy et al. 1993). Two kinds of nodes called "places" and "transitions" comprise a standard Petri net. Places are shown graphically as circles and represent distinct species of molecules. The number of "tokens" inside a place represents the number of discrete molecules of that species in existence at any time. Transitions, shown graphically as rectangles, represent reactions and they are connected to places by arrows. The stoichiometry of a reaction is indicated by a weighting where the arrow meets the transition. Transitions are said to be activated (or "fire") at some time step when the number of tokens available at all its input places are greater than the stoichiometric weights on the arrows. Upon firing, tokens are removed from input places and created in output places, representing the consumption and production of molecular species. Although Petri nets are basically deterministic, stochastic extensions have been made to more realistically model very low concentration species (Goss and Peccoud 1998; Kurtz 1972).

Agent based models (ABMs) are a third discrete stochastic approach to modeling and they extend the power and flexibility of DCAs even further. In ABMs, macromolecular and small molecular species are defined as "agents" with arbitrarily complex behavior depending on the specific implementation (Emonet et al. 2005). ABM simulations usually take place on a lattice-free grid representing real spatial dimensions and agent interaction is specified by a set of rules which may be deterministic or stochastic in nature. ABMs have been used to model such complex phenomena as calcium-dependent cell migrations in wound healing (Walker et al. 2004) or clinical trials of anti-cytokine treatments of sepsis (An 2004). The relationship between epidermal growth factor receptor (EGFR) overexpression and tumor growth and metastasis has recently been studied with a multi-scale hybrid ABM (Athale, Mansury, and Deisboeck 2005; Athale and Deisboeck 2006; Zhang, Athale, and Deisboeck 2007). The model utilized differential equations to determine intracellular and extracellular concentrations of small molecules and then discretized cellular decision making for division or migration across a lattice free grid based on programmed threshold values. Recently, an ABM has been analyzed to propose a novel therapeutic approach to tumor growth, whereby modified cancer cells (presumably with metastatic regulators) would be injected into a tumor to out-compete the harmful cancer cells (Deisboeck and Wang 2008).

16.2.3 Continuous Deterministic or Stochastic Approaches

Continuous models estimate the mass behavior of individual molecules through experimentally accessible parameters such as concentrations of molecular species and reaction rates. Because much more experimental data is available about these bulk parameters than about the single molecule parameters required in discrete approaches, it is often easier to find appropriate numerical values using continuous approaches. Systems of ordinary differential equations (ODEs) and partial differential equations (PDEs) are by far the most common approaches to modeling biological processes (de Jong 2002; Kitano 2002). For many researchers, converting a simple chemical reaction like $A + B \xrightarrow{k_1} C$ into a corresponding ODE ($dC/dt = k_1 AB$) is obvious and natural, and derivations of ODEs for many biochemical reactions can be found in most introductory biochemistry textbooks. ODEs assume that all reactions occur within a single compartment. This means that changes in concentrations are given as a function of time only. PDEs, on the other hand, explicitly incorporate spatial dimensions into their formulation, reflecting variations in molecular concentration within the compartment. Nevertheless, explicit transport functions need to be written with either approach to accurately model species movement between compartments.

While simple ODEs (e.g., for the spontaneous molecular degradation of a single species) may have exact integral solutions, most applicable ODEs in biology must be solved numerically. Using the Euler method, integration programs divide the area under a continuous function into approximate rectangular or trapezoidal regions whose areas can be exactly computed and then summed. These methods have relatively large errors and have mostly been replaced by fourth-order Runge–Kutta algorithms (Butcher 2003). Many variations for specific ODE cases have been derived from these equations. For example, delay differential equations (DDEs) can implicitly incorporate time delay factors into ODE solutions to more accurately model signal transduction cascades with reduced computational effort (Srividhya, Gopinathan, and Schnell 2007). DDEs permit the modeler to simplify a model by replacing intermediate reactions with a time delay term without general loss of information or accuracy. In addition, solutions of so-called "stiff" ODE systems (those with slow changes in most variables) can be solved faster without loss of accuracy using modified algorithms.

ODE models have been used extensively to simulate tumor development. For example, a simple model with 17 equations compared the relative effects of genetic instability, avoidance of apoptosis, increased growth rate, and angiogenic signaling on tumor progression and showed that increases in the mutational rate were only important in late-stage, sporadic tumors (Spencer et al. 2004). A more complex ODE model with 80 equations was used to explore the CD95-inducible apoptotic pathway. This study demonstrated that derepression of the c-FLIP inhibition of the death-inducing signaling complex (DISC) was dependent on activation of the CD-95 signal above a threshold value (Bentele et al. 2004). An ODE simulation incorporating effects of both angiogenic factor VEGF and vessel-maturation factor Ang2 led to a suggestion that combining antagonistic drugs might be more efficacious than monotherapeutic treatments (Arakelyan, Vainstein, and Agur 2002). This prediction was actually borne out in phase 3 clinical trials of the anti-angiogenic drug Avastin (Garber 2002).

In contrast to ODEs, which model temporal processes only, systems of PDEs have been used to model several spatially dependent processes in cancers, including chemotactically directed tumor growth (Castro, Molina-Paris, and Deisboeck 2005), growth factor-stimulated glioblastoma development (Khain and Sander 2006), tumor-immune system interactions (Matzavinos, Chaplain, and Kuznetsov 2004), and tumor growth along tubular structures (Marciniak-Czochra and Kimmel 2007). Due to the incorporation of spatial variables, PDE solvers are more computationally expensive than ODE solvers and are not found in many software packages (Alves, Antunes, and Salvador 2006).

Deterministic numerical integrators do not accurately model the stochastic behavior of molecular species with very low concentrations. To overcome this important limitation, Gillespie developed the Stochastic Simulation Algorithm (SSA), a method incorporating a probabilistic "master equation" into standard integrators (Gillespie 1976). The SSA ranks the probabilities that any one of the reactions in a system will be the next one to occur in some infinitesimal time interval and solves the ODE for that reaction before others in the system (see Gillespie, 1976, for a complete derivation).

SSA methods have been used to model many biological processes, including PKC signal transduction (Manninen, Linne, and Ruohonen 2006) and *Hox* gene expression (Kastner, Solomon, and Fraser 2002). Because SSA methods are computationally expensive, Chatterjee and colleagues have developed methods to accelerate computation by two to three orders of magnitude and demonstrated this improvement in a MAPK cascade simulation (Chatterjee et al. 2005). This "tau-leap" method computes transition probabilities per unit time for a reaction system and then allows a "bundle" of events sampled from a binomial distribution to occur simultaneously in the next time interval. This bypasses the dominating effects of fast kinetics reactions in the SSA model (such as explicit diffusion) and emphasizes slower reactions that are likely of more interest to the modeler. Gillespie has also recently introduced modifications to the original SSA method that result in similar improvements in processing speed when solving stiff (numerically unstable) ODE systems (Gillespie 2007).

Standard Petri nets have also recently been extended to permit the incorporation of continuous quantities of tokens and stoichiometric weightings through Hybrid Petri net and Functional Hybrid Petri net (FHPNs) models (Matsuno et al. 2003). An FHPN model of p53 tumor suppression has been developed which suggested that the protein displays transcriptional activation in a heterotrimer with p19-ARF and MDM2 (Doi et al. 2006). Colored Petri nets (CPNs) also allow the definition of arbitrary mathematical formulas inside transitions and have recently been used to quantitatively model EGF signaling with results similar to ODE simulations (Lee et al. 2006).

16.2.4 Network Approaches

While the lack of detailed concentration and kinetic data may make it difficult to use discrete or continuous modeling approaches, data describing networks of interacting genes and proteins are frequently available. A substantial number of network visualization and annotation systems have arisen to deal with the flood of network information available from microarray, ChIP-chip, RNAi, and two-hybrid experiments, including Cytoscape, Pathway Studio, and VisANT (Suderman and Hallett 2007). Although, these tools generally don't support simulations, some genetic regulatory network (GRN) tools, such as BioTapestry, permit the creation of databases that describe changes in the network over time (Longabaugh, Davidson, and Bolouri 2008).

The simplest biological network descriptions only identify interacting species (through nodes and directed or undirected arcs) without concern for the relative activating or inhibiting contributions of each species. However, two recent extensions of Boolean networks (BNs) used piecewise-linear differential equations (PLDEs) to enable realistic simulations of genetic networks (de Jong et al. 2003; R. Zhang et al. 2008).

A standard BN uses directed arcs to encode input and output relationships between nodes (representing genes) in the network. Input arcs may indicate that one node is activating or inhibiting to another. Furthermore, these inputs can be combined through standard Boolean logical operations such as AND, OR, NOT. PLDEs provide weightings to both input and output arcs so that each input is considered only if it surpasses a defined threshold value. The resulting output from an activated node is described by an ODE which takes into account both the concentration of the output node and its decay rate (Albert et al. 2008). BooleanNet and the Genetic Network Analyzer (de Jong et al. 2003) are two computer programs that provide PLDE genetic network analysis. A PLDE-based network model of apoptotic escape in large granular lymphocyte leukemia, with potential use in screening new therapeutic agents, has recently been developed (Zhang et al. 2008).

16.3 CONCLUSIONS AND FUTURE DEVELOPMENT

Considerable progress has been made over the past few decades in representing and simulating biological processes related to tumor development. With the advent of organizations such as the Center for the Development of a Virtual Tumor (http://www.cvit.org) and the National Resource for Cell Analysis and Modeling (http://www.vcell.org), along with other centers for quantitative biology or systems biology, the utility of simulating complex biological systems is becoming more apparent. Agent based models capable of spanning multiple spatial and temporal scales (Zhang et al. 2009) and hybrid modeling systems, such as COPASI (Hoops et al. 2006), which combine the best features of discrete and continuous approaches, are under development. Further, software that integrates multiple databases with a variety of analytic tools is being created by the open source, international GAGGLE community (Bare et al. 2007; Shannon et al. 2006).

Nevertheless, challenges still remain. For example, while graphical notation systems for molecular-level biological processes are readily available, no comparable system for describing tissue-, organ-, or organism-level phenomena has been developed. In addition, no single approach has yet been selected as a definitive standard and model development continues along competing lines. Initiatives such as the System Biology Graphical Notation project demonstrate that standard development is also under active development as a variety of approaches move toward an effective consensus. Simulating the development of cancer has already led to an increased understanding of the disease and predicted clinical efficacy of therapeutic approaches. Researchers hope that improved modeling techniques with greater accuracy will have even greater importance in the future.

REFERENCES

Alberghina, L., Chiaradonna, F., and Vanoni, M. 2004. Systems biology and the molecular circuits of cancer. *ChemBioChem* 5:1322–1333.
Albert, I., Thakar, J., Li, S., Zhang, R., and Albert, R. 2008. Boolean network simulations for life scientists. *Source Code for Biology and Medicine* 3:16.

Alon, U. 2007. *Introduction to Systems Biology: Design Principles of Biological Circuits*. Boca Raton, FL: Chapman & Hall/CRC Press.

Alves, R., Antunes, F., and and Salvador, A. 2006. Tools for kinetic modeling of biochemical networks. *Nature Biotechnol* 24:667–672.

An, G. 2004. In silico experiments of existing and hypothetical cytokine-directed clinical trials using agent-based modeling. *Crit Care Med* 32:2050–2060.

Ander, M., Beltrao, P., Di Ventura, B. et al. 2004. SmartCell, a framework to simulate cellular processes that combines stochastic approximation with diffusion and localisation: analysis of simple networks. *Syst Biol (Stevenage)* 1:129–138.

Arakawa, K., Yamada, Y., Shinoda, K., Nakayama, Y., and Tomita, M. 2006. GEM System: automatic prototyping of cell-wide metabolic pathway models from genomes. *BMC Bioinformatics* 7:168.

Arakelyan, L., Vainstein, V., and Agur, Z. 2002. A computer algorithm describing the process of vessel formation and maturation, and its use for predicting the effects of anti-angiogenic and anti-maturation therapy on vascular tumor growth. *Angiogenesis* 5:203–214.

Araujo, R. P. and McElwain, D. L. 2004. A history of the study of solid tumour growth: the contribution of mathematical modelling. *Bull Math Biol* 66:1039–1091.

Athale, C., Mansury, Y., and Deisboeck, T. S. 2005. Simulating the impact of a molecular "decision-process" on cellular phenotype and multicellular patterns in brain tumors. *J. Theor. Biol.* 233:469–481.

Athale, C. A. and Deisboeck, T. S. 2006. The effects of EGF-receptor density on multiscale tumor growth patterns. *J. Theor. Biol.* 238:771–779.

Bare, J. C., Shannon, P., Schmid, A., and Baliga, N. 2007. The Firegoose: two-way integration of diverse data from different bioinformatics web resources with desktop applications. *BMC Bioinformatics* 8:456.

Bentele, M., Lavrik, I., Ulrich, M. et al. 2004. Mathematical modeling reveals threshold mechanism in CD95-induced apoptosis. *J. Cell. Biol.* 166:839–851.

Bhalla, U.S. and Iyengar, R. 1999. Emergent properties of networks of biological signaling pathways. *Science* 283:381–387.

Bolouri, H. 2008. *Computational Modeling of Gene Regulatory Networks: A Primer*. London: Imperial College Press.

Broderick, G., Ru'aini, M., Chan, E., and Ellison, M. J. 2005. A life-like virtual cell membrane using discrete automata. *In Silico Biol* 5:163–178.

Bugrim, A., Nikolskaya, T., and Nikolsky, Y. 2004. Early prediction of drug metabolism and toxicity: systems biology approach and modeling. *Drug Discov Today* 9:127–135.

Burkert, J., Wright, N. A., and Alison, M. R. 2006. Stem cells and cancer: an intimate relationship. *J. Pathol.* 209:287–297.

Burton, A. C. 1966. Rate of growth of solid tumours as a problem of diffusion. *Growth* 30:157–176.

Butcher, J. C. 2003. *Numerical Methods for Ordinary Differential Equations*. New York: John Wiley & Sons.

Castro, M., Molina-Paris C., and Deisboeck T.S. 2005. Tumor growth instability and the onset of invasion. *Phys Rev E Stat Nonlin Soft Matter Phys* 72:041907.

Chatterjee, A., Mayawala, K., Edwards, J. S., and Vlachos, D. G. 2005. Time accelerated Monte Carlo simulations of biological networks using the binomial {tau}-leap method. *Bioinformatics* 21:2136–2137.

Clegg, J. S. 1984. Properties and metabolism of the aqueous cytoplasm and its boundaries. *Am. J. Physiol. Regul. Integr. Comp. Physiol.* 246:R133–R151.

de, Jong H. 2002. Modeling and simulation of genetic regulatory systems: a literature review. *J. Comput. Biol.* 9:67–103.

de, Jong H., Geiselmann, J., Hernandez, C., and Page, M. 2003. Genetic Network Analyzer: qualitative simulation of genetic regulatory networks. *Bioinformatics* 19:336–344.

Deisboeck, T. S. and Wang, Z. 2008. A new concept for cancer therapy: out-competing the aggressor. *Cancer Cell Int* 8:19.

Demin, O. and Goryanin, I. 2009. *Kinetic Modelling in Systems Biology*. Boca Raton, FL: Chapman & Hall/CRC Press.

Dhar, P., Meng, T. C., Somani, S. et al. 2004. Cellware: a multi-algorithmic software for computational systems biology. *Bioinformatics* 20:1319–1321.

Doi, A., Nagasaki, M., Matsuno, H., and Miyano, S. 2006. Simulation-based validation of the p53 transcriptional activity with hybrid functional Petri net. *In Silico Biol* 6:1–13.

Ellis, R. J. 2001. Macromolecular crowding: obvious but underappreciated. *Trends Biochem Sci* 26:597–604.

Emonet, T., Macal, C. M., North, M. J ., Wickersham, C. E., and Cluzel P. 2005. AgentCell: a digital single-cell assay for bacterial chemotaxis. *Bioinformatics* 21:2714–2721.

Feist, A. M., Herrgard, M. J., Thiele, I., Reed, J. L., and Palsson B.O. 2009. Reconstruction of biochemical networks in microorganisms. *Nature Rev Microbiol* 7:129–143.

Finney, A. and Hucka, M. 2003. Systems Biology Markup Language: Level 2 and beyond. *Biochem Soc Trans* 31:1472–1473.

Galle, J., Loeffler, M., and Drasdo, D. 2005. Modeling the effect of deregulated proliferation and apoptosis on the growth dynamics of epithelial cell populations *in vitro*. *Biophys. J.* 88:62–75.

Garber, K. 2002. Angiogenesis inhibitors suffer new setback. *Nature Biotechnol* 20:1067–1068.

Garny, A., Nickerson, D. P., Cooper, J. et al. 2008. CellML and associated tools and techniques. *Philos Trans R Soc A Math Phys Eng Sci* 366:3017–3043.

Gillespie, D. T. 1976. Exact stochastic simulation of coupled chemical reactions. *J. Phys. Chem.* 81:2340–2361.

Gillespie, D. T. 2007. Stochastic simulation of chemical kinetics. *Annu Rev Phys Chem* 58: 35–55.

Goss, P. J. and Peccoud, J. 1998. Quantitative modeling of stochastic systems in molecular biology by using stochastic Petri nets. *Proc Natl Acad Sci USA* 95:6750–6755.

Hanahan, D. and Weinberg, R. A. 2000. The hallmarks of cancer. *Cell* 100:57–70.

Hollywood, K., Brison, D. R., and Goodacre, R. 2006. Metabolomics: current technologies and future trends. *Proteomics* 6:4716–4723.

Hoops, S., Sahle, S., Gauges, R. et al. 2006. COPASI: a COmplex PAthway SImulator. *Bioinformatics* 22:3067–3074.

Hucka, M., Finney, A., Sauro, H. M. et al. 2003. The Systems Biology Markup Language (SBML): a medium for representation and exchange of biochemical network models. *Bioinformatics* 19:524–531.

Hunter, P., Robbins, P., and Noble, D. 2002. The IUPS human physiome project. *Pflugers Arch* 445:1–9.

Ideker, T., Galitski, T., and Hood, L. 2001a. A new approach to decoding life: systems biology. *Annu Rev Genomics Hum Genet* 2:343–372.

Ideker, T., Thorsson, V., Ranish, J.A. et al. 2001b. Integrated genomic and proteomic analyses of a systematically perturbed metabolic network. *Science* 292:929–934.

Jares, P. 2006. DNA microarray applications in functional genomics. *Ultrastruct Pathol* 30:209–219.

Kastner, J., Solomon, J., and Fraser, S. 2002. Modeling a hox gene network in silico using a stochastic simulation algorithm. *Dev Biol* 246:122–131.

Khain, E. and Sander, L. M. 2006. Dynamics and pattern formation in invasive tumor growth. *Phys Rev Lett* 96:188103.

Kier, L. B., Cheng, C. K., Testa B., and Carrupt P. A. 1996a. A cellular automata model of enzyme kinetics. *J. Mol. Graph.* 14:227–231, 26.

Kier, L. B., Cheng, C. K., Testa, B., and Carrupt P. A. 1996b. A cellular automata model of micelle formation. *Pharm Res* 13:1419–1422.

Kier, L. B., Cheng, C .K., Testa B., and Carrupt P. A. 1997. A cellular automata model of diffusion in aqueous systems. *J. Pharm. Sci.* 86:774–778.

Kitano H. 2002. Computational systems biology. *Nature* 420:206–210.

Kitano H., Funahashi A., Matsuoka Y., and Oda K. 2005. Using process diagrams for the graphical representation of biological networks. *Nature Biotechnol* 23:961–966.

Kohn, K. W., Aladjem, M. I., Kim, S., Weinstein, J. N., and Pommier, Y. 2006a. Depicting combinatorial complexity with the molecular interaction map notation. *Mol Syst Biol* 2.

Kohn, K. W., Aladjem, M. I., Weinstein, J. N., and Pommier, Y. 2006b. Molecular interaction maps of bioregulatory networks: a general rubric for systems biology. *Mol Biol Cell* 17:1–13.

Kolpakov, F., Poroikov, V., Sharipov, R. et al. 2007. CYCLONET: an integrated database on cell cycle regulation and carcinogenesis. *Nucleic Acids Res* 35:D550–D556.

Kunkel, E. J., Dea, M., Ebens, A. et al. 2004. An integrative biology approach for analysis of drug action in models of human vascular inflammation. *FASEB J.*:04-1538fje.

Kurata, H., Inoue, K., Maeda, K. et al. 2007. Extended CADLIVE: a novel graphical notation for design of biochemical network maps and computational pathway analysis. *Nucleic Acids Res* 35:e134-.

Kurtz, T. G. 1972. The relationship between stochastic and deterministic models for chemical reactions. *J. Chem. Phys.* 57:2976–2978.

Lee, D.-Y., Zimmer, R., Lee, S. Y., and Park, S. 2006. Colored Petri net modeling and simulation of signal transduction pathways. *Metab Eng* 8:112–122.

Lloyd, C. M., Halstead, M. D. B., and Nielsen, P. F. 2004. CellML: its future, present and past. *Progr Biophys Mol Biol* 85:433–450.

Loew, L. M. and Schaff, J. C. 2001. The Virtual Cell: a software environment for computational cell biology. *Trends Biotechnol* 19:401–406.

Longabaugh, W. J., Davidson, E. H., and Bolouri, H. 2008. Visualization, documentation, analysis, and communication of large-scale gene regulatory networks. *Biochim Biophys Acta* 1789: 363–374.

Manninen, T., Linne, M.-L., and Ruohonen, K. 2006. Developing Ito stochastic differential equation models for neuronal signal transduction pathways. *Comput Biol Chem* 30:280–291.

Marciniak-Czochra, A. and Kimmel, M. 2007. Reaction-diffusion approach to modeling of the spread of early tumors along linear or tubular structures. *J. Theor. Biol.* 244:375–387.

Materi, W. and Wishart, D. S. 2007. Computational system biology in cancer: modeling methods and applications. *Gene Regul Syst Biol* 1:91–110.

Matsuno, H., Tanaka, Y., Aoshima, H. et al. 2003. Biopathways representation and simulation on hybrid functional Petri net. *In Silico Biol* 3:389–404.

Matzavinos, A., Chaplain, M. A. J., and Kuznetsov, V. A. 2004. Mathematical modelling of the spatiotemporal response of cytotoxic T-lymphocytes to a solid tumour. *Math Med Biol* 21:1–34.

Moodie, S., Sorokin, A., Goryanin, I., and Ghazal, P. 2006. A graphical notation to describe the logical interactions of biological pathways. *J. Integrative. Bioinform.* 3:36.

Oda, K., Matsuoka, Y., Funahashi, A., and Kitano, H. 2005. A comprehensive pathway map of epidermal growth factor receptor signaling. *Mol Syst Biol* 1: 1–17.

Peleg, M., Rubin, D., and Altman, R. B. 2005. Using Petri Net tools to study properties and dynamics of biological systems. *J. Am. Med. Inform. Assoc.* 12:181–199.

Ramsey, S., Orrell, D., and Bolouri, H. 2005. Dizzy: stochastic simulation of large-scale genetic regulatory networks. *J. Bioinform. Comput. Biol.* 3:415–436.

Raza, S., Robertson, K., Lacaze, P. et al. 2008. A logic-based diagram of signalling pathways central to macrophage activation. *BMC Systems Biol* 2:36.

Reddy, V. N., Smith, A., Smith, B. et al. 1993. Petri net representations in metabolic pathways. *Proc Int Conf Intell Syst Mol Biol* 1:328–336.

Ridgway, D., Broderick, G., and Ellison, M. J. 2006. Accommodating space, time and randomness in network simulation. *Curr Opin Biotechnol* 17:493–498.

Rucker, R., and Waller, J. 1997. CelLab user guide: exploring cellular automata [http://www.fourmilab.ch/cellab/manual/chap5.html]

Sanford, C., Yip, M. L. K., White, C., and Parkinson, J. 2006. Cell++: simulating biochemical pathways. *Bioinformatics* 22:2918–2925.

Sauro, H. M., Hucka, M., Finney, A. et al. 2003. Next generation simulation tools: The Systems Biology Workbench and BioSPICE integration. *OMICS: J. Integrative. Biol.* 7:355–372.

Schnell, S. and Turner, T. E. 2004. Reaction kinetics in intracellular environments with macromolecular crowding: simulations and rate laws. *Progr Biophys Mol Biol* 85:235–260.

Shannon, P., Reiss, D., Bonneau, R., and Baliga, N. 2006. The Gaggle: An open-source software system for integrating bioinformatics software and data sources. *BMC Bioinformatics* 7:176.

Spencer, S. L., Berryman, M. J., Garcia, J. A., and Abbott, D. 2004. An ordinary differential equation model for the multistep transformation to cancer. *J. Theor. Biol.* 231:515–524.

Spencer, S. L., Gerety, R. A., Pienta, K. J., and Forrest S. 2006. Modeling somatic evolution in tumorigenesis. *PLoS Comput Biol* 2:e108.

Srere, P., Jones, M. E., and Mathews, C. 1989. *Structural and Organizational Aspects of Metabolic Regulation*. New York: Alan R. Liss.

Srividhya, J., Gopinathan, M. S., and Schnell, S. 2007. The effects of time delays in a phosphorylation-dephosphorylation pathway. *Biophys Chem* 125:286–297.

Suderman, M. and Hallett, M. 2007. Tools for visually exploring biological networks. *Bioinformatics* 23:2651–2659.

Thiery, J. P. 2002. Epithelial-mesenchymal transitions in tumor progression. *Nature Rev Cancer* 2:442–454.

Von, Neumann J. 1966. *Theory of Self-Reproducing Automata*. Chicago: University of Illinois Press.

Walker, D. C., Hill, G., Wood, S. M., Smallwood, R. H., and Southgate, J. 2004. Agent-based computational modeling of wounded epithelial cell monolayers. *IEEE Trans Nanobioscience* 3:153–163.

Wang, E., Lenferink, A., and O'Connor-McCourt, M. 2007. Cancer systems biology: exploring cancer-associated genes on cellular networks. *Cell Mol Life Sci* 64:1752–1762.

Webb, K. and White, T. 2005. UML as a cell and biochemistry modeling language. *Biosystems* 80:283–302.

Wilkinson, D. J. 2006. *Stochastic Modelling for Systems Biology*. London: Chapman & Hall/CRC Press.

Wishart, D. S., Yang, R., Arndt, D., Tang, P., and Cruz, J. 2005. Dynamic cellular automata: an alternative approach to cellular simulation. *In Silico Biol* 5:139–161.

You, L., Hoonlor, A., and Yin, J. 2003. Modeling biological systems using Dynetica: a simulator of dynamic networks. *Bioinformatics* 19:435–436.

Zhang, L., Athale, C. A., and Deisboeck, T. S. 2007. Development of a three-dimensional multiscale agent-based tumor model: simulating gene-protein interaction profiles, cell phenotypes and multicellular patterns in brain cancer. *J. Theor. Biol.* 244:96–107.

Zhang, L., Wang, Z., Sagotsky, J. A., and Deisboeck, T. S. 2009. Multiscale agent-based cancer modeling. *J. Math. Biol.* 58:545–559.

Zhang, R., Shah, M. V., Yang, J. et al. 2008. Network model of survival signaling in large granular lymphocyte leukemia. *Proc Natl Acad Sci USA* 105:16308–16313.

CHAPTER 17

Advanced Visualization, Analysis, and Inference of Biological Networks Using VisANT

Zhenjun Hu

CONTENTS

17.1 INTRODUCTION

VisANT is a free, online software platform used for integrative visualization, modeling, and analysis of biological networks. VisANT was developed in an effort to model the cell as an interconnected information network, with molecular components linked with one another in topologies that can encode and represent many features of cellular function. This networked view of biology brings the potential for systematic understanding of living molecular systems (Hu, Mellor, and DeLisi 2004; Hu et al. 2004, 2005, 2007b). Important features that VisANT offers to the research community are (1) exploratory navigation of integrated database-driven interaction and association networks; (2) multi-scale visualization, manipulation, and storage of known and/or user-defined networks with integrated hierarchical knowledge; (3) the ability to perform data mining and basic graph operations on arbitrary networks and subnetworks, including loop detection, degree distribution (the distribution of edges per node), and exhaustive shortest path identification between various component genes or proteins; and (4) expression visualization and analysis in the network context. Unlike other network visualization tools, VisANT provides fundamental data integration services driven by the Predictome database (Mellor et al. 2002), such as name resolution, which greatly lessens the burden of integrating data from various sources (Hu et al. 2008). For example, expression data can automatically be mapped to the appropriate nodes in a network once the Name Resolution function has been applied to its nodes, regardless of the different naming conventions used in expression and network data, respectively.

A metagraph (Hu et al. 2005, 2007a, 2007b) is an advanced graph type developed in our lab to integrative inclusive or partially inclusive relationships and the adjacent relationships into one single network, as illustrated in Figure 17.1. The inclusive relationship in a metagraph is represented by a metanode, which is a special type of node that contains associated subnodes, much as a Gene Ontology (GO) term contains its subterms or associated genes. A metanode has two states, expanded or collapsed; the expanded state manifests the internal subgraph (that is, places all descendent nodes with their connections into the graph) while the collapsed state replaces this subgraph with the single node. Networks represented by a metagraph are usually termed metanetworks, and such visualization technology is often referred to as multi-scale visualization because information at different abstraction scales is presented in one network. A detailed mathematical definition of the metagraph can be found in Section 17.5.1.

In this chapter we focus on the VisANT functions associated with the work flows (Figure 17.2) that will potentially be useful for cancer systems biology. Users are advised to visit http://visant.bu.edu for other functions of VisANT. Both work flows shown in Figure 17.2 are usually aimed at finding network modules that may account for the differential RNA expression patterns (e.g., tumor vs. normal) determined by genome-wide

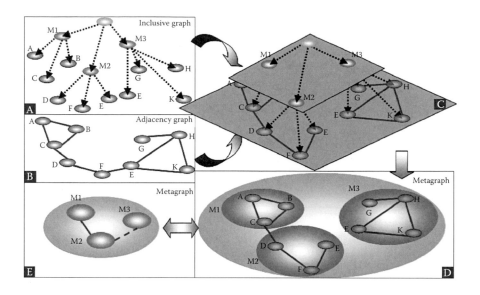

FIGURE 17.1 Illustration of the multi-scale visualization using a metagraph. Note that node E has two instances in the inclusive tree. (A) A network where an edge represents the inclusive relationship such as F belongs to M2, E is part of M2 and M3. (B) A network with adjacency relations. (C) Integration of inclusive relations (dashed lines) and adjacency relations (solid lines). (D) The integrated network using a metagraph (also referred to as meta-network) where node E belongs to both metanode M2 and M3. (E) The same meta-network with three metanodes (M1 to M3) collapsed; the dashed line between M2 and M3 indicates there is a shared node between two metanodes.

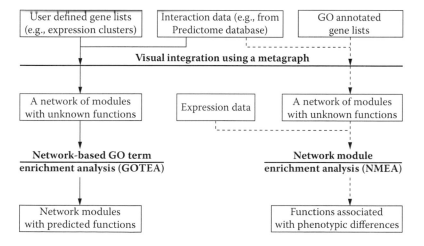

FIGURE 17.2 Different work flows focused on in this chapter. The solid lines represent the work flow of functional profiling where GO annotations are used to interpret the roles of a given gene set. The dashed lines represent the work flow of the gene set/network module enrichment analysis, where GO terms and associated genes may be used to construct the functional modules.

association studies. The first work flow starts with the modules whose functions are unknown; therefore, the task is to determine their functions, whereas the other starts with the modules whose functions are already known and the task is to determine whether they are enriched in the expression pattern. Finally, we illustrate the construction of the cancer network using the metagraph and the automatic creation of the cancer gene network based on it using the built-in VisANT function "Create the co-metanode network."

VisANT is implemented in Java and can run on any platform where Java is supported. The VisANT applet has been tested on many popular browsers, including Internet Explorer, FireFox, Chrome, and Netscape. VisANT relies on the Java Script to communicate between the web pages and the applet. VisANT can also be downloaded and run as a local application or through Java's web start technology for desktop use on Windows, Mac OS, Linux, or Unix computers where Java is supported. Please reference Section 17.5.2 for more information.

All instructions, examples, and figures used in the chapter are based on VisANT online version 3.62 (the latest version as of May 3, 2007). Because the Predictome database is updated weekly, the queried interactions used in the examples may vary; please reference http://visant.bu.edu/statistics_visant/ for more information.

17.1.1 Necessary Resources

17.1.1.1 Hardware
- Computer with 1 GHz CPU or higher, 512M free memory recommended, screen resolution of 1024 × 768 or higher recommended

- Internet connection to obtain network data from online databases

17.1.1.2 Software
- Operating System: Windows, Mac OS, Linux, or any other platform that supports Java

- Java 2 Platform, Standard Edition, version 1.4 or higher (http://java.sun.com/javase/downloads/index.jsp)

- Web browser: e.g., Microsoft Internet Explorer (http://www.microsoft.com), Mozilla Firefox (http://www.mozilla.org/firefox), or Apple Safari (http://www.apple.com/safari), etc.

17.2 NETWORK-BASED FUNCTIONAL PROFILING

Functional profiling (Rhee et al. 2008), or GO term enrichment analysis, aims to determine whether particular GO terms inform the difference of molecular phenotypes in any set of user-specified genes, typically the co-expression modules (Figure 17.2, solid lines). In a network context, the goal is to identify biological functions for a given subnetwork, or for a network module. Although many algorithms and tools (Alibes et al. 2008; Antonov et al. 2008; Antonov, Tetko, and Mewes 2006; Brohee et al. 2008; Draghici et al. 2003; Huang et al. 2007; Khatri et al. 2004, 2007; Lee et al. 2008; Reimand et al. 2008; Salomonis et al. 2007; Zhang et al. 2008; Zhu et al. 2007) have been developed for GO term enrichment analysis, they generally omit correlations based on disparate and varied datasets, such as yeast two hybrid, genetic interaction, mass spectrometry (MS), and so on. Such relations may help to overcome some drawbacks in the current enrichment analysis. For example, one drawback is that all terms are weighted equally (Khatri and Draghici 2005), while in a network module, terms annotated for highly connected genes will have more weight than those annotated for loosely connected genes. Accuracy may also be improved if network type is considered; for example, for a regulatory network, we probably can exclude those annotations of metabolic processes. From this perspective, flexible annotation schema will be needed to enable users to select subsets of GO annotations. Such flexibility could help determine the functions of genes in a specified network (see Section 17.5.3, Annotate Gene Functions Using Flexible Schema).

17.2.1 Construct a Network of Modules

1. Open the web browser and load the page http://visant.bu.edu (Figure 17.3). If the tool-tip text "Click to Start VisANT" does not appear with the mouse over the link "Start VisANT" at the right-upper corner of the Web page (Figure 17.3), you will need to turn on the support of javascript in the browser.

   Click the "Start VisANT" link; the VisANT main window will appear (Figure 17.4). The center of the main window is the network panel and a default startup network is loaded to demonstrate some basic capabilities of VisANT. The startup network also provides the default configurations for VisANT. The default startup network may

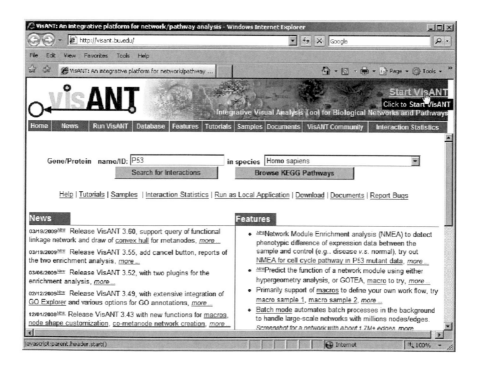

FIGURE 17.3   The VisANT homepage.

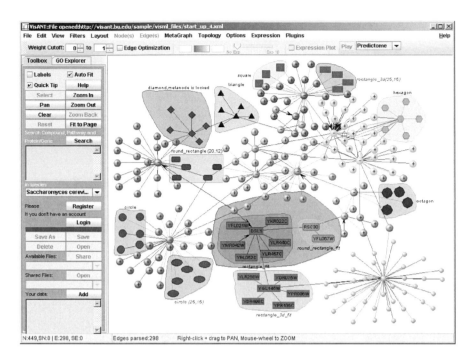

FIGURE 17.4 (See color insert following page 332.) The VisANT main window with a startup network.

vary at different times to illustrate the latest development. When VisANT is run as a local application, the startup network can be customized through the Files→Save as Startup Network menu, or through the key combination CTRL-T.

At the top of the main window is the toolbar containing some controls with the tooltips, such as the control for the minimum and maximum edge weight cutoff, as well as the current database to query the interactions (Predictome database in Figure 17.4). At the bottom of the main window is the status bar where some information on the current network, such as total number of edges and nodes, will be shown. At the left of the main window is the control panel, which contains two tabs: toolbox and GO explorer. The default tab toolbox provides many important controls that will be used in the chapter, and GO explorer is used to navigate the GO hierarchy structure. Please reference http:// visant.bu.edu/vmanual/ver3.50.htm for more information about GO explorer, such as search of interactions using the key words.

2. Clear the network by clicking the button "Clear" on the toolbox, and change the current species to *Homo sapiens* through the drop-down list for the species. Press the key "h" to speed up the search of *Homo sapiens* in the drop-down list. It is important to set the correct species because all the integrated knowledge is directly associated with a specific species, including the gene-GO associations that will be used later.

3. Assume we have three co-expression clusters named CLUSTER_A, CLUSTER_B, and CLUSTER_C, each containing a number of genes as listed below. Copy and paste

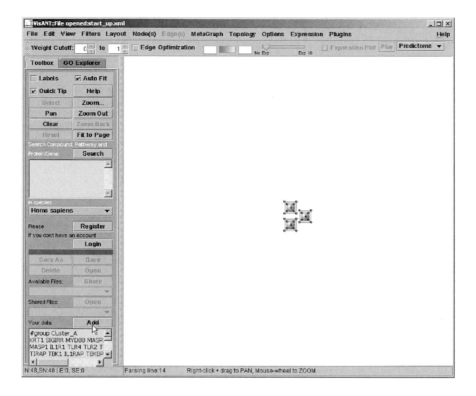

FIGURE 17.5   Use the extended edge-list to create three metanodes representing three co-expression clusters.

(either through the pop-out menu, or the key combination CTRL-C and CTRL-V) the following text into the Add text box of VisANT's toolbox and click the Add button (left-bottom corner indicated by the mouse cursor in Figure 17.5). Three metanodes will be created (Figure 17.5). The Add textbox can be used to add any type of data whose format is supported by VisANT (Table 17.1).

```
#group Cluster_A
KRT1 SIGIRR MYD88 MASP2 C1QA
MASP1 IL1R1 TLR4 TLR2 TLR1
TIRAP TBK1 IL1RAP TBKBP1 MBL2
SERPING1 CR2 C1S C1R

#group Cluster_B
NA SNAPAP BLOC1S3 BLOC1S2 DTNBP1 BLOC1S1
MUTED SNAP25 PLDN TRPV1 EBAG9 STX12

#group Cluster_C
HYAL2 CLP1 TEP1 RPP40 TSEN15
ERVWE1 RPP38 POP1 LOC100128314 TSEN34
RPP30 TSEN2 TSEN54 TERT
```

TABLE 17.1   File Formats Supported by VisANT

File Format	Description & Identification	Related URL
Edge List	Default tab-delimited text format for non-XML-based data in VisANT.	http://visant.bu.edu/import#Edge
VisML (VisANT XML file)	Default VisANT file format for XML-based data, containing all the network information. Network stored as VisML format can be safely replayed as it was stored.	http://visant.bu.edu/misi/visML.htm
Expression (expression matrix file)	Expression matrix file; the first line must start with "#!Expression". Optional parameter addNewNode to determine whether to abandon the nodes that are not in the current network, e.g., addNewNode=false.	http://visant.bu.edu/vmanual/expression.htm http://visant.bu.edu/vmanual/ver3.50.htm#Expression
Macro/Batch file	The file format to store a list of commands for VisANT to carry out. The first line must start with "#!batch commands".	http://visant.bu.edu/vmanual/cmd.htm
ID-Mapping (ID mapping file)	This file format is designed to allow the user to add various database IDs, as well as alias and functional descriptions, to the nodes in a network.	http://visant.bu.edu/vmanual/ver2.60.htm
KGML (KEGG XML file)	The KEGG Markup Language (KGML) is an exchange format of the KEGG graph objects, especially the KEGG pathway maps that are manually drawn and updated.	http://www.genome.jp/kegg/xml/
GML (Graph Markup Language)	A common graph file format supported by several network software packages	http://www.infosun.fim.uni-passau.de/Graphlet/GML/
PSI-MI (Proteomics Standards Initiative-Molecular Interaction format)	XML standard format for molecular interactions supported by molecular interaction databases.	http://www.psidev.info/index.php?q=node/60
BioPAX 1.0 (Biological PAthway eXchange)	Standard format for pathway information supported by multiple pathway databases.	http://www.biopax.org

The above text uses VisANT's extended edge-list format (Table 17.1) to create the network, which is the simplest format supported in VisANT. It can also be used to easily add nodes (each line with the name of one single node) or edges (each line with the name of the two nodes separated by space or tab).

Alternatively, users can load this edge-list from the URL through the File→Open URL menu and enter the URL http://visant.bu.edu/other_formats/edge_list_3_clusters.txt (depending on the type of browser, you may be able to paste the above URL using the key combination CTRL-V), and follow the instructions to achieve the same result.

When you are uncertain about the format of edge-list, you can always export the network in the format of edge-list with the menu File→Export as Tab-Delimited File→All and follow the exported examples.

4. Layout the network. Apply the Layout→Circle* menu four times, and click the button "Fit to Page." The network will look similar to the one shown in Figure 17.6.

The layout will be carried recursively for all embedded metanodes when there is "*" shown for the corresponding layout menu. If only a set of metanodes needs to be laid out, simply select these metanodes only (reference http://visant.bu.edu/docs.htm for complete instructions on node selection), and apply the corresponding menu. The circle size of the metanode will be increased each time the layout is applied. Alternatively, users can move the component nodes of the metanode around to achieve the desired rectangle size, and the circle size will be determined by the minimal size of the rectangle's width and height.

FIGURE 17.6   Layout network of three clusters.

17.2.2 Predict the Functions of Three Co-Expression Clusters Using the Hyper-Geometric Test

1. Annotate the genes of three clusters using the menu MetaGraph→GO Annotation of All Nodes→Using Most Specific GO Terms.

   VisANT automatically resolves the node names when annotating the nodes.

   Please reference Section 17.5.3 for other annotation options. The same annotation menus are also available under the menu Nodes which are only used to annotate the selected nodes. In the case where you have collapsed metanodes, such as for KEGG pathways, always use the annotation options under the MetaGraph menu.

2. Uncheck the menu Options→Open Link Using Same Browser so that we can compare the prediction reports produced by VisANT for different algorithms.

3. Make sure that no node is selected by mouse-clicking on the empty space of the network. Predict the functions of the clusters using hyper-geometric-based analysis through the menu MetaGraph→ Predict Functions of Metanodes Using GO → Detect Over-represented GO Terms Using Hypergeometric Test→ Start Hypergeometric Test over GO Database. VisANT will perform the prediction for all non-embedded metanodes. For more information, please reference the manual at http://visant.bu.edu/vmanual/ver3.50.htm#hyper.

The prediction results will be added to the metanode as part of its description that is available as tooltips when the mouse is over the node (Figure 17.7). Table 17.2 lists all predictions of three clusters based on the report created by VisANT at this writing: http://visant.bu.edu/misi/hyper_3_cluster.htm.

The Predictome database maintains a local copy of the GO database and the gene-GO associations are extracted from the Entrez Gene database. Both datasets are being updated constantly; therefore, the actually prediction results may be a little different from the results shown in the link above. This also applies to the GOTEA algorithm that will be illustrated later because the interactions are also being updated from a list of interaction databases.

17.2.3 Predict the Functions of Three Co-Expression Clusters Using GO Term Enrichment Analysis (GOTEA)

GOTEA uses a permutation-based method to predict the functions of a given network module with the evaluation of the association between genes in the module. The algorithm detail can be found at 5.4 GO term enrichment analysis (GOTEA).

1. Return to the VisANT home page and click on the link Interaction Statistics as shown at the upper right corner in Figure 17.3. The opened page lists the number of interactions available for a total of 108 organisms supported in the Predictome database. Find the species *Homo Sapiens* and click the link of the number; it will bring you to a page similar to the one shown in Figure 17.8:

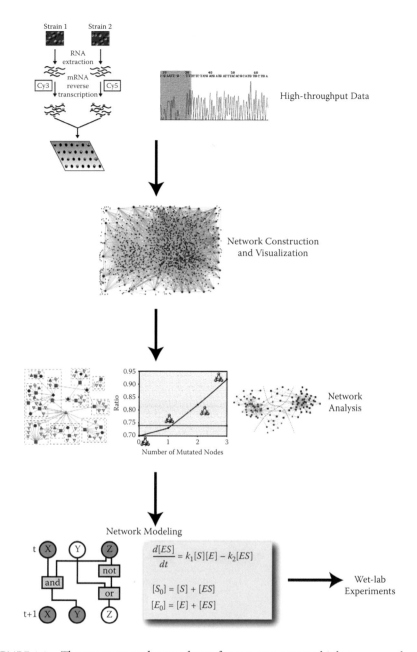

COLOR FIGURE 1.1   The strategy and procedures for cancer systems biology research.

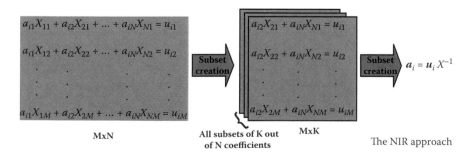

COLOR FIGURE 3.2   Network identification by multiple regression algorithm.

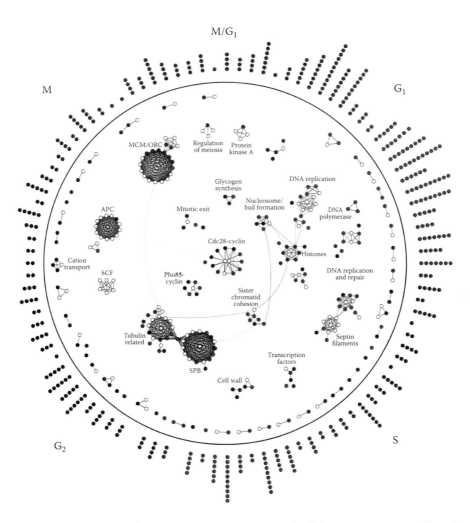

COLOR FIGURE 4.1 Temporal protein interaction network of the yeast mitotic cell cycle. Circle nodes represent cell cycle proteins, which are members of protein complexes or other physically interacting partners. White nodes represent static proteins, while colored nodes represent the time of peak expression of the proteins. The proteins without interactions are outside the circle, and are positioned and colored according to their peak time. (From de Lichtenberg, U. et al. 2005. *Science* 307: 724–727. With permission.)

COLOR FIGURE 4.2 Protein communities of cancer metastasis. The communities were identified by *k*-clique analysis performed on the predicted genome-wide rat protein network. The communities are distinguished by different colors and labeled by the overall function or the dominating protein class. Note that proteins, particularly at community edges, can belong to more than two communities. (From Jonsson, P.F. et al. 2006. *BMC Bioinformatics* 7: 2.)

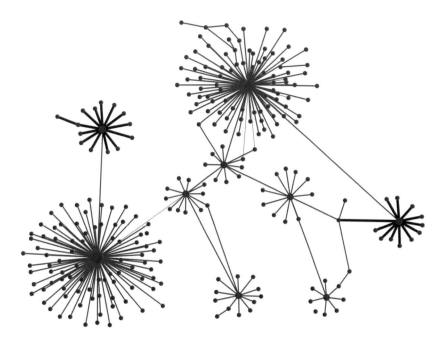

COLOR FIGURE 4.3   A global picture of the cancer protein network encodes the driver-mutating information, network rewiring, and module dynamics. The cancer protein network architecture contains many small network modules containing intramodular hubs which are connected by intermodular hubs in which cancer driver-mutating genes are dominantly enriched. The expression levels of the intramodular and intermodular hub genes are not significantly changed between metastatic and nonmetastatic tumors; however, the coexpressions between intramodular hubs and their interacting partners are significantly changed between metastatic and nonmetastatic tumors. Nodes represent proteins while links represent physical interactions. Red, orange, and blue nodes represent intermodular and intramodular hubs, and nonhub nodes, respectively. Square nodes represent cancer driver-mutating genes while black links represent the co-expression changes between the two linked genes.

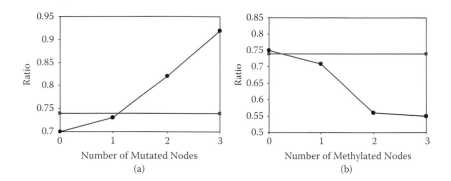

COLOR FIGURE 5.1 Enrichment of mutated and methylation genes in network motifs. (a) Relations between the fractions of positive links in all 3-node size network motifs and the fractions of mutated genes in these motifs. (b) Relations between the fractions of positive links in all 3-node size network motifs and the fractions of methylated genes in these motifs. All network motifs were classified into subgroups based on the number of nodes that are either mutated genes or methylated genes, respectively. The ratio of positive links to total positive and negative links in each subgroup was plotted. The horizontal lines indicate the ratio of positive links to the total positive and negative links in all network motifs. (Adapted from Cui et al. 2007a.)

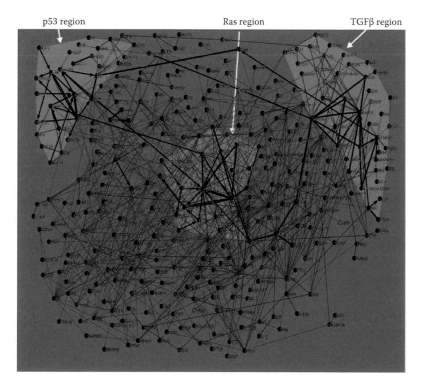

COLOR FIGURE 5.2 Human oncogenic signaling map. The human cancer signaling map was extracted from the human signaling network, which was mapped with cancer mutated and methylated genes. The map shows three "oncogenic dependent regions" (background in light grey), in which genes of the two regions are also heavily methylated. Nodes represent genes, while the links with and without arrows represent signal and physical relations, respectively. Nodes in red, purple, brown, cyan, blue, and green represent the genes that are highly mutated but not methylated, both highly mutated and methylated, lowly mutated but not methylated, both lowly mutated and methylated, methylated but not mutated, and neither mutated nor methylated, respectively. (Adapted from Cui et al. 2007a.)

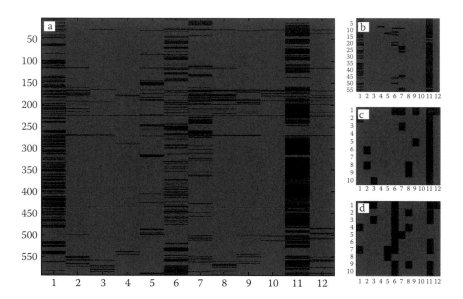

COLOR FIGURE 5.3   Heatmaps of the gene mutation distributions in oncogenic signaling blocks. Twelve topological regions or oncogenic signaling blocks have been identified based on the gene connectivity of the human oncogenic signaling map. A heatmap was generated from a matrix, which was built by querying the oncogenic signaling blocks using tumor samples, in which each sample has at least two mutated genes. If a gene of a particular signaling block (b) gets mutated in a tumor sample (s), we set Ms,b to 1; otherwise we set Ms,b to 0. (a) A heatmap generated using the gene mutation data of the 592 tumor samples. (b) A heatmap generated using the gene mutation data of the NCI-60 cancer cell lines. (c) and (d) Heatmaps generated using the output from the genome-wide sequencing of breast and colon tumor samples, respectively. Rows represent samples, while columns represent oncogenic signaling blocks. Blocks with gene mutations are marked in red. (Adapted from Cui et al. 2007a.)

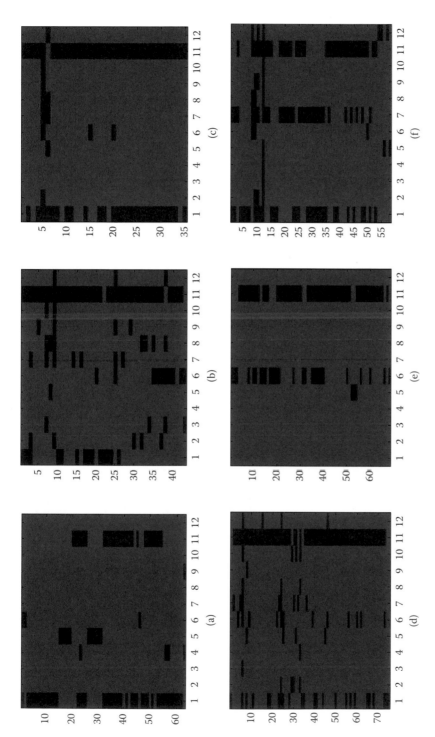

COLOR FIGURE 5.4   Heatmaps of the gene mutation distributions in oncogenic signaling blocks for six representative cancer types. Twelve topological regions or oncogenic signaling blocks have been identified based on the gene connectivity of the human oncogenic signaling map. A heatmap was generated from a matrix, which was built by querying the oncogenic signaling blocks using tumor samples, in which each sample has at least two mutated genes. If a gene of a particular signaling block (b) gets mutated in a tumor sample (s), we set $M_{s,b}$ to 1; otherwise we set $M_{s,b}$ to 0. Heatmaps for (a) blood, (b) breast, (c) central nervous system, (d) lung, (e) pancreas, and (f) skin tumors were built using tumor samples of these cancer types, respectively. Rows represent samples, while columns represent oncogenic signaling blocks. Blocks with gene mutations are marked in red. (Figure is adapted from Cui et al. (2007a).

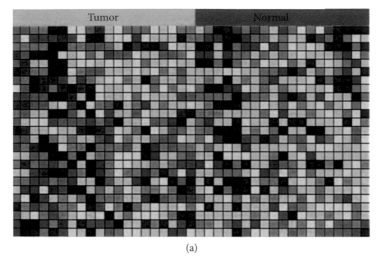

(a)

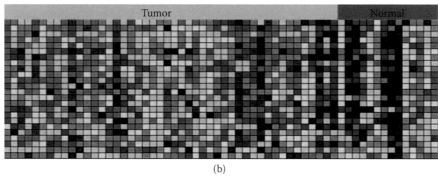

(b)

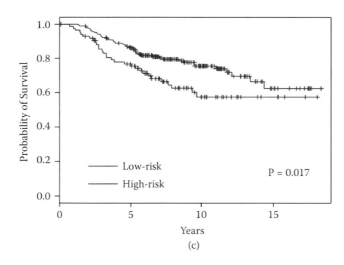

(c)

COLOR FIGURE 6.2 The association of the 26S proteasome genes with tumor progression and metastasis. Heat map generated by GSEA based on the gene expression values of the 26S proteasome genes: (a) lung tumor, GSE2514, (b) bladder tumor, GSE3167. Kaplan–Meier survival analysis of breast cancer patient groups stratified by the 26S proteasomes genes' expression in tumors: (c) breast tumor, 295-set, (d) breast tumor, GSE349. (Adapted from Fu, Li, and Wang 2009.)

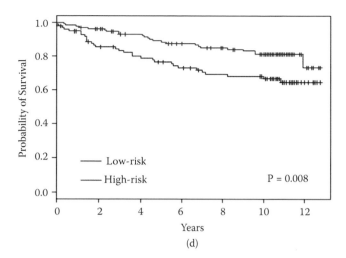

COLOR FIGURE 6.2   (Continued)

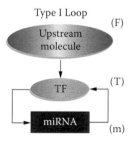

COLOR FIGURE 7.1   Type-1 circuits of transcription factor and miRNA regulatory motifs (F), (T), and (m) represent upstream factor, target gene, and miRNA, respectively.

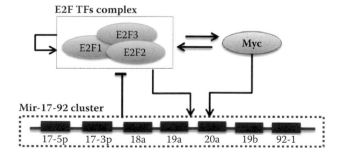

COLOR FIGURE 7.3   The regulatory loop of c-Myc/E2Fs and miR-17-92. Arrows represent active regulation while T signs represent negative regulation.

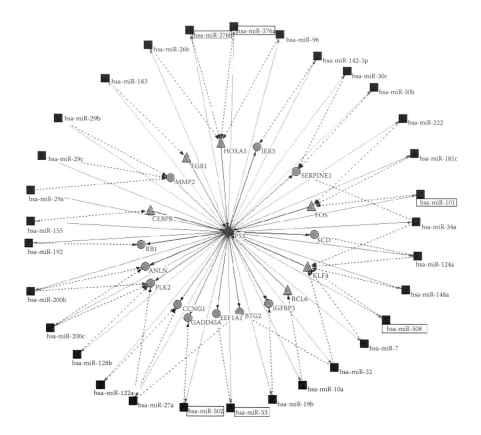

COLOR FIGURE 7.4 The p53-centered miRNA regulatory cancer network. Red boxes represent miRs. Red dotted lines represent the regulation of miRs to TFs which either regulate p53 or are regulated by p53. Yellow triangles represent the TFs which regulate p53. Yellow circles represent the TFs which are regulated by p53. Red filled lines represent repression. Blue filled lines represent activation. Black filled lines represent repression/activation. Green filled lines represent uncharacterized regulation. Differentially expressed miRs in cancer are boxed. miRs whose expression 1.5-fold ($p < 0.05$) in cancer is triggered by the p53 activation are underlined. All of the TFs in this network were predicted as miRNA targets by two or more algorithms (picTar, miRtarget, microT, miRanda, TargetScanS).

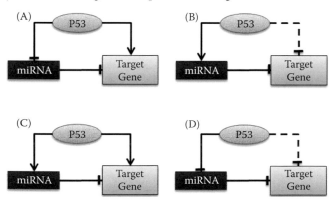

COLOR FIGURE 7.5 Types of representative miRNA-TF regulatory circuits in the p53-miRNA regulatory cancer network. (A) and (B): downstream incoherent models; (C) and (D): downstream coherent models.

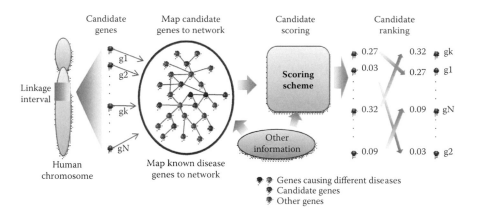

COLOR FIGURE 11.1   Sketch map of network-based candidate gene prioritization and prediction. A list of candidate genes such as those in a linkage interval or all the human genes are mapped onto a human gene/protein network, and if applicable, known disease genes and other information (such as sequence characteristics and mRNA expression) are also mapped onto the network. A scoring scheme is used to score each candidate gene based on current data and outputs a rank list of all candidate genes. Genes ranked above a certain position are predicted as disease causative.

COLOR FIGURE 12.4   Heatmap demonstrating associations of cancer-linked GO terms (linked to cancer hallmarks) in various signaling pathways. Significance value q < 0.05 (red in color) implies more significant association whereas q > 0.15 (green in color) implies less significant association.

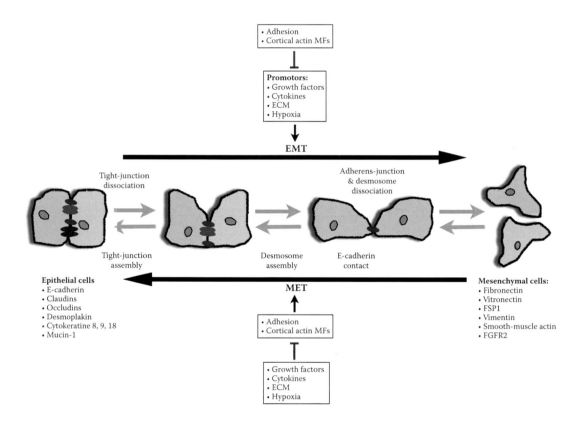

COLOR FIGURE 13.1   Simplified representation of the balance between the mechanisms driving epithelial-to-mesenchymal transition (EMT) and mesenchymal-to-epithelial transition (MET).

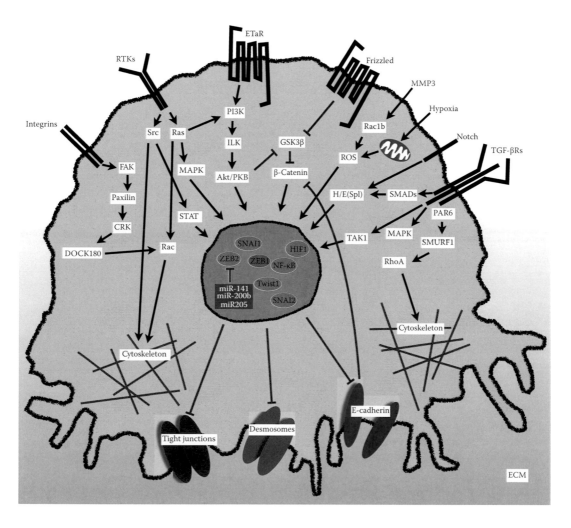

COLOR FIGURE 13.2   Schematic overview of the major signal transduction pathways that induce epithelial-to-mesenchymal transition (EMT).

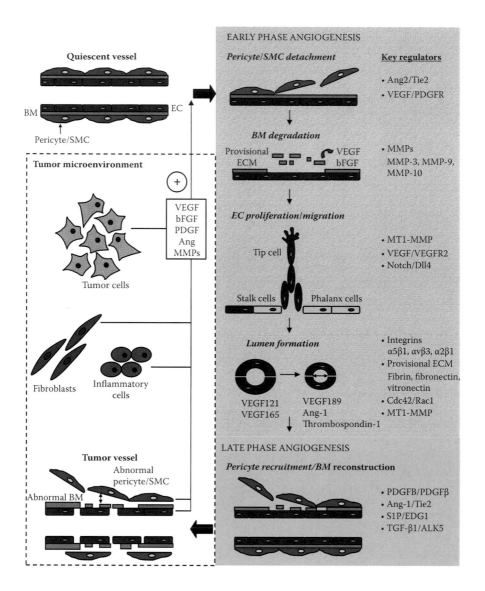

COLOR FIGURE 15.1 Schematic illustration of the different steps and key modulators involved in the process of tumor angiogenesis. The transition from a pre-vascular to a vascularized tumor phenotype is tightly regulated by the balance between pro- and anti-angiogenic factors secreted by the tumor cells and/or stromal cells—including endothelial cells, pericytes, smooth muscle cells and fibroblasts, and by infiltrating cells of the immune system. Each phase during this process is specifically modulated by temporal and spatial interactions between the tumor and vascular cell types and the mediators released into the tumor microenvironment. **BM**: basement membrane; **EC**: endothelial cells; **SMC**: smooth muscle cells; **ECM**: extracellular matrix; MMPs: matrix metalloproteinases.

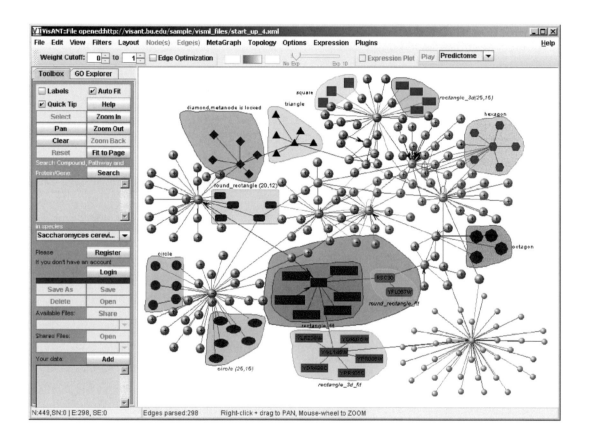

COLOR FIGURE 17.4   The VisANT main window with a startup network.

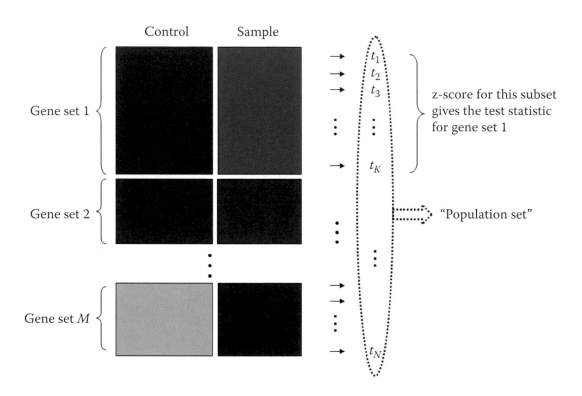

COLOR FIGURE 18.1   An example of a competitive GSA method: PAGE. (From Kim S.W. and Volsky D.J. 2005. *BMC Bioinformatics* 6:144.)

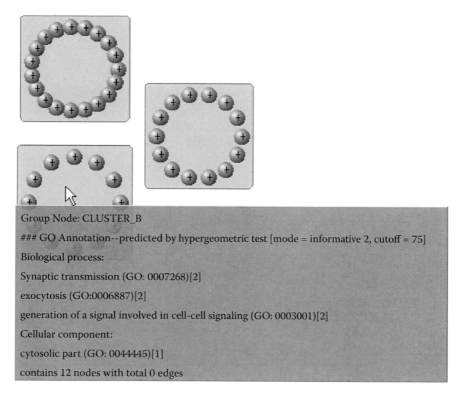

FIGURE 17.7 Prediction results are available as tooltips of the corresponding metanode nodes.

TABLE 17.2   Cluster Functions Predicted with the Hyper-Geometric-Based Test

	Molecular Function	Biological Process	Cellular Component
Cluster_A	Cytokine binding(GO:0019955) growth factor binding(GO:0019838) serine-type endopeptidase activity(GO:0004252)	Positive regulation of immune response(GO:0050778) innate immune response(GO:0045087) acute inflammatory response(GO:0002526)	
Cluster_B		Synaptic transmission(GO:0007268) exocytosis(GO:0006887) generation of a signal involved in cell-cell signaling(GO:0003001)	
Cluster_C	Endonuclease activity(GO:0004519) nucleotidyltransferase activity(GO:0016779)	tRNA metabolic process(GO:0006399)	Nucleolus(GO:0005730)

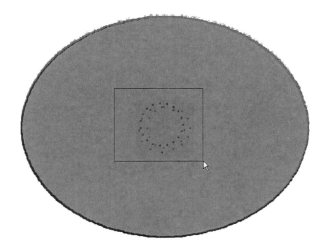

Start VisANT

Integrative Visual Analysis Tool for Biological Networks and Pathways

| Home | News | Run VisANT | Database | Features | Tutorials | Samples | Documents | VisANT Community | Interaction Statistics |

Homo sapiens

affinity technology(M0045):	10672	anti tag coimmunoprecipitation(M0065)	3090
Biochemical Activity(M0083):	70	chromatography technology(M0085):	4
Co-fractionation(M0079):	167	coimmunoprecipitation(M0010):	4488
colocalization by immunostaining(M0025):	276	colocalization/visualisation technologies(M0063):	316
Competition binding(M0012):	121	copurification(M0013):	688
cosedimentation(M0033):	91	cross-linking studies(M0014):	3901
electron microscopy(M0062):	1	elisa:enzyme-linked immunosorbent assay(M0051):	79
enzymatic study(M0066):	118	far western blotting(M0060):	303
filter binding(M0053)	165	fluorescence technology(M0052):	114

FIGURE 17.8   Total interactions available in the Predictome database for *Homo sapiens*.

2. Clicking the number of affinity technology (10672 at the time of writing) shown in Figure 17.8 will load a total of 10,672 interactions detected by affinity technology in VisANT changes the network shown in Figure 17.6 to one similar to Figure 17.9.

    VisANT automatically adjusts the global zoom level when loading a large interaction set. To resume the zoom level, simply click first the Zoom Out button and then the Reset button in VisANT's toolbox.

3. Select all nodes of the three clusters by dragging a rectangle using the left mouse button, then apply Edit→Invert Node Selection menu, remove those selected nodes by either pressing the Delete key, or through the Edit→Delete Selected Nodes menu. Then click the Fit to Page button on VisANT's toolbox. A network similar to the one shown in Figure 17.10 appears with nodes of the three clusters being connected.

4. Use GO → Network-based GO Term Enrichment Analysis (GOTEA) → Configure GOTEA menu to change the number of interactions to 20,000.

FIGURE 17.9   An integrated network of three clusters with 10,672 interactions detected by affinity technology (M0045).

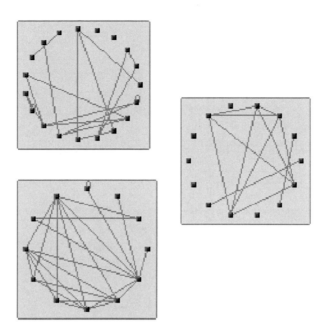

FIGURE 17.10   Network modules for the three clusters using integrated interactions of M0045.

5. Perform GOTEA analysis through the MetaGraph→ Predict Functions of Metanodes. Using GO → Network-based GO Term Enrichment Analysis (GOTEA) → Fast GOTEA menu.

The prediction results will be added to the metanode as part of its description that is available as tooltips. Table 17.3 lists the top three GO terms that resulted from

TABLE 17.3   Cluster Functions Predicted by GO with Integrated Interaction of M0045

	Molecular Function	**Biological Process**	**Cellular Component**
Cluster_A	Cytokine binding(GO:0019955) growth factor binding(GO:0019838) sugar binding (GO:0005529) …	Cytokine biosynthetic process(GO:0042089) positive regulation of immune response (GO:0050778) innate immune response (GO:0045087) …	Extracellular space(GO:0005615) receptor complex(GO:0043235) secretory granule(GO:0030141) …
Cluster_B	Calmodulin binding(GO:0005516) ATP binding(GO:0005524) calcium channel activity(GO:0005262)	Synaptic transmission(GO:0007268) neurotransmitter transport(GO:0006836) generation of a signal involved in cell-cell signaling(GO:0003001) …	Clathrin-coated vesicle(GO:0030136) neuron projection(GO:0043005) cytoplasmic vesicle membrane(GO:0030659) …
Cluster_C	Endonuclease activity(GO:0004519) nucleotidyltransferase activity(GO:0016779) ATP binding (GO:0005524) …	tRNA metabolic process(GO:0006399) DNA recombination(GO:0006310) cellular carbohydrate catabolic process(GO:0044275) …	Nucleolus(GO:0005730) anchored to membrane(GO:0031225) soluble fraction(GO:0005625)

N:48,5N:48 | E:56, 5E:0   GOTEA for CLUSTER_C:scan GO term-->GO:0000171[8 of 206]-->permutation test[1396 of 20000]   ⬤

FIGURE 17.11   Use the cancel button to cancel the computational heavy analysis.

GOTEA analysis for three clusters. The complete report can be found at http://visant.bu.edu/misi/gotea_M0045_3_cluster.htm.

It is obvious that GOTEA finds more enriched GO terms for each cluster than the hyper-geometric test, which is mainly because GOTEA uses a fuzzy searching algorithm to find those GO terms that are semantically similar. As a result, GOTEA is much slower than the hyper-geometric test, and takes about half an hour to finish the analysis of three clusters. From this perspective, VisANT provides a red cancel button at the right end of the status bar to cancel the analysis, as shown in Figure 17.11.

More information about GOTEA in VisANT can be found at http://visant.bu.edu/vmanual/ver3.50.htm#gotea.

6. Save the network. Click on the toolbox tab on VisANT's control panel. If you have a VisANT account, login to VisANT and click the Save as button to save the network. If you do not have a VisANT account, you can register for one for free; otherwise, you can use the key combination CTRL+Y or the File→Copy to Add TextBox menu to export the current network to Add textbox in the format of VisML, then copy/paste VisML into any text editor and save it as a local file. The network can be restored by copy/paste VisML into the Add Textbox and click the Add button.

17.3 NETWORK-BASED EXPRESSION ENRICHMENT ANALYSIS

Another typical appplication of enrichment analysis is the study of differential RNA expression patterns (e.g., tumor versus normal) determined by genome-wide association studies, to determine if one or more specified gene sets (e.g., KEGG pathways) might account for some of the differences (Figure 17.2, dashed lines; Barry, Nobel, and Wright 2005; Mootha et al. 2003; Subramanian et al. 2005; Volinia et al. 2004). Gene set enrichment analysis (GSEA; Subramanian et al. 2005) is probably the most used algorithm in such analysis which does not take account of prior network knowledge. Here we introduce network module enrichment analysis (NMEA) to test whether the modules are enriched with transcriptional changes between the control and the sample. NMEA is basically an extension of GSEA but takes advantage of the extra information provided by network connectivity. In VisANT, a network can be constructed using the data from any combination of 60-odd methods (e.g., Y2H, ChIP-Chip, MS, knock-outs, etc.) for the gene lists of interest. Modules can be easily constructed as metanodes through corresponding menus, simple drag and drop operations from GO explorer, and extended edge-list (http://visant.bu.edu/import#Edge) of the user's own data.

17.3.1 NMEA with GO Modules

1. Repeat steps 1–2 in Section 17.2.1 to have an empty network for *Homo sapiens*.

2. Resume the zoom level by clicking first the Zoom Out button and then Reset button in VisANT's toolbox.

3. Click on the GO Explorer tab in VisANT's control panel, enter GO:0000077 in the search box at the bottom of the GO explorer, and click the Search button to search for the GO term. Drag and drop the highlighted term DNA damage checkpoint to the network to create the metanode for GO:0000077 (Figure 7.12).

4. Repeat step 2 above for GO:0051320, GO:0007127, and GO:0051318. All three meta-nodes have overlaps with the first metanode of GO:0000077. Move the overlapped genes to the center of each metanode, and a metanetwork similar to the one shown in Figure 17.12 will appear, except there is no edge.

5. Map the expression profiles by opening the expression data from the following address: http://visant.bu.edu/sample/exp/p53_visant.dat using File→Open URL menu.

The expression data shown in the above link contains 22 microarray samples with mutations in P53 and 17 wild-type samples. The data is downloaded from the

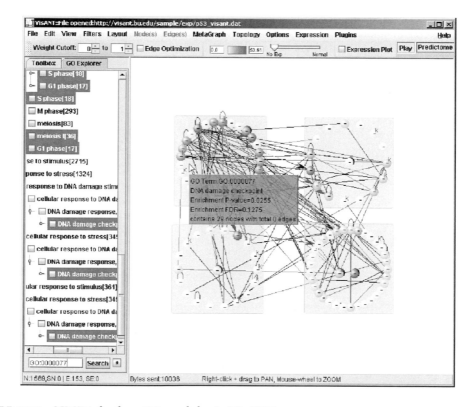

FIGURE 17.12 NMEA for four GO modules in VisANT.

GSEA website (http://www.broad.mit.edu/gsea/). Please reference http://visant.bu.edu/vmanual/ver3.50.htm#Expression for the format of expression data supported by VisANT.

An alternative way to load the expression data is to copy and paste expression data in the Add textbox of the toolbox.

6. Change the color mapped for the minimal and maximum expression values to the light green and darker green, respectively, by clicking the left or right side of the color map shown in the toolbar (Figure 17.12). The color map will also be used to indicate the relative contribution to the enrichment score within each metanode.

7. Select all nodes using the Edit→Select All Nodes menu.

8. Query the interactions between selected nodes from the Predictome database using the Node(s)→Query Internal Interactions menu. The edges between the nodes appear as in Figure 17.12.

In comparison with steps 2 and 3 in Section 17.2.3 where only a portion of the interactions are used to construct the network modules, here we query all possible interactions in the Predictome database.

9. Clear all selections by left-mouse clicking on an empty space of the network panel.

10. Start NMEA using the Expression→Network Module Enrichment Analysis (NMEA)→ Start NMEA Analyze menu. Once finished, the P-value and FDR score will be added to each metanode's description (Figure 17.12) and an HTML report will be generated similar to the one at: http://visant.bu.edu/misi/nmea_go_modules.htm.

From the report it is clear that only the process DNA damage checkpoint (GO:0000077) exhibits the phenotypic difference in the expression of genes between mutated and wild-type samples, probably due to the fact that P53 plays a role in the process. As mentioned in step 6 above, nodes with the darker color have more contribution to the enrichment score.

17.3.2 NMEA with KEGG Pathways

Pathway databases provide a valuable resource of network modules to apply the enrichment analysis although the number of available pathways is still limited. VisANT is closely integrated with the KEGG pathway and provides several different ways to load KGML-based KEGG pathways (Hu et al. 2007b).

1. Click the toolbox tab on the control panel.

2. Clear the network by clicking the Clear button on the toolbox. Alternatively, you can clear the network using the Edit→Clear menu so that you will not need to change back and forth between the toolbox and GO explorer.

3. Enter map04110 in the Search box above the drop-down list of the species in VisANT's toolbox, and click the Search button to load the KEGG cell cycle pathway. VisANT will search for KEGG pathways for the current species using the number (e.g., 04110) as the pathway ID, if the search term starts with "map" followed by the number.

4. Resolve all the node names using the MetaGraph→Resolve All Nodes Name menu. This step is necessary because the KGML of the KEGG pathway may use different naming systems for genes.

5. Repeat step 5 in the previous section to load the same expression data.

6. Repeat steps 9 and 10 above to carry out the NMEA analysis of the cell cycle pathway. The result is shown in Figure 17.13.

Since P53 mutants influence the behavior of the cell cycle, it is expected that cell cycle-related modules should be enriched. NMEA reported the cell cycle pathway as significantly enriched with a p-value = 0.04 and therefore supports this point (Figure 17.13).

More information about NMEA in VisANT can be found at http://visant.bu.edu/vmanual/ver3.50.htm#nmea.

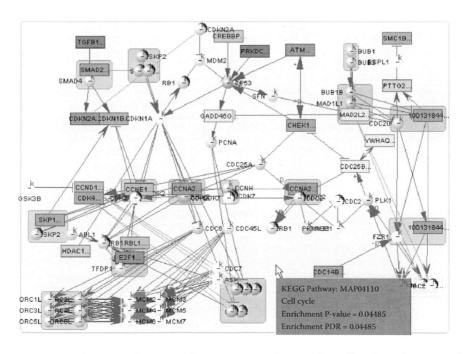

FIGURE 17.13 Visualization of NMEA of P53 mutation data on the cell cycle pathway. Nodes with a darker color have more contributions to the enrichment score.

17.4 USING A METAGRAPH TO MODEL THE CANCER NETWORK

In this section we will illustrate how to use a metagraph to build a network of cancers based on the simple cancer-gene association, and how this cancer network can be used to create the cancer gene network.

17.4.1 Construct the Cancer Network

1. Clear the network by clicking the Clear button.

2. Load the edge-list for the cancer network from http://visant.bu.edu/other_formats/edge_list_cancers.txt using the File→Open URL menu. Once finished, click the Fit to Page button on the toolbox.

   The data shown in the above URL is extracted from the work of Goh and co-workers (2007). The disease is represented by the disease ID and is not very informative. From this perspective, we use the ID-Mapping format (Table 17.1) to add an informative description for each cancer. The first few lines of the file are shown below:

   ```
   #!ID Mapping AddNewNode=false
   #VisANT_ID   description
   DOR2212      Rhabdomyosarcoma, alveolar, 268220 (3) [DOR2212]
   DOR2211      Rhabdomyosarcoma, 268210 (3) [DOR2211]
   DOR2210      Rhabdoid tumors (3) [DOR2210]
   DOR1804      Nasopharyngeal carcinoma, 161550 (3) [DOR1804]
   ```

3. Similar to the above step, load the ID-Mapping file from the URL: http://visant.bu.edu/other_formats/IDMapping_cencers.txt.

4. Collapse all metanodes using the MetaGraph→MetaNode→Collapse All menu. A dashed edge between two cancers will be created automatically if they share at least one gene.

5. Click the Zoom Out button on the toolbox six times and then click the Fit to Page button to reduce the node size and make it easier to examine the connections between diseases.

6. Lay out the cancer network using the Layout→Spring Embedded Relaxing menu. Click the Stop Animation button whenever appropriate (Figure 17.14). The cancer network will look similar to the one shown in Figure 17.14.

17.4.2 Construct the Cancer Gene Network

Apply a concept similar to the "disease gene network" (Goh et al. 2007); that is, two genes are connected if they are associated with the same disorder. We can easily create a cancer gene network in VisANT.

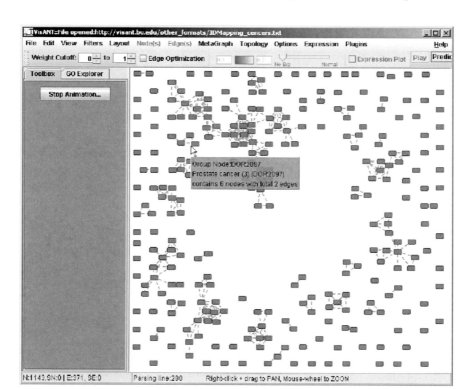

FIGURE 17.14   Network of cancers rebuilt in VisANT using a metagraph with a subset of data of cancer extracted from the work of Goh and co-workers (2007). Each metanode (grey box) represents one type of cancer. The correlations between cancers are evaluated based on the number of shared genes.

1. Use the MetaGraph→Create Co-Metanode Network menu to create the cancer gene network.

2. Repeat step 6 above to apply the spring embedded relaxing layout.

3. Change the node shape, color, and size of the cancer gene network by copying and pasting the following macros into the Add textbox (clear the textbox if necessary using the key CTRL-A and then Backspace):

```
#!batch commands
select_all_node
set_node_property=node_size:7
set_node_property=node_shape:circle
clear_selection
```

Please reference http://visant.bu.edu/vmanual/cmd.htm for more information about macros.

4. The cancer gene network will look similar to Figure 17.15.

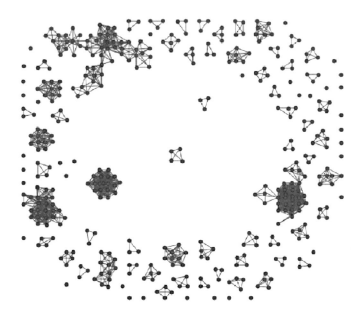

FIGURE 17.15 Cancer gene network automatically created by VisANT based on the cancer network. The dark dot represents the gene and the edge represents the two genes that are associated with the same cancer.

17.5 COMMENTARY

17.5.1 Mathematical Definition of Metagraph

A metagraph $G_m = \{V, E\}$ consists of a finite set V of nodes and a finite set E of edges. Nodes in a metagraph can be denoted as $V = \{V_s, V_m\}$ where V_s represents simple nodes as generally defined in a simple graph and V_m represents the metanodes. The subscript s represents the simple node/edge and the subscript m represents metanode/metaedge. Each metanode $v_m \in V_m$ contains a subgraph consisting of child nodes and connected edges. In addition, each node $v \in V$ represents a set of its instance nodes, that is, $v = \{V_i | i > 0\}$ where v_i is the instance nodes of v. Instance nodes have the same identity between them but can have individual specific properties. The statement that two metanodes share a node implies that each metanode contains an instance of the same node.

A metanode v_m has two states, *expanded* or *contracted*; the expanded state manifests the internal subgraph (that is, places all children nodes with their connections into the graph) while the contracted state replaces this subgraph with the single node. The combination of different states of the metanodes for a given metagraph results in multiple *views* that are abstract representations of the same underlying data. The change of views for a given metagraph is defined as the dynamics of the metagraph, as shown in Figures 17.1D and E.

Edges in a metagraph can be denoted as $E = (E_s, E_m)$ where E_s represents simple edges that are generally defined in the simple graph and E_m represents metaedges. Each metanode edge $e_m \in E_m = e_{v_m, v}$ is associated with at least one contracted metanode v_m and is transient: it appears when the metanode is contracted and disappears when one or two connected metanode nodes are expanded, that is, the metaedge is derived from the properties of two connected nodes. The most common derivation of the metaedge is the connection transfer. For example, when metanodes M1 and M2 are contracted in Figure 17.1E, the connection between C and E is transferred to M1 and M2. However, the metaedge can also be derived from other properties of the metanode. The metaedge shown in Figure 17.1E is derived because two metanodes M2 and M3 share the same node E. The derivation of the metaedge can be generalized as $e_{v_m,v} = g(v_m, v)$, where g is the aggregation function and $v \in V$ can either be a metanode node or a simple node.

17.5.2 Download and Run VisANT as a Local Application

VisANT has four running modes and two of them require a local copy of VisANT. Please visit http://visant.bu.edu and click the link "Run VisANT" for detailed instructions on other modes. It is recommended to run VisANT as a local application when handling large-scale networks, such as a network with more than 100,000 nodes and edges because you will have the option to specify the memory size that VisANT can use. In addition, a local application allows VisANT to access local resources, such as load/save network files directly; it also allows the user to develop VisANT plugins, as well as run a list of batch commands in the background without any user interface (batch mode).

The only drawback to running VisANT as a local application is that it easily becomes out of date because VisANT is under active development. Fortunately, VisANT provides a function to check for updates automatically and an icon will be shown near the Help menu if update is available. Users can either click the icon or corresponding menu to upgrade an VisANT to the latest version, as shown in Figure 17.16.

1. If not already installed, download and install the Java 2 Platform, Standard Edition, version 1.4 or higher (http://java.sun.com/javase/downloads/index.jsp).

2. Go to http://visant.bu.edu and click on the link "Download", then click the link "Latest Version of VisANT".

3. Select a directory to save the file "VisAnt.jar". The VisAnt.jar is only about 400K in size and the download takes less than one minute to finish. No installation is needed to run VisANT.

4. To launch VisANT, double-click VisAnt.jar.

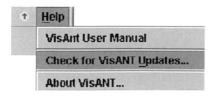

FIGURE 17.16 VisANT upgrade.

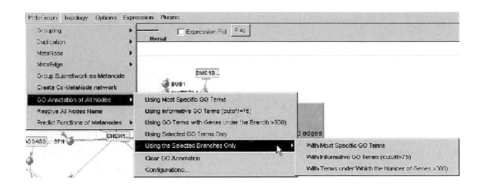

FIGURE 17.17 Menus for GO annotations under the MetaGraph menu. The menu will annotate all the genes, including those hidden in the collapsed metanodes. The same list of menus is also available under the Nodes menu, which should be used to annotate the selected nodes.

5. To launch VisANT by an alternative means, open a Dos window in Win OS, or a shell window in other operating systems, and go to the directory where VisAnt.jar is located, and run the command:

java -Xmx512M -classpath VisAnt.jar cagt.bu.visant.VisAntApplet

where 512M indicates the maximum size of the memory that VisANT can use. Increase this number if you have a large network or you get the "run out of memory" error.

6. The VisANT main window will appear (Figure 17.4).

7. To exit VisANT, close the VisANT main window, or use the File→Exit menu option, or press the key combination ALT+X.

17.5.3 Annotate Gene Functions Using Flexible Schema

VisANT provides four basic options to annotate genes using GO annotations with corresponding menus, shown in Figure 17.17. Options 1 to 3 listed below can also be applied to the selected branches. These options provide users great flexibility to test various hypotheses. To save space, we use the human gene ACN9, which that is involved in the predisposition to alcohol dependence to illustrate these options.

1. *Using the Most Specific GO Terms*: Genes are annotated with the most specific functional descriptions available at Entrez Gene database. The table below lists the GO annotation of ACN9 with this option.

Biological Process	Cellular Component
Gluconeogenesis(GO:0006094)[ISS]	Mitochondrion(GO:0005739)[IEA]
	Mitochondrial intermembrane space(GO:0005758) [ISS]

2. *Using Informative GO Terms*: Genes are annotated using GO terms (1) having more than a user-specified number of genes and (2) each of whose descendent terms have less than the specified number of genes. Let's use 145 as the cutoff (click the button near Search button of GO explorer, enter 145 in the corresponding field, and press the Enter key). The informative GO annotations for ACN9 are shown below:

Biological Process	Cellular Component
Hexose metabolic process(GO:0019318)	

3. *Using GO Terms with Genes under the Branch > cutoff*: A term must have more than a user-specified number of genes. Let's again use 145 as the cutoff, and here are the results:

Biological Process	Cellular Component
Hexose metabolic process(GO:0019318)	
Cellular alcohol metabolic process(GO:0006066)	Mitochondrion(GO:0005739)
Cellular biosynthetic process(GO:0044249)	Mitochondrial envelope(GO:0005740)
Biosynthetic process(GO:0009058)	Mitochondrial part(GO:0044429)
Carbohydrate metabolic process(GO:0005975)	Intracellular organelle part(GO:0044446)
Monosaccharide metabolic process(GO:0005996)	Membrane-enclosed lumen(GO:0031974)
Monocarboxylic acid metabolic process(GO:0032787)	Organelle envelope(GO:0031967)

4. *Using Selected GO Terms Only*: Genes are annotated using only selected GO terms. Figure 17.18 shows the selected terms and resulting annotation for ACN9.

Options 2 and 3 are frequently used when predicting gene functions using functional linkages. Annotations resulting from different options can coexist as node descriptions in VisANT for comparison purposes.

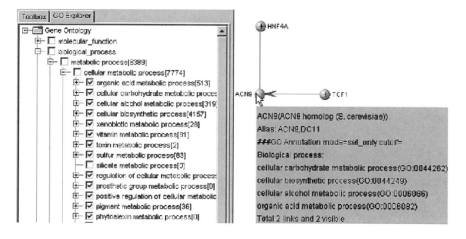

FIGURE 17.18 Annotate the gene using the selected GO terms only. Four among the total thirteen terms are annotated for ACN9 because the GO term hexose metabolic process (GO:0019318) is the child term, which will be very clear when the hierarchy of GO:0019318 is shown in the GO explorer. Please reference http://visant.bu.edu/vmanual/ver3.50.htm for information about GO hierarchy visualization.

17.5.4 GO Term Enrichment Analysis (GOTEA)

The four steps here describe how GOTEA works in VisANT. For illustration purposes, the following steps only take one metanode, G, into account and only calculate the enrichment score of one target GO term, T.

Step 1: Fully annotate all of the nodes in G with gene names and GO terms.

Step 2: Calculate density scores for each node based on the topology and the GO term similarity to T. A vector D^G of density scores of each gene in G is computed, with the element of D^G for the ith gene denoted D_i. The density score is used to evaluate the impact of other genes in G on the ith gene, according to both the GO term similarity and the topological distance to the ith gene. D_i is defined as:

$$D_i = \sum_{j \in G} \log_2 \left[\left(\frac{M_j}{\alpha} \right) \Theta(M_j - \alpha) + \Theta(\alpha - M_j) \right] e^{-\beta d_{ij}},$$

where the step function,

$$\Theta(x - y) = \begin{cases} 1 & x \geq y \\ 0 & x < y, \end{cases}$$

ensures that $D_i \geq 0$. M_j is a measure of the GO term similarity calculated based upon the graph structure of the GO term hierarchy (Wang et al. 2007). A significance threshold, α, is used to control the contribution that gene j makes to D_i. For larger α, a greater number of less statistically significant (with $M_j < \alpha$) genes are filtered and do not contribute to D_i. The shortest distance between genes i and j given the topology of G is denoted d_{ij} and was calculated with the Floyd–Warshall algorithm. We assume that shorter distances make an exponentially greater contribution to the density than do longer distances, with the steepness of the exponential determined by the parameter β When a bigger β is chosen, more distant genes can contribute to the density. Taken together, the parameters α and β are used to control the sensitivity and selectivity of the density.

Step 3: Another vector of density scores, D^{NG}, is computed based on a randomly chosen subset of genes representative of the background distribution. The background consists of all genes annotated by NCBI.

Step 4: Statistical significance for rejecting the null hypothesis is determined by a permutation test. For statistical robustness, step 3 is repeated n times. The number of times the average density score of randomly chosen genes is found to be larger than the average density score of genes in G is counted after n iterations and used to compute the final p-value.

These four steps can be carried out for multiple testing by using multiple metanodes and multiple targeting GO terms. In this case, the p-values are corrected using FDR methods (Benjamini et al. 2001). Specifically, $FDR = p \times m/k$, where m is the total number of GO terms tested and k is the rank of the GO terms under consideration. There is also an option for GOTEA to identify representative GO terms from all its discoveries based on approaches that identify the most informative GO term (Zhou, Kao, and Wong 2002).

17.5.5 Network Module Enrichment Analysis (NMEA)

NMEA is implemented in a manner similar to GOTEA. Where GOTEA used GO term similarities, NMEA uses p-values from T-tests on the expression values of two phenotypes.

Step 1: Fetch the expression profile of each gene in a given module (i.e., metanode, denoted M in the following context) from formatted user input. The input should include an adequate number of samples with comparable phenotypes (e.g., normal and disease).

Step 2: A vector D^M of density scores of each gene is computed, with the element of D^M for the ith gene denoted as D_i. D_i is defined as:

$$D_i = \sum_{j \in G} \log_2 \left[\left(\frac{\alpha}{M_j} \right) \Theta(\alpha - M_j) + \Theta(M_j - \alpha) \right] e^{-\beta d_{ij}},$$

where the step function,

$$\Theta(x - y) = \begin{cases} 1 & x \geq y \\ 0 & x < y, \end{cases}$$

ensures that $D_i \geq 0$. M_j is the p-value from a two-tailed T-test of differential expression between two phenotypes (for example, normal and disease). The parameters α and β are used to control the sensitivity and selectivity of the density, as described in the previous section.

The density score is used to evaluate the impact of other genes in M on the ith gene, according to both the p-value calculated by T-test (an indicator of differential expression) and their topological distances to the ith gene.

Step 3: Another vector of density scores, D^{NM}, is computed by randomly shuffling the phenotypes to obtain a representative sampling of the background distribution.

Step 4: Statistical significance for rejecting the null hypothesis is determined by a permutation test. For statistical robustness, step 3 is repeated n times. The number of times the average density score of randomly chosen genes is found to be larger than the average density score of genes in M is counted after n iterations and used to compute the final p-value.

When applying NMEA to multiple metanodes, the p-value must be corrected by FDR in a manner similar to what was described above for GOTEA. In this case, $FDR = p \times m/k$ as before, but m is the total number of metanodes and k is the rank of the metanodes under consideration.

ACKNOWLEDGMENTS

We thank colleagues, students, and other users of VisANT for their valuable comments and suggestions in improving all facets of the system. We appreciate the guidelines provided by the VisANT advisory board and many of our collaborators. This work is supported by the National Institutes of Health (1R01RR022971-01A1 and 1R21CA135882-01).

REFERENCES

Alibes, A., Canada, A., and Diaz-Uriarte, R. 2008. PaLS: filtering common literature, biological terms and pathway information. *Nucleic Acids Res* 36:W364–W367.

Antonov, A. V., Schmidt, T., Wang, Y., and Mewes, H. W. 2008. ProfCom: a web tool for profiling the complex functionality of gene groups identified from high-throughput data. *Nucleic Acids Res* 36:W347–W351.

Antonov, A. V., Tetko, I. V., and Mewes, H. W. 2006. A systematic approach to infer biological relevance and biases of gene network structures. *Nucleic Acids Res* 34:e6.

Barry, W. T., Nobel, A. B., and Wright, F. A. 2005. Significance analysis of functional categories in gene expression studies: a structured permutation approach. *Bioinformatics* 21:1943–1949.

Benjamini, Y., Drai, D., Elmer, G., Kafkafi, N., and Golani, I. 2001. Controlling the false discovery rate in behavior genetics research. *Behav Brain Res* 125:279–284.

Brohee, S., Faust, K., Lima-Mendez, G. et al. 2008. NeAT: a toolbox for the analysis of biological networks, clusters, classes and pathways. *Nucleic Acids Res* 36:W444–W451.

Draghici, S., Khatri, P., Bhavsar, P. et al. 2003. Onto-Tools, the toolkit of the modern biologist: Onto-Express, Onto-Compare, Onto-Design and Onto-Translate. *Nucleic Acids Res* 31:3775–3781.

Goh, K. I., Cusick, M. E., Valle, D. et al. 2007. The human disease network. *Proc Natl Acad Sci USA* 104:8685–8690.

Hu, Z., Mellor, J., and DeLisi, C. 2004. Analyzing networks with VisANT. In: *Current Protocols in Bioinformatics*. Edited by Baxevanis A., Davison D., Page R., Petsko G., Stein L., and Stormo G. New York: John Wiley & Sons.

Hu, Z., Mellor, J., Wu, J., and DeLisi, C. 2004. VisANT: an online visualization and analysis tool for biological interaction data. *BMC Bioinformatics* 5:17.

Hu, Z., Mellor, J., Wu, J. et al. 2005. VisANT: data-integrating visual framework for biological networks and modules. *Nucleic Acids Res* 33:W352–W357.

Hu, Z., Mellor, J., Wu, J. et al. 2007a. Towards zoomable multidimensional maps of the cell. *Nature Biotechnol* 25:547–554.

Hu, Z., Ng, D. M., Yamada, T. et al. 2007b. VisANT 3.0: new modules for pathway visualization, editing, prediction and construction. *Nucleic Acids Res* 35:W625–W632.

Hu, Z., Snitkin, E. S., and DeLisi, C. 2008. VisANT: an integrative framework for networks in systems biology. *Brief Bioinform* 9:317–325.

Huang, D. W., Sherman, B. T., Tan, Q. et al. 2007. The DAVID Gene Functional Classification Tool: a novel biological module-centric algorithm to functionally analyze large gene lists. *Genome Biol* 8:R183.

Khatri, P., Bhavsar, P., Bawa, G., and Draghici, S. 2004. Onto-Tools: an ensemble of web-accessible, ontology-based tools for the functional design and interpretation of high-throughput gene expression experiments. *Nucleic Acids Res* 32:W449–W456.

Khatri, P. and Draghici, S. 2005. Ontological analysis of gene expression data: current tools, limitations, and open problems. *Bioinformatics* 21:3587–3595.

Khatri, P., Voichita, C., Kattan, K. et al. 2007. Onto-Tools: new additions and improvements in 2006. *Nucleic Acids Res* 35:W206–W211.

Lee, T., Desai, V. G., Velasco, C., Reis, R. J., and Delongchamp R.R. 2008. Testing for treatment effects on gene ontology. *BMC Bioinformatics* 9 (Suppl 9):S20.

Mellor, J. C., Yanai, I., Clodfelter, K. H., Mintseris, J., and DeLisi, C. 2002. Predictome: a database of putative functional links between proteins. *Nucleic Acids Res* 30:306–309.

Mootha, V. K., Lindgren, C. M., Eriksson, K. F. et al. 2003. PGC-1alpha-responsive genes involved in oxidative phosphorylation are coordinately downregulated in human diabetes. *Nature Genet* 34:267–273.

Reimand, J., Tooming, L., Peterson, H., Adler, P., and Vilo, J. 2008. GraphWeb: mining heterogeneous biological networks for gene modules with functional significance. *Nucleic Acids Res* 36:W452–W459.

Rhee, S. Y., Wood, V., Dolinski, K., and Draghici S. 2008. Use and misuse of the gene ontology annotations. *Nature Rev Genet* 9:509–515.

Salomonis, N., Hanspers, K., Zambon, A. C. et al. 2007. GenMAPP 2: new features and resources for pathway analysis. *BMC Bioinformatics* 8:217.

Subramanian, A., Tamayo, P., Mootha, V. K. et al. 2005. Gene set enrichment analysis: a knowledge-based approach for interpreting genome-wide expression profiles. *Proc Natl Acad Sci USA* 102:15545–15550.

Volinia, S., Evangelisti, R., Francioso, F. et al. 2004. GOAL: automated Gene Ontology analysis of expression profiles. *Nucleic Acids Res* 32:W492–W499.

Wang, J. Z., Du, Z., Payattakool, R., Yu, P. S., and Chen, C. F. 2007. A new method to measure the semantic similarity of GO terms. *Bioinformatics* 23:1274–1281.

Zhang, M., Ouyang, Q., Stephenson, A. et al. 2008. Interactive analysis of systems biology molecular expression data. *BMC Syst Biol* 2:23.

Zhou, X., Kao, M. C., and Wong, W. H. 2002. Transitive functional annotation by shortest-path analysis of gene expression data. *Proc Natl Acad Sci USA* 99:12783–12788.

Zhu, J., Wang, J., Guo, Z. et al. 2007. GO-2D: identifying 2-dimensional cellular-localized functional modules in Gene Ontology. *BMC Genomics* 8:30.

CHAPTER 18

Gene Set and Pathway-Based Analysis for Cancer Omics

Dougu Nam and Seon-Young Kim

CONTENTS

18.1 INTRODUCTION

Biological processes, especially on the molecular level, are modulated by a complex network of functionally related cell components. Deregulation of the components in specific pathways results in the progression of diseases such as cancer. Therefore, in order to decipher disease mechanisms, it is of primary importance to identify pathways or specific groups of genes that exhibit unusual behavior.

Such group-wise patterns are readily investigated on the transcriptional level by combining the vast amounts of microarray expression data with predefined gene sets derived from biological databases. Indeed, since the inspiring work of Mootha et al. (Mootha et al. 2003), group-wise expression pattern analysis, designated here as gene set analysis (GSA), has received great attention, and various GSA methods have subsequently been developed and used intensively for microarray analysis (Al-Shahrour et al. 2008; Dinu et al. 2009; Huang, Sherman, and Lempicki 2009; Nam and Kim 2008).

The typical microarray analysis, as we call frequency-based analysis (FBA), first determines a list of differentially expressed genes (DEGs) using a cutoff value and then identifies frequently observed members of gene groups or pathways from the list. The use of a cutoff value, however, causes significant loss of information and statistical power. GSA methods act in the opposite direction of FBA; GSA assesses the significance of each gene set directly, and then looks into the significant gene sets to identify responsive members and their roles in pathways.

The GSA approach is coherent with biology because it analyzes the modular behavior of functionally related gene groups. Moreover, GSA methods exhibit higher statistical power than FBA methods, and hence have revealed many important group-wise patterns that FBA could not detect (Mootha et al. 2003). The high statistical power of GSA originates from the utilization of every member's information in a gene set and the group-wise approach, such that even genes with moderate expression changes taken together can represent significant patterns. Despite these advantages, the search for optimal GSA methods has generated a number of debates because of the different statistical hypotheses employed and disagreement about the concept of *differentially expressed gene sets*.

In this chapter, we introduce three different GSA methods and discuss their pros and cons. Then we provide examples of the application of these methods to the genomic analysis of disease, and introduce some widely used tools and databases. This is not an exhaustive review, but is aimed to offer a practical guide for analyzers and developers by clarifying the concepts of different GSA methods, showing recent applications of GSA, and introducing useful GSA tools. See the reviews by Nam and Kim (2008), Dopazo (2009), Huang, Sherman, and Lempicki (2009), and Dinu et al. (2009) for extensive coverage of GSA methods and tools.

18.2 DESCRIPTION OF FBA

The typical FBA procedure for microarray analysis is shown as follows:

Step 1. Compile biologically predefined gene sets derived from Gene Ontology, KEGG, or other pathway databases.

Step 2. Evaluate individual statistics for each gene: t_i, $i = 1, 2, \ldots N$, between the two sample groups compared. t_i can be mean difference, two sample t-statistic, SAM (modified t-statistic), Wilcoxson rank sum, and so on.

Step 3. Using a cutoff value, choose a list of genes (the DEG list) that have some level of significance. Then evaluate the p-value for enrichment of the members of each gene set in the list using hypergeometric distribution, Fisher's exact test, the binomial test, and so on. Adjustment for multiple hypotheses follows.

FBA methods, though simple and widely used, have several drawbacks and limitations. First, FBA methods lack sensitivity in detecting relevant gene groups, which may not be clearly identified from expression changes of only a few genes found in the gene list (Ben-Shaul, Bergman, and Soreq 2005). Second, the use of a cutoff value to determine DEGs causes information loss both for the used and the discarded genes; that is, equal weight is assigned to all DEGs irrespective of the different levels of association signals in each gene, and many of the discarded genes can still have moderate but meaningful expression changes. Third, the biological conclusions derived from FBA are often altered substantially under different choices of the cutoff value (Pan, Lih, and Cohen 2005). Last, the enrichment test of a biological annotation is based on the assumption of independent gene sampling, which usually increases false positive predictions (Goeman and Buhlmann 2007). We will investigate how GSA methods can address these issues.

Most GSA methods, except for multivariate and regression-based methods, share the first two steps of FBA by making use of individual statistics and predefined gene sets. On the other hand, GSA exploits all the individual statistics without using a cutoff value, and aggregates those contained in each gene set to assess the significance of the gene set. However, the method of aggregation and the hypothesis to be tested differ among GSA methods, which are described below.

18.3 THREE DIFFERENT APPROACHES FOR GSA

The purpose of GSA is twofold: to assess the association of gene sets with a given phenotype, and the enrichment of such association signals in each gene set. Depending on their purpose and their method of computing p-values, GSA methods can be classified into three categories: competitive, self-contained, and hybrid. The first two categories were termed and extensively discussed by Goemann and Buhlmann (2007).

18.3.1 Competitive Methods

The competitive methods test the null hypothesis that a gene set and its complement have the same level of association with the phenotype (say Q1). In other words, competitive methods are interested in whether the association signal distributed over the genome is relatively "concentrated" or "enriched" in each gene set. To assess the enrichment of DEG signals in a gene set, competitive methods use gene randomization to compute p-values. For this reason, competitive methods are widely applicable even to a small number of data samples. One competitive GSA method (Kim and Volsky 2005) is illustrated in Figure 18.1 and described by replacing Step 3 of FBA with the following step:

Step 3′ Regarding $P = \{t_1, \ldots, t_N\}$ as the population set, assume each gene set G_j to be a random collection from P and compute "summary statistic" T_j using t_i's contained in G_j, which provides the test statistic for the gene set.

The summary statistic can be a Z-score (Kim and Volsky 2005), average t (Tian et al. 2005), or SAM-statistic, or average of their pth moment (Dinu et al. 2007), which reflects the significance of each gene set. We can test its significance by randomizing gene labels. The main drawback of competitive methods that is in common with FBA is the invalid assumption of independent gene sampling. A conceptual issue with the competitive approach, called zero sum game (Allison et al. 2006), may also arise from the concept of enrichment itself. In other words, the significance of a gene set is relatively determined by its background distribution. For an extreme example, even if 70% of the members of each

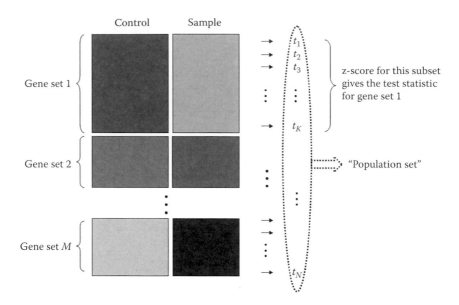

FIGURE 18.1 (See color insert following page 332.) An example of a competitive GSA method: PAGE. (From Kim S.W. and Volsky D.J. 2005. *BMC Bioinformatics* 6:144.)

gene set are DEGs, most of the gene sets will not be declared "enriched," because of the equal portion of DEGs in the background genes.

18.3.2 Self-Contained Methods

While competitive methods test the relative enrichment of the association signal with the phenotype in a gene set, self-contained methods test the existence of such association signals in a gene set. Self-contained methods test the strictest null hypothesis that no gene in the given gene set is associated with the phenotype, which, therefore, is relatively easily rejected to yield many significant gene sets. Indeed, repeated tests in the literature report that self-contained methods exhibit the highest statistical power (Ackermann and Strimmer 2009; Dinu et al. 2007). They employ sample randomization to compute p-values for each gene set, and hence cause no statistical problem with gene sampling. Self-contained methods focus on the phenotypic difference between two sample groups and do not compare the association signal with that outside a gene set. A typical self-contained method is described by replacing the third step of FBA with the following step:

Step 3″. Compute a summary statistic for a gene set, and permute sample labels to assess the significance of the gene set.

In addition to the summary statistic listed for competitive methods, the multivariate statistic of Hotelling's T^2 (Kong, Pu, and Park 2006) or the regression-based global test (Goeman et al. 2004) and ANCOVA (Hummel, Meister, and Mansmann 2008), have also been used for some self-contained methods. However, Ackermann and Strimmer (2009) reported the use of a simpler statistic provided sufficiently good performance. Although statistically legitimate, a conceptual issue appears with the self-contained approach: one or only a few DEGs can make the whole gene set significant. We may not say a pathway with 100 genes has altered its expression pattern for only one DEG in the pathway. One useful way to address the problem is the gene set reduction method (Dinu et al. 2009) that identifies the core DEGs in each gene set. As shown by Dinu et al. (2009), the most highly significant gene sets in their self-contained approach had only one or a few core DEGs. We suggest applying additional filtering steps to remove redundant and irrelevant gene sets as follows:

1. For core DEGs identified for each gene set, assign only the smallest gene set that contains all the core DEGs (say, the core gene set) and discard other larger gene sets.

2. Among the core gene sets, discard those sets that contain a smaller portion of core genes than a threshold, for example, 20%.

With these reduction and filtering processes, we expect self-contained methods will provide coherent and more useful biological information from expression data.

18.3.3 Hybrid Methods: GSEA

The third class of methods, known as GSEA (Mootha et al. 2003), takes a hybrid approach. GSEA aims to find enriched gene sets, but avoids the problem of gene sampling by comparing whole populations of gene set scores generated under sample permutations. GSEA regards the whole dataset (the set of gene sets) as the analysis unit, and tests the null hypothesis that no gene set contained in the dataset is associated with the phenotype (say Q3). The GSEA procedure (Subramanian et al. 2005) can be described by replacing the third step of FBA with the following steps:

Step 3‴. Compute the summary statistic $T_j, j = 1, 2, \dots, S$ for each gene set, where S is the number of gene sets considered. Let $T = \{T_j\}_{1 \le j \le S}$.

Step 4. Permute the sample labels and repeat *Step 3‴* to generate the $S \times P$ matrix of randomized summary statistics $T_{perm} = \{T_j^k, \ 1 \le j \le S, \ 1 \le k \le P\}$, where T_j^k is the score of the jth gene set for the kth permutation, and P is the number of permutations executed. Then normalize $[T \ T_{perm}]$ by dividing its elements by the corresponding means of the rows of T_{perm}, which we denote $[NT \ NT_{perm}]$.

Step 5. To assess the significance of NT_j, compute the FDR q value as follows:

$$FDR(NT_j) = \frac{\#\{NT_u^k \ge NT_j\}/\#\{NT_u^k \ge 0\}}{\#\{NT_v \ge NT_j\}/\#\{NT_v \ge 0\}}, \quad \text{for } NT_j \ge 0,$$

and similarly for $NT_j \le 0$.

The original GSEA (Mootha et al. 2003) used a random-walk-like Kolmogorov–Smirnov (K-S) statistic for the summary statistic. However, Subramanian et al. (2005) revised the summary statistic by weighting each gene by its level of association with the phenotype to increase statistical sensitivity. They suggested computing the FDR q value for each gene set to derive multiple significant gene sets.

Although GSEA compares the scores of different gene set scores to each other, there is a key difference from ordinary competitive methods. The ordinary competitive methods compare each gene set score with "randomly" categorized gene sets of the same size, while GSEA compares already-categorized existing gene sets with possibly different sizes. This is where the main problem of GSEA arises. As Damian and Gorfine (2004) have indicated, in original GSEA, gene sets with larger sizes can have higher significance scores than those with small sizes and intensive signals. Moreover, gene sets have different correlation structures, and hence should not be compared on equal terms. To address these problems, Subramanian et al. (2005), in their revised version of GSEA, divided each gene set score by the mean of the sample-permuted gene set score to adjust for different gene set sizes (and correlation structures). Wang and co-workers, in the context of SNP gene set analysis, applied GSEA and suggested Z-normalization for each gene set score to adjust for the different number of SNPs in each gene (Wang, Li, and Bucan 2007). This kind of normalization may also be incorporated into standard GSEA instead of dividing by the mean of each row as follows:

Step 4′. Use $T_j' = \frac{T_j - \mu_j}{v_j}$ and $T_{\text{perm}}' = \{\frac{T_j^k - \mu_j}{v_j}, \ 1 \leq j \leq S, \ 1 \leq k \leq P\}$ in place of T_j and T_{perm}, respectively, where $\{\mu_j, v_j\}, j = 1, 2, \ldots, S$ are the mean and standard deviation of the jth row of T_{perm} which takes into account the variability of permuted gene set scores.

In addition to FDR, Subramanian et al. also suggested a simplified p-value computation for GSEA, called the nominal p-value, as well as its pooled version in score when the number of samples are small, in which each gene set score is compared with the sample-permuted scores corresponding to the gene set. Efron and Tibshirani (2007) also used the simplified p-value computation, and tested five summary statistics: mean, mean.abs, maxmean, K-S. abs, and K-S in simulation tests. Among them, the maxmean statistic exhibited consistently low p-values in each test. They also suggested applying a "restandardization" procedure for general gene set statistics. Restandardization makes the test statistic reflect the background distribution by applying gene randomization to the statistic.

Since GSEA uses sample permutation for computing p-values, it may not be reliable for data with small samples. For such cases, a preranked version of GSEA was also developed, where T_{perm} is obtained by permuting gene labels (Subramanian et al. 2005).

The three GSA approaches are briefly summarized in Table 18.1. See also Table 18.2 for some widely used GSA tools.

18.3.4 Simulation Studies

We now introduce simulation studies that demonstrate the characteristics of each GSA approach (Dinu et al. 2008; Nam and Kim 2009). We commonly used the average of the absolute t-statistic in a gene set for the summary statistic and newly included restandardized GSA results Efron and Tibshirani (2007). The p-value for restandardized GSA is computed as follows (pooled version):

Step R1. Compute

$$g(T_{perm}) = \left\{ \frac{g(T_j^k) - \mu^\star}{v^\star}, \ 1 \leq j \leq S, \ 1 \leq k \leq P \right\}$$

TABLE 18.1   Comparison of GSA Methods

Methods	Competitive	Self-Contained	Hybrid: GSEA
Null hypothesis	Q1: A gene set and its complement have the same level of association with the phenotype	Q2: No gene in a gene set is associated with the phenotype	Q3: No gene set in the whole dataset is associated with the phenotype
Nature	Enrichment of association signal in a gene set	Existence of association signal in a gene set	Enrichment of association signal in a gene set/Existence of associated gene set in the full dataset
P-value computation	Gene randomization	Sample randomization	Sample randomization
Advantages	1. Applicable to dataset with a small number of samples	1. Highly sensitive 2. Statistically legitimate	1. Use statistically sound sample randomization 2. Able to compute FDR using relatively small number of permutations
Weaknesses	1. Invalid assumption of gene-gene independence 2. Zero-sum game	1. Many samples are required 2. Too many gene sets are detected that have only a small number of core genes	1. Not sensitive 2. Zero-sum game 3. Compares different gene sets on equal terms

where $g(T_j^k)$ represents a general gene set statistic and $\{\mu^*, \upsilon^*\}$ are their mean and standard deviation of randomly drawn gene sets over all the permutations performed.

Step R2. For each gene set, compute the generalized statistic $g(T_j)$ as well as its mean and variance $\{\mu, \upsilon\}$, then count the number of permutations that satisfy

$$\frac{g(T_j) - \mu}{\upsilon} \leq \frac{g(T_j^k) - \mu^*}{\upsilon^*}.$$

g represents a general function that summarizes individual gene scores. If g is a linear transform of individual scores such as mean of absolute values, pth moment, or maxmean statistic, $\{\mu^*, \upsilon^*\}$ and $\{\mu, \upsilon\}$ can simply be replaced by those mean and variance of the individual gene scores without randomly drawing gene sets.

In the first test, we considered 2,000 genes and divided them into 100 gene sets of 20 genes each. We generated 40 samples and divided them into two groups of 20 samples each. All expression values were sampled from a standard normal distribution. We randomly chose 200 genes (10%) and added a constant 2 to the second sample group to generate DEGs. Since the DEGs were chosen uniformly at random, most gene sets are not enriched with DEGs. Indeed, the p-values for competitive and restandardization methods were uniformly distributed (Figure 18.2a). On the other hand, the self-contained method detected most of the gene sets (76) to be significant with the p-value <0.05, because most

TABLE 18.2   List of Representative Gene Set Analysis Tools

Name	Statistical Model	Application Type	URL	Reference
		Univariate		
ASSESS	Q2	Octave/Java standalone	http://people.genome.duke.edu/~jhg9/assess/	(Edelman et al. 2006)
Babelomics	Q1, Q2	Web Server	http://www.babelomics.org	(Al-Shahrour et al. 2008)
ErmineJ	Q1	Java standalone	http://www.bioinformatics.ubc.ca/ermineJ/	(Lee et al. 2005)
Gazer	Q1, Q2	Web server	http://integromics.kobic.re.kr/GAzer/index.faces	(Kim et al. 2007)
GeneTrail	Q1, Q3	Web server	http://genetrail.bioinf.uni-sb.de	(Backes et al. 2007)
GSA	Q3	R package	http://www-stat.stanford.edu/~tibs/GSA/	(Efron and Tibshirani 2007)
GSEA	Q3	Java standalone, R package	http://www.broad.mit.edu/gsea/	(Subramanian et al. 2005)
PLAGE	Q2	Web server	http://dulci.biostat.duke.edu/pathways/	(Tomfohr et al. 2005)
SAM-GS	Q2	Windows Excel add-in	http://www.ualberta.ca/~yyasui/homepage.html	(Dinu et al. 2007)
SAFE	Q2	R package	http://bioconductor.org/packages/bioc/html/safe.html	(Barry et al. 2005)
sigPathway	Q1, Q2	R package	http://bioconductor.org/packages/bioc/html/sigPathway.html	(Tian et al. 2005)
		Global and Multivariate		
GlobalANCOVA	Q2	R package	http://bioconductor.org/packages/bioc/html/GlobalAncova.html	(Hummel et al. 2008)
Global test	Q2	R package	http://bioconductor.org/packages//bioc/html/globaltest.html	(Goeman et al. 2005)

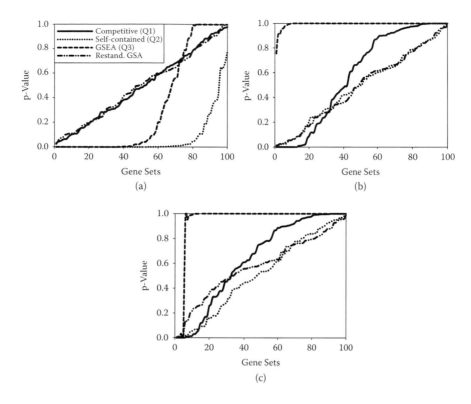

FIGURE 18.2 The p-value distributions of 100 gene sets for the three GSA approaches as well as restandardized GSA on simulated data. In each example, 2,000 permutations were performed for gene or sample randomizations on the average absolute *t*-score. (a) Randomly assigned DEGs for 10% of the genes. (b) Each gene set was sampled from a multi-dimensional normal distribution and no DEG was included. (c) Half of the members of the first ten gene sets were replaced by DEGs from the second example.

gene sets contained at least one DEG. This test example clearly contrasts the competitive and self-contained methods that target enrichment and existence of DEGs, respectively.

In the second scenario, we used the same number of genes and samples, but sampled data from a 20-dimensional normal distribution with positive correlation in each gene set, and no DEG was included. The correlation-covariance matrix was generated to have constant off-diagonal terms that were sampled from U[0,1]. The test result is shown in Figure 18.2b. Although no DEG was included, the competitive method showed a tendency to detect some correlated gene sets as significant (18) with p-value <0.05. By contrast, the self-contained method showed uniform p-value distribution. The restandardization method was not affected by correlation structures so that it also exhibited uniform p-value distribution although it used a gene-randomization score. This was an unexpected result, and we infer that the sample permutation automatically adjusted for the correlation structures.

In the third example, we replaced 10 genes (half of the members) with DEGs for the first 10 gene sets in the second example. With the p-value cutoff 0.05, all the methods correctly detected the 10 DEG sets. However, the competitive and self-contained methods detected five and six additional gene sets, respectively, while the restandardization method detected no false positive gene sets. On the other hand, for a lower cutoff value of 0.01, the true negatives increased for the restandardization (seven), while the false positives of the former two methods decreased to three and zero, respectively.

Overall, competitive and self-contained methods are highly sensitive, but the competitive method tends to additionally detect some correlated gene sets. On the other hand, the restandardization method provided the most conservative predictions and yielded few false positives. These features may not be exactly realized in real situations due to the complexity of gene set structures (e.g., different sizes, overlaps, and correlations) but it will be helpful to understand the basic properties of each algorithm.

18.4 GENE SET AND PATHWAY-BASED APPROACHES IN CANCER OMICS DATA MINING

Cancer is a disease caused by multiple genetic and epigenetic variations in many genes over a long period of time. Cancer-causing genes include many oncogenes and tumor suppressor genes that interact in multiple pathways, and whose deregulation is critical in all stages of carcinogenesis, including initiation, progression, and distant metastasis. Numerous gene expression and genetic mutation studies have shown repeatedly that cancer is a pathway disease, thus highlighting the importance of pathway-level understanding of carcinogenesis in each stage. We discuss several examples in which novel insights and knowledge were obtained by successful application of gene set and pathway-based analysis to cancer omics data.

18.4.1 The Gene Set Approach to Understanding Carcinogenesis

One of the representative gene set-level studies for carcinogenesis was the cancer module map conducted by Segal et al. (2004). With 1,975 samples, including 22 tumor types and 2,849 predefined gene sets, they first identified 456 gene sets that were activated or repressed in specific tumor groups. Then, by associating each module with each sample's clinical information, they constructed a global cancer module map that identifies significantly activated or repressed genes under specific clinical conditions. The module map covers diverse processes, from general tumorigenic processes (such as the cell cycle) to those relevant to a specific tissue or tumor type (such as growth inhibition in acute lymphoblastic leukemia.) Segal et al.'s work illustrates the power of the gene set approach to gain novel insights when applied to multiple cancer gene expression datasets.

Similarly, Tomlins et al. (2007) applied gene set and pathway analysis to their prostate cancer gene expression data obtained from laser capture microdissection (LCM)-dissected prostate samples, and identified key gene sets for each of the prostate cancer tumorigenic processes. Edelman et al. (2008) recently suggested a method for pathway-level modeling of cancer progression. First, relevant pathways in each step of cancer progression were identified by applying GSEA to each step, and then a pathway interaction network was constructed by measuring dependencies among relevant pathways. Their approach enables a more detailed understanding of pathway deregulation during each step of carcinogenesis.

18.4.2 Gene Set Analysis for Personalized Cancer Treatment

Cancer consists of molecularly heterogeneous disease types; many morphologically and histopathologically similar tumors show various clinical outcomes and responses to chemotherapeutic drugs. Recent successful targeted therapeutics such as Herceptin and Gleevec well illustrate the importance of selecting a subset of patients likely to respond to the drug, but patient selection is also important in using traditional cytotoxic anticancer drugs. Gene set and pathway analyses are useful methods for selecting patients likely to respond to specific chemotherapeutic treatments.

By infecting cells with adenovirus expressing Myc, Ras, E2F3, Src, or beta-catenin, Bild et al. (2006) generated five oncogenic pathway signatures and showed that clinical outcome depends on the deregulation level of these five signatures in breast, lung, and ovarian cancer patients. They also showed that the deregulation pattern of the five oncogenic pathways indicates the sensitivity of a cell line to drugs that target each pathway. Potti et al. (2006) proved the feasibility of chemotherapy-guiding gene sets by showing that gene sets defined from *in vitro* sensitivity tests to chemotherapeutic drugs could successfully predict patients' responses to chemotherapeutic drugs.

To address the necessity of patient selection for individualized treatment, Edelman et al. (2006) suggested a GSA method, called analysis of sample set enrichment analysis (ASSESS), to identify the activation status of oncogenic pathways in individual patients. After normalization of gene expression values by suitable reference samples (for example, normal tissues), gene label permutation can also be applied to get sample-level gene set values.

18.4.3 Gene Set Approaches for Other Types of Omics Data

Until now, we have discussed the application of GSA to gene expression data analysis, but it is equally applicable to the analysis of other types of omics data, such as genome-wide association studies (GWAS) and genome-wide mutation analysis.

GWAS is an efficient method for identifying genetic variants with small to moderate disease risks that has been widely used to identify risk variants in complex diseases such as diabetes, hypertension, and cancer. Most of the complex diseases are the result of complex interactions of many small- to moderate-risk genetic variants with each other and with environmental factors as well as gene-environment interactions. In this regard, GSA is an efficient approach for aggregating small to moderate genetic risks and identifying significant group-wise patterns in GWAS data analysis.

Wang and co-workers, by applying GSEA to three GWAS datasets (two Parkinson's disease and one acute macular degeneration), showed that novel insights can be derived by pathway-based analysis of GWAS data (Wang, Li, and Bucan 2007). To adjust GSEA for the SNP categorical data, they first assigned the most significant SNP to each gene and then applied a sample permutation to infer the significance of each gene set. Two Parkinson's disease datasets were compared, but no commonly significant SNPs were found between them. However, GSA revealed that the two datasets are congruent in selecting the "gluta-mate receptor activity" gene set, a well-known Parkinson's disease susceptibility pathway. Such increased congruence by GSA between independent datasets is commonly observed in microarray gene expression analysis. Torkamani and co-workers applied gene set and pathway analysis to the Wellcome Trust Case-Control Consortium dataset (2007) of seven diseases (bipolar disease, coronary artery disease, Crohn's disease, hypertension, rheuma-toid arthritis, type 1 diabetes, and type 2 diabetes) and found significant pathways in each disease (Torkamani, Topol, and Schork 2008). Holden et al. (2008) provided GSEA-SNP, a modified version of GSEA software for gene set analysis of GWAS data.

Due to the recent advances in next-generation sequencing technology that dramatically reduce time and cost, more and more research groups are expected to produce genome-scale mutation data that have previously been available only from a few laboratories (Shendure and Ji 2008). GSA is also suitable for genome-scale mutation analysis, because genetic changes in cancer ultimately converge into several core pathways. For example, Lin et al. (2008) identified many novel as well as well-known pathways in breast and colon cancers through a multidimensional mutation analysis of genes. This shows that gene set- and pathway-level understanding of genome-scale genetic mutations will be a promising approach to understand the complexity of cancer genomes.

18.5 AVAILABLE GENE SET AND PATHWAY DATABASES AND VISUALIZATION TOOLS

Gene set databases are as important as GSA algorithms. Previously, we listed several downloadable gene set databases on the web (Nam and Kim 2008). Here, we provide a revised list with a few additional gene set and pathway databases as well as some pathway manipulation and visualization tools (Table 18.3). We note that only a small portion of pathway databases is included here. For further pathway-related information, the Pathway Commons web site (http://www.pathwaycommons.org/pc/) is a good resource.

In addition to gene set-level analysis, direct visual inspection of gene expression changes in the context of biological pathways can substantially increase our understanding of the biological mechanism of interest. GenMAPP is one of the most famous software tools for visualizing gene expression, and it provides many useful tools for pathway drawing, manipulation, and statistical analysis (Dahlquist et al. 2002). PathVisio is newer software for pathway visualization that provides editing in the GenMAPP pathway markup lan-guage (GPML) format, as well as increased flexibility in visualizing many different types of data (van Iersel et al. 2008). Pathways in the GPML format are easily converted from and to GenMAPP pathways. Cytoscape is the most famous software for biological network analysis, and provides many useful plug-ins, including visualization of gene expression data (Shannon et al. 2003). Advanced Pathway Painter is a freeware program for the visu-alization of pathways in the context of genomic data. It automatically imports BioCarta, GenMAPP, and KEGG pathways from the web and offers easy-to-use functions for data visualization.

TABLE 18.3   List of Gene Set and Pathway Databases and Visualization Tools

Name	Organism	Gene Sets	Web Address	Reference
		Databases		
ASSESS	H	Cytogenetic, pathway, motif	http://people.duke.edu/~jhg9/assess/genesets.shtml	(Edelman et al. 2006)
DAVID Knowledgebase	H, M, R	GO, cytogenetic, pathways, InterPro,	http://david.abcc.ncifcrf.gov/home.jsp	(Sherman et al. 2007)
ErmineJ	H, M, R	GO	http://www.bioinformatics.ubc.ca/ermineJ/	(Lee et al. 2005)
Gazer	H, M, R, Y	Go, composite GO, InterPro, Pathways, TFBS	http://integromics.kobic.re.kr/GAzer/documents.jsp	(Kim et al. 2007)
GSA	H	Tissue, cellular processes, cytobands, chromosome arms, cancer module	http://www-stat.stanford.edu/~tibs/GSA/	(Efron and Tibshirani 2007)
MSigDB	H	Positional, curated pathways motif, computed, GO	http://www.broad.mit.edu/gsea/msigdb/index.jsp	(Subramanian et al. 2005)
Pathway Commons	H, M, R and others	Pathways	http://www.pathwaycommons.org/pc/	
PLAGE	H, M	KEGG and BioCarta pathways	http://dulci.biostat.duke.edu/pathways/misc.html	(Tomfohr et al. 2005)
WikiPathways	H, M, R, and others	Pathways	http://www.wikipathways.org/	(Pico et al. 2008)
		Visualization Tools		
Advanced Pathway Painter	H, M, R	GO, BioCarta, GenMAPP, KEGG	http://www.gsa-online.de/eng/app.html	
Cytoscape	H, M, R, and others	GO, PPI, Pathways	http://www.cytoscape.org/	(Shannon et al. 2003)
GenMAPP	H, M, R, and others	GO, KEGG, GenMAPP	http://www.genmapp.org/	(Dahlquist et al. 2002)
PathVisio	H, M, R, and others	GO, KEGG, GenMAPP	http://www.pathvisio.org/	(van Iersel et al. 2008)

18.6 CONCLUSIONS AND FUTURE DIRECTIONS

Gene set or pathway-based analysis has become a powerful approach for deriving biological information from microarray data, either gene expression or SNP. It utilizes gene sets or pathways as the basic units of analysis and substantially broadens our understanding of biological processes via newly found group-wise patterns. The group-wise approach of GSA is well suited to and very efficient for identifying the deregulation of specific pathways in disease, and thus represents a primary step toward deciphering disease mechanisms from omics data.

GSA methods generalize different aspects of the widely used FBA. Competitive and GSEA methods generalize the enrichment analysis of FBA (Step 3), while self-contained methods generalize the detection of phenotypic difference in individual genes (Step 2). In particular, the reduction analysis for self-contained methods covers the role of gene-level analysis by analyzing the behavior of individual core genes or small groups of them. Since different GSA approaches reveal different aspects of group-wise patterns, we recommend using different GSA methods simultaneously as follows: apply the most powerful self-contained method to identify any gene set with a significant phenotypic difference, and then apply competitive or GSEA methods to identify gene sets relatively enriched with DEGs. Significant sets identified by competitive methods, but not by self-contained methods, are likely to be false positives detected by correlation structures. And those identified by self-contained methods should be reduced and filtered to obtain more relevant gene sets.

As Huang and colleagues appropriately compared GSA to a Google search, human discretion is the most important factor in GSA, because the cutoff for statistical significance has some level of arbitrariness and the multiple hypothesis correction is far from trivial due to gene-gene interactions and the complex organization of gene sets (Huang, Sherman, and Lempicki 2009).

The utility of GSA will increase in proportion with the advance of biological databases. Because the development of new GSA algorithms may be becoming saturated, enhancing the quality of information within biological databases will be more critical for GSA. For example, ADGO suggests using multidimensional (composite) gene sets across different databases for GSA, which will easily and considerably enrich the biological information of the databases in use (Nam et al. 2006).

ACKNOWLEDGMENTS

This work was supported by grants from KRIBB Research initiative grant (KGM5160911) and C-RESEARCH-08-09-NIMS from Korea Research Council of Fundamental Science and Technology.

REFERENCES

Ackermann, M. and Strimmer, K. 2009. A general modular framework for gene set enrichment analysis. *BMC Bioinformatics* 10:47.
Al-Shahrour, F., Carbonell, J., Minguez, P. et al. 2008. Babelomics: advanced functional profiling of transcriptomics, proteomics and genomics experiments. *Nucleic Acids Res* 36:W341–W346.

Allison, D. B., Cui, X., Page, G. P., and Sabripour, M. 2006. Microarray data analysis: from disarray to consolidation and consensus. *Nature Rev Genet* 7:55–65.

Backes, C., Keller, A., Kuentzer, J. et al. 2007. GeneTrail: advanced gene set enrichment analysis. *Nucleic Acids Res* 35:W186–W192.

Barry, W. T., Nobel, A. B., and Wright, F. A. 2005. Significance analysis of functional categories in gene expression studies: a structured permutation approach. *Bioinformatics* 21:1943–1949.

Ben-Shaul, Y., Bergman, H., and Soreq, H. 2005. Identifying subtle interrelated changes in functional gene categories using continuous measures of gene expression. *Bioinformatics* 21:1129–1137.

Bild, A. H., Yao, G., Chang, J. T. et al. 2006. Oncogenic pathway signatures in human cancers as a guide to targeted therapies. *Nature* 439:353–357.

Dahlquist, K. D., Salomonis, N., Vranizan, K., Lawlor, S. C., and Conklin, B. R. 2002. GenMAPP, a new tool for viewing and analyzing microarray data on biological pathways. *Nature Genet* 31:19–20.

Damian, D. and Gorfine, M. 2004. Statistical concerns about the GSEA procedure. *Nature Genet* 36:663; author reply 63.

Dinu I., Potter J.D., Mueller T. et al. 2007. Improving gene set analysis of microarray data by SAM-GS. *BMC Bioinformatics* 8:242.

Dinu, I., Potter, J. D., Mueller, T. et al. 2009. Gene-set analysis and reduction. *Brief Bioinformatics*.

Dopazo, J. 2009. Formulating and testing hypotheses in functional genomics. *Artificial Intelligence in Medicine* 45:97–107.

Edelman, E., Porrello, A., Guinney, J. et al. 2006. Analysis of sample set enrichment scores: assaying the enrichment of sets of genes for individual samples in genome-wide expression profiles. *Bioinformatics* 22:e108–e116.

Edelman, E. J., Guinney, J., Chi, J. T., Febbo, P. G., and Mukherjee, S. 2008. Modeling cancer progression via pathway dependencies. *PLoS Comput Biol* 4:e28.

Efron, B. and Tibshirani, R. 2007. On testing the significance of sets of genes. *Ann Appl Stat* 1:107–129.

Goeman J. J. and Buhlmann P. 2007. Analyzing gene expression data in terms of gene sets: methodological issues. *Bioinformatics* 23:980–987.

Goeman, J. J., Oosting, J., Cleton-Jansen, A. M., Anninga, J. K., and van Houwelingen, H. C. 2005. Testing association of a pathway with survival using gene expression data. *Bioinformatics* 21:1950–1957.

Goeman, J. J., van de Geer, S. A., de Kort, F., and van Houwelingen, H. C. 2004. A global test for groups of genes: testing association with a clinical outcome. *Bioinformatics* 20:93–99.

Holden, M., Deng, S., Wojnowski, L., and Kulle, B. 2008. GSEA-SNP: applying gene set enrichment analysis to SNP data from genome-wide association studies. *Bioinformatics* 24:2784–2785.

Huang, D. A. W., Sherman, B. T., and Lempicki, R. A. 2009. Bioinformatics enrichment tools: paths toward the comprehensive functional analysis of large gene lists. *Nucleic Acids Res* 37:1–13.

Hummel, M., Meister, R., and Mansmann, U. 2008. GlobalANCOVA: exploration and assessment of gene group effects. *Bioinformatics* 24:78–85.

Kim, S. B., Yang, S., Kim, S. K. et al. 2007. GAzer: gene set analyzer. *Bioinformatics* 23:1697–1699.

Kim S.Y., and Volsky D.J. 2005. PAGE: parametric analysis of gene set enrichment. *BMC Bioinformatics* 6:144.

Kong, S. W., Pu ,W. T., and Park, P. J. 2006. A multivariate approach for integrating genome-wide expression data and biological knowledge. *Bioinformatics* 22:2373–2380.

Lee, H. K., Braynen, W., Keshav, K., and Pavlidis, P. 2005. ErmineJ: tool for functional analysis of gene expression data sets. *BMC Bioinformatics* 6:269.

Lin, R., Dai, S., Irwin, R. D. et al. 2008. Gene set enrichment analysis for non-monotone association and multiple experimental categories. *BMC Bioinformatics* 9:481.

Mootha, V. K., Lindgren, C. M., Eriksson, K. F. et al. 2003. PGC-1alpha-responsive genes involved in oxidative phosphorylation are coordinately downregulated in human diabetes. *Nature Genet* 34:267–273.

Nam, D. and Kim, S. B., Kim, S. K. et al. 2006. ADGO: analysis of differentially expressed gene sets using composite GO annotation. *Bioinformatics* 22:2249–2253.

Nam, D. and Kim, S. Y. 2008. Gene-set approach for expression pattern analysis. *Brief Bioinform* 9:189–197.

Pan, K. H., Lih, C. J., and Cohen, S. N. 2005. Effects of threshold choice on biological conclusions reached during analysis of gene expression by DNA microarrays. *Proc Natl Acad Sci USA* 102:8961–8965.

Pico, A. R., Kelder, T., van Iersel, M. P. et al. 2008. WikiPathways: pathway editing for the people. *PLoS Biol* 6:e184.

Potti, A., Dressman, H. K., Bild, A. et al. 2006. Genomic signatures to guide the use of chemotherapeutics. *Nature Med* 12:1294–1300.

Segal, E., Friedman, N., Koller, D., and Regev, A. 2004. A module map showing conditional activity of expression modules in cancer. *Nature Genet* 36:1090–1098.

Shannon, P., Markiel, A., Ozier, O. et al. 2003. Cytoscape: a software environment for integrated models of biomolecular interaction networks. *Genome Res* 13:2498–2504.

Shendure, J. and Ji, H. 2008. Next-generation DNA sequencing. *Nature Biotechnol* 26:1135–1145.

Sherman, B. T., Huang, D. A. W., Tan, Q. et al. 2007. DAVID Knowledgebase: a gene-centered database integrating heterogeneous gene annotation resources to facilitate high-throughput gene functional analysis. *BMC Bioinformatics* 8:426.

Subramanian, A., Tamayo, P., Mootha, V. K. et al. 2005. Gene set enrichment analysis: a knowledge-based approach for interpreting genome-wide expression profiles. *Proc Natl Acad Sci USA* 102:15545–15550.

Tian, L., Greenberg, S. A., Kong, S. W. et al. 2005. Discovering statistically significant pathways in expression profiling studies. *Proc Natl Acad Sci USA* 102:13544–13549.

Tomfohr, J., Lu, J., and Kepler, T. B. 2005. Pathway level analysis of gene expression using singular value decomposition. *BMC Bioinformatics* 6:225.

Tomlins, S. A., Mehra, R., Rhodes, D. R. et al. 2007. Integrative molecular concept modeling of prostate cancer progression. *Nature Genet* 39:41–51.

Torkamani, A., Topol, E. J., and Schork, N. J. 2008. Pathway analysis of seven common diseases assessed by genome-wide association. *Genomics* 92:265–272.

van Iersel, M. P., Kelder, T., Pico, A. R. et al. 2008. Presenting and exploring biological pathways with PathVisio. *BMC Bioinformatics* 9:399.

Wang, K., Li, M. and Bucan, M. 2007. Pathway-based approaches for analysis of genomewide association studies. *Am. J. Hum. Genet.* 81.

Wellcome Trust Case-Control Consortium. 2007. Genome-wide association study of 14,000 cases of seven common diseases and 3,000 shared controls. *Nature* 447:661–678.

CHAPTER 19

SH2 Domain Signaling Network and Cancer

Shawn S.-C. Li and Thamara K.J. Dayarathna

CONTENTS

19.1   PHOSPHOTYROSINE BINDING MODULES

The covalent modification of a protein on a Tyr residue serves several important functions. First, tyrosine phosphorylation of an enzyme may regulate its activity. This is typified by the phosphorylation of the activating loop Tyr residue in a PTK, which leads to activation of the kinase domain, and by the phosphorylation of a C-terminal Tyr residue by Csk, which functions to shut down the kinase (Levinson et al. 2008). Second, phosphorylation may alter the subcellular localization of a protein. For instance, the activation and nuclear retention of the STAT transcription factors are regulated by homodimerization mediated by tyrosine phosphorylation (Wenta et al. 2008). Third, phosphorylation may lead to endocytosis or degradation of a protein. In this regard, the phosphorylation of a C-terminal tyrosine residue on the EGF receptor triggers the binding of the E3 ligase Cbl and subsequent ubiquitination and degradation of the receptor (Sweeney and Carraway 2004). And last, and perhaps most relevant to this chapter, phosphorylation creates docking sites for proteins harboring a phosphotyrosine-recognition domain such that a signal initiated at a PTK may be transduced to downstream molecules efficiently and with high fidelity. A prototypical example is found on insulin receptor substrate 1 (IRS-1), which is phosphorylated on multiple tyrosine residues by the insulin receptor following its activation. This effectively creates multiple docking sites for the recruitment of downstream signaling molecules such as Grb2, SHP2, and PI3K (Ogawa, Matozaki, and

Kasuga 1998). Together with other modular domains, pTyr-binding modules provide an effective means by which to form elaborate and highly regulated pathways and networks for signal integration and diversification (Kaneko, Li, and Li 2008; Li 2005; Pawson, Gish, and Nash 2001; Pawson and Scott 2005; Schlessinger and Lemmon 2003; Wiggin, Fawcett, and Pawson 2005).

Selective protein-protein interactions, frequently mediated by interaction modules such as the SH2 domain, are important in organizing and regulating cellular processes. The SH2 domain, first identified as a highly conserved noncatalytic region in the Src family of cytoplasmic kinases, serves as a prototypical example of a modular interaction domain. It is the largest family of phosphotyrosine-binding modules, with 120 members found in the human genome (Sadowski, Stone, and Pawson 1986). The importance of the SH2 domain in normal cellular functions and tumorigenesis is underscored by the fact that mutation in an SH2 domain or deletion of a gene encoding an SH2 protein often leads to aberrant cellular behavior, such as transformation (Johnson and Hunter 2005; Liu et al. 2006). An SH2 domain typically binds to a specific phosphotyrosine (pTyr)-containing motif and thereby couples an activated PTK to intracellular pathways that regulate many aspects of cellular communication in metazoans (Huang et al. 2008; Liu et al. 2006; Pawson, Gish, and Nash 2001; Pawson 2004). The intimate relationship between the tyrosine kinase and SH2 domain is supported by their coordinated emergence during eukaryotic evolution. Evolutionary analysis of the phosphotyrosine signaling machinery suggests concurrent expansion of tyrosine kinases (PTK), protein tyrosine phosphatases (PTPs), and SH2 domains that function, respectively, as "writers," "erasers," and "readers" of phosphotyrosine modifications (Pincus et al. 2008). Besides SH2, phosphotyrosine-binding or PTB domains are capable of binding to NPxY motifs, where x represents any amino acid. However, among the approximately 60 PTB domains found in a mammalian cell, only a small portion (<25%) bind to their cognate ligands in a phosphorylation-dependent manner, while the majority do not require tyrosine phosphorylation for binding (Uhlik et al. 2005). Curiously, the C2 domain of PKCθ was recently shown to recognize phosphotyrosine residues on target proteins (Benes et al. 2005). However, this may be an idiosyncratic rather than a general phenomenon as no other C2 domain has since been found to bind tyrosine phosphorylated proteins or peptides.

19.2 SH2 DOMAIN STRUCTURE AND SPECIFICITY

SH2 domains are a group of structurally conserved protein modules of 100 amino acids, which, in general, bind selectively to phosphotyrosine-containing sequences (Songyang et al. 1993, 1995). Different SH2 domains have distinct preferences for residues C-terminal to the pTyr. All SH2 domains share the same structure characterized by a central β-sheet flanked by two α-helices (Kuriyan and Cowburn 1997; Waksman et al. 1992; Waksman, Kumaran, and Lubman 2004). Apart from a highly conserved phosphotyrosine-binding pocket (Figure 19.1A) that is used to engage the phosphotyrosine residue in a ligand, the remaining binding surface on different SH2 domains is much more variable. As shown in Figure 19.1A for the NCK SH2 domain, some SH2 domains contain another pocket

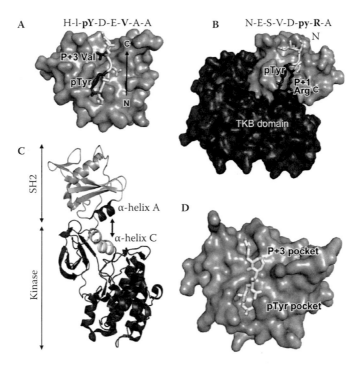

FIGURE 19.1 Structural basis of SH2-ligand and SH2-inhibitor interactions. SH2 domains are in light grey. (A) A representative SH2-ligand complex structure. The crystal structure of the NCK2 SH2 domain in complex with a peptide derived from the enteropathogenic *E. coli* protein Tir (translocated intimin receptor) (PDB code: 2Cia) is shown. The phosphotyrosine and P+3 Val of the peptide are the two major specificity determinants (PDB: 1z3k) (Ran et al. 2005). (B) An unconventional SH2 domain, residing within the larger tyrosine kinase-binding (TKB) domain of Cbl, in complex (PDB; 1jpa) with a peptide from c-Met, both of which are proto-oncogene products. Note that the direction of peptide binding is opposite to that found in a conventional SH2-ligand interaction (Peschard et al. 2004). (C) The SH2 and kinase domains of the proto-oncogenic kinase Fes. Electrostatic interactions between the SH2 α-helix A and the kinase α-helix C play an important role in kinase activation (PDB: 1WQU) (Filippakopoulos et al. 2008). Figures were drawn in the same scale and with identical SH2 domain orientation throughout (A) to (C). (D) The Src SH2 domain in complex with an inhibitor RU84687 (PDB; 1O45) that mimics phosphopeptide binding. The IC_{50} is 0.25 uM (Xu et al. 2009). The program PyMol was used for the preparation of the figures.

for the third residue (i.e., P+3), usually hydrophobic, downstream of the phosphotyrosine (Boggon and Eck 2004; Engen et al. 2008; Huang et al. 2008; Machida and Mayer 2005; Roskoski 2004; Shen et al. 2005; Waksman et al. 1992). This P+3-binding pocket is defined by a hydrophobic cleft molded between the EF and BG loops. The two-pronged (i.e., pTyr and P+3) binding mode is seen in the majority of SH2-ligand interactions (Waksman et al. 1993). Nevertheless, variations from this conventional mode of ligand recognition are found in certain SH2 domains. For instance, the SAP SH2 domain is capable of binding to either a phosphorylated or a nonphosphorylated peptide. The binding of a nonphosphorylated ligand, however, requires an N-2 Thr or Ser residue (Hwang et al. 2002; Li et al. 1999; Ostrakhovitch and Li 2006; Zwahlen et al. 2000). Together with a binding pocket for P+3 hydophobic residues, the SAP SH2 domain recognizes its cognate ligand in a "three-pronged," instead of a "two-pronged," mode of binding (S.C. Li et al. 1999). In another special case, the proto-oncogenic Cbl protein features an expanded tyrosine kinase binding (TKB) domain that engages a ligand peptide in an orientation opposite to the conventional mode of binding (Figure 19.1B; Peschard et al. 2004). Binding of the SH2 domain with the kinase domain in the Fes (Figure 19.1C) and Abl, mediated by electrostatic interactions between the α-helix A in the former and the α-helix C in the latter, is coupled to kinase activation (Filippakopoulos et al. 2008).

The distinct features of different SH2 domains binding to their cognate phosphotyrosyl sequences provide a general mechanism for the formation of unique protein complexes in PTK-mediated intracellular signal transduction (Birge and Hanafusa 1993; Campbell and Jackson 2003; Huang et al. 2008; Pawson 2004). Such specificity is engendered by the differences in architecture for different SH2-ligand interfaces despite the conserved overall structure for all SH2 domains (Machida et al. 2007). The specificity and affinity of an SH2 domain are important contributors to the specificity and regulation of cellular signal transduction pathways involving the corresponding SH2 proteins. SH2 domains bind to their cognate ligands with affinities in the submicromolar to micromolar range (De Fabritiis et al. 2008; Zhou et al. 1995). For a typical high-affinity phosphopeptide-SH2 interaction, greater than 50% of the binding free energy comes from the pTyr residue (Grucza et al. 1999). This makes phosphorylation and dephosphorylation of a tyrosine, mediated by a PTK and a PTP, respectively, an important binary switch in signal transduction (Figure 19.2A). An SH2 domain often co-occurs with other modular domains in regulatory proteins and kinases, suggesting a combinatorial mechanism of regulation. In the case of cytosolic tyrosine kinases such as the Src family PTK, the coexistence of an SH2 domain and a kinase domain serves an important regulatory function (Moniakis et al. 2001; Pincus et al. 2008). In resting cells, Src is kept in an inactive conformation by an intramolecular interaction involving the SH2 domain and a C-terminal phosphotyrosine. Occupancy of the SH2 domain by a high affinity ligand would then open up this inhibitory conformation, leading to activation of Src (Figure 19.2B). The recruitment of Src to a substrate by its SH2 domain would also allow processive tyrosine phosphorylation of a substrate. This apparent coupling of SH2 domain binding to kinase signaling stems from the fact that both the Src SH2 and kinase domains of an Src kinase recognize the same peptide motif C-terminal to the pTyr or Tyr residue (Songyang and Cantley 1995). Additional mechanisms exist that pertain to the control of SH2-pTyr signaling. Structural analysis of the tandem SH2 domains of human ZAP-70 in complex with a doubly phosphorylated peptide drawn from the zeta-subunit of the T-cell receptor reveals a cooperative behavior of the two SH2 domains in conferring high affinity and high specificity binding (Hatada et al. 1995; Figure 19.2C). Moreover, sequential interactions of SH2 domains to create active sites for other proteins have also been observed. For example, c-Cbl, an adapter protein for receptor protein tyrosine kinases (RPTKs), regulates a RPTK ubiquitination by binding to a pTyr site via its SH2 domain and promoting receptor

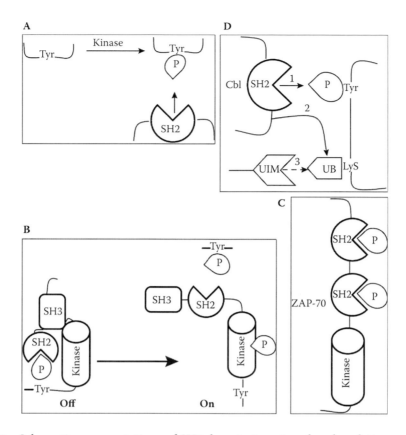

FIGURE 19.2   Schematic representations of SH2 domain-tyrosine phosphorylation-mediated regulatory modes. (A) Inducible interaction. Upon kinase activation, phosphorylated tyrosine residues serve as docking sites for SH2 domains. (B) Intramolecular interaction. Autoinhibition of a Src kinase through intramolecular interactions between its SH2 domain and a phosphotyrosine residue at the C-terminus. When the tail is dephosphorylated or when the SH2 and SH3 domains are engaged in intermolecular interactions, the kinase is activated. (C) Cooperative interaction. Tandem SH2 domains of Zap-70 (zeta-chain associated protein kinase) interacts with doubly phosphorylated ITAMs (immunoreceptor tyrosine-based activation motifs) in a cooperative manner. (D) Sequential interactions. Phosphotyrosine of a receptor tyrosine kinase on a specific Tyr residue results in the recruitment of Cbl through its SH2 domain. The RING finger domain in Cbl functions as an E3 ligase to promote the ubiquitination of the receptor on a specific Lys residue, which ultimately leads to downregulation of the receptor (Joazeiro et al. 1999).

ubiquitination on Lys residues through its RING finger domain that serves as an E3 ligase (Joazeiro et al. 1999; Figure 19.2D).

19.3 SH2 DOMAIN IN CANCER

Cancers are frequently caused by perturbations in signal transduction pathways that are regulated by protein kinases (Hunter 2000, Pawson and Nash 2000). Certain classes of signaling proteins, including receptor tyrosine kinases, cytosolic tyrosine kinases, and phosphotyrosine phosphatases, are frequently targeted for overexpression and/or amplification, creation of an autocrine stimulation loop, point mutations, deletions, or chromosomal rearrangements in cancer cells (Hanahan and Weinberg 2000; Hunter 2000). For example, untimely expression of the RTK Ret is associated with papillary tyroid carcinomas (Jenkins et al. 2002; Sagartz et al. 1997). Furthermore, specific point mutations in Ret are found responsible for multiple endocrine neoplasia types 2A and 2B and familial medullary thyroid carcinomas (Carlomagno et al. 1995). Not surprisingly, over-expression of an RTK leads to enhanced kinase activity as is the case with overexpression of the Neu/ErbB2 receptor in breast and lung cancers (Harari and Yarden 2000; Levkowitz et al. 2000; Mohi and Neel 2007; Neel, Gu, and Pao 2003; Ooms et al. 2009; Tsui et al. 2006). A number of cytoplasmic PTKs have been identified in either mutated or overexpressed forms in human malignancies (Blume-Jensen and Hunter 2001; Rikova et al. 2007). Bcr-Abl, an oncogene fusion protein produced by the Philadelphia chromosome, is associated with chronic myeloid leukemia (Konopka, Watanabe, and Witte 1984). A recent global survey of the phosphotyrosine profiles in cancer cell lines and tumor tissue samples identified an array of kinases activated in lung cancer (Rikova et al. 2007). These include RTKs such as the anaplastic lymphoma kinase (Alk), the proto-oncogenic tyrosine kinase ROS, insulin receptor (INSR), fibroblast growth factor receptor 1 (FGFR1), EGFR family members ErbB2 and ErbB3, PDGFR, and the EPH receptor family. Many nonreceptor tyrosine kinases, including Fyn, Lyn, HCK, Lck, Fer, and Fak, have also been found aberrantly activated. The activation of such a wide spectrum of PTKs suggests that lung cancer, and likely other types of cancer also, is caused by an amplified kinase signaling program rather than the aberrant activation of a single kinase (Guo et al. 2008; Hynes and MacDonald 2009; Yeh and Der 2007).

Because a primary function of PTK is to create phosphorylation sites for the recruitment of SH2 domain-containing proteins, it is reasonable to expect aberrant SH2 domain signaling to have a detrimental effect on normal cell physiology which, in some cases, may lead to cellular transformation. Indeed, mutations of certain SH2 domain containing proteins are associated with various human cancers and cancer subtypes (Lappalainen et al. 2008; Waksman, Kumaran, and Lubman 2004). For instance, missense and nonsense mutations of the SH2 adaptor protein SAP cause the X-linked lymphoproliferative syndrome (Erdos et al. 2005; Hare et al. 2006). Nonsense mutations in the SH2 domain of RASA1/RasGAP are associated with basal cell carcinoma which leads to the formation of tumors in the chest (Friedman et al. 1993). Borges and co-workers showed that cytokine-inducible SH2 containing (Carlomagno et al. 1995) protein expression is increased in human breast cancer cells with a high level of growth hormone synthesis (Borges et al. 2008). Mutations in a number of other SH2 domain proteins, including Lck, Hck, Grb2, Grb7, Src, Shc, PI3K, Grb2, GAP, Crkl, Btk, and Tec, directly or indirectly cause cancer (Waksman, Kumaran, and Lubman 2004).

These and other studies suggest that the SH2 domain represents an attractive target for cancer therapy development. In exploring this avenue of drug design, peptides, peptidomimetics, and nonpeptidic compounds have been developed that target the SH2 domain of Stat-3, a protein constitutively activated by aberrant upstream tyrosine kinase activities in a broad spectrum of human cancers (Costantino and Barlocco 2008). Screening of chemical libraries led to the identification of Stattic, a nonpeptidic small molecule that selectively inhibits Stat-3 activation *in vitro* by blocking the SH2 domain (Schust and Berg 2004). Importantly, Stattic selectively inhibits the dimerization and subsequent nuclear translocation of Stat-3 and thereby sensitizes Stat-3-dependent breast cancer cell lines to apoptosis (Turkson et al. 2001). The growing collection of structures of SH2 domains in complex with physiological ligands or inhibitors (Figure 19.1D), respectively, will not only provide a view on the SH2-ome at atomic resolution but also guide the design of more selective inhibitors for cancer treatment. A virtual screen strategy based on docking a compound onto the active site of an SH2 domain was recently developed by Xu et al. (2009) and used to screen 920,000 small drug-like compounds for inhibitors of the Stat-3 SH2 domain, from which three were subsequently verified to competitively inhibit Stat-3 binding to its physiological phosphopeptide and IL-6 induced phosphorylation of Stat-3.

19.4 SH2 DOMAIN AND THE HUMAN PROTEIN-PROTEIN INTERACTOME

One of the main challenges in the post-genomic era is to decipher how proteins interact with one another in a cell and how this "interactome" fluxes dynamically with the cellular state (Rual et al. 2005). A need to map all possible protein-protein interactions has driven the development of a number of high-throughput technologies, including mass spectrometry (MS), yeast two-hydrid (Y2H), protein arrays, and LUMIER (Barrios-Rodiles et al. 2005; Krysiak, Marek, and Okopien 2009; Kung and Snyder 2006; Olsen and Macek 2009; Phizicky et al. 2003). Although all these methods can be used to uncover protein-protein interactions in a high-throughput manner, each has advantages and limitations. Mass spectrometry, when combined with affinity purification, provides a robust platform to identify protein complexes (Macek, Mann, and Olsen 2009; Oppermann et al. 2009). However, a protein-protein interaction network identified from AP-MS is generally of low resolution because it is impossible to infer, without further experimentation, whether an identified interaction is a direct binder of the bait or not. Y2H, in contrast, maps binary protein-protein interactions (Phizicky et al. 2003; Rual et al. 2005; Yu et al. 2008). Nevertheless, current Y2H protocols are not suitable for identifying interactions mediated by posttranslational modifications. Therefore, the current databases of protein-protein interactions are critically lacking in interactions mediated by posttranslational modifications such as phosphorylation. Protein arrays, especially arrays of antibodies specific for physiological phosphoproteins, bear the potential to address direct phosphorylation-regulated interactions (Blackburn and Hart 2005; Combaret et al. 2005; Engelman et al. 2007; Hamelinck et al. 2005; Stommel et al. 2007). However, a challenging aspect in protein microarray technology development is the difficulty in maintaining the native state of the protein following purification and surface immobilization. To avoid the laborious purification step and to minimize the loss of protein activity incurred during purification and immoblization, Ramachandran et al. (2004, 2008) developed a high-density self-assembling protein micro array that displays thousands of proteins produced and captured *in situ* from immobilized cDNA templates. However, it remains to be seen whether this strategy can be adapted for mapping PTM-mediated protein-protein interactions. LUMIER is a LUminescence-based Mammalian IntERactome assay that can potentially be used for mapping phosphorylation

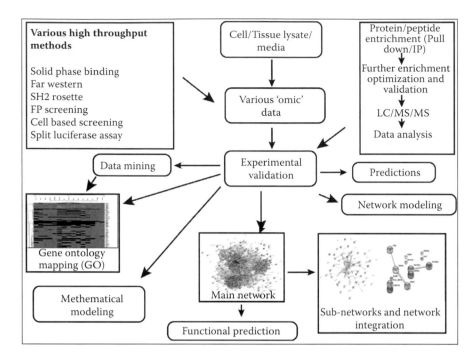

FIGURE 19.3 An overview of strategies used to derive SH2 domain-mediated protein-protein interaction networks. Different high-throughput experimental methods can be used to produce 'omic' data that involve SH2 domains. Computational methods can aid in the prediction and/or understanding of functional SH2 interaction networks.

modification and phosphorylation-mediated protein-protein interactions (Barrios-Rodiles et al. 2005). An important issue to address in a LUMIER assay is the controlled activation of a specific kinase and the development of specific antibodies that may be used to identify phosphorylation sites generated by the kinase. To overcome limitations associated with a specific method, there is a growing interest in using a combination of different methods to map signaling cascades involving PTKs and/or SH2 domains (Bork, Schultz, and Ponting 1997; Giovannone et al. 2000; Monteiro, Arai, and Travassos 2008). As outlined in Figure 19.3, combinations of various approaches, which include far-Western blot analysis, oligonucleotide-tagged multiplex assay (OTM), SH2 rosette assay, high-throughput fluorescence polarization screening, cell-based screening using co-transfection, microfluidic assay, multiplexed fluorescent microsphere assay, fluorometric microvolume assay, solid-phase binding assay, split luciferase assay, and phage-display-based binding assay, as well as other modeling, prediction, computational, and data mining methods, have been developed to map these signaling cascades and interactions mediated by PTKs and/or SH2 domains (Dierck et al. 2009; Gordus and MacBeath 2006; Huang et al. 2008; Lawrence 2005; Machida et al. 2007; Muller, Schust, and Berg 2008; Schust and Berg 2004; Yaoi et al. 2006; V. Zhou et al. 2009).

Relatively simple domain arrays have been developed to map interactions mediated by the PDZ (postsynaptic density 95, PSD-85, discs large, Dlg, zonula occludens-1, ZO-1), SH3 (src homology 3), and SH2 domains in which purified domains are immobilized on a membrane, a multiwell plate, or a glass slide followed by probing with a protein or peptide ligand (Chamnongpol and Li 2004; Day and Kobilka 2006; X. Jiang et al. 2008; MacBeath 2002). Using protein microarrays containing most human SH2 and PTB domains, MacBeath and associates generated quantitative protein interaction maps for the Erb receptors and three receptor tyrosine kinases, namely, EGFR, FGFR1, and IGF1R, by probing the arrays with phosphopeptides drawn from intracellular tyrosine residues on these receptors (Jones et al. 2006; Kaushansky et al. 2008). These systematic and quantitative analyses of SH2 domain-mediated interactions provide evidence suggesting that the biological and pathological effects of receptor tyrosine kinases arise in part from quantitative differences in the specific set of SH2 or PTB proteins they recruit and in the corresponding affinities.

As a complementary strategy to domain arrays, peptide arrays and peptide library arrays have been used to map SH2-phosphopeptide interactions and to charter SH2 domain specificity. Cesareni and colleagues have synthesized an array of phosphotyrosyl peptides representing over 6000 physiological phosphorylation sites and probed the array for binding to purified SH2 domains. This type of proteome-wide SH2-pTyr interaction analysis provides valuable information on the specificity and signaling potential of the SH2-ome (Miller et al. 2008). Recently, an oriented peptide array library (OPAL) approach was used to identify specific sequence motifs recognized by kinases or SH2 domains (Huang et al. 2008; Rodriguez et al. 2004; Shigaki et al. 2007). When applied to the complement of human SH2 domains, systematic OPAL screens allowed for the determination of the specificity for the majority of the 120 human SH2 domains (Huang et al. 2008; L. Li et al. 2008). The unique binding profile produced by each SH2 domain yields quantitative information on specificity, which, in turn, could be harnessed for the prediction of the SH2 interactome by methods such as SMALI (scoring matrix-assisted ligand identification) and for generating a virtual phosphorylation-dependent signaling network using NetPhorest (Huang et al. 2008; L. Li et al. 2008; Miller et al. 2008).

As pointed out by Seong and Choi (2003), the use of highly stable, inexpensive peptide array methods also has disadvantages. Since peptides have small molecular masses they are not easily accessible when immobilized on a solid support. Moreover, since peptides lack three-dimensional structure, they need to be correctly oriented for interactions with the protein targets. Another major concern of peptide array is the high false positive rate likely owing to the chemistry of the support material used (Min and Mrksich 2004; Min, Su, and Mrksich 2004). Finally, interactions identified from peptide array screening represent *in vitro* potential only and need to be validated in cells using the corresponding intact proteins.

19.5 SYSTEMS BIOLOGY OF THE PHOSPHOPROTEOME

In contrast to the paucity in pTyr-mediated interactions in current PPI databases, recent advances in mass spectrometry have permitted the identification of thousands of *in vivo* phosphorylation sites (Olsen et al. 2006; Rikova et al. 2007). The challenge remains to elucidate the associated phosphorylation and interaction network in order to understand the signaling network employed by a cancer cell. In this regard, computational algorithms such as SMALI, NetworKIN, and NetPhorest have been developed to enhance the accuracy of modeling kinase signaling or interaction networks based on substrate specificity of protein kinases, consensus motifs recognized by phosphorylation-binding modules, and network contextual information (L. Li et al. 2008; Linding et al. 2007, 2008; Miller et al. 2008). Despite the usefulness of these computer programs in predicting phosphotyrosine signaling and/or interaction networks, experimental approaches are needed to gauge the phosphoproteome and its dynamic flux from normal to disease states.

Mass spectrometry provides a powerful means to observe the entire proteome of a cell under a defined condition (Aebersold and Mann 2003; de Hoog and Mann 2004; Kuster and Mann 1998; Mann, Hendrickson, and Pandey 2001; Mann and Jensen 2003; Rappsilber and Mann 2002; Vermeulen, Hubner, and Mann 2008). Peptide mass fingerprinting and tandem mass spectrometry (MS/MS) can be used to detect, identify, and/or verify phosphorylation sites *in vivo* (Olsen et al. 2006; Rush et al. 2005). However, sample preparation for the detection of low abundant phosphorylated peptides or proteins is still a significant challenge for MS mapping of protein-protein interactions mediated by phosphorylation (Aebersold and Mann 2003; Machida, Mayer, and Nollau 2003; Mann, Hendrickson, and Pandey 2001; Mann et al. 2002; Mann and Jensen 2003; Vermeulen, Hubner, and Mann 2008). In this regard, phosphotyrosine (pY)-specific antibodies and metal affinity chromatography (IMAC) have been used to enrich phosphorylated proteins or peptides. Using the IMAC approach, Salomon and co-workers have profiled a considerable number of known as well as novel phosphorylated sites in different proteins from human hematopoietic cells (Salomon et al. 2003). Moreover, a large-scale analysis of GRB2-mediated epidermal growth factor receptor protein-protein interactors by Mann and co-workers reveals the significance of the SH2 domain as a bait in mapping phosphoproteome in normal and cancer cells (Blagoev et al. 2003). In addition to MS phosphorylation-dependent interactions by MS, Machida and colleagues employed a large-scale far-Western analysis and developed SH2 Rossete assays to profile the global tyrosine phosphorylation state of the cell (Dierck et al. 2009; Machida et al. 2007). Reverse-phase protein arrays were used to generate comprehensive, quantitative SH2 binding profiles for phosphopeptides, recombinant proteins, and entire proteomes (Machida et al. 2007). It can be envisaged that the combination of SH2 domain array screening and reverse screening of cell lysate arrays would provide a comprehensive readout of the phosphoproteome (R. Jiang et al. 2006; Nishizuka et al. 2003; Tangrea et al. 2004). When applied to cancer cell lines and cancer tissues, high-throughput comprehensive SH2 profiling could provide valuable mechanistic insights into tyrosine kinase signaling pathways associated with tumorigenesis and/or metastasis.

19.6 CONCLUDING REMARKS

Monitoring the continuum of phosphoproteome and relevant system trajectories to identify vectors of cell communication and translation of these identifications into models of cellular functions remain a significant challenge in systems biology. Efficient and systematic identification of tyrosine phosphorylation-mediated protein-protein interactions and monitoring the dynamic flux of the phosphoproteome and the phosphoprotein interactome demand a combination of approaches and assay platforms. SH2 domains and the underlying interactome constitute a significant part of the phosphoprotein interactome, and it can be envisaged that, by systematically mapping SH2 domain interactions and by comparing SH2 signaling profiles between a normal and a cancer cell, new avenues for cancer diagnosis and treatment may be created. Future efforts in drug design that target the SH2 domain should consider the impact of the drug on the targeted SH2 domain as well as on the entire SH2 interactome.

REFERENCES

Aebersold, R. and Mann, M. 2003. Mass spectrometry-based proteomics. *Nature* 422: 198–207.

Barrios-Rodiles, M., Brown, K. R., Ozdamar, B. et al. 2005. High-throughput mapping of a dynamic signaling network in mammalian cells. *Science* 307: 1621–1625.

Benes, C. H., Wu, N., Elia, A. E. et al. 2005. The C2 domain of PKCdelta is a phosphotyrosine binding domain. *Cell* 121: 271–280.

Birge, R. B. and Hanafusa, H. 1993. Closing in on SH2 specificity. *Science* 262: 1522–1524.

Blackburn, J. M. and Hart, D. J. 2005. Fabrication of protein function microarrays for systems-oriented proteomic analysis. *Methods Mol Biol* 310: 197–216.

Blagoev, B., Kratchmarova, I., Ong, S. E. et al. 2003. A proteomics strategy to elucidate functional protein-protein interactions applied to EGF signaling. *Nature Biotechnol* 21: 315–318.

Blume-Jensen, P. and Hunter, T. 2001. Oncogenic kinase signalling. *Nature* 411: 355–365.

Boggon, T. J. and Eck, M. J. 2004. Structure and regulation of Src family kinases. *Oncogene* 23: 7918–7927.

Borges, S., Moudilou, E., Vouyovitch, C. et al. 2008. Involvement of a JAK/STAT pathway inhibitor: cytokine inducible SH2 containing protein in breast cancer. *Adv Exp Med Biol* 617: 321–329.

Bork, P., Schultz, J., and Ponting, C. P. 1997. Cytoplasmic signalling domains: the next generation. *Trends Biochem Sci* 22: 296–298.

Campbell, S. J. and Jackson, R. M. 2003. Diversity in the SH2 domain family phosphotyrosyl peptide binding site. *Protein Eng Design Selection* 16: 217–227.

Carlomagno, F., Salvatore, D., Santoro, M. et al. 1995. Point mutation of the RET proto-oncogene in the TT human medullary thyroid carcinoma cell line. *Biochem Biophys Res Commun* 207: 1022–1028.

Chamnongpol, S. and Li, X. 2004. SH3 domain protein-binding arrays. *Methods Mol Biol* 264: 183–189.

Combaret, V., Bergeron, C., Brejon, S. et al. 2005. Protein chip array profiling analysis of sera from neuroblastoma patients. *Cancer Lett* 228: 91–96.

Costantino, L. and Barlocco, D. 2008. STAT 3 as a target for cancer drug discovery. *Curr Med Chem* 15: 834–843.

Day, P. and Kobilka, B. 2006. PDZ-domain arrays for identifying components of GPCR signaling complexes. *Trends Pharmacol Sci* 27: 509–511.

De Fabritiis G., Geroult, S., Coveney, P. V., and Waksman, G. 2008. Insights from the energetics of water binding at the domain-ligand interface of the Src SH2 domain. *Proteins* 72: 1290–1297.

de Hoog, C. L. and Mann, M. 2004. Proteomics. *Annu Rev Genomics Hum Genet* 5: 267–293.

Dierck, K., Machida, K., Mayer, B. J., and Nollau, P. 2009. Profiling the tyrosine phosphorylation state using SH2 domains. *Methods Mol Biol* 527: 131–155, ix.

Engelman, J. A., Zejnullahu, K., Mitsudomi, T. et al. 2007. MET amplification leads to gefitinib resistance in lung cancer by activating ERBB3 signaling. *Science* 316: 1039–1043.

Engen, J. R., Wales, T. E., Hochrein, J. M. et al. 2008. Structure and dynamic regulation of Src-family kinases. *Cell Mol Life Sci* 65: 3058–3073.

Erdos, M., Uzvolgyi, E., Nemes, Z. et al. 2005. Characterization of a new disease-causing mutation of SH2D1A in a family with X-linked lymphoproliferative disease. *Hum Mutat* 25: 506.

Filippakopoulos, P., Kofler, M., Hantschel, O. et al. 2008. Structural coupling of SH2-kinase domains links Fes and Abl substrate recognition and kinase activation. *Cell* 134: 793–803.

Friedman, F., Gejman, P. V., Martin, G. A., and McCormick, F. 1993. Nonsense mutations in the C-terminal SH2 region of the GTPase activating protein (GAP) gene in human tumours. *Nature Genet* 5: 242–247.

Giovannone, B., Scaldaferri, M. L., Federici, M. et al. 2000. Insulin receptor substrate (IRS) transduction system: distinct and overlapping signaling potential. *Diabetes Metab Res Rev* 16: 434–441.

Gordus, A. and MacBeath, G. 2006. Circumventing the problems caused by protein diversity in microarrays: implications for protein interaction networks. *J. Am. Chem. Soc.* 128: 13668–13669.

Grucza, R. A., Bradshaw, J. M., Futterer, K., and Waksman, G. 1999. SH2 domains: from structure to energetics, a dual approach to the study of structure-function relationships. *Med Res Rev* 19: 273–293.

Guo, A., Villen, J., Kornhauser, J. et al. 2008. Signaling networks assembled by oncogenic EGFR and c-Met. *Proc Natl Acad Sci USA* 105: 692–697.

Hamelinck, D., Zhou, H., Li, L. et al. 2005. Optimized normalization for antibody microarrays and application to serum-protein profiling. *Mol Cell Proteomics* 4: 773–784.

Hanahan, D. and Weinberg, R. A. 2000. The hallmarks of cancer. *Cell* 100: 57–70.

Harari, D. and Yarden, Y. 2000. Molecular mechanisms underlying ErbB2/HER2 action in breast cancer. *Oncogene* 19: 6102–6114.

Hare, N. J., Ma, C. S., Alvaro, F., Nichols, K. E., and Tangye, S. G. 2006. Missense mutations in SH2D1A identified in patients with X-linked lymphoproliferative disease differentially affect the expression and function of SAP. *Int Immunol* 18: 1055–1065.

Hatada, M. H., Lu, X., Laird, E. R. et al. 1995. Molecular basis for interaction of the protein tyrosine kinase ZAP-70 with the T-cell receptor. *Nature* 377: 32–38.

Huang, H., Li, L., Wu, C. et al. 2008. Defining the specificity space of the human SRC homology 2 domain. *Mol Cell Proteomics* 7: 768–784.

Hunter, T. 2000. Signaling: 2000 and beyond. *Cell* 100: 113–127.

Hwang, P. M., Li, C., Morra, M. et al. 2002. A "three-pronged" binding mechanism for the SAP/SH2D1A SH2 domain: structural basis and relevance to the XLP syndrome. *EMBO. J.* 21: 314–323.

Hynes, N. E. and MacDonald, G. 2009. ErbB receptors and signaling pathways in cancer. *Curr Opin Cell Biol* 21: 177–184.

Jenkins, L. W., Peters, G. W., Dixon, C. E. et al. 2002. Conventional and functional proteomics using large format two-dimensional gel electrophoresis 24 hours after controlled cortical impact in postnatal day 17 rats. *J. Neurotrauma.* 19: 715–740.

Jiang, R., Mircean, C., Shmulevich, I. et al. 2006. Pathway alterations during glioma progression revealed by reverse phase protein lysate arrays. *Proteomics* 6: 2964–2971.

Jiang, X., Roth, L., Han, S., and Li, X. 2008. SH2 domain-based tyrosine phosphorylation array. *Methods Mol Biol* 441: 153–161.

Joazeiro, C. A., Wing, S. S., Huang, H. et al. 1999. The tyrosine kinase negative regulator c-Cbl as a RING-type, E2-dependent ubiquitin-protein ligase. *Science* 286: 309–312.

Johnson, S. A. and Hunter, T. 2005. Kinomics: methods for deciphering the kinome. *Nature Methods* 2: 17–25.

Jones, R. B., Gordus, A., Krall, J. A., and MacBeath, G. 2006. A quantitative protein interaction network for the ErbB receptors using protein microarrays. *Nature* 439: 168–174.

Kaneko, T., Li, L., and Li, S. S. 2008. The SH3 domain: a family of versatile peptide- and protein-recognition module. *Front Biosci* 13: 4938–4952.

Kaushansky, A., Gordus, A., Chang, B., Rush, J., and MacBeath, G. 2008. A quantitative study of the recruitment potential of all intracellular tyrosine residues on EGFR, FGFR1 and IGF1R. *Mol Biosyst* 4: 643–653.

Konopka, J. B., Watanabe, S. M., and Witte, O. N. 1984. An alteration of the human c-abl protein in K562 leukemia cells unmasks associated tyrosine kinase activity. *Cell* 37: 1035–1042.

Krysiak, R., Marek, B., and Okopien, B. 2009. [Inhibitory effect of carbamazepine on hypotensive action of spironolactone in primary aldosteronism]. *Endokrynol Pol* 60: 52–55.

Kung, L. A. and Snyder, M. 2006. Proteome chips for whole-organism assays. *Nature Rev Mol Cell Biol* 7: 617–622.

Kuriyan, J. and Cowburn, D. 1997. Modular peptide recognition domains in eukaryotic signaling. *Annu Rev Biophys Biomol Struct* 26: 259–288.

Kuster, B. and Mann, M. 1998. Identifying proteins and post-translational modifications by mass spectrometry. *Curr Opin Struct Biol* 8: 393–400.

Lappalainen, I., Thusberg, J., Shen, B., and Vihinen, M. 2008. Genome wide analysis of pathogenic SH2 domain mutations. *Proteins* 72: 779–792.

Lawrence, D. S. 2005. Signaling protein inhibitors via the combinatorial modification of peptide scaffolds. *Biochim Biophys Acta* 1754: 50–57.

Levinson, N. M., Seeliger, M. A., Cole, P. A., and Kuriyan, J. 2008. Structural basis for the recognition of c-Src by its inactivator Csk. *Cell* 134: 124–134.

Levkowitz, G., Oved, S., Klapper, L. N. et al. 2000. c-Cbl is a suppressor of the neu oncogene. *J. Biol. Chem.* 275: 35532–35539.

Li, L., Wu, C., Huang, H. et al. 2008. Prediction of phosphotyrosine signaling networks using a scoring matrix-assisted ligand identification approach. *Nucleic Acids Res* 36: 3263–3273.

Li, S. C., Gish, G., Yang, D. et al. 1999. Novel mode of ligand binding by the SH2 domain of the human XLP disease gene product SAP/SH2D1A. *Curr Biol* 9: 1355–1362.

Li, S. S. 2005. Specificity and versatility of SH3 and other proline-recognition domains: structural basis and implications for cellular signal transduction. *Biochem. J.* 390: 641–653.

Linding, R., Jensen, L. J., Ostheimer, G. J. et al. 2007. Systematic discovery of *in vivo* phosphorylation networks. *Cell* 129: 1415–1426.

Linding, R., Jensen, L. J., Pasculescu, A. et al. 2008. NetworKIN: a resource for exploring cellular phosphorylation networks. *Nucleic Acids Res* 36: D695–D699.

Liu, B. A., Jablonowski, K., Raina, M. et al. 2006. The human and mouse complement of SH2 domain proteins: establishing the boundaries of phosphotyrosine signaling. *Mol Cell* 22: 851–868.

MacBeath, G. 2002. Protein microarrays and proteomics. *Nature Genet* 32 (Suppl): 526–532.

Macek, B., Mann, M., and Olsen, J. V. 2009. Global and site-specific quantitative phosphoproteomics: principles and applications. *Annu Rev Pharmacol Toxicol* 49: 199–221.

Machida, K. and Mayer, B. J. 2005. The SH2 domain: versatile signaling module and pharmaceutical target. *Biochim Biophys Acta* 1747: 1–25.

Machida, K., Thompson, C. M., Dierck, K. et al. 2007. High-throughput phosphotyrosine profiling using SH2 domains. *Mol Cell* 26: 899–915.

Machida, K., Mayer, B. J., and Nollau, P. 2003. Profiling the global tyrosine phosphorylation state. *Mol Cell Proteomics* 2: 215–233.

Mann, M., Hendrickson, R. C., and Pandey, A. 2001. Analysis of proteins and proteomes by mass spectrometry. *Annu Rev Biochem* 70: 437–473.

Mann, M. and Jensen, O. N. 2003. Proteomic analysis of post-translational modifications. *Nature Biotechnol* 21: 255–261.

Mann, M., Ong, S. E., Gronborg, M. et al. 2002. Analysis of protein phosphorylation using mass spectrometry: deciphering the phosphoproteome. *Trends Biotechnol* 20: 261–268.

Miller, M. L., Jensen, L. J., Diella, F. et al. 2008. Linear motif atlas for phosphorylation-dependent signaling. *Sci Signal* 1: ra2.

Min, D. H. and Mrksich, M. 2004. Peptide arrays: towards routine implementation. *Curr Opin Chem Biol* 8: 554–558.

Min, D. H., Su, J., and Mrksich, M. 2004. Profiling kinase activities by using a peptide chip and mass spectrometry. *Angew Chem Int Ed Engl* 43: 5973–5977.

Mohi, M. G. and Neel, B. G. 2007. The role of Shp2 (PTPN11) in cancer. *Curr Opin Genet Dev* 17: 23–30.

Moniakis, J., Funamoto, S., Fukuzawa, M. et al. 2001. An SH2-domain-containing kinase negatively regulates the phosphatidylinositol-3 kinase pathway. *Genes Dev* 15: 687–698.

Monteiro, H. P., Arai, R. J., and Travassos, L. R. 2008. Protein tyrosine phosphorylation and protein tyrosine nitration in redox signaling. *Antioxid Redox Signal* 10: 843–889.

Muller, J., Schust, J., and Berg, T. 2008. A high-throughput assay for signal transducer and activator of transcription 5b based on fluorescence polarization. *Anal Biochem* 375: 249–254.

Neel, B. G., Gu, H., and Pao, L. 2003. The 'Shp'ing news: SH2 domain-containing tyrosine phosphatases in cell signaling. *Trends Biochem Sci* 28: 284–293.

Nishizuka, S., Charboneau, L., Young, L. et al. 2003. Proteomic profiling of the NCI-60 cancer cell lines using new high-density reverse-phase lysate microarrays. *Proc Natl Acad Sci USA* 100: 14229–14234.

Ogawa, W., Matozaki, T., and Kasuga, M. 1998. Role of binding proteins to IRS-1 in insulin signalling. *Mol Cell Biochem* 182: 13–22.

Olsen, J. V., Blagoev, B., Gnad, F. et al. 2006. Global, *in vivo*, and site-specific phosphorylation dynamics in signaling networks. *Cell* 127: 635–648.

Olsen, J. V. and Macek, B. 2009. High accuracy mass spectrometry in large-scale analysis of protein phosphorylation. *Methods Mol Biol* 492: 131–142.

Ooms, L. M., Horan, K. A., Rahman, P. et al. 2009. The role of the inositol polyphosphate 5-phosphatases in cellular function and human disease. *Biochem. J.* 419: 29–49.

Oppermann, F. S., Gnad, F., Olsen, J. V. et al. 2009. Large-scale proteomics analysis of the human kinome. *Mol Cell Proteomics* 8: 1751–1764.

Ostrakhovitch, E. A. and Li, S. S. 2006. The role of SLAM family receptors in immune cell signaling. *Biochem Cell Biol* 84: 832–843.

Pawson, T. 2004. Specificity in signal transduction: from phosphotyrosine-SH2 domain interactions to complex cellular systems. *Cell* 116: 191–203.

Pawson, T., Gish, G. D., and Nash, P. 2001. SH2 domains, interaction modules and cellular wiring. *Trends Cell Biol* 11: 504–511.

Pawson, T. and Nash, P. 2000. Protein-protein interactions define specificity in signal transduction. *Genes Dev* 14: 1027–1047.

Pawson, T. and Scott, J. D. 2005. Protein phosphorylation in signaling: 50 years and counting. *Trends Biochem Sci* 30: 286–290.

Peschard, P., Ishiyama, N., Lin, T., Lipkowitz, S., and Park, M. 2004. A conserved DpYR motif in the juxtamembrane domain of the Met receptor family forms an atypical c-Cbl/Cbl-b tyrosine kinase binding domain binding site required for suppression of oncogenic activation. *J. Biol. Chem.* 279: 29565–29571.

Phizicky, E., Bastiaens, P. I., Zhu, H., Snyder, M., and Fields, S. 2003. Protein analysis on a proteomic scale. *Nature* 422: 208–215.

Pincus, D., Letunic, I., Bork, P., and Lim, W. A. 2008. Evolution of the phospho-tyrosine signaling machinery in premetazoan lineages. *Proc Natl Acad Sci USA* 105: 9680–9684.

Ramachandran, N., Hainsworth, E., Bhullar, B. et al. 2004. Self-assembling protein microarrays. *Science* 305: 86–90.

Ramachandran, N., Raphael, J. V., Hainsworth, E. et al. 2008. Next-generation high-density self-assembling functional protein arrays. *Nature Methods* 5: 535–538.

Rappsilber, J. and Mann, M. 2002. Is mass spectrometry ready for proteome-wide protein expression analysis? *Genome Biol* 3: COMMENT2008.

Ran, X. and Song, J. 2005. Structured insight into the binding diversity between the Tyr-phosphorylated human EphrinBs and Nck SH$_2$ domain. *J. Biochem.* 280:19205–19212.

Rikova, K., Guo, A., Zeng, Q. et al. 2007. Global survey of phosphotyrosine signaling identifies oncogenic kinases in lung cancer. *Cell* 131: 1190–1203.

Rodriguez, M., Li, S. S., Harper, J. W., and Songyang, Z. 2004. An oriented peptide array library (OPAL) strategy to study protein-protein interactions. *J. Biol. Chem.* 279: 8802–8807.

Roskoski, R., Jr. 2004. Src protein-tyrosine kinase structure and regulation. *Biochem Biophys Res Commun* 324: 1155–1164.

Rual, J. F., Venkatesan, K., Hao, T. et al. 2005. Towards a proteome-scale map of the human protein-protein interaction network. *Nature* 437: 1173–1178.

Rush, J., Moritz, A., Lee, K. A. et al. 2005. Immunoaffinity profiling of tyrosine phosphorylation in cancer cells. *Nature Biotechnol* 23: 94–101.

Sadowski, I., Stone, J. C., and Pawson, T. 1986. A noncatalytic domain conserved among cytoplasmic protein-tyrosine kinases modifies the kinase function and transforming activity of Fujinami sarcoma virus P130gag-fps. *Mol Cell Biol* 6: 4396–4408.

Sagartz, J. E., Jhiang, S. M., Tong, Q., and Capen, C. C. 1997. Thyroid-stimulating hormone promotes growth of thyroid carcinomas in transgenic mice with targeted expression of the ret/PTC1 oncogene. *Lab Invest* 76: 307–318.

Salomon, A. R., Ficarro, S. B., Brill, L. M. et al. 2003. Profiling of tyrosine phosphorylation pathways in human cells using mass spectrometry. *Proc Natl Acad Sci USA* 100: 443–448.

Schlessinger, J. and Lemmon, M. A. 2003. SH2 and PTB domains in tyrosine kinase signaling. *Sci STKE* 2003: RE12.

Schust, J. and Berg, T. 2004. A high-throughput fluorescence polarization assay for signal transducer and activator of transcription 3. *Anal Biochem* 330: 114–118.

Seong, S. Y. and Choi, C. Y. 2003. Current status of protein chip development in terms of fabrication and application. *Proteomics* 3: 2176–2189.

Shen, K., Hines, A. C., Schwarzer, D., Pickin, K. A., and Cole, P. A. 2005. Protein kinase structure and function analysis with chemical tools. *Biochim Biophys Acta* 1754: 65–78.

Shigaki, S., Yamaji, T., Han, X. et al. 2007. A peptide microarray for the detection of protein kinase activity in cell lysate. *Anal Sci* 23: 271–275.

Songyang, Z. and Cantley, L. C. 1995. Recognition and specificity in protein tyrosine kinase-mediated signalling. *Trends Biochem Sci* 20: 470–475.

Songyang, Z., Gish, G., Mbamalu, G., Pawson, T., and Cantley, L. C. 1995. A single point mutation switches the specificity of group III Src homology (SH) 2 domains to that of group I SH2 domains. *J. Biol. Chem.* 270: 26029–26032.

Songyang, Z., Shoelson, S. E., Chaudhuri, M. et al. 1993. SH2 domains recognize specific phosphopeptide sequences. *Cell* 72: 767–778.

Stommel, J. M., Kimmelman, A. C., Ying, H. et al. 2007. Coactivation of receptor tyrosine kinases affects the response of tumor cells to targeted therapies. *Science* 318: 287–290.

Sweeney, C. and Carraway, K. L., III. 2004. Negative regulation of ErbB family receptor tyrosine kinases. *Br. J. Cancer.* 90: 289–293.

Tangrea, M. A., Wallis, B. S., Gillespie, J. W. et al. 2004. Novel proteomic approaches for tissue analysis. *Expert Rev Proteomics* 1: 185–192.

Tsui, F. W., Martin, A., Wang, J., and Tsui, H. W. 2006. Investigations into the regulation and function of the SH2 domain-containing protein-tyrosine phosphatase, SHP-1. *Immunol Res* 35: 127–136.

Turkson, J., Ryan, D., Kim, J. S. et al. 2001. Phosphotyrosyl peptides block Stat3-mediated DNA binding activity, gene regulation, and cell transformation. *J Biol Chem* 276: 45443–45455.

Uhlik, M. T., Temple, B., Bencharit, S. et al. 2005. Structural and evolutionary division of phosphotyrosine binding (PTB) domains. *J. Mol. Biol.* 345: 1–20.

Vermeulen, M., Hubner, N. C., and Mann, M. 2008. High confidence determination of specific protein-protein interactions using quantitative mass spectrometry. *Curr Opin Biotechnol* 19: 331–337.

Waksman, G., Kominos, D., Robertson, S. C. et al. 1992. Crystal structure of the phosphotyrosine recognition domain SH2 of v-src complexed with tyrosine-phosphorylated peptides. *Nature* 358: 646–653.

Waksman, G., Kumaran, S., and Lubman, O. 2004. SH2 domains: role, structure and implications for molecular medicine. *Expert Rev Mol Med* 6: 1–18.

Waksman, G., Shoelson, S. E., Pant, N., Cowburn, D., and Kuriyan, J. 1993. Binding of a high affinity phosphotyrosyl peptide to the Src SH2 domain: crystal structures of the complexed and peptide-free forms. *Cell* 72: 779–790.

Wenta, N., Strauss, H., Meyer, S., and Vinkemeier, U. 2008. Tyrosine phosphorylation regulates the partitioning of STAT1 between different dimer conformations. *Proc Natl Acad Sci USA* 105: 9238–9243.

Wiggin, G. R., Fawcett, J. P., and Pawson, T. 2005. Polarity proteins in axon specification and synaptogenesis. *Dev Cell* 8: 803–816.

Xu, X., Kasembeli, M. M., Jiang, X., Tweardy, B. J., and Tweardy, D. J. 2009. Chemical probes that competitively and selectively inhibit Stat3 activation. *PLoS One* 4: e4783.

Yaoi, T., Chamnongpol, S., Jiang, X., and Li, X. 2006. Src homology 2 domain-based high throughput assays for profiling downstream molecules in receptor tyrosine kinase pathways. *Mol Cell Proteomics* 5: 959–968.

Yeh, J. J. and Der, C. J. 2007. Targeting signal transduction in pancreatic cancer treatment. *Expert Opin Ther Targets* 11: 673–694.

Yu, H., Braun, P., Yildirim, M. A. et al. 2008. High-quality binary protein interaction map of the yeast interactome network. *Science* 322: 104–110.

Zhou, M. M., Harlan, J. E., Wade, W. S. et al. 1995. Binding affinities of tyrosine-phosphorylated peptides to the COOH-terminal SH2 and NH2-terminal phosphotyrosine binding domains of Shc. *J. Biol. Chem.* 270: 31119–31123.

Zhou, V., Gao, X., Han, S. et al. 2009. An intracellular conformational sensor assay for Abl T315I. *Anal Biochem* 385: 300–308.

Zwahlen, C., Li, S. C., Kay, L. E., Pawson, T., and Forman-Kay, J. D. 2000. Multiple modes of peptide recognition by the PTB domain of the cell fate determinant Numb. *EMBO. J.* 19: 1505–1515.

CHAPTER 20

Data Sources and Computational Tools for Cancer Systems Biology

Yun Ma, Pradeep Kumar Shreenivasaiah, and Edwin Wang

CONTENTS

20.1 DATA SOURCES AND QUALITY FOR CANCER SYSTEMS BIOLOGY

Cancer systems biology studies often integrate many datasets representing different facets of cancer cells (i.e., interaction data, gene expression data, gene silencing data, etc.). Public data are available via the literature or collections of high-throughput datasets. Generally speaking, the data quality of small-scale studies is high, whereas high-throughput datasets have low quality. However, manual curation of data from small-scale studies might introduce errors (Cusick et al. 2009), whereas new technologies may improve the data quality of high-throughput datasets. For example, RNA-seq may generate digit reading of transcripts, providing high quality gene expression profiles.

20.1.1 Cancer Datasets

Cancer driver-mutating genes: literature-mined cancer genes are available (Futreal et al. 2004). Recently, as genome sequencing technology has become cheaper, tumor genome sequencing has generated more information about cancer genes (Cui et al. 2007). The COSMIC database (http://www.sanger.ac.uk/genetics/CGP/cosmic/) collects and assembles cancer genes derived from literature and tumor genome sequencing efforts. The Cancer Genome Atlas (TCGA) also collects tumor genome sequencing data. In the future, the *International Cancer Genome Consortium* (ICGC, http://www.icgc.org/)

will host a data repository for tumor genome sequencing data. The ICGC and the TCGA have worked together and plan to sequence more than 50 cancer types (250 tumors for each type) in the future.

Cancer methylation genes: such genes have been determined for certain cancer stem cells (Ohm et al. 2007; Schlesinger et al. 2007; Widschwendter et al. 2007) and NCI-60 cell lines (Ehrich et al. 2008) from high-throughput studies. Some databases have been built to collect cancer methylation data, for example, PubMeth (http://www.pubmeth.org/).

Tumor gene expression profiles: a great deal of gene expression data has been generated using microarray technology over the past 10 years. These data can be downloaded and queried from the Gene Expression Omnibus (GEO) database (http://www.ncbi. nlm.nih.gov/geo/) or from the tumor-specific gene expression profile database, Oncomine (http://www.oncomine.org/).

RNAi knockout of cancer cells: genome-wide RNAi knockout in cancer cell lines and tumor samples has been performed (Baldwin et al. 2008; Bommi-Reddy et al. 2008; Grueneberg et al. 2008a, 2008b; Manning 2009; Schlabach et al. 2008; Silva et al. 2008).

Profiling of drugs and small molecules on NCI-60 cell lines: more than 100,000 small molecules have been used to examine the growth of NCI-60 cell lines by a group from the National Cancer Institute (NCI), National Institutes of Health (NIH). These data are available at http://dtp.nci.nih.gov/. Furthermore, the Connectivity Map database (http://www.broad.mit.edu/node/305) collects genome-wide transcriptional expression data from human cancer cells treated with bioactive small molecules.

Phospoproteomic profiling of cancer cells: a large-scale survey of kinase activities in cancer cells and tumor samples has been performed (Du et al. 2009; Rikova et al. 2007; Wolf-Yadlin et al. 2006). These data are suitable for cancer signaling network studies.

Cancer protein atlas: large-scale survey of protein expression patterns in cancer cell lines, tumor samples, and normal tissues using an immunohistochemistry-based approach. Data containing 5 million images of immunohistochemically stained tissues and cells, based on 6122 antibodies representing 5011 human proteins, are available at the Human Protein Atlas (http://www.proteinatlas.org).

Tumor clinical data: some large-scale genome analysis of tumor samples is accompanied by patient clinical data, such as drug treatment, survival, and tumor recurrence, etc. It is critical for personalized medicine to be able to link clinical information and genome data at a systems level. Some databases exist that aim to collect clinical information and genome information of cancer patients. Specifically, Rembrandt (https://caintegrator.nci.nih.gov/rembrandt/menu.do) contains genomic and clinical data for brain tumor patients.

20.1.2 Molecular Interaction Datasets

Public databases collect and assemble literature-mined datasets describing human protein interactions, and metabolic and signaling pathways. Some examples of this type of database are the human protein interaction database HPRD (http://www.hprd.org/), IntAct (http://www.ebi.ac.uk/intact/site/index.jsf), MINT (http://mint.bio.uniroma2.it/mint/Welcome.do), and DIP (http://dip.doe-mbi.ucla.edu/); the signaling pathway databases BioCarta (http://www.biocarta.com/) and Reactome (http://reactome.org/). Additional databases are listed in Table 20.1.

TABLE 20.1   Public Data Resources for Systems Biology

	Name	Data Source (Manual/ Predicted)	Types of Data
1	4DXpress http://ani.embl.de/4DXpress	Automatically integrates from several other databases and those submitted by researchers.	Gene expression data during development of multiple model organisms.
2	ArrayExpress http://www.ebi.ac.uk/arrayexpress	Manually curated, re-annotated subsets of data from the archives.	Functional genomic data.
3	MGED http://www.mged.org	Manually curated.	Ontology for gene expression.
4	OMG http://www.omwg.org/	Manually curated.	Ontology management tools distributed through their sites.
5	BioGRID http://www.thebiogrid.org	Manually curated.	Protein-protein interaction data.
6	BioThesaurus http://pir.georgetown.edu/iprolink/ biothesaurus/data/thesaurus	Predicted.	Protein and gene names to uniprot knowledge accession mapping.
7	CancerGenes http://cbio.mskcc.org/CancerGenes/Select. action	Gene lists are annotated by experts/Information from other databases is added automatically.	Cancer gene database.
8	Cellmap.org http://cancer.cellmap.org/cellmap/	Manually curated.	Cancer related signaling pathways.
9	Entrez query http://www.ncbi.nlm.nih.gov/sites/entrez	Predicted and user submitted.	Provides information from discrete databases related to health sciences.
10	Sanger COSMIC Database http://www.sanger.ac.uk/genetics/CGP/cosmic/	Manually curated and predicted.	Cancer gene database.
11	Cancer Chromosomes http://www.ncbi.nlm.nih.gov/sites/ entrez?db=cancerchromosomes	Predicted.	Database of chromosome aberrations in cancer.
12	Mitelman Database of Chromosome Aberrations in Cancer http://cgap.nci.nih.gov/Chromosomes/Mitelman	Manually curated.	Database of chromosome aberrations in cancer.

(continued)

TABLE 20.1 Public Data Resources for Systems Biology (Continued)

	Name	Data Source (Manual/ Predicted)	Types of Data
13	*Haematology* http://www.infobiogen.fr/services/chromcancer/	Manually curated.	Database for genes involved in cancer, cytogenetics, and clinical entities involved in cancer and cancer prone diseases.
14	CGH data: Charite http://amba.charite.de/cgh/	Manually curated.	Database of tumor collectives.
15	Progenetix http:// www.progenetix.net/	Manually curated.	This database provides an overview of copy number abnormalities in human cancer from comparative genomic hybridization.
16	Laboratory of Cytomolecular Genetics (CMG) http://www.helsinki.fi/cmg/	Raw and processed data from the experimental pipeline are distributed from their site.	Information from the experiments using several techniques.
17	CGH Data Base http://www.cghtmd.jp/cghdatabase/index_e.htm	Manually curated.	Molecular cancer cytogenetics data obtained using comparative genomic hybridization technique.
18	Chromosome Rearrangements in Carcinomas http://www.path.cam.ac.uk/~pawefish/	Manually curated.	A collection of SKY and molecular cytogenetics data on cell lines mostly from epithelial cancers.
19	Cell Line NCI60 Drug Discovery Panel http://home.ncifcrf.gov/CCR/60SKY/new/ demo1.asp	Manually curated.	Molecular cytogenetics data in various tissues.
20	ChemBank http://chembank.broad.harvard.edu/	Manually curated.	Small-molecule screening and cheminformatics resource database.
21	DIP™ database http://dip.doe-mbi.ucla.edu/	Manually curated and predicted.	Database of experimentally determined interactions between proteins.
22	DrugBank database http://www.drugbank.ca/	Manually curated.	Database of drug data with drug target information.
23	Evola http://www.h-invitational.jp/evola/	Manually curated.	Ortholog database of human genes.
24	GenomeRNAi http://rnai2.dkfz.de/GenomeRNAi/	Manually curated.	Database for cell-based RNAi phenotypes

TABLE 20.1   Public Data Resources for Systems Biology (Continued)

	Name	Data Source (Manual/ Predicted)	Types of Data
25	GEO http://www.ncbi.nlm.nih.gov/geo/	Deposited by the community/automated.	Gene expression database.
26	GLIDA http://pharminfo.pharm.kyoto-u.ac.jp/services/ glida/	Manually curated.	GPCR-Ligand database.
27	Het-PDB Navi http://daisy.bio.nagoya-u.ac.jp/golab/ hetpdbnavi.html	Manually curated.	Protein-small molecule interaction database.
28	Genew, the Human Gene Nomenclature Database http://www.gene.ucl.ac.uk/cgi-bin/ nomenclature/searchgenes.pl	Manually curated.	Human gene database.
29	HPTAA http://www.hptaa.org	Automated collection.	Human potential tumor associated antigen database.
30	Human Proteinpedia http://www.humanproteinpedia.org/	Manual curated.	Integration of human protein data.
31	Human Protein Reference Database (HPRD) http://www.hprd.org/	Manually curated.	Database for human protein interactions.
32	CLDB http://www.biotech.ist.unige.it/interlab/cldb. html	Manually curated/ automated.	Database of cell lines.
33	I2D - Interologous Interaction Database http://ophid.utoronto.ca/ophidv2.201/	Manually curated and predicted.	Protein interaction database.
34	IMGT-GENE-DB http://www.imgt.org/IMGT_GENE-DB/ GENElect	Manually curated.	Database for human and mouse immunoglobulin and T cell receptor genes.
35	IntAct http://www.ebi.ac.uk/intact/site/index.jsf	Manually curated.	Protein interaction database.
36	Oncomine http://www.oncomine.org/	Manually curated.	Cancer gene expression database.
37	Phospho.ELM http://phospho.elm.eu.org/	Manually curated.	Database of serine, threonine, and tyrosine sites in eukaryotic proteins.
38	NetworKIN http://networkin.info/search.php	Predicted.	Consensus motifs with context for kinases and phosphoproteins

Efforts are ongoing to perform large-scale determination of protein interactions and signaling relationships in normal and cancer cells. For example, the human Src homology 2 domain (SH2 domain) protein interactions have been determined at a genome scale (Huang et al. 2008). This effort extends the human signaling map. More details regarding the extension of current signaling networks are discussed in Chapter 19.

There are also efforts to manually curate signaling relationships from research articles. Useful datasets can be found in research articles that have manually curated data from the literature (Oda et al. 2005; Oda and Kitano 2006). For instance, we have manually curated a large human signaling network containing more than 1600 proteins and 5000 signaling relationships (Cui et al. 2007). Our group is accumulating this type of curated signaling network. At present, the human signaling network contains more than 4000 proteins and 22,000 signaling relationships.

When using these public datasets, the quality of data should be carefully examined. For example, false positives are present in the protein interaction data derived from high-throughput studies. Relevant computational methods have been developed to eliminate these false positives as much as possible (Braun et al. 2009; Venkatesan et al. 2009). However, dealing with these problems is still a challenging task. In addition to false positives, public datasets are often incomplete. To overcome these problems, sensitivity analysis can be applied. False positives and false negatives can be mimicked by randomly adding or removing an extra 10% or 20% of the network nodes and the analysis is then performed on the modified network (Cui et al. 2006).

20.2 COMPUTATIONAL TOOLS FOR NETWORK CONSTRUCTION, ANALYSIS, AND MODELING

Many computational tools have been developed for visually and numerically exploring biological networks, including well-known examples such as Cytoscape, VisANT, and Pajek. These tools play an important role in systems biology, integration of data sources, and bioinformatics. These computational tools assist in network construction, visualization, and analysis.

Some tools, such as Cytoscape and VisANT, are used for many aspects of network analysis. Other tools are designed for specific purposes of network analysis. For instance, Mfinder (http://www.weizmann.ac.il/mcb/UriAlon/groupNetworkMotifSW.html), FANMOD (http://www.minet.uni-jena.de/~wernicke/motifs/index.html), and MAVisto (http://mavisto.ipk-gatersleben.de/) have been specifically designed to find network motifs. CFinder (http://www.cfinder.org/) can be used to define network communities.

Additional descriptions of these tools are provided in Table 20.2. Furthermore, Chapter 17 provides an in-depth explanation of how to use VisANT to perform network visualization and analysis. It also provides a discussion of some new network concepts, such as meta networks. Finally, network modeling tools are reviewed extensively in Chapter 16.

TABLE 20.2   Useful Tools for Network Analysis and Systems Biology

	Tools	Functions	Application
1	NeAT http://rsat.ulb.ac.be/ neat/	The network analysis tools include: Graph manipulation tools: covert-graph (graph format interconversions), alter-graph (adding and removing of nodes and edges), and random-graph (generates random graphs either from existing graph or from scratch). Network analysis tools: comparing graphs (supports set operations such as computing union, intersection, and difference between two networks), graph-topology (calculates the degree, betweenness, and closeness of each node), pathfinder (finds k-shortest path between nodes). Network visualization tools: display-graph, which draws a network graphical representation, random-graph. The Network cluster tools include: MCL and RNSC (finding the densely connected subsets of the graph). Graph-clique and graph-neighbors extractor (extracting all the cliques of a graph and neighborhood of a node/set of seed node, respectively). Graph-cluster-membership (mapping a cluster onto a graph and computing the membership degree between each node and each cluster). Graph-get-clusters (comparing graphs with clusters. Extracting the intra-clusters edges of map the clusters on the network). Cluster tools: Compare-classes (comparing query file and reference file). Contingency-stats (studying a contingency table). Roc-stats (calculating and draws ROC curves).	Neighborhood analysis can be applied to predict the function of an unknown polypeptide by collecting its neighbors with known functions in a protein interaction network ("guilt by association"). Network comparison is typically applicable to estimate the relevance of a protein-protein interaction network obtained by some high-throughput experiments, by comparing it with a manually curated network such as BioGrid or MIPs database. Path finding tools can be applied to uncover signal transduction pathways from protein-protein interaction networks. Clusters predicted by NeAt can be used in comparing classes to extract some overlap with biologically relevant classes (i.e., gene ontology classes). Further, the program helps to create a contingency table that can be analyzed via the contingency-statistical applications.
2	GraphWeb http://biit.cs.ut.ee/ graphweb/	Clustering algorithms: Markov cluster (MCL) algorithm and Betweenness Centrality Clustering (BCC). Basic graph algorithms: connected components, strongly connected components, biconnected components, maximal cliques. Node grouping: hub-based modules, input graph-based module, weight graph. Node filtering (i.e., keep N% of highest degree nodes), network neighborhood. Edge filtering (i.e., keep N% of heaviest edges).	Methods to analyze directed and undirected, weighted and unweighted heterogeneous networks of genes, proteins, and microarray probesets for many eukaryotic genomes.

(*continued*)

TABLE 20.2   Useful Tools for Network Analysis and Systems Biology (Continued)

	Tools	Functions	Application
		Module filtering (i.e., hide modules with less than N nodes, show N largest modules, hide insignificant modules).	Help to integrate multiple diverse datasets into global networks. Help to incorporate multispecies data using gene orthology mapping. Extract customized networks using filters for nodes and edges based on dataset support, edge weight, and node annotation. Analysis and detecting of gene modules from networks using various algorithms from the collection. Functional interpretation of predicted modules using Gene Ontology, pathways, and cis-regulatory motifs.
3	DAVID http://david.abcc. ncifcrf.gov/	Identify enriched biological themes, particularly GO terms and functionally related genes. Visualize genes on BioCarta & KEGG pathway maps. Display related many-genes-to-many-terms on 2-D view. Search for other functionally related genes in genome, but not in the list and search other annotations functionally similar to one of interest. List interacting proteins. Link gene-disease associations. Highlight protein functional domains and motifs. Redirect to related literature. Convert gene identifiers from one type to another. Cluster redundant and heterozygous annotation terms. Read all annotation contents associated with a gene. All these can be done for a single gene and also in batches.	DAVID's design provides automated solutions that enable researchers to rapidly discover biological themes in lists of genes from large experimental datasets. The tools and analysis algorithms have been applied to various studies. Identify enriched annotation terms associated with user's gene list. Cluster functionally similar terms associated with user's gene list into groups. Query associated terms like disease, heterozygous annotation terms.

TABLE 20.2   Useful Tools for Network Analysis and Systems Biology (Continued)

	Tools	Functions	Application
4	iHOP http://www.pdg.cnb. uam.es/UniPub/ iHOP/	Allow literature investigation starting with a gene or protein of interest. Gene-name-serves has hyperlinks to their corresponding pages. Ranking systems to emphasize the information with high experimental evidence. All the sentences/phrases displayed as result of a query are likened to their corresponding abstracts.	Text mining and easy reference search. Visualize gene network based on their co-occurrence in scientific literature.
5	VisANT http://visant.bu.edu	Provide a visual interface for combining and annotating network data and support for very large networks. Provide supporting functional annotation for different genomes from the Gene Ontology and KEGG databases. Provide various statistical and analytical tools that could be used to extract network topological properties of the user-defined networks. Provide network-drawing capabilities. Advanced iconic representation pertaining to biological entities such as protein complexes or pathways allowing exuberant visualizations.	Can be extensively used for sophisticated visualization and analysis of many types of networks of biological interactions and associations including cellular pathways and functional modules.
6	Hub Objects Analyzer http://hub.iis.sinica. edu.tw/Hubba	Find the degree of the network nodes. Find the bottleneck in the network. Find the edge percolation component (EPC). Find the Subgraph Centrality (SC). Identify Maximum Neighborhood Component (MNC). Identify Density of Maximum Neighborhood Component (DMNC). Perform Double Screening Scheme (DSS).	Helps to find the most essential nodes in a protein-protein interaction network. Helps to elucidate roles of a protein in a cell.
7	bioNMF http://bionmf. dacya.ucm.es.	A web-based tool for nonnegative matrix factorization in biology. Bicluster analysis using a sparse variant of the NMF model.	Sample classification with an unsupervised classification method that uses NMF to classify experimental samples.
8	Cytoscape http://www. cytoscape.org/	Basic network analysis tools for global features of networks. Many plug-ins for specific topics of the network analyses, such as finding active modules, enrichment analysis of functions in some of the network components, network inferring from functional genomic data, comparing networks, and so on.	An open source bioinformatics software platform for visualizing molecular interaction networks and integrating these interactions with gene expression profiles and other state data.

(continued)

TABLE 20.2   Useful Tools for Network Analysis and Systems Biology (Continued)

	Tools	Functions	Application
9	CellNetAnalyzer / FluxAnalyzer (CNA) http://www. mpi-magdeburg. mpg.de/projects/ cna/cna.html	Facilitate the analysis of metabolic (stoichiometric) as well as signaling and regulatory networks solely on their network topology, i.e., independent of kinetic mechanisms and parameters. Provide a powerful collection of tools and algorithms for structural network analysis which can be started in a menu-controlled manner within interactive network maps. Enable interested users to call algorithms of CNA from external programs. Compute paths and cycles.	A package for MATLAB and provides a comprehensive and user-friendly environment for structural and functional analysis of biochemical networks. Applications of CNA can be found in systems biology, biotechnology, metabolic engineering, and chemical engineering.
10	SYCAMORE http://sycamore. eml.org/sycamore/	Allow building a draft model of your system of interest in such a way that kinetic expressions and parameters are as close to reality as possible. Build, view, edit, refine, and analyze the models.	SYCAMORE is a system that facilitates access to a number of tools and methods in order to build models of biochemical systems; view, analyze, and refine them; as well as perform quick simulations. SYCAMORE is not intended to substitute for expert simulation and modeling software packages, but might interact with those. It is rather intended to support and guide system biologists when doing computational research.
11	ChemChains http://www. bioinformatics.org/ chemchains/wiki/	Provide a Boolean network-based simulation and analysis. Combine the advantages of the parameter-free nature of logical models while providing the ability for users to interact with their models in a continuous manner. Allow users to simulate models in an automatic fashion under tens of thousands of different external environments, as well as perform various mutational studies.	ChemChains combines the advantages of logical and continuous modeling and provides a way for laboratory biologists to perform *in silico* experiments on mathematical models easily for systems biology.
12	Nested effects models (NEMs) http://bioconductor. org/packages/2.4/ bioc/html/nem. html	Allow reconstruction of features of pathways from the nested structure of perturbation effects. Take input data: a set of pathway components, which were perturbed, and high-dimensional phenotypic readout of these perturbations (i.e., gene expression or morphological profiles).	Nested effects models (NEMs) are a class of probabilistic models introduced to analyze the effects of gene perturbation screens visible in high-dimensional phenotypes like microarrays or cell morphology.

TABLE 20.2   Useful Tools for Network Analysis and Systems Biology (Continued)

	Tools	Functions	Application
			NEMs reverse engineer upstream and downstream relations of cellular signaling cascades. NEMs take as input a set of candidate pathway genes and phenotypic profiles of perturbing these genes. NEMs return a pathway structure explaining the observed perturbation effects.
13	Signaling pathway impact analysis (SPIA) http://vortex. cs.wayne.edu/ ontoexpress/	Provide a bootstrap procedure used to assess the significance of the observed total pathway perturbation using microarray data. Provide increased sensitivity as well as improved specificity and better pathway ranking.	Signaling pathway impact analysis (SPIA) combines the evidence obtained from the classical enrichment analysis with a novel type of evidence, which measures the actual perturbation on a given pathway under a given condition.
14	MetNetAligner http://alla.cs.gsu. edu:8080/ MinePW/pages/ gmapping/ GMMain.html	Provide aligning metabolic networks (similar to sequence alignment), taking into account the similarity of network topology and the enzymes' functions. Allow or forbid enzyme deletion and insertion. Provide measurement of enzyme-to-enzyme functional similarity and a fast algorithm to find optimal mappings from a directed graph with restricted cyclic structure to an arbitrary directed graph.	MetNetAligner can be used for predicting unknown pathways, comparing and finding conserved patterns, and resolving ambiguous identification of enzymes.
15	JClust http://jclust.embl. de/	Implemented the procedures: (1) density, (2) haircut, (3) best neighbor, and (4) cutting edge operation. Provide k-Means, Affinity Propagation, Spectral Clustering, Markov Clustering (MCL), Restricted Neighborhood Search Cluster (RNSC), MULIC. Provide filtering procedures as haircut, outside– inside, best neighbors, and density control operations. Provide visualization tool for data analysis and information extraction.	JClust provides a collection of clustering algorithms that can be applied to various data (i.e., the datasets of networks and microarrays) to find network clusters, or cluster chemicals, and clusters of heterogeneous data to see connections between clusters.

REFERENCES

Baldwin, A., Li, W., Grace, M. et al. 2008. Kinase requirements in human cells. II. Genetic interaction screens identify kinase requirements following HPV16 E7 expression in cancer cells. *Proc Natl Acad Sci USA* 105: 16478–16483.

Bommi-Reddy, A., Almeciga, I., Sawyer, J. et al. 2008. Kinase requirements in human cells. III. Altered kinase requirements in VHL-/- cancer cells detected in a pilot synthetic lethal screen. *Proc Natl Acad Sci USA* 105: 16484–16489.

Braun, P., Tasan, M., Dreze, M. et al. 2009. An experimentally derived confidence score for binary protein-protein interactions. *Nature Methods* 6: 91–97.

Cui, Q., Ma, Y., Jaramillo, M. et al. 2007. A map of human cancer signaling. *Mol Syst Biol* 3: 152.

Cui, Q., Yu, Z., Purisima, E. O., and Wang, E. 2006. Principles of microRNA regulation of a human cellular signaling network. *Mol Syst Biol* 2: 46.

Cusick, M. E., Yu, H., Smolyar, A. et al. 2009. Literature-curated protein interaction datasets. *Nature Methods* 6: 39–46.

Du, J., Bernasconi, P., Clauser, K. R. et al. 2009. Bead-based profiling of tyrosine kinase phosphorylation identifies SRC as a potential target for glioblastoma therapy. *Nature Biotechnol* 27: 77–83.

Ehrich, M., Turner, J., Gibbs, P. et al. 2008. Cytosine methylation profiling of cancer cell lines. *Proc Natl Acad Sci USA* 105: 4844–4849.

Futreal, P. A., Coin, L., Marshall, M. et al. 2004. A census of human cancer genes. *Nature Rev Cancer* 4: 177–183.

Grueneberg, D. A., Degot, S., Pearlberg, J. et al. 2008a. Kinase requirements in human cells. I. Comparing kinase requirements across various cell types. *Proc Natl Acad Sci USA* 105: 16472–16477.

Grueneberg, D. A., Li, W., Davies, J. E. et al. 2008b. Kinase requirements in human cells. IV. Differential kinase requirements in cervical and renal human tumor cell lines. *Proc Natl Acad Sci USA* 105: 16490–16495.

Huang, H., Li, L., Wu, C. et al. 2008. Defining the specificity space of the human SRC homology 2 domain. *Mol Cell Proteomics* 7: 768–784.

Manning, B. D. 2009. Challenges and opportunities in defining the essential cancer kinome. *Sci Signal* 2: e15.

Oda, K. and Kitano, H. 2006. A comprehensive map of the toll-like receptor signaling network. *Mol Syst Biol* 2: 2006.

Oda, K., Matsuoka, Y., Funahashi, A., and Kitano, H. 2005. A comprehensive pathway map of epidermal growth factor receptor signaling. *Mol Syst Biol* 1: 2005.

Ohm, J. E., McGarvey, K. M., Yu, X. et al. 2007. A stem cell-like chromatin pattern may predispose tumor suppressor genes to DNA hypermethylation and heritable silencing. *Nature Genet* 39: 237–242.

Rikova, K., Guo, A., Zeng, Q. et al. 2007. Global survey of phosphotyrosine signaling identifies oncogenic kinases in lung cancer. *Cell* 131: 1190–1203.

Schlabach, M. R., Luo, J., Solimini, N. L. et al. 2008. Cancer proliferation gene discovery through functional genomics. *Science* 319: 620–624.

Schlesinger, Y., Straussman, R., Keshet, I. et al. 2007. Polycomb-mediated methylation on Lys27 of histone H3 pre-marks genes for *de novo* methylation in cancer. *Nature Genet* 39: 232–236.

Silva, J. M., Marran, K., Parker, J. S. et al. 2008. Profiling essential genes in human mammary cells by multiplex RNAi screening. *Science* 319: 617–620.

Venkatesan, K., Rual, J. F., Vazquez, A. et al. 2009. An empirical framework for binary interactome mapping. *Nature Methods* 6: 83–90.

Widschwendter, M., Fiegl, H., Egle, D. et al. 2007. Epigenetic stem cell signature in cancer. *Nature Genet* 39: 157–158.

Wolf-Yadlin, A., Kumar, N., Zhang, Y. et al. 2006. Effects of HER2 overexpression on cell signaling networks governing proliferation and migration. *Mol Syst Biol* 2: 54.

Index